Recent Advances in Weed Management

Bhagirath S. Chauhan · Gulshan Mahajan
Editors

Recent Advances in Weed Management

Springer

Editors
Bhagirath S. Chauhan
Queensland Alliance for Agriculture and Food Innovation (QAAFI)
The University of Queensland
Toowoomba
Queensland
Australia

Gulshan Mahajan
Department of Plant Breeding and Genetics
Punjab Agricultural University
Ludhiana
Punjab
India

ISBN 978-1-4939-1018-2 ISBN 978-1-4939-1019-9 (eBook)
DOI 10.1007/978-1-4939-1019-9
Springer New York Heidelberg Dordrecht London

Library of Congress Control Number: 2014940295

Printed on acid-free paper

Springer is part of Springer Science+Business Media (www.springer.com)

Preface

Agriculture will remain the mainstay to feed the teeming millions in the years to come, which is indeed a tremendous and tough task. The untiring efforts and unflinching zeal of research scientists have transformed agricultural production from mere sustenance into commercial farming. An influx of technologies has transformed the very outlook of the farmers who look toward scientists for support in diverting their farming into profitable enterprises. Efficient weed management approaches are expected to contribute significantly in sustaining and increasing the profitability of agriculture. Advanced research in weed science provides knowledge to the weed science community in formulating research planning as well as developing guidelines for the farmers to save their crops from the menace of weeds.

Weed problems have turned into a continuing struggle for farmers on account of pressure to raise crops and increasing their productivity to meet the ever-growing demands of a fast-growing human population. As per the requirements of various crops, starting from hand weeding, weed control has gone through a number of changes with the advent of new technologies. Herbicide use is increasing globally as agriculture labor is becoming not only scarce, but also costly and not available at the right times. The growth of chemical weed control is attracting scientists and industries to work on herbicides that are eco-friendly and required in low doses. The new molecules that can be used in small quantities help in reducing the herbicide load in the environment, but may create some residue problems and pose high selection pressure. Research, therefore, is now focused on new methods of weed control, such as the use of cultural, biological, and biotechnological approaches that could be integrated with chemical weed control to reduce the herbicide load in the environment.

In this book, an attempt has been made to highlight the emerging weed management issues and to suggest measures to tackle these issues through advanced methods of weed control and better understanding of the ecology and biology of weeds. The authors of each chapter of this book were invited to contribute based on their experience and respective areas of expertise. To our knowledge, no book exists that summarizes the advanced methods of weed control to handle the emerging issues of weed science, and that too in the current changing scenario.

In this book, the thrust areas requiring immediate attention of weed scientists are covered: biology and ecology of weeds, new challenges in weed science and research priorities, development of resistance to herbicides in weeds, control of aquatic and parasitic weeds, weed management in conservation agriculture, role of allelopathy in weed management, and integrated approaches for weed management in important crops. Through this book, the message has been given that to make an integrated weed management program a success, it would require improved information and technical assistance to growers in choosing correct methods for controlling the complexes of weeds. The main goal of this book is to provide comprehensive knowledge that will enable the weed scientists and policy makers—in careful planning, designing, and orientation of research and development of weed management—to ensure sustainability in agriculture. We expect that this book will provide sound guidelines for future weed management strategies to boost agricultural production by allowing the readers to benefit from the collective experience of others instead of learning through "the hard way."

Bhagirath S. Chauhan
Gulshan Mahajan

Contents

Contributors

Narayana Rao Adusumilli Resilient Dryland Systems and International Rice Research Institute (IRRI), International Crop Research Institute for the Semi-Arid Tropics (ICRISAT), Hyderabad, India

Radi Aly Department of Plant Pathology and Weed Research, Newe-Yaar Research Center, Haifa, Israel

K. K. Barman Directorate of Weed Science Research, Indian Council of Agricultural Research, Jabalpur, Madhya Pradesh, India

Hugh J. Beckie Saskatoon Research Centre, Agriculture and Agri-Food Canada, Saskatoon, SK, Canada

Bhagirath S. Chauhan Queensland Alliance for Agriculture and Food Innovation (QAAFI), The University of Queensland, Queensland, Australia

Sardar Alam Cheema Allelopathy and Eco-physiology Lab, Department of Agronomy, University of Agriculture, Faisalabad, Pakistan

Zahid Ata Cheema Allelopathy and Eco-physiology Lab, Department of Agronomy, University of Agriculture, Faisalabad, Pakistan

David R. Clements Department of Biology, Trinity Western University, Langley, BC, Canada

Antonio DiTommaso Department of Crop and Soil Sciences, Cornell University, Ithaca, NY, USA

Anil Dixit Directorate of Weed Science Research, Indian Council of Agricultural Research, Jabalpur, Madhya Pradesh, India

Mehmet Nedim Doğan Department of Plant Protection, Adnan Menderes University, Aydin, Turkey

Neeraj Kumar Dubey Unit of Weed Science, Newe-Yaar Research Center, Haifa, Israel

R. P. Dubey Directorate of Weed Science Research, Indian Council of Agricultural Research, Jabalpur, Madhya Pradesh, India

Robert M. Durborow Division of Aquaculture, College of Agriculture, Food Science and Sustainable Systems, Kentucky State University, Aquaculture Research Center, Frankfort, KY, USA

Seyed Vahid Eslami Agronomy Department, Faculty of Agriculture, University of Birjand, Birjand, South Khorasan, Iran

Muhammad Farooq Allelopathy and Eco-physiology Lab, Department of Agronomy, University of Agriculture, Faisalabad, Pakistan

Eric Gallandt School of Food and Agriculture, University of Maine, Orono, ME, USA

Zahoor A. Ganie Department of Agronomy and Horticulture, University of Nebraska–Lincoln, Lincoln, NE, USA

Gurjeet S. Gill Department of Agriculture and Animal Science, University of Adelaide, Adelaide, SA, Australia

Terho Hyvönen MTT Agrifood Research Finland, Plant Production Research, Planta, Jokioinen, Finland

Khawar Jabran Department of Plant Protection, Adnan Menderes University, Aydin, Turkey

Amit J. Jhala Department of Agronomy and Horticulture, University of Nebraska–Lincoln, Lincoln, NE, USA

Stevan Z. Knezevic Northeast Research and Extension Center, Department of Agronomy and Horticulture, Haskell Agricultural Laboratory, University of Nebraska–Lincoln, Concord, NE, USA

Vivek Kumar Department of Plant Breeding and Genetics, College of Agriculture, Punjab Agricultural University, Ludhiana, Punjab, India

J. K. Ladha International Rice Research Institute (IRRI)/India Office, New Delhi, India

Gulshan Mahajan Department of Plant Breeding and Genetics, Punjab Agricultural University, Ludhiana, Punjab, India

R. K. Malik Cereal Systems Initiative for South Asia (CSISA) Hub for Eastern U.P. and Bihar, Patna, India

Ahmad Nawaz Allelopathy and Eco-physiology Lab, Department of Agronomy, University of Agriculture, Faisalabad, Pakistan

Probir Kumar Pal Natural Plant Products Division, Council of Scientific and Industrial Research (CSIR)-Institute of Himalayan Bioresource Technology, Palampur, Himachal Pradesh, India

Darren E. Robinson Department of Plant Agriculture, University of Guelph, Ridgetown, ON, Canada

A. R. Sharma Directorate of Weed Science Research, Indian Council of Agricultural Research, Jabalpur, Madhya Pradesh, India

Megh Singh Citrus Research and Education Center, University of Florida, Lake Alfred, FL, USA

P. K. Singh Directorate of Weed Science Research, Indian Council of Agricultural Research, Jabalpur, Madhya Pradesh, India

Rakesh Deosharan Singh Department of Biodiversity, Council of Scientific and Industrial Research (CSIR)-Institute of Himalayan Bioresource Technology, Palampur, Himachal Pradesh, India

Samunder Singh Department of Agronomy, CCS Haryana Agricultural University, Hisar, India

V. P. Singh Directorate of Weed Science Research, Indian Council of Agricultural Research, Jabalpur, Madhya Pradesh, India

Rakesh Kumar Sud Hill Area Tea Science Division, Council of Scientific and Industrial Research (CSIR)-Institute of Himalayan Bioresource Technology, Palampur, Himachal Pradesh, India

Aydin Unay Department of Field Crops, Agricultural Faculty, Adnan Menderes University, Aydin, Turkey

Ashok Yadav Weed Science, Department of Agronomy, Chaudhary Charan Singh Haryana Agricultural University, Hisar, India

Chapter 1
Ecologically Based Weed Management Strategies

Bhagirath S. Chauhan and Gurjeet S. Gill

Introduction

Weeds are one of the most important biological constraints in agricultural production systems. They negatively affect crop growth and yield by competing with crops for nutrient, sunlight, space, and water. In some regions, especially in developing countries, weeds are controlled by using hand weeding. However, manual weeding is becoming less common due to labor scarcity on farms and high labor wages [1]. In other regions, herbicide use has allowed a massive release of labor from agriculture [2]. The increased use of herbicides, however, has been accompanied by concerns over the evolution of herbicide resistance in weeds, weed species population shifts, increased costs of herbicides, surface-water pollution, and effects on nontarget organisms [3–5]. Therefore, a heavy reliance on chemical weed control is considered objectionable in some regions [6, 7]. Water, as flooding, is used to manage weeds in crops such as rice (*Oryza sativa* L.). However, farmers in many areas, especially in Asia, are expected to experience economic and physical water scarcity [8], which may make it unfeasible for them to flood rice fields to ensure sufficient weed control [4]. These concerns have increased the interest of weed scientists around the globe to develop ecologically based weed management strategies [4, 9–12].

To develop ecologically based weed management strategies, however, knowledge of weed ecology and biology is essential. Even in the era of herbicides, understanding the biology of weeds remains essential for developing effective weed management tactics [12]. In this chapter, we discuss ecologically based strategies to

B. S. Chauhan (✉)
Queensland Alliance for Agriculture and Food Innovation (QAAFI),
The University of Queensland, Toowoomba 4350, Queensland, Australia
e-mail: b.chauhan@uq.edu.au

G. S. Gill
Department of Agriculture and Animal Science, University of Adelaide,
Waite Building, GN 11, Adelaide, SA, Australia

B. S. Chauhan, G. Mahajan (eds.), *Recent Advances in Weed Management*,
DOI 10.1007/978-1-4939-1019-9_1, © Springer Science+Business Media New York 2014

reduce the weed seed bank before crop sowing and to reduce weed emergence and growth in crops.

Strategies to Reduce the Weed Seed Bank

Weed seed banks are the reserves of viable weed seeds present on the soil surface and in the soil. These are the primary source of annual weed infestation in most crop production systems [13–15]. Farmers would benefit from management practices that reduce weed seed input, increase weed seed losses, and reduce the probability that residual weed seeds establish [16]. Weed seed banks are usually depleted through germination, predation, or death. Before discussing specific strategies, there is a need to better understand the effect of light and seed coat on weed seed germination.

Light plays an important role in weed seed germination. However, the germination of different weeds in light and darkness varies [1, 15]. Seeds of some species (e.g., *Avena fatua* L., *Malva parviflora* L., and *Mimosa invisa* Mart. ex Colla) germinate equally in light and dark; seeds of some species (e.g., *Eclipta prostrata* [L.] L. and *Cyperus difformis* L.) do not germinate in the dark at all; seeds of some species (e.g., *Galium tricornutum* Dandy) do not germinate in the light; and, for some weed species (e.g., *Echinochloa crus-galli* [L.] P. Beauv. and *Sisymbrium orientale* L.), light is not an absolute requirement for germination, but light stimulates germination [4, 17–21]. In the field, light conditions differ for weed seeds present on the soil surface, beneath the crop residue cover, or buried in the soil.

Seeds of some weed species (e.g., *Malva parviflora, Mimosa invisa, Abutilon theophrasti* Medik, *Urena lobata* L.) have a hard seed coat, which imposes dormancy due to the impermeability of the seed coat to water or gases [19, 22–25]. Germination of such seeds is generally low unless they are scarified. In some species (e.g., *Rottboellia cochinchinensis* [Lour.] W.D. Clayton and *Raphanus raphanistrum* L.), dormancy is largely due to the pod surrounding the seeds [26, 27]. Mechanisms that increase breakdown of the pod will increase germination of species with such seeds. Possible factors that may account for a dormancy break in hard-seeded species and seeds surrounded by the pod are microbial and fungi attack, changes in temperature and moisture regimes, and fire [15, 28].

Seed Predation and Decay

One way to reduce the size of weed seed banks is through mortality of newly produced weed seeds by predators [16, 29]. Seed predation has been recognized as an important means of seed mortality, particularly after seed shed [6, 30]. Weed seeds are most prone to seed predators while on the soil surface and burial makes seeds largely unavailable to most seed predators [31]. Furthermore, weed seeds present on the soil surface are also prone to rapid decay due to unfavorable weather conditions, such as extreme changes in temperature and moisture fluctuations [32]. Therefore, the use of no-till systems, in which most of the weed seeds remain on the soil surface, may expose weed seeds to seed predators. By delaying tillage operations or creating an additional time lapse between seedbed preparation and seeding, the first weed flush can be easily controlled [6, 33]. The number of seed predators can be increased by creating better opportunities for shelter and additional food [34]. For example, the management of field bunds, through creating favorable environments for seed predators by accumulating crop and weed residues on bunds rather than burning them, could provide a promising opportunity to encourage weed seed predation [29]. Similarly, organic cropping practices, especially cover cropping, may increase the activity of weed seed predators [16].

Different studies suggest that seed predation can cause a substantial reduction in the number of weed seeds entering the seed bank, and therefore could contribute to ecologically based weed management in different crops. Seed predation could be achieved with no additional costs, and it could easily be integrated into existing management practices, which could increase adoption by farmers.

The Stale Seedbed Technique

In the stale seedbed technique, weeds are allowed to germinate after a light irrigation or rainfall and are then killed by using a nonselective herbicide (e.g., glyphosate or paraquat) or a shallow tillage operation. As most of the weed seeds remain on the soil surface after crop harvest, this practice may help to reduce the weed seed bank. Most weed species sensitive to the stale seedbed practice are those that require light to germinate (as discussed in a previous section), have low initial dormancy, and are present on or near the soil surface. Some of these weed species are *Digitaria ciliaris* (Retz.) Koel., *Leptochloa chinensis* (L.) Nees, *Eclipta prostrata,* and *Cyperus iria* L. Therefore, knowledge about the effect of light on the germination of different weeds may help to make the decision regarding the use of the stale seedbed practice. The feasibility of this practice, however, should be assessed by farmers themselves, especially when the period between the harvest of the previous crop and the sowing of the subsequent crop is short.

Strategies to Reduce Weed Emergence and Growth in Crops

Various strategies—such as tillage practices, the use of crop residue as mulch, cultivars with weed competitiveness and allelopathy, and agronomic practices aimed at early canopy closure with the use of a high seeding density and narrow row spacing—can be used to reduce weed seedling emergence and weed growth in crops.

Tillage Systems

Weeds emerging in a crop can be reduced by using different tillage practices. However, the effect of tillage practices on weed emergence depends on the intensity and timing of tillage; type, speed, and depth of the tillage or seeding equipment; and the extent that the soil environment is modified by the tillage [15]. Tillage and seeding operations determine the vertical seed distribution of weeds in the soil profile, and this distribution affects weed seed germination and seedling emergence through the influence of seed predation, seed decay, seed dormancy, seed longevity, seed size, light requirement for germination, and potential of a seedling to emerge from a given depth [14, 35]. An earlier study, for example, reported that a no-till system retained 56 % of the weed seeds in the top 1-cm soil layer, whereas a conventional tillage system buried 65 % of the seeds to a depth of 1–5 cm and only 5 % of the seeds remained in the top 1-cm soil layer [14]. In another study, about 85 % of all weed seeds were found in the top 5-cm soil layer in a reduced tillage system and only 28 % of the weed seeds were found in this soil layer in the conventional tillage system [36]. These studies suggest that no-till or reduced tillage systems leave most of the weed seeds on or near the soil surface. In some species (e.g., *Lolium rigidum* Gaud.), weed seeds present on the soil surface under no-till and zero-till germinate and emerge at a slower rate than seeds buried to a shallow depth by tillage. Weeds emerging later and after the crop are likely to be at a competitive disadvantage against the crop in no-till than those emerging before or with the crop under conventional tillage systems [14].

As discussed in a previous section, seeds present on or near the soil surface are prone to seed predation and decay. Therefore, adopting no-till systems for some crops may help to enhance seed predation and deplete the seed bank, resulting in fewer weed seedlings in the crop. In no-till systems, most of the weed seeds are present on the soil surface, where light may stimulate germination and help in reducing the seed bank through germination. In some situations, a large weed seed bank may accumulate on the soil surface. In such situations, a deep inversion tillage operation could be used to bury weed seeds below the maximum depth of their emergence. Most weed seedlings cannot emerge from depths more than 10 cm. A previous study also suggested that the success of *Alopecurus myosuroides* Huds. in reduced tillage systems could be overcome by plowing once every 5 years [37].

Rotation of tillage or crop establishment systems may also help to reduce weed problems in crops. In rice, for example, the built-up population of *Ischaemum rugosum* Salisb. in wet-seeded rice was reduced by using a no-till system [38]. Similarly, the increasing population of *Echinochloa colona* in no-till rice could be managed by shifting to wet-seeded rice [38, 39].

In conservation agriculture, permanent residue for soil cover has been advocated as this improves soil and moisture conservation [40, 41]. The presence of crop residue on the soil surface can also help suppress weed seed germination and seedling emergence; however, the extent of suppression depends on the quantity and allelopathic potential of the residue and the weed species [15, 42, 43]. The presence of large amount of crop residue on soil surface can substantially reduce and delay weed seedling emergence by preventing light penetration, decreasing thermal amplitude, and increasing the time needed for seedlings to emerge through the residue cover. Crop residues may also reduce weed seed germination through their chemical effect, such as allelopathy and toxic microbial products. The Turbo seeder has been found effective in India to plant wheat (*Triticum aestivum* L.) and rice under high residue amounts because it diverts straw in front of the tines and places it in between two crop rows [44]. Straw mulch placed between the two crop rows inhibits the emergence of weeds and also adds organic matter to the soil. In a recent study in the Philippines, a residue amount of 6 t ha^{-1} significantly reduced seedling emergence and biomass of *Dactyloctenium aegyptium* (L.) Willd., *Eclipta prostrata, Eleusine indica* (L.) Gaertn., and *L. chinensis* as compared to a no-residue situation in a sprinkler-irrigated zero-till dry-seeded rice system [45]. In some crops (e.g., corn [*Zea mays* L.], soybean [*Glycine max* L.], etc.), cover crops and their residues are used to suppress weeds [46, 47]. The presence of rye mulch, for example, was reported to reduce weed biomass in corn, without any detrimental effect on corn yield [48]. Therefore, integrating the use of residue as mulch with other weed management strategies could help in reducing weed pressure in crops.

The Role of Cultivars in Suppressing Weed Emergence and Growth in Crops

The use of weed-competitive cultivars and cultivars having allelopathy can help in providing supplemental weed control when herbicide inputs decrease [49]. Weed competitiveness has been investigated for several crops, such as sugar beet (*Beta vulgaris* L.), soybean, corn, wheat, and rice [50–55]. Tall and traditional crop cultivars with droopy leaves are generally more competitive, but they are often lower in yield potential than short-statured modern cultivars with erect leaves. In Australia, Vandeleur and Gill showed that there was a significant positive linear relationship between the year of wheat cultivar release and crop yield loss from weed competition, indicating the inferior competitive ability of the modern cultivars related to their shorter stature [56]. Therefore, there is a trade-off between yield potential and competitive ability. In the future, the use of nitrogen fertilizers may rise in some crop production systems to meet the increasing food demand, and high nitrogen doses are known to cause lodging in tall cultivars [1]. Therefore, by selecting traits other than tall plant type, the trade-off between yield potential and competitive ability may be minimized.

High genetic correlations between leaf area index of wheat and its yield loss ($r = -0.81$) as well as suppression of *L. rigidum* ($r = -0.91$) indicate that traits contributing to early ground cover would be important for developing weed competitive wheat genotypes [57]. In another study, wheat cultivars with early canopy cover and greater biomass were found to shade grass weeds [58]. Similarly, rice cultivars having high seedling vigor suppressed weeds to a greater extent, especially in rainfed and upland environments, where dry seeding is practiced [59]. In an earlier study, shoot length of rice was reported to have a positive correlation with fresh and dry biomass of seedlings, and vigor index [60]. Therefore, seedling vigor could play a critical role in dry-seeded rice as it helps in better crop emergence and offers greater crop competition with weeds [49]. In general, the traits associated with weed competitiveness in rice are early canopy cover, high tiller density, droopy leaves, high biomass at the early stage, high leaf area index and high specific leaf area during vegetative growth, and early vigor. In herbicide-dominant systems, using weed-competitive cultivars to suppress weeds may substantially reduce herbicide use, selection pressure for herbicide resistance, and labor costs. Most efforts to select for improved weed competitive ability have focused on aboveground traits and little is known about the importance of root competition, especially in low-input production systems. Fofana and Rauber undertook one of the few competition studies in which crop varietal differences in root growth was investigated in rice (*O. sativa* and *O. glaberrima*) [61]. They concluded that rice varieties with greater root lengths were able to cause larger suppression of weed biomass. Therefore, there is a need to quantify variation in root growth in research aimed at improving weed competitive ability of field crops.

Allelopathic crop cultivars can also be used to suppress weed seedling emergence, as they release chemical compounds through living and intact roots, and these compounds affect the growth of other plant species [49, 62]. Some progress has been made in determining the role of allelopathy in rice. Field experiments by Olofsdotter et al. revealed allelopathy accounted for 34 % of overall competitive ability in rice [63]. They have argued that optimizing allelopathy in combination with breeding for other weed competitive traits (e.g., early vigor) could result in crop cultivars with superior weed-suppressive ability. However, the benefits of allelopathy for weed management in field crops, including rice, still remain largely conjectural at this stage and much research work needs to occur before these benefits can be realized by farmers.

In crops, such as rice, flooding is used to suppress weeds as most weed species cannot germinate and emerge under flooded conditions [4]. In the USA, rice is seeded in standing water (water seeding), mainly to suppress weeds. In Asia, however, flooding can be introduced only after the rice seedlings have emerged as rice cultivars capable of germinating under anaerobic conditions are not widely available. Work on such cultivars is in progress at the International Rice Research Institute (IRRI) and such cultivars will be available to farmers in the near future. Rice cultivars having tolerance of anaerobic conditions during germination are increasingly required because of the shift of rice establishment methods in many areas from transplanting to direct seeding [49]. Direct-seeded rice fields can be easily submerged immediately after crop sowing if such cultivars are available and this could provide economical and environmentally friendly weed control. However, the feasibility of such systems needs to be examined in water-limited environments.

Role of Crop Density and Row Spacing in Suppressing Weed Emergence and Growth in Crops

The impact of weeds on crops can be reduced by agronomic manipulations, such as increased crop density and reduced row spacing. Increasing crop competitiveness through the use of high crop density is a possible technique for weed management, especially in low-input and organic production systems or when herbicide resistance develops in weeds. At low crop density, crop cover early in the growing season is usually low and a large amount of resources are available for the weeds [64]. These conditions enable weeds to establish and grow quickly.

In an earlier study in wheat, doubling the crop density of several cultivars from 100 to 200 plants m^{-2} halved *L. rigidum* biomass from 100 to 50 g m^{-2} [10]. In another study, increasing wheat density from 75 to 200 plants m^{-2} reduced the biomass of *L. rigidum* and increased wheat grain yield [65]. *L. rigidum* biomass declined by 43 % when the wheat-seeding rate doubled from 55 to 110 kg ha^{-1} [51]. In a later study, increasing wheat density from 50 to 200 plants m^{-2} in the presence of 200 plants m^{-2} of *Avena* spp. almost doubled the gross margin [66].

In Asia, rice is generally grown after transplanting of seedlings into puddled soil. Weeds are not a big problem in these establishment systems. However, there is a trend toward direct seeding (wet and dry seeding). In these systems, weeds are the number-one biological constraint. Recently, several studies reported the effect of increased seeding rates on weed suppression in direct-seeded rice systems. In one study, reducing the seeding rate from 80 to 26 kg ha^{-1} increased weed biomass significantly and therefore a seeding rate of 80 kg ha^{-1} was needed to avoid a large yield loss because of weeds [55]. Results from another study in India and the Philippines showed that the maximum grain yield of an inbred cultivar was achieved at 95–125 kg seed ha^{-1} when grown in the presence of weeds; however, seeding rates from 15 to 125 kg ha^{-1} had little effect on yield in weed-free conditions [67]. In the same study, increasing the rice seeding rate from 25 to 100 kg ha^{-1} reduced weed biomass by 47–59 %.

No-till farmers in many countries have widened crop row spacing to enable their seeders to cope with the large amounts of crop residues present in the field. However, wider row spacing provides more interrow space for weeds to establish and proliferate. In many crops, it is well known that reduced row spacing suppresses weed emergence and growth. Narrow row spacing improves crop competitiveness by developing faster canopy closure and allowing less light penetration to the ground. In wheat, it was shown that reducing crop row spacing from 23 to 7.5 cm decreased the seed production of *Bromus secalinus* L. [68]. Another study suggested the possibility of using narrow row spacing in corn to minimize the addition of weed seeds to the soil seed bank and to progressively deplete weed seeds in the long term [69]. In direct-seeded rice, 15–45-cm row spacing had little effect on rice grain yield in weed-free conditions; however, in weedy conditions, the widest spacing resulted in lower grain yield [70]. The critical periods for weed control can also be shorter for a crop grown in narrow rows than in wider rows. For example, the critical periods to achieve 95% of maximum yield for weed control in dry-seeded rice were fewer in 15-cm rows (18–52 days) than in 30-cm rows (15–58 days) [71]. In another similar study at IRRI, the seedlings of *Echinochloa colona* and *Echinochloa crus-galli* emerging up to 2 months after crop emergence in dry-seeded rice produced less shoot biomass and fewer seeds in 20-cm rows than in 30-cm rows [72].

Conclusion

In summary, weeds are the major constraint to crop production systems. Various ecologically based weed management strategies, such as the adoption of practices that enhance seed predation and seed decay, the use of a stale seedbed technique and appropriate tillage systems, retention of crop residue on the soil surface, and the use of crop cultivars with weed competitiveness and allelopathy, high crop density, and narrow row spacing, need to be integrated to achieve effective and sustainable weed control.

References

1. Chauhan BS (2012) Weed ecology and weed management strategies for dry-seeded rice in Asia. Weed Technol 26:1–13
2. Nelson RJ (1996) Herbicide use in Asian rice production: perspectives from economics, ecology and the agricultural sciences. Herbicides in Asian rice: transitions in weed management: Palo Alto (California): Institute for International Studies, Stanford University and Manila (Philippines), International Rice Research Institute, pp 3–26
3. Buhler DD, Liebman M, Obrycki JJ (2002) Review: theoretical and practical challenges to an IPM approach to weed management. Weed Sci 48:274–280
4. Chauhan BS, Johnson DE (2010) The role of seed ecology in improving weed management strategies in the tropics. Adv Agron 105:221–262
5. Primot S, Valentin-Morrison M, Makowski D (2006) Predicting the risk of weed infestation in winter oilseed rape crops. Weed Res 46:22–33
6. Bastiaans L, Paolini R, Baumann DT (2008) Focus on ecological weed management: what is hindering adoption? Weed Res 48:481–491
7. Liebman M (2001) Managing weeds with insects and pathogens. In: Liebman M, Mohler CL, Staver CP (eds) Ecological management of agricultural weeds. Cambridge: Cambridge University Press, pp 375–408
8. Bouman BAM, Tuong TP (2003) Growing rice with less water. Issues of water management in agriculture: compilation of essays comprehensive assessment secretariat. Colombo: International Irrigation Management Institute, pp 49–54
9. Bhowmik PC (1997) Weed biology: importance to weed management. Weed Sci 45:349–356
10. Lemerle D, Cousens RD, Gill GS, Peltzer SJ, Moerkerk M, Murphy CE et al (2004) Reliability of higher seeding rates of wheat for increased competitiveness with weeds in low rainfall environments. J Agric Sci 142:395–409
11. Mortensen DA, Bastiaans L, Sattin M (2000) The role of ecology in the development of weed management systems: an outlook. Weed Res 40:49–62
12. Van Acker RC (2009) Weed biology serves practical weed management. Weed Res 49:1–5

13. Buhler DD, Hartzler RG, Forcella F (1997) Implications of weed seed bank dynamics to weed management. Weed Sci 45(3):329–336
14. Chauhan BS, Gill G, Preston C (2006) Influence of tillage systems on vertical distribution, seedling recruitment and persistence of rigid ryegrass (*lolium rigidum*) seed bank. Weed Sci 54:669–676
15. Chauhan BS, Gill G, Preston C (2006) Tillage system effects on weed ecology, herbicide activity and persistence: a review. Aust J Exp Agric 46:1557–1570
16. Gallandt ER (2006) How can we target the weed seedbank? Weed Sci 54:588–596
17. Boyd NS, Van Acker RC (2004) Seed germination of common weed species as affected by oxygen concentration, light, and osmotic potential. Weed Sci 52:589–596
18. Chauhan BS, Gill G, Preston C (2006) Factors affecting seed germination of threehorn bedstraw (*Galium tricornutum*) in Australia. Weed Sci 54:471–477
19. Chauhan BS, Gill G, Preston C (2006) Factors affecting seed germination of little mallow (*Malva parviflora*) in southern Australia. Weed Sci 54:1045–1050
20. Chauhan BS, Gill G, Preston C (2006) Influence of environmental factors on seed germination and seedling emergence of oriental mustard (*Sisymbrium orientale*). Weed Sci 54:1025–1031
21. Chauhan BS, Johnson DE (2011) Ecological studies on *Echinochloa crus-galli* and the implications for weed management in direct-seeded rice. Crop Prot 30:1385–1391
22. Cardina J, Sparrow DH (1997) Temporal changes in velvetleaf (*Abutilon theophrasti*) seed dormancy. Weed Sci 45:61–66
23. Chauhan BS, Johnson DE (2008) Seed germination and seedling emergence of giant sensitiveplant (*Mimosa invisa*). Weed Sci 56:244–248
24. Foley ME (2001) Seed dormancy: an update on terminology, physiological genetics, and quantitative trait loci regulating germinability. Weed Sci 49:305–317
25. Wang J, Ferrell J, MacDonald G, Sellers B (2009) Factors affecting seed germination of cadillo (*Urena lobata*). Weed Sci 57:31–35
26. Cheam AH (1986) Seed production and seed dormancy in wild radish (*Raphanus raphanistrum* l.) and some possibilities for improving control. Weed Res 26:405–413
27. Mercado BL (1978) Biology, problems and control of *Rottboellia exaltata* l.F. Biotrop Bull 14:5–38
28. Baskin CC, Baskin JM (1998) Seeds: ecology, biogeography, and evolution of dormancy and germination. San Diego, Academic
29. Chauhan BS, Migo T, Westerman PR, Johnson DE (2010) Post-dispersal predation of weed seeds in rice fields. Weed Res 50:553–560
30. Westerman PR, Hofman A, Vet LEM, Van Der Werf W (2003) Relative importance of vertebrates and invertebrates in epigaeic weed seed predation in organic cereal fields. Agric Ecosyst Environ 95:417–425
31. Hulme PE (1994) Post-dispersal seed predation in grassland: its magnitude and sources of variation. J Ecol 81:645–652
32. Mohler CL (1993) A model of the effects of tillage on emergence of weed seedlings. Ecol Appl 3:53–73
33. Chauhan BS (2012) Can knowledge in seed ecology contribute to improved weed management in direct-seeded rice? Curr Sci 103:486–489
34. Landis DA, Manalled FD, Costamagna AC, Wilkinson TK (2005) Manipulating plant resources to enhance beneficial arthropods in agricultural landscapes. Weed Sci 53:902–908
35. Yenish JP, Fry TA, Durgan BR, Wyse DL (1996) Tillage effects on seed distribution and common milkweed (*Asclepias syriaca*) establishment. Weed Sci 44(4):815–820
36. Pareja MR, Staniforth DW, Pareja GP (1985) Distribution of weed seeds among soil structural units. Weed Sci 33:182–189
37. Cussans GW, Moss SR (1982) Population dynamics of annual grass weeds. In: Austin RB (ed) Decision making in the practice of crop protection. London, BCPC, pp 91–98
38. Singh VP, Singh G, Singh Y, Mortimer M, Johnson DE (2008) Weed species shifts in response to direct seeding in rice. In: Singh Y, Singh VP, Chauhan B, Orr A, Mortimer AM,

Johnson DE et al (eds) Direct seeding of rice and weed management in the irrigated rice-wheat cropping system of the indo-gangetic plains: Los Baños (Philippines): International Rice Research Institute, and Pantnagar (India), Directorate of Experiment Station, G.B. Pant University of Agriculture and Technology, pp 213–219.
39. Chauhan BS, Johnson DE (2009) Influence of tillage systems on weed seedling emergence pattern in rainfed rice. Soil Tillage Res 106:15–21
40. Hobbs PR, Sayre K, Gupta R (2008)The role of conservation agriculture in sustainable agriculture. Philos Trans R Soc Lond B Biol Sci 363:543–555
41. Locke MA, Bryson CT (1997) Herbicide-soil interactions in reduced tillage and plant residue management systems. Weed Sci 45:307–320
42. Chauhan BS, Singh RG, Mahajan G (2012) Ecology and management of weeds under conservation agriculture: a review. Crop Prot 38:57–65
43. Locke MA, Reddy KN, Zablotowicz RM (2002) Weed management in conservation crop production systems. Weed Biol Manage 2:123–132
44. Singh S (2007) Role of management practices on control of isoproturon-resistant littleseed canarygrass (*Phalaris minor*) in India. Weed Technol 21:339–346
45. Chauhan BS, Abugho SB (2013) Effect of crop residue on seedling emergence and growth of selected weed species in a sprinkler-irrigated zero-till dry-seeded rice system. Weed Sci 61:403–409
46. Ateh CM, Doll JD (1996) Spring-planted winter rye (*Secale cereale*) as a living mulch to control weeds in soybean (*Glycine max*). Weed Technol 10:347–353
47. Johnson GA, DeFelice MS, Helsel ZR (1993) Cover crop management and weed control in corn (*Zea mays*). Weed Technol 7:425–430
48. Mohler CL (1991) Effects of tillage and mulch on weed biomass and sweet corn yield. Weed Technol 5:545–552
49. Mahajan G, Chauhan BS (2013) The role of cultivars in managing weeds in dry-seeded rice production systems. Crop Prot 49:52–57
50. Callaway MB, Forcella F (1993) Crop tolerance to weeds. In: Callaway MB, Francis CA (eds) Crop improvement for sustainable agriculture. Lincoln, University of Nebraska Press, pp 100–131
51. Lemerle D, Verbleek B, Cousens RD, Coombes NE (1996) The potential for selecting wheat cultivars strongly competitive against weeds. Weed Res 36:505–513
52. Lindquist JL, Mortensen DA, Johnson BE (1998) Mechanisms of corn tolerance and velvetleaf suppressive ability. Agron J 90:787–792
53. Lotz LAP, Groeneveld RMW, De Groot NAMA (ed) (1991) Potential for reducing herbicide input in sugar beet by selecting early closing cultivars. Proceedings 1991 Brighton crop protection conference—weeds, Brighton, UK
54. Namuco OS, Cairns JE, Johnson DE (2009) Investigating early vigor in upland rice (*Oryza sativa* l.). Part i. Seedling growth and grain yield in competition with weeds. Field Crops Res 113:197–206
55. Zhao D (2006) Weed competitiveness and yielding ability of aerobic rice genotypes: PhD thesis, Wageningen University, The Netherlands
56. Vandeleur RK, Gill GS (2004) The impact of plant breeding on the grain yield and competitive ability of wheat in Australia. Aust J Agric Res 55:855–861
57. Coleman RK, Gill GS, Rebetzke GJ (2001) Identification of quantitative trait loci for traits conferring weed competitiveness in wheat (*Triticum aestivum* L.). Aust J Agric Res 52:1235–1246
58. Singh S (2007) Role of management practices on control of isoproturon-resistant littleseed canarygrass (*Phalaris minor*) in India. Weed Technol 21:339–346
59. Kanbar A, Janamatti M, Sudheer E, Vinod MS, Shashidhar HE (2006) Mapping qtls underlying seedling vigour traits in rice (*Oryza sativa* l.). Current Sci 90:24–26
60. Shashidhar HE (1990) Root studies and stability of quantitative traits in rice (*Oryza sativa* l.) under punji (dry cum wet cultivation) and irrigated habitats. PhD thesis. Bangalore, India: University of Agricultural Sciences

61. Fofana B, Rauber R (2000) Weed suppression ability of upland rice underlow-input conditions in West Africa. Weed Res 40:271–280
62. Belz RG (2007) Allelopathy in crop/weed interactions-an update. Pest Manage Sci 63:308–326
63. Olofsdotter M, Jensen LB, Courtois B (2002) Improving crop competitive ability using allelopathy–an example from rice. Plant Breed 121:1–9
64. Kristensen L, Olsen J, Weiner J (2008) Crop density, sowing pattern, and nitrogen fertilization effects on weed suppression and yield in spring wheat. Weed Sci 56:97–102
65. Medd RW, Auld BA, Kemp DR, Murison RD (1985) The influence of wheat density and spatial arrangement on annual grass, *lolium rigidum* gaudin, competition. Aust J Agric Res 36:361–371
66. Murphy C, Lemerle D, Jones R, Harden S (2002) Use of density to predict crop yield loss between variable seasons. Weed Res 42:377–384
67. Chauhan BS, Singh VP, Kumar A, Johnson DE (2011) Relations of rice seeding rates to crop and weed growth in aerobic rice. Field Crops Res 121:105–115
68. Justice GG, Peeper TF, Solie JB, Epplin FM (1993) Net returns from cheat (*Bromus secalinus*) control in winter wheat (*Triticum aestivum*). Weed Technol 7:459–464
69. Mashingaidze AB, van der Werf W, Lotz LAP, Chipomho J, Kropff MJ (2009) Narrow rows reduce biomass and seed production of weeds and increase maize yield. Ann Appl Biol 155:207–218
70. Akobundu IO, Ahissou A (1985) Effect of interrow spacing and weeding frequency on the performance of selected rice cultivars on hydromorphic soils of west Africa. Crop Prot 4:71–76
71. Chauhan BS, Johnson DE (2011) Row spacing and weed control timing affect yield of aerobic rice. Field Crops Res 121:226–231
72. Chauhan BS, Johnson DE (2010) Implications of narrow crop row spacing and delayed *Echinochloa colona* and *Echinochloa crus-galli* emergence for weed growth and crop yield loss in aerobic rice. Field Crops Res 117:177–182

Chapter 2
Ecology and Management of Weeds in a Changing Climate

David R. Clements, Antonio DiTommaso and Terho Hyvönen

Introduction

The annual economic cost of weeds throughout the world is estimated at US$ 400 billion [1]. In the USA alone, the cost of invasive plants was estimated to be US$ 34.7 billion per year [2]. Oerke estimated that 34 % of potential crop losses throughout the globe are due to weeds, as compared to 16 % for pathogens and 18 % for animal pests [3]. Traditionally, the cost of weed management has been principally accounted for within the agricultural sector, but in the last decade weeds, or more inclusively invasive plants, have been increasingly recognized for their negative impact on a broad array of human enterprises in addition to agriculture, including forestry, transportation, human health, recreation, and tourism [2, 4]. These collective economic influences are difficult to estimate reliably, but the estimates that have been done indicate that these are threats to be taken seriously. In the UK, for example, more than 175 million euros is used annually to control *Fallopia japonica* (Houtt.) Ronse Decr. (Japanese knotweed; Fig. 2.1) [4]. Taken together, these impacts on the economy are issues that require urgent action, particularly because weeds are a dynamic threat—they evolve in response to management practices [5, 6].

D. R. Clements (✉)
Department of Biology, Trinity Western University,
7600 Glover Road, Langley, BC V2Y 1Y1, Canada
e-mail: clements@twu.ca

A. DiTommaso
Department of Crop and Soil Sciences, Cornell University,
903 Bradfield Hall, 306 Tower Road, Ithaca, NY 14853, USA
e-mail: ad97@cornell.edu

T. Hyvönen
MTT Agrifood Research Finland, Plant Production Research,
Planta, 31600 Jokioinen, Finland
e-mail: terho.hyvonen@mtt.fi

B. S. Chauhan, G. Mahajan (eds.), *Recent Advances in Weed Management*,
DOI 10.1007/978-1-4939-1019-9_2, © Springer Science+Business Media New York 2014

Fig. 2.1 Illustrations of invasive plants. (**a**) *Heracleum mantegazzianum* Sommier and Levier (giant hogweed) with author David Clements. (**b**) *Fallopia japonica* (Houtt.) Ronse Decr. (Japanese knotweed). (**c**) *Impatiens glandulifera* Royle (Himalayan balsam). (**d**) *Ambrosia artemisiifolia* L. (common ragweed) infesting corn (*Zea mays* L.). (**e**) *Sorghum halepense* (L.) Pers. (Johnsongrass). (**f**) *Lantana camara* L. (lantana). (Photo credits: (**a**) Vincent Clements; (**b**, **d**, **e**) Antonio DiTommaso; (**c**, **f**) David Clements)

The history of weeds and how they became weeds is tightly interwoven with that of our cropping practices [5, 7, 8], or for that matter, our horticultural, forestry, and numerous other practices that have inadvertently fostered weeds to flourish [9–11]. Thus, even aside from the specter of climate change, our response to threats caused by weeds must be as proactive as possible. Our environmental history is rife with narratives of plant invasions and weed infestations that have worsened because the actions taken were insufficient or too late [12]. Presently, we find ourselves in the midst of an increasingly worldwide dilemma in which numerous weed species are developing resistance to the nonselective herbicide glyphosate that had been considered to be a "silver bullet" for managing a vast array of weeds, including difficult-to-control perennial weed species. However, because policies governing its use were too lenient and its economic benefits within glyphosate-tolerant cropping systems were so lucrative, the selection pressure for the development of glyphosate resistance is unprecedented and creating "super weeds" [13, 14]. Until now, populations from 24 different weed species have developed resistance to glyphosate, including *Amaranthus palmeri* S. Watson (palmer amaranth), *Amaranthus tuberculatus* (Moq.) Sauer var. *rudis* (Sauer; waterhemp), *Conyza canadensis* (L.) Cronquist (horseweed), *Lolium multiflorum* Lam. (Italian ryegrass), and *Lolium rigidum* Gaudin (rigid ryegrass) [15].

The dynamic nature of weeds frequently involves expansion of their distributions [5, 16, 17]. With the increasing attention to climate change in the past several decades and impacts on biota, researchers and managers have attempted to map how the distribution of weeds might change or is already changing with climate warming and other climatic changes anticipated in temperate regions [18–22]. With these weed distribution changes, there is the prospect of increased economic damage due to weeds, either in newly infested areas or through more favorable conditions in their current ranges [23]. At the same time, historically problematic species may become less damaging in certain regions with climate change [23] but, obviously, much study is needed to ascertain which scenarios are most likely to occur. Failure to adequately predict potential impacts of weed distribution changes on agriculture, forestry, and conservation lands could have serious consequences for human sustainability. As mentioned earlier, the costs of weed management are already extremely high. Furthermore, it has been shown repeatedly that the best course of action in dealing with invasive species is "early detection and rapid response" [12]. Thus, taking proactive steps based on information on the potential spread of invasive weeds in an era of climate change should be a high priority.

In this chapter, we review current knowledge of climate change and its effects on weeds, examining regional patterns of recent range expansions across the globe, the influence of climate change on cropping systems, and biological and evolutionary responses of weeds to climate change, including resultant research and management priorities.

Regional Patterns of Recent Weed Range Expansions

The developed world has historically been centered in Europe and North America, and, by extension, these regions have featured the most intensive weed management efforts, including the monitoring of potential expansion of weed distributions

[5, 21, 24–27]. Parts of Oceana, particularly Australia and New Zealand, have also received considerable attention in this regard [28, 29]. Economic development and weeds have also tended to go hand in hand, as illustrated by the widespread introductions of Eurasian weeds to North America [30]. Interestingly, with recent large-scale economic development in other regions, such as Southeast Asia, these regions have also experienced drastic increases in weed species introductions as a result of expanded commerce and trade [31, 32]. As accurately predicted in 1958 by Elton, in his seminal book on invasion biology, the alarming tendency arising from increased globalization is towards homogenization of the world's flora and fauna [33]. In the following section, we characterize recent weed range expansions on a regional basis, including the potential influences of climate change and other factors on these distributional shifts.

North America

As noted previously, the weed flora of North America is largely a product of its colonial history with European settlement and the ever-expanding cultivation of land, particularly during the nineteenth and twentieth centuries, with many weeds introduced through the seed trade [30]. There are some weeds native to North America that cause economic damage but the magnitude of their impact tends to be much lower than that for nonnative species. Weed managers and researchers have tended to focus efforts on weed species in North America established during the colonial period rather than "invasive weeds," but careful survey work reveals that new weed problems are emerging due to novel species, either moving from other subregions of North America or as recent invaders from outside the continent [34]. The weed science community cannot afford to be complacent in treating weed problems based on the status quo, particularly if it can be shown that climate change and other drivers of regional weed distribution may increase the incidence of novel weeds and associated management problems.

As shown in Table 2.1 [21, 35–45], relatively recent northward expansion of weed ranges has been well documented for numerous weed species in North America, and doubtless, there are many more examples of this phenomenon [46]. This northward weed migration is of particular concern in areas where traditionally many weeds have been unable to establish because of severe winter conditions, such as in the northeastern USA [47] or Canada [48]. Of course, some weeds are predicted to decline in some regions as climate changes [23, 49]. Ziska and Runion demonstrated how *Cirsium arvense* (L.) Scop. or *Panicum miliaceum* L. (proso millet) is likely to decline in the southern parts of their ranges with climate warming [49]. Interestingly, some weed species exhibit increased tolerance to lower temperatures with increasing CO_2 availability [50, 51], so an increase in CO_2 levels even if unaccompanied by warming could stimulate poleward weed distribution extensions [52].

Table 2.1 Recent expansion of ranges for selected weed species in North America

Weed species	Range expansion	Attributed mechanism(s)	References
Centaurea stoebe L. (spotted knapweed)	More northerly latitudes than in native Europe	Shift in the climatic niche due to lack of natural enemies, adaptation to drier/colder climates	Broennimann et al. [35]
Datura stramonium L. (jimsonweed)	Northward invasion of Canadian and northeastern US cropland since 1950s	Selection for heavier seeds, earlier growth	Weaver et al. [36]; Warwick [37]
Echinochloa crus-galli (L.) P. Beauv (barnyardgrass)	Northward invasion of Quebec from the USA in the nineteenth century	More rapid maturation at each life cycle stage	Potvin [38]
Fallopia japonica (Houtt.) Ronse Decr. (Japanese knotweed)	Northward range expansion in both Ontario and British Columbia, Canada	Genotypes with different temperature thresholds and potential hybridization	Bourchier and Van Hezewijk [21]
Panicum miliaceum L. (proso millet)	Northward invasion into Canadian cropland by early 1970s	Modified seed germination and dispersal characteristics	Bough et al. [39]; McCanny et al. [40]; McCanny and Cavers [41]
Setaria faberi Herrm. (giant foxtail)	Northward expansion into Canadian cropland by the 1970s	Modified life history traits	Warwick et al. [42]
Setaria viridis (L.) P. Beauv. (green foxtail)	Survival at Churchill, Manitoba, at nearly 60° N latitude (normal range 45–55° N)	Leaf production at low temperatures	Douglas et al. [43]; Swanton et al. [44]
Sorghum halepense (L.) Pers. (Johnsongrass)	Northward expansion by 5° latitude between 1926 and 1979	Northern populations annual (vs. perennial southern population)	Warwick et al. [45]

As indicated in Table 2.1, *Fallopia japonica* has recently shown rapid range expansion in both Ontario and British Columbia, Canada [21]. *Fallopia japonica* is one of the most aggressive invasive plants in Europe, having invaded large areas of the UK and other countries after its introduction from Asia [53]. Bourchier and Van Hezewijk compared the distribution of *Fallopia japonica* between 1971–2000 and 2000–2008 weather normals, and found an increase of 53 % in suitable habitats for this invasive plant in Southern Ontario for the period 2000–2008, when temperatures were warmer than for the 1971–2000 period [21]. In contrast, only 35 % of the habitat in Southern Ontario was suitable for *Fallopia japonica* for the 1971–2000 period. Similarly, with approximately half of the potentially suitable regions in British Columbia, encompassing 12.3 % of the total territory invaded by *Fallopia japonica*, there is much more potential for future expansion. Costs of *Fallopia japonica* control in both Europe and North America are considerable (e.g., between £ 1 and 8 m^{-2} in the UK), and thus climate change scenarios threaten to increase these costs [21]. There are also indications that *Fallopia japonica* could develop increased frost tolerance [54] and genetic diversity through hybridization with *Fallopia sachalinense* (F. Schmidt) Ronse Decraene (giant knotweed), as has been documented in Washington State [55] and British Columbia [56].

It is likely that the actual range expansions already observed in North America (examples in Table 2.1 and additional examples) are just harbingers of a much larger-scale expansion of weed distributions in response to climate change and other factors, given the high dispersal characteristics of many of these weeds and their ability to respond to climate change. By simply examining eight species with the potential for range expansion (Table 2.2) [23, 57–64], it is evident that there are many ways a species can achieve this expansion.

Pueraria lobata (Willdenow) Ohwi (kudzu) is one of the world's worst invasive plants [65], largely known for spreading through large areas of the southeastern USA [62]. It is restricted to fairly warm environments. Sasek and Strain noted that its range is limited by low winter temperatures of −15 °C [61]. Thus, its potential northward advance in response to warming temperatures, as predicted by Sasek and Strain [61], is of great concern. In 2009, a patch of *Pueraria lobata* was found growing near Leamington, Ontario, the first verified occurrence in Canada [66]. Leamington has one of the warmest climates in Canada, but winter temperatures occasionally fall below −15 °C, such as in 1937, when a record low of −32 °C was recorded.

Another well-documented invasive plant that threatens to expand its range and impact in North America is *Sorghum halepense* (L.) Pers. (Johnsongrass) (Fig. 2.1; Table 2.2). *Sorghum halepense* is a perennial C_4 grass native to Eurasia that was initially adapted to the warm, humid conditions of Mediterranean Europe and Africa, and originally introduced to North America as a forage crop in the southern USA [67]. Increasingly broad climatic tolerance among new ecotypes found in North America includes increasing cold tolerance in rhizomes [45]. Furthermore, although southern populations in North America are perennial, northern populations generally have an annual life history [45]. Utilizing a damage niche model to project the potential change in the distribution of *Sorghum halepense* under a "business as usual" greenhouse gas emissions scenario, McDonald et al. predicted that the damage niche in maize could move 200–650 km northward by 2100 [23]. This would result in a much greater impact on US maize-growing regions (e.g., Midwestern USA) for this weed, which has historically had greater impacts in the southern USA [23]. In addition to this predicted increased negative impact on maize production, *Sorghum halepense* is also an increasing threat to native tallgrass prairie ecosystems under climate change [68]. When *Sorghum halepense* invades native tallgrass prairies, its rhizomatous growth allows it to advance at rates of 0.45 m $year^{-1}$ in addition to the deleterious effects of its allelopathic leachates on native vegetation [68].

Table 2.2 Potential range expansion for selected weed species in North America due to climate change and adaptive traits possessed by these particular weed species

Weed species	Potential range expansion	Critical adaptive weed traits	References
Abutilon theophrasti Medik. (velvetleaf)	Damage niche could move 200–650 km northward in North America	Coadaptation with crops (especially maize)	McDonald et al. [23]
Bromus tectorum L. (cheatgrass)	Greater expansion of populations within Canada	De novo creation of weedy genotypes among ecotypes already present	Valliant et al. [57]
Buddleja davidii Franch. (ornamental butterfly bush)	Capable of northward movement in North America	Lack of local adaptation; current range well within climatic requirements	Ebeling et al. [58]
Impatiens glandulifera Royle (Himalayan balsam)	Potential for northward range expansion in North America	Differences in flowering phenology among populations	Kollmann and Bañuelos [59]; Clements et al. [60]
Pueraria lobata (Willdenow) Ohwi (kudzu)	Capable of expanding northward to the −15 °C (low winter temperature) isocline	Rapid growth rate and ability to establish extensive systems of vines and respond to CO_2 enrichment	Sasek and Strain [61]; Lindgren et al. [62]
Phalaris arundinacea L. (reed canarygrass)	Capable of more rapid evolution at edges of range in response to climate change	Greater genetic variation and greater biomass of introduced populations	Lavergne and Molofsky [63]
Sorghum halepense (L.) Pers. (johnsongrass)	Damage niche could move 200–600 km northward in North America	Coadaptation with crops (especially maize)	McDonald et al. [23]
Tamarix ramosissima Ledeb. (saltcedar)	North of Montana in North America	Increased investment in seedling root growth	Sexton et al. [64]

Europe

Europe has a long history of nonnative species introductions, and is reported to have as many as 2843 plant species of non-European origin [69]. Most of these species possess narrow ranges and do not cause notable management problems. However, the most alarming examples of rapid range expansion of plant species in Europe are exemplified by nonnatives (Table 2.3) [59, 70–72], suggesting that their potential range has not yet been attained. Unlike for North America, documented examples of weed range expansions due especially to climate change are limited for Europe, although range expansion limited by temperature is evident for many species. A study across altitudinal gradients in Italy found that life-form was strongly linked to

Table 2.3 Recent expansion of ranges for selected nonnative weed species in Europe

Weed species	Range expansion	Attributed mechanism(s)	References
Ailanthus altissima (P. Mill) Swingle (tree of heaven)	Range expansion in southern and central Europe	Effective wind dispersal	DAISIE [70]
Ambrosia artemisiifolia L. (common ragweed)	Range expansion in central Europe	Niche expansion from ruderal to agricultural habitats	DAISIE [70]; Essl et al. [71]
Fallopia japonica (Houtt.) Ronse Decr. (Japanese knotweed)	Range expansion in central Europe	Hybridization	Hollingsworth and Bailey [72]; DAISIE [70]
Heracleum mantegazzianum Sommier and Levier (giant hogweed)	In northern and central Europe	Niche expansion	DAISIE [70]
Impatiens glandulifera Royle (Himalayan balsam)	Range expansion throughout Europe	Differences in flowering phenology among populations	Kollmann and Bañuelos [59]; DAISIE [70]
Robinia pseudoacacia L. (black locust)	Range expansion throughout Europe	Nitrogen fixation	DAISIE [70]
Rosa rugosa Thunb. ex Murray (rugosa rose)	Range expansion throughout Europe	Effective dispersal by floating seeds	DAISIE [70]

temperature for native species but not alien species [73]. The implication was that alien plants in Europe are less limited by temperature and depend more on anthropogenic factors for their spread, which does not preclude the influence of climate change but does highlight other important factors, such as land use.

Among the most troublesome nonnative species in Europe, *Heracleum mantegazzianum* Sommier and Levier (giant hogweed) and *Impatiens glandulifera* Royle (Himalayan balsam) have been successful invaders in most of northern Europe (Fig. 2.1) [17, 59], suggesting that climate is not limiting their northern distributional limit [53]. Adaptation to northern climate conditions has resulted in northern populations of *I. glandulifera* flowering earlier and producing less biomass compared with southern populations [59]. In Finland, both of these species are continuously expanding their ranges and are considered to be the most important nonnative species that should be targeted for control [74].

The distribution of two other notable nonnatives—*Fallopia japonica* and *Ambrosia artemisiifolia* L. (common ragweed; Fig. 2.1)—is evidently limited by temperature [26, 53] in Europe. The core of their ranges is situated in central Europe [70], and even though they are regularly found further north (*Ambrosia artemisiifolia* as a contaminant of sunflower [*Helianthus annuus* L.] seeds used as bird feed and *Fallopia japonica* as an ornamental), they are currently not able to establish permanent populations there. Rapid range expansion of *Ambrosia artemisiifolia* has been reported from France [75], Austria [71], and Hungary [76], whereas *Fallopia japonica* has been especially problematic in the UK [4]. A key factor in the range expansion of *Ambrosia artemisiifolia* has been a niche shift from ruderal to agricultural habitats, whereas for *Fallopia japonica* hybridization has been the most important (Table 2.3). Because the distribution of both of these species is limited by temperature, it can be assumed that they may take advantage of climate warming to expand their ranges northwards in the future.

In arable habitats, several weed species are regarded to have potential for range expansion in the future [25]. Many of these species are found in several cropping systems and are difficult to control, making them economically important weeds to manage [77]. In Europe, the number of arable weed species declines from south to north following a climate gradient [78], suggesting that the climate warming may result in the movement of weedy species towards northern limits of the climate zones [79]. Indeed, predictions of future changes in suitable climate conditions for weed species have provided evidence for this (Table 2.4) [27]. For example, *Amaranthus retroflexus* L. (redroot pigweed) has been predicted to successfully establish about 500 km further north in the future than under current climate conditions. It is also notable that for several weed species, including *Chenopodium rubrum* L. (red goosefoot), *Papaver argemone* L. (long prickly head poppy), and *Sinapis arvensis* L. (wild mustard), a 60% decline in suitable climate conditions in the future is predicted. These findings highlight species-specific responses to climate change and subsequent effects on their ranges, which should, however, be confirmed with field experiments that assess reproductive success outside the current ranges [80, 81]. Importantly, the combined effects of northward extension of crop regions and potential climate warming trends [82] will have significant consequences for weed management in Europe, as in other continents [18, 20, 27].

Oceana

The position of Australia and New Zealand in the southern hemisphere makes the potential trend of southward changes in weed distribution of greater interest than northern regions, from the standpoint of climate change. As with North America and Europe, such distributional changes (southward) have been documented for a number of weed species (Table 2.5) [22, 28, 29, 83, 84].

Gallagher et al. showed how 11 species of alien perennial grasses that were either shortlisted or listed as weed threats of national importance in Australia could undergo alterations in distribution with climate change [29]. As is seen worldwide, these grasses can have devastating impacts on crops and rangeland in Australia, even though most of these were deliberately introduced as forage grasses. However, because these grasses are already near the edge of their climate optima in Australia, if temperatures along with drought conditions increase through climate change as predicted, these grasses may not be able to maintain their current extent [29]. For example, the range of *Cortaderia selloana* (J.A. and J.H. Schultes) Aschers. and Graebn. (pampas grass) is predicted to decline by 68% by 2050, according to a climate change scenario for Australia generated from four models. Likewise, the other grass species listed in Table 2.5 that have hitherto featured rapidly expanding ranges are predicted by Gallagher et al. [29] to decline by 2050: *Eragrostis curvula* (Schrad.) Nees (African lovegrass), *Nassell aneesiana* (Trin, and Rupr.) Barkworth

Table 2.4 Potential range expansion from 2051 to 2080 for selected weed species in Europe due to climate change under two scenarios developed from climate land-use (CLU) models. (From Hyvönen et al. [27])

Species	Distribution	Percent change with less severe scenario	Percent change with more severe scenario
Amaranthus graecizans L. (Mediterranean amaranth)	Southern	19.3	25.9
Amaranthus retroflexus L. (redroot pigweed)	Southern and central	26.2	44.6
Cardaria draba (L.) Desv.	Southern and central	15.3	21.6
Chenopodium vulvaria L. (stinking goosefoot)	Southern and central	20.1	26.1
Consolida regalis Gray (royal knight's-spur)	Central	5.7	11.7
Coronopus squamatus (Forssk.) Asch. (greater swinecress)	Southern and central	13.8	14.5
Fumaria parviflora Lam. (fineleaf fumitory)	Southern and western	26.3	39.6
Fumaria vaillantii Loisel (earth smoke)	Central	3.6	0.4
Neslia paniculata (L.) Desv. (ball mustard)	Southern, central, and eastern	10.8	17.6
Papaver hybridum L. (round prickly head poppy)	Southern and western	26.5	41.1
Papaver rhoeas L. (common poppy)	Throughout Europe	18.3	22.8
Portulaca oleracea var. oleracea L. (common purslane)	Southern and central	23.2	30.9
Ranunculus arvensis L. (corn buttercup)	Southern and central	21.3	31.3

(Chilean needle grass), *Nassella trichotoma* (Nees) Hack. (serrated tussock), *Sporobolus africanus* (Poir.) Robyns and Tourn. (Parramatta grass), *Sporobolus pyramidalis* Beauv. (giant rat's tail grass), and *Themeda quadrivalvis* (L.) Kuntze (grader grass). Interestingly, *Eragrostis curvula* and *Themeda quadrivalvis* are also found in areas in Australia where their global niches do not predict them, whereas the other species have not yet fully expanded to occupy all areas in Australia where the climate is suitable [29]. These findings suggest that some grasses could defy range predictions based on global climate niche modeling. Certainly, in the case of *Nasella* spp. (needle grass species), there is a high likelihood that their current high levels of infestation in southeast Australia will be reduced by anticipated warming trends in the region [29]. The case of *Nassella* spp. is further complicated by the fact that these grasses still have suitable areas that are yet to be colonized based on their potential climate niches; hence, there are still many parts of Australia that are vulnerable to new invasions by *Nassella* spp.

Table 2.5 Recent expansion of ranges for selected weed species in Oceana

Weed species	Range expansion	Attributed mechanism(s)	References
Aira cupaniana Guss. (silvery hairgrass)	Expansion into more arid parts of New South Wales	Decreased plant height	Buswell et al. [83]
Cortaderia selloana (J.A. and J.H. Schultes) Aschers. and Graebn. (pampas grass)	Spread widely throughout southern Australia since 1901	Wind dispersal, superior competitor	Gallagher et al. [29]
Eragrostis curvula (Schrad.) Nees (African lovegrass)	Spread widely throughout southern, eastern, and western Australia since 1914	Superior competitor, low palatability to grazers; drought tolerant	Gallagher et al. [29]
Facelis retusa (Lam.) Sch. Bip. (trampweed)	Expansion into more arid parts of New South Wales	Decreased plant height	Buswell et al. [83]
Hyparrhenia hirta (L.) Stapf. (Coolatai grass)	Rapid spread to all Australian states from New South Wales since the 1890s	High population growth in areas with 175–600 mm of rain; ability to germinate in a broad range of temperatures and moisture levels	Chejara et al. [84]
Nassella neesiana (Trin, and Rupr.) Barkworth (Chilean needle grass)	Rapid spread in southeastern Australia since 1941	Transport in sheep wool, mowing equipment, and by natural means (wind, water)	Gardener et al. [28]; Bourdöt et al. [22]; Gallagher et al. [29]
Nassella trichotoma (Nees) Hack. (serrated tussock)	Rapid spread in southeastern Australia since 1937	A variety of seed dispersal mechanisms: wind, animals, and contaminated feed	Gallagher et al. [29]
Polycarpon tetraphyllum (L.)L. (fourleaf allseed)	Expansion into more arid parts of New South Wales	Decreased plant height	Buswell et al. [83]
Silene gallica L. (French catchfly)	Expansion into more arid parts of New South Wales	Decreased plant height	Buswell et al. [83]
Sporobolus africanus (Poir.) Robyns and Tourn. (Parramatta grass)	Spread widely throughout Australia since 1802	Seed dispersal via wind, water, and machinery	Gallagher et al. [29]
Sporobolus pyramidalis Beauv. (giant rat's tail grass)	Spread rapidly across northern and eastern Australia since 1921	Seeds carried on animal fur	Gallagher et al. [29]
Themeda quadrivalvis (L.) Kuntze (grader grass)	Spread rapidly across northern and eastern Australia since 1935	Seed dispersal via wind, water, and machinery	Gallagher et al. [29]
Trifolium glomeratum L. (cluster clover)	Expansion into more arid parts of New South Wales	Decreased plant height	Buswell et al. [83]

As with weeds throughout the world [6, 48], questions remain as to how stable the fundamental niches of invasive weeds are and as assumed by the modeling performed by Gallagher et al. [29]. Buswell et al. studied a variety of invasive species to determine whether their morphology had changed under environmental conditions experienced since their introduction to Australia [83]. Seventy percent of the species examined using historical herbarium records showed changes in at least one trait, with the most commonly observed modification being a change in height. Changes in height were found in 8 of the 21 species, including *Facelis retusa* (Lam.) Sch. Bip. (trampweed), *Polycarpon tetraphyllum* (L.)L. (four leaf allseed), *Silene gallica* L. (French catchfly), and *Trifolium glomeratum* L. (cluster clover; Table 2.5), species now inhabiting the arid western portion of New South Wales. The relatively large number of cases where height decreased is consistent with the view that these species underwent selection for decreased height to cope with the more arid conditions, and the implication is that this kind of selection could enable invasive plants like these to thrive and spread under climate change. Buswell et al. did not eliminate the possibility that much of the observed height differences might be due to phenotypic plasticity, but through tracking the trend in the same location through time, there were strong indications that genetic differences were represented in the results [83].

As with North America and Europe, there is a considerable amount of research being done to predict further spread of invasive species. For example, *Senna obtusifolia* L. (sicklepod) is currently found in northern Australia but could move into southern regions, as ecotypes have been identified that are adapted to a range of temperatures [85].

Asia

For Asia, the north–south axis does not carry the same significance in terms of climate change and floral distribution as seen in North America, Europe, and Oceana. However, recent modeling work by Qin et al. suggests dramatic impacts of predicted climate change on the distribution of two highly invasive alien annual herbs, *Ambrosia artemisiifolia* and *Ambrosia trifida* L. (giant ragweed), in China [86]. The projected distribution under future climatic change scenarios suggests an overall increase in *Ambrosia artemisiifolia* distribution with further expansion to climatically favorable locations in southeastern China and northern Taiwan. The models reveal a significant progressive northward and northeastward contraction in *Ambrosia trifida*'s range in China, with southeastern Tibet and northern Taiwan as novel and potentially suitable climate habitats.

Japan experiences a substantial range in climate along its north–south axis and its overall climate has changed in the past few decades as indicated by earlier flowering of *Ginkgo biloba* L. trees in the spring [87]. Tsutsumi modeled the potential range expansion of *Senecio madagascariensis* Poir. (fireweed) in Japan using a maximum entropy ecological niche modeling approach (Maxent) [88]. Tsutsumi predicted that the northern extent of *Senecio madagascariensis* could shift from where it is currently at 36.9° N in southern Tohoku to 39.1° N in central Tohoku, with temperature in the warmest quarter of the year identified as the key variable predicting its range [88]. Thus, if temperatures continue to rise, invasive plants like *Senecio madagascariensis* that require higher temperatures to complete their life cycle would be predicted to move northward on the Japanese archipelago.

Africa and South America

As with Oceana, the southward expansion of weed distributions is important to monitor in Africa and South America under climate change. *Lantana camara* L. (lantana) is a troublesome invasive species native to South America, now thriving in many subtropical habitats throughout the world (Fig. 2.1). It was first collected by Dutch explorers in the 1640s, who introduced it to European gardens where hybrids were produced and distributed throughout the globe [89–91]. In Kenya, *Lantana camara* has spread over large areas, threatening wildlife habitat [92]. It has likewise spread rapidly in other African countries. For example, the area infested by *Lantana camara* increased by roughly sixtyfold in South Africa between 1962 and 2000 [90, 93, 94]. As in other parts of the world, *Lantana camara* affects South African ecosystems in many deleterious ways, including impacts on water availability and biodiversity [94]. Vardien et al. utilized correlative bioclimatic models to demonstrate that *Lantana camara* threatens to expand its South African range even further within a few decades [94]. A climate scenario for the year 2050 predicted increased habitat suitability in areas where *Lantana camara* was already present and in a few additional areas as well, thus forecasting greater impacts of the plant on South African ecosystems unless it is effectively managed. Taylor and Kumar modeled potential changes in climate suitability for *Lantana camara* in Queensland, Australia, and found that although further potential invasion was predicted for 2030, for future climate projected to 2070 and 2100, a dramatic decline in available niche space in Queensland was forecasted [95]. Once temperatures increase beyond a specific threshold and combine with reduced rainfall, this subtropical plant has physiological challenges [96]. These predictions from northeastern Australia are likely to apply to more equatorial regions of Africa and other areas near the equator around the globe as well.

Lygodium microphyllum (Cav.) R. Br. (the Old World climbing fern), native to the Old World wet tropics and subtropics of Africa, Asia, Australia, and Oceania [97], has been recently causing serious problems in a variety of habitats in the New World, in particular, Florida where it overtops trees and smothers plant communities with its extensive growth [98]. Goolsby modeled its potential spread in the New World and found that much of Florida as far north as Tampa was vulnerable to further invasion, based on current climatic conditions [99]. Furthermore, Goolsby's model [99] suggested that large areas within Central America, the Caribbean, and South America could be colonized by *Lygodium microphyllum* in addition to Jamaica and

Guyana, which have already been colonized [97]. Aggressive Old World subtropical weeds, such as *Lygodium microphyllum,* are likely to benefit from changing climates in terms of spread through the New World subtropics, but little information is available on South American weed invasions.

Influence of Climate Change on Weed Competition in Cropping Systems

There have been various predictions made about the impact of climate change on the world's crops; some predictions have indicated an increase in crop yields by as much as 13 % by 2050 due to increased CO_2 production [100], but other factors such as increased drought severity, increased temperature during the growing cycle, and changes in monsoon patterns may result in lower yield gains or worse [101, 102]. For example, the Fourth Assessment Report of the World Meteorological Organization/The United Nations Environment Programme (WMO/UNEP) Intergovernmental Panel on Climate Change (IPCC) released in 2007 predicted that decreases in moisture availability will accompany rising temperatures in semiarid regions of Asia, Africa, and Latin America and will become even more severe during the twenty-first century [103]. Longer growing seasons in temperate regions could provide wider windows of opportunity for infestations of weeds and other pests [104]. Weeds figure prominently in the uncertainty surrounding the impact of climate change on cropping systems, because of differential impacts of changing moisture regimes, temperatures, and CO_2 levels on weeds versus crops [101]. In general, weeds have an advantage in making use of increased CO_2 levels [101, 105] with some possible exceptions [106], indicating that without adequate weed management, climate change could lead to increased yield losses due to weeds. Furthermore, many weed species may grow better than crops under warmer conditions. In the USA, many of the invasive plants infesting warm season crops originated in tropical or warm temperate regions, and warming would foster even greater northward movement of such crops [46]. If the crops themselves can be grown further north under climate change, the weeds are likely to move with them, as has frequently been observed in the past when a crop is grown outside its normal range [5].

It is clear that prediction of changes in weed distribution often requires tracking potential changes in the distribution of the cropping systems with which the weeds have coevolved [5]. As Marini et al. point out, disentangling the relationship between alien and native plant communities is challenging because human settlement and economic activity is related to climate [73, 107]. Types and relative intensity of management dictate weed distribution at local and regional scales [34]. For example, intensification of cereal production practices dramatically altered weed community composition in Spain between 1976 and 2007 [108]. Geographic shifts in the areas where particular crops are grown have been predicted based on projected changes in climate [103, 109]. Another management response already being carefully considered and implemented is strategically changing the varieties and/or types of crops being grown within certain regions in anticipation of climate change [100, 109, 110].

Biological and Evolutionary Responses of Weeds to Climate Change

Actual and predicted ecophysiological changes in weed species in response to climate change are complex, due to the range of impacts that climate change has on plant biology. Four important components to consider are: responses to increased CO_2 levels; responses to increased temperatures predicted in temperate regions; responses to increased climate variability such as changes in moisture availability; and, finally, actual and predicted evolutionary adaptation by weeds under climate change. These four components are examined in detail as follows.

Responses to Increased CO_2 Levels

Plants in general are expected to exhibit an increased growth rate in response to enhanced CO_2 levels due to the obvious impact of increased carbon available for fixation via photosynthesis, when increased CO_2 level is considered independent of other climatic factors. As mentioned previously, weed species often exhibit differential responses to CO_2 levels compared with crop species. Ziska examined the responses of six invasive species to past, present, and projected future CO_2 levels (284, 380, and 719 μmol mol^{-1}, respectively) and observed an average increase in plant biomass of 46 % among the species tested, with the greatest response of 72 % by *Cirsium arvense* L. (Canada thistle) [111]. The growth response from past to present was significantly higher at 110 %, with *Cirsium arvense* once again exhibiting the greatest response (180 %). The remaining species *Convolvulus arvensis* L. (field bindweed), *Euphorbia esula* L. (leafy spurge), *Sonchus arvensis* L. (perennial sowthistle), *Centaurea stoebe* L. (spotted knapweed), and *Centaurea solstitialis* L. (yellow star thistle) also demonstrated increased biomass with CO_2 enrichment [111]. Given the threefold greater response of these species compared with other plant species tested, Ziska concluded that increased CO_2 levels could result in increased selection of these weed species over other plants, including crop plants [111]. However, it should be noted that in some cases, crop plants could produce greater biomass relative to weeds. For example, Ziska observed that when soybeans (*Glycine max* L.) were grown in competition with the well-known agronomic weed, *Abutilon theophrasti* Medik. (velvetleaf), with elevated CO_2 levels, competition favored the soybeans whereby soybean yield components, including pod numbers $plant^{-1}$, were higher under increased CO_2 [106]. Similarly, when another widespread agronomic weed *Chenopodium album* L. (common lamb's-quarters) was grown in a Canadian pasture community, CO_2 enrichment failed to elicit increased growth in *Chenopodium album* [112]. Thus, although increased CO_2 levels clearly promote enhanced weed growth in general, weed–crop competition relationships should be evaluated on a case-by-case basis.

Photosynthetic pathway is a critical factor to consider, particularly since many of the world's most problematic weeds are C_4 plants, which tend to photosynthesize more efficiently at higher temperatures, and therefore would likely be able to utilize increased CO_2 levels compared to C_3 plants, including crops [113]. Alberto et al. observed that the C_4 weed *Echinochloa glabrescens* Munro ex Hook. f. gained a photosynthetic advantage over rice (*Oryza sativa* L.) at elevated temperatures [113].

Of course, it is unrealistic to consider increased CO_2 in isolation of other factors, given the multiple dimensions of climate change. For example, Nonhebel found that although enhanced CO_2 was predicted to increase wheat (*Triticum aestivum* L.) yields in Europe, these effects would be neutralized by reduced growth due to an elevated temperature [114]. However, if drought occurred due to limited water availability under climate change, wheat yields would be depressed [114]. Similar three-way interactions should apply to other plant species, including many weed species, highlighting the value of studying the impact of multiple climatic factors on weeds and their interactions with crops or native plants.

Responses to Increased Temperatures Predicted in Temperate Regions

Examples of potential responses of weeds to increased temperatures, as well as the actual observed trends due to recent climate warming, are presented in Tables 2.1–2.5 and discussed with respect to specific global regions. Fundamentally, warmer temperatures provide plants with an opportunity to complete their life cycles within a shorter time period, and thus allow certain weeds to occupy areas where they formerly could not reproduce, or at least reproduce efficiently enough to be successful. For example, recent trends towards warmer temperatures in temperate regions have promoted earlier flowering in a variety of ecosystems and over a wide range of plant species [115, 116]. However, there are also upper temperature thresholds that impact plant growth; and thus in areas such as parts of Australia where climate change is predicted to result in very high temperatures accompanied by moisture shortage, even very prolific weed species may decline in such extreme conditions [29]. Woody plants, particularly shrubs and other weeds physiologically adapted to high temperatures and moisture stress, are expected to prevail in regions experiencing extreme impacts of rising temperatures [117]. Similarly, C_4 plants are generally predicted to fare better in regions experiencing warming. For example, *Rottboellia cochinchinensis* (Lour.) W.D. Clayton (itchgrass), a C_4 plant that currently occupies a fairly restricted range in the southern USA, is expected to expand its range and become troublesome over a much larger area of the country [118]. Established latitudinal niches for particular weed species could be dramatically altered if mean temperatures in temperate arable regions increase by just a few degrees Celsius as anticipated under even relatively conservative climate change scenarios [23]. In Europe, the distribution of C_4 weeds is limited to a great extent by temperature [79]. Climate warming could enable their expansion from temperate to boreal region [27].

Responses to Increased Climate Variability

Simply accounting for the impacts of rising levels of greenhouse gasses on a single parameter, such as temperature, is not sufficient to account for climate change; and in fact, it is not appropriate to refer to climate change as "global warming" because the extreme variation expected from region to region may lead to cooler temperatures in certain regions. Thus, efforts to account for an array of potential climate variables affected by global climate change are vital and such efforts have been underrepresented in the literature to date [119]. More variable precipitation, particularly, when accompanied by warmer temperatures, will likely lead to increased drought; some studies have addressed the potential impact of such conditions on plant life, with once again weeds expected to fare better than native plants or crop plants [120–122]. This projected scenario is of concern and will require improved weed management strategies in regions especially affected by this climatic variability [123].

Considerable research on how annual plant communities are adapted to variable rainfall levels in arid and semiarid ecosystems has been performed [124, 125]. This knowledge may provide some clues as to the ability of annual weeds to thrive in currently more mesic environments but predicted to experience more variable rainfall as the climate changes [126]. Robinson and Gross studied how increased variability in precipitation would impact two common annual weeds in the USA, *Chenopodium album* and *Setaria faberi* Herrm. (giant foxtail) [126]. As might be expected, the impact of periodic droughts varied, depending on plant life stage, but overall, *Chenopodium album* showed greater resilience under prolonged dry periods than *Setaria faberi*. The authors concluded that predicted changes in the occurrence and severity of precipitation events are likely to alter relative abundances of agriculturally important weeds [126].

Actual and Predicted Evolutionary Adaptation by Weeds under Climate Change

Although considerable progress has been made in predicting potential weed distributions using bioclimatic models, the accuracy of these predictions may always be compromised to some extent by the ability of weeds to adapt [6]. There are an increasing number of examples illustrating how weed species are effectively adapting to selection pressures as the climate continues to change, such as extremes of heat or moisture deprivation (Tables 2.2 and 2.5) [6, 48]. Therefore, as in all areas of weed science, there is a need for more "evolutionary thinking" [127] in assessing the risk of increased weed invasions under climate change. Weed species, in general, can be viewed as a complex set of ecotypes occurring both in their native and introduced ranges, with genetic variation and the potential for natural selection even among species exhibiting a predominantly selfing strategy [5, 127]. Climate change introduces additional selection pressures that add to the dynamic environments that weeds must already adapt to, including adaptation to normally variable climate, heterogeneous environments, herbicide applications, and various other anthropogenic factors. Recent studies have examined actual and potential evolutionary responses to climate changes, such as the work by Buswell et al. in Australia that revealed changes in weed attributes [83], or the evolutionary adaptations of *Fallopia japonica* or *Sorghum halepense* observed in North America [21, 45, 56].

Many examples of weed evolutionary responses are related to adaptations to increased temperatures, but Franks et al. demonstrated that *Brassica rapa* L. (field mustard) exhibited an adaptive response to multiyear droughts that may come to typify modified climates in more arid areas [128]. They compared pre-drought genotypes to genotypes exposed to growing seasons shortened by drought and found first flowering was advanced by between 1.9 and 8.6 days, depending on the population. There is a need to account for such interpopulation variability in weed responses to be able to better model and respond to such evolutionary changes under the various predicted climate change scenarios [6, 127, 129].

Use of Predictive Models to Develop Early Warning Systems

As indicated throughout this chapter, although much research has been done to gain a better understanding of the impact of climate change on weeds, there are still many gaps in our knowledge. As with climate change in general, the use of predictive models is critical and as climate change modeling is undergoing continual fine-tuning, predictive models for weed responses must be concomitantly updated. However, improved predictive modeling requires better empirical understanding of weed evolution and ecology [130]. Weed ecology is a multifaceted discipline requiring extensive knowledge specific to the individual weed species. Taken together, predictive modeling and improved understanding of weed biology and ecology should provide more effective early warning systems to track changes in weed distributions and their impact under climate change.

An increasing number of modeling approaches and techniques have been developed in recent years to predict and map the expected ranges of habitat suitability of various invasive weeds. These models have helped to assess the potential geographic distributions of these species in response to different factors, such as climate change and land-use type. Among these modeling approaches, niche-based species distribution models [131–133] have been used for assessing and identifying regions with a high invasion potential. These models use known species distributions combined with a set of environmental variables to develop a correlative model of the environmental conditions that meet a species' ecological requirements [133]. This approach makes it possible to project modeled niches into new regions and under future climate change scenarios, and ultimately to estimate the geographical distribution of suitable conditions.

One concern with distribution predictions for invasive species is the type of range information (native, invaded, or full ranges) that is used to develop the niche-based models. A comprehensive list of concerns with modeling plant species' distributions is provided in Thuiller et al. [134]. Models based on data from the native range assume that the same environmental factors determine the distribution of the species in the adventive range. However, these predictions may not adequately reflect the distribution of an invasive species in its introduced range [135]. On the other hand, "climatic niche shifts are rare among terrestrial plant invaders" giving support for the use of ecological niche models [136].

Models using data from a previously invaded range may be more appropriate and accurate because the fundamental niche is likely to be more fully realized in invaded ranges than in the native range, where the species may be constrained by such factors as competition and dispersal barriers. Combining native and alien distribution records in models (i.e., discriminative correlative models) may be most insightful [137], but has been shown not to consistently improve model projections [138]. However, given niche differences across native and introduced ranges, distribution models using introduced range data alone may be more valid in some cases [35, 139]. Despite contrasting views about the robustness of these methods, niche-based models and the presence of species in a region can provide insight into those factors that may favor or restrict the expansion of invasive species over a large scale. For example, Qin et al. used ecological niche maximum entropy (Maxent) modeling based on occurrence records of *Ambrosia artemisiifolia* and *Ambrosia trifida* in their native and introduced ranges to predict the potential distribution of these two invasive congeners in China under current climate conditions and under future climate projections [86].

The recent development of a large range of modeling tools for predicting potential weed distributions and to determine what areas are vulnerable to further weed invasion with or without climate change is very promising [140]. However, this ability to predict possible future changes in weed distribution must be accompanied by sufficient resources, to develop these tools, utilize them, and then take active measures to prevent new weed infestations.

Conclusion

As with the other issues discussed with respect to climate change, there is an urgent need to address the threats posed by weeds under climate change, and formulate better management approaches fostered by comprehensive research efforts. One distinction between problems associated with weed management and climate change, not necessarily true for other climate change issues, is that with or without climate change, worsening of economic and ecological impacts of weeds is fairly predictable. Invasive plants will continue to spread regardless of climate change, particularly in an era of ever-increasing globalization and movement of goods [31]. As seen in this review, many plant invasions will progress farther and faster if assisted by climate change, and climate change is likely to increase evolutionary adaptation of weeds to climate extremes. However, encouraging progress is being made in our understanding of many of these, often, complex dynamics. Prospects for tracking and mitigating anticipated impacts of weeds in a changing climate are good, provided sufficient resources continue to be made available.

References

1. Oerke E-C, Dehne HW, Schönbeck F, Weber A (1994) Crop production and crop protection—estimated losses in major food and cash crops. Elsevier, Amsterdam
2. Pimentel D, Lach L, Ziniga R, Morrison D (2000) Environmental and economic costs associated with non-indigenous species in the United States. BioScience 50:53–65
3. Oerke E-C (2006) Crop losses to pests. J Agric Sci 144:31–43
4. EEA (2012) The impacts of invasive alien species in Europe. EEA Technical report No 16/2012. European Environment Agency, Copenhagen. doi:10.2800/65864
5. Clements DR, DiTommaso A, Jordan N, Booth B, Murphy SD, Cardina J et al (2004) Adaptability of plants invading North American cropland. Agric Ecosyst Environ 104:379–398
6. Clements DR, DiTommaso A (2011) Climate change and weed adaptation: can evolution of invasive plants lead to greater range expansion than forecasted? Weed Res 51:227–240
7. Altieri MA, Liebman M (eds) (1988) Weed management in agroecosystems. CRC, Boca Raton
8. Ghersa CM, Roush ML, Radosevich SR, Corday SM (1994) Coevolution of agroecosystems and weed management. BioScience 44:85–94
9. Mack RN (1981) The invasion of *Bromus tectorum* L. into western North America: an ecological chronicle. Agroecosystems 7:145–165
10. Mack RN (1986) Alien plant invasion into the Intermountain West: a case history. In: Mooney HA, Drake JA (eds) Ecology of biological invasions of North America and Hawaii. Springer-Verlag, New York, pp 191–213
11. Mack RN, Simberloff D, Londsdale WM, Evans H, Cloutand M, Bazzaz FA (2000) Biotic invasions: causes, epidemiology, global consequences and control. Ecol Appl 10:689–710
12. Westbrooks RG (2004) New approaches for early detection and rapid response to invasive plants in the United States. Weed Technol 18:1468–1471
13. Gressel J (1999) Tandem constructs: preventing the rise of superweeds. Trends Biotechnol 17:361–366
14. Harker KN, Clayton GW, Blackshaw RE, O'Donovan JT, Lupwayi NZ, Johnson EN et al (2005) Glyphosate-resistant spring wheat production system effects on weed communities. Weed Sci 53:451–464
15. Heap I (2013) The international survey of herbicide resistant weeds. http://www.weedscience.org. Accessed 25 June 2013
16. Pyšek P, Prach K (1995) Invasion dynamics of *Impatiens glandulifera*—a century of spreading reconstructed. Biol Conserv 74:41–48
17. Pyšek P, Jarošík V, Müllerová J, Pergl J, Wild J (2008) Comparing the rate of invasion by *Heracleum mantegazzianum* at continental, regional and local scales. Divers Distrib 14:355–363
18. Kriticos DJ, Yonow T, McFadyen RE (2005) The potential distribution of *Chromolaena odorata* (Siam weed) in relation to climate. Weed Res 45:246–254
19. Dukes JS, Pontius J, Orwig D, Garnas JR, Rodgers VL, Brazee N et al (2009) Responses of insect pests, pathogens, and invasive plant species to climate change in the forests of northeastern North America: what can be predicted? Can J For Res 39:231–248
20. Potter KJB, Kriticos DJ, Watt MS, Leriche A (2009) The current and future potential distribution of *Cytisus scoparius*: a weed of pastoral systems, natural ecosystems and plantation forestry. Weed Res 49:271–282

21. Bourchier RS, Van Hezewijk BE (2010) Distribution and potential spread of Japanese knotweed (*Polygonum cuspidatum*) in Canada relative to climatic thresholds. Invasive Plant Sci Manage 3:32–39
22. Bourdöt GW, Lamoureaux SL, Watt MS, Manning LK, Kriticos DJ (2012) The potential global distribution of the invasive weed *Nassella neesiana* under current and future climates. Biol Invasions 14:1545–1556
23. McDonald A, Riha S, DiTommaso A, Degaetano A (2009) Climate change and geography of weed damage: analysis of US maize systems suggests the potential for significant range transformation. Agric Ecosyst Environ 130:131–140
24. Peterson AT, Papes M, Kluza DA (2003) Predicting the potential invasive distributions of four alien plant species in North America. Weed Sci 51:863–868
25. Weber E, Gut D (2005) A survey of weeds that are increasingly spreading in Europe. Agron Sustain Dev 25:109–121
26. Beerling DJ, Huntley B, Bailey JP (2009) Climate and the distribution of *Fallopia japonica*: use of an introduced species to test the predictive capacity of response surfaces. J Veg Sci 6:269–282
27. Hyvönen T, Luoto M, Uotila P (2012) Assessment of weed establishment risk in a changing European climate. Agric Food Sci (Finland) 21:348–360
28. Gardener MR, Whalley RDB, Sindel BM (2003) Ecology of *Nassella neesiana*, Chilean needle grass, in pastures on the northern Tablelands of New South Wales. I. Seed production and dispersal. Aust J Agric Res 54:613–619
29. Gallagher RV, Duursma DE, O'Donnell J, Wilson PD, Downey PO, Hughes L et al (2013) The grass may not always be greener: projected reductions in climatic suitability for exotic grasses under future climates in Australia. Biol Invasions 15:961–975
30. Mack RN (1991) The commercial seed trade: an early disperser of weeds in the United States. Econ Bot 45:257–273
31. Ding J, Mack RN, Lu P, Ren M, Huang H (2008) China's booming economy is sparking and accelerating biological invasions. BioScience 58:317–324
32. Weber E, Sun S-G, Li B (2008) Invasive alien plants in China: diversity and ecological insights. Biol Invasions 10:1411–1429
33. Elton CS (1958) The ecology of invasions by animals and plants. Methuen, London
34. Thomas AG, Leeson JY (2007) Tracking long-term changes in the arable weed flora of Canada. Pages 43–69 In: Invasive plants: inventories, strategies and action. Topics in Canadian Weed Science, Volume 5. Sainte Anne de Bellevue, Québec: Canadian Weed Science Society—Société canadienne de malherbologie. 165 pp
35. Broennimann O, Treier UA, Müller-Shärer H, Thuiller W, Peterson AT, Guisan A (2007) Evidence of climatic niche shift during biological invasion. Ecol Lett 10:701–709
36. Weaver SE, Dirks VA, Warwick SI (1985) Variation and climatic adaptation in northern populations of *Datura stramonium*. Can J Bot 63:1303–1308
37. Warwick SI (1990) Allozyme and life history variation in five northwardly colonizing North American weedy species. Plant Syst Evol 169:41–54
38. Potvin C (1986) Biomass allocation and phonological differences among southern and northern populations of the C4 grass *Echinochloa crus-galli*. J Ecol 74:915–923
39. Bough MA, Colosi JC, Cavers PB (1986) The major weedy biotypes of proso millet (*Panicum miliaceum* L.) in Canada. Can J Bot 64:1188–1198
40. McCanny SJ, Bough M, Cavers PB (1988) Spread of prosomillet (*Panicum miliaceum* L.) in Ontario, Canada. I. Rate of spread and crop susceptibility. Weed Res 28:59–65
41. McCanny SJ, Cavers PB (1988) Spread of proso millet (*Panicum miliaceum* L.) in Ontario, Canada. II. Dispersal by combines. Weed Res 28:67–72
42. Warwick SI, Thompson BK, Black LD (1987) Life history variation in populations of the weed species *Setaria faberi*. Can J Bot 65:1396–1402
43. Douglas BJ, Thomas AG, Morrison IN, Maw MG (1985) The biology of Canadian weeds 70. *Setaria viridis* (L.) Beauv. Can J Plant Sci 65:669–690

44. Swanton CJ, Huang JZ, Deen W, Tollenaar M, Shrestha A, Rahimian H (1999) Effects of temperature and photoperiod on *Setaria viridis*. Weed Sci 47:446–453
45. Warwick SI, Phillips D, Andrews C (1986) Rhizome depth: the critical factor in winter survival of *Sorghum halepense* (L.) Pers. (Johnsongrass). Weed Res 26:381–387
46. Patterson DT (1993) Implications of global climate change for impact of weeds, insects and plant diseases. Int Crop Sci 1:273–280
47. Wolfe DW, Ziska L, Petzoldt C, Seaman A, Chase L, Hayhoe K (2008) Projected change in climate thresholds in the Northeastern U.S.: implications for crops, pests, livestock, and farmers. Mitig Adapt Strateg Glob Change 13:555–575
48. Clements DR, DiTommaso A (2012) Predicting weed invasion in Canada under climate change: evaluating evolutionary potential. Can J Plant Sci 92:1013–1020
49. Ziska LH, Runion GB (2007) Future weed, pest and disease problems for plants. In: Newton PCD et al (ed) Agroecosystems in a changing climate. CRC, Boston, pp 262–279
50. Potvin C, Strain BR (1985) Effects of CO_2 enrichment and temperature on growth in two C_4 weeds, *Echinochloa crus-galli* and *Eleusine indicn*. Can J Bot 63:1495–1499
51. Boese SR, Wolfe DW, Melkonian JJ (1997) Elevated CO_2 mitigates chilling-induced water stress and photosynthetic reduction during chilling. Plant Cell Environ 20:625–632
52. Bunce JA, Ziska LH (2000) Crop ecosystem responses to climate change: crop/weed interactions. In: Reddy KR, Hodge RF (eds) Climate change and global crop productivity. CABI, Wallingford, pp 333–354
53. Beerling DJ (1993) The impact of temperature on the northern distribution limits of the introduced species *Fallopia japonica* and *Impatiens glandulifera* in north-west Europe. J Biogeogr 20:45–53
54. Maruta E (1983) Growth and survival of current-year seedlings of *Polygonum cuspidatum* at the upper distribution limit on Mt. Fuji. Oecologia 60:316–320
55. Urgenson LS, Reichard SH, Halpern CB (2009) Community and ecosystem consequences of giant knotweed (*Polygonum sachalinense*) invasion into riparian forests of western Washington, USA. Biol Conserv 142:1536–1541
56. Gillies SL (2011) Japanese knotweed: invasion of the clones? In: Biology on the cutting edge. Pearson Education, Toronto
57. Valliant MT, Mack RN, Novak SJ (2007) Introduction history and population genetics of the invasive grass *Bromus tectorum* (Poaceae) in Canada. Am J Bot 94(7):1156–1169
58. Ebeling SK, Welk E, Auge H, Bruelheide H (2008) Predicting the spread of an invasive plant: combining experiments and ecological niche models. Ecography 31:709–719
59. Kollmann J, Bañuelos MJ (2004) Latitudinal trends in growth and phenology of the invasive alien plant *Impatiens glandulifera* (Balsaminaceace). Divers Distrib 10:377–385
60. Clements DR, Feenstra KR, Jones K, Staniforth R (2008) The biology of invasive alien plants in Canada. 9. *Impatiens glandulifera* Royle. Can J Plant Sci 88(2):403–417
61. Sasek TW, Strain BR (1990) Implications of atmospheric CO_2 enrichment and climatic change for the geographical distribution of two introduced vines in the USA. Clim Change 16:31–51
62. Lindgren CJ, Castro KL, Coiner HA, Nurse RE, Darbyshire SJ (2013) The biology of invasive alien plants in Canada. 12. *Pueraria montana* var. *lobata* (Willd.) Sanjappa & Predeep. Can J Plant Sci 93:71–95
63. Lavergne S, Molofsky J (2007) Increased genetic variation and evolutionary potential drive the success of an invasive grass. Proc Natl Acad Sci U S A 104(10):3883–3888
64. Sexton JP, McKay JK, Sala A (2002) Plasticity and genetic diversity may allow saltcedar to invade cold climates in North America. Ecol Appl 12(6):1652–1660
65. Global Invasive Species Database (2007) 100 of the World's worst weeds. Invasive species specialist group. I.U.C.N.—The World Conservation Union Species Survival Commission. University of Auckland, New Zealand. http://www.issg.org/database/welcome. Accessed 25 June 2013
66. Waldron GE, Larson BMH (2012) Kudzu vine, *Pueraria montana*, adventive in southern Ontario. Can Field Nat 126(1):31–33

67. Warwick SI, Black LD (1983) The biology of Canadian weeds. 61. *Sorghum halapense* (L.) Pers. Can J Plant Sci 63:997–1114
68. Rout ME, Chrzanowski TH, Smith WK, Gough L (2013) Ecological impacts of the invasive grass *Sorghum halepense* on native tallgrass prairie. Biol Invasions 15:327–339
69. Lambdon PW, Pyšek P, Basnou C, Hejda M, Arianoutsou M, Essl F et al (2008) Alien flora of Europe: species diversity, temporal trends, geographical patterns and research needs. Preslia 80:101–149
70. DAISIE (2009) Handbook of alien species in Europe. Springer, Dordrecht, 399 pp
71. Essl F, Dullinger S (2009) Kleimnbauer I. Changes in the spatio-temporal patterns and habitat preferences of *Ambrosia artemisiifolia* during its invasion of Austria. Preslia 81:119–133
72. Hollingsworth ML, Bailey JP (2000) Hybridisation and clonal diversity in some introduced *Fallopia* species (Polygonaceae). Watsonia 23:111–121
73. Marini L, Gaston KJ, Prosser F, Hulme PE (2012) Alien and native plant life-forms respond differently to human and climate pressures. Glob Ecol Biogeogr 21:534–544
74. MAFF (2012) Finland's national strategy on invasive alien species. Ministry of Agriculture and Forestry in Finland, Helsinki, 125 pp
75. Chauvel B, Dessaint F, Cardinal-Legrand C, Bretagnolle F (2006) The historical spread of *Ambrosia artemisiifolia* L. in France from herbarium records. J Biogeogr 33:665–673
76. Novák R, Dancza I, Szentey L, Karamán J (2009) Arable weeds of hungary. Fifth national weed survey (2007–2008). Ministry of Agricuture and Rural Development, Budapest
77. Schroeder D, Mueller-Schaerer H, Stinson CSA (1993) European weed survey in 10 major crop systems to identify targets for biological control. Weed Res 33:449–458
78. Glemnitz M, Radics L, Hoffmann J, Czimber G (2006) Weed species richness and species composition of different arable field types.—A comparative analysis along a climate gradient from south to north Europe. J Plant Dis Prot 20:577–586
79. Hyvönen T, Glemnitz M, Radics L, Hoffmann J (2011) Impact of climate and land use type on the distribution of Finnish casual arable weeds. Eur Weed Res 51:201–208
80. Milberg P, Andersson L (2006) Evaluating the potential northward spread of two grass weeds in Sweden. Acta Agric Scand B Soil and Plant Sci 56:91–95
81. Hyvönen T (2011) Impact of temperature and germination time on the success of a C_4 weed in a C_3 crop: *Amaranthus retroflexus* and spring barley. Agric Food Sci 20:183–189
82. Olesen JE, Trnka M, Kersebaum KC, Skjelvåg AO, Seguin B, Peltonen-Sainio P et al (2011) Impacts and adaptation of European crop production systems to climate change. Eur J Agron 34:96–112
83. Buswell JM, Moles AT, Hartley S (2011) Is rapid evolution common in introduced plant species? J Ecol 99:214–224
84. Chejara VK, Kriticos DJ, Kristiansen P, Sindel BM, Whalley DB, Nadolny C (2010) The current and future potential geographical distribution of *Hyparrhenia hirta*. Weed Res 50:174–184
85. Dunlop EA, Wilson JC, MacKey AP (2006) The potential geographic distribution of the invasive weed *Senna obtusifolia* in Australia. Weed Res 46:404–413
86. Qin Z, DiTommaso A, Zhang JE (2013) Potential distribution of two *Ambrosia* species in China under projected climate change. Weed Res (under review)
87. Matsumoto K, Ohta T, Irasawa M, Tsutomu N (2003) Climate change and extension of the *Ginkgo biloba* L. growing season in Japan. Glob Change Biol 9:1634–1642
88. Tsutsumi M (2011) Current and potential distribution of *Senecio madagascariensis* Poir. (fireweed), an invasive alien plant in Japan. Grassl Sci 57:150–157
89. Howard RA (1970) *Lantana camara*—a prize and a peril. Am Hortic Mag 49:31
90. Stirton CH (1977) Some thoughts on the polyploid *Lantana camara* L, (Verbenaccae). In: Proceedings of the second national weeds conference. Balkema, Cape Town, pp 321–340
91. Morton JF (1994) Lantana, or red sage (*Lantana camara* L. (Verbenaceae)), notorious weed and popular garden flower; some cases of poisoning in Florida. Econ Bot 48:259–270
92. Nanjappa HV, Saravanane P, Ramachandrappa BK (2005) Biology and management of *Lantana camara* L.-a review. Agric Rev 26:272–280
93. Le Maitre DC, Versfeld DB, Chapman RA (2000) The impact of invading alien plants on water resources in South Africa: a preliminary assessment. Water SA 26:397–408

94. Vardien W, Richardson DM, Foxcroft LC, Thompson GD, Wilson JRU, Le Roux JJ (2012) Invasion dynamics of *Lantana camara* L. (sensulato) in South Africa. S Afr J Bot 81:81–94
95. Taylor S, Kumar L (2012) Potential distribution of an invasive species under climate change scenarios using CLIMEX and soil drainage: a case study of *Lantana camara* L. in Queensland. Aust J Environ Manage 114:414–422
96. Day MD, Wiley CJ, Playford J, Zalucki MP (2003) Lantana: current management status and future prospects. ACIAR Monograph. Australian Centre for International Agricultural Research, Canberra
97. Pemberton RW (1998) The potential of biological control to manage Old World climbing fern (*Lygodium microphyllum*), an invasive weed in Florida. Am Fern J 88:176–182
98. Pemberton RW, Ferriter AP (1998) Old world climbing fern (*Lygodium microphyllum*), a dangerous invasive weed in Florida. Am Fern J 88:165–175
99. Goolsby JA (2004) Potential distribution of the invasive old world climbing fern, *Lygodium microphyllum* in North and South America. Nat Areas J 24:351–353
100. Jaggard KW, Qi A, Ober ES (2010) Possible changes to arable crop yields by 2050. Philos Trans R Soc B 365:2835–2851
101. Hatfield JL, Boote KJ, Kimball BA, Ziska LH, Izaurralde RC, Ort D et al (2011) Climate impacts on agriculture: implications for crop production. Agron J 103:351–370
102. Mahajan G, Singh S, Chauhan BS (2012) Impact of climate change on weeds in the rice-wheat cropping system. Curr Sci 102:1254–1255
103. Sivakumar MVK, Stefanski R (2009) Climate change mitigation, adaptation, and sustainability in agriculture. Idojaras 113:89–102
104. Hakala K, Hannukkala A, Huusela-Veistola E, Jalli M, Peltonen-Sainio P (2011) Pests and diseases in a changing climate a major challenge for Finnish crop production. Agric Food Sci (Finland) 20:3–14
105. Ziska LH (2004) Rising carbon dioxide and weed ecology. In: Inderjit (ed) Weed biology and management. Kluwer, The Netherlands, pp 159–176
106. Ziska L (2013) Observed changes in soyabean growth and seed yield from *Abutilon theophrasti* competition as a function of carbon dioxide concentration. Weed Res 53:140–145
107. Marini L, Battisti A, Bona E, Federici G, Martini F, Pautasso M, Hulme PE (2009) Contrasting response of native and alien plant species richness to environmental energy and human impact along alpine elevation gradients. Glob Ecol Biogeogr 18:652–661
108. Cirujeda A, Zaragoza C (2011) Remarkable changes of weed species in Spanish cereal fields from 1976 to 2007. Agron Sustain Dev 31:675–688
109. Singh RP, Prasad PVV, Reddy KR (2013) Impacts of changing climate and climate variability of seed production and seed industry. Adv Agron 118:49–110
110. Bennett RG, Ryan MH, Colmer TD, Real D (2011) Prioritisation of novel pasture species for use in water-limited agriculture: a case study of Cullen in the Western Australian wheatbelt. Genet Resour Crop Evol 58:83–100.
111. Ziska LH (2003) Evaluation of the growth response of six invasive species to past, present and future atmospheric carbon dioxide. J Exp Bot 381:395–405
112. Taylor K, Potvin C (1997) Understanding the long-term effect of CO_2 enrichment on a pasture: the importance of disturbance. Can J Bot 75:1621–1627
113. Alberto AMP, Ziska LH, Cervancia CR, Manalo PA (1996) The influence of increasing carbon dioxide and temperature on competitive interactions between a C_3crop, rice (*Oryza sativa*) and a C_4 weed (*Echinochloa glabrescens*). Aust J Plant Physiol 23:795–802
114. Nonhebel S (1996) Effects of temperature rise and increase in CO_2 concentration on simulated wheat yields. Eur Clim Change 34:73–90
115. Parmesan C, Yohe GA (2003) A globally coherent fingerprint of climate change impacts across natural systems. Nature 421:37–42
116. Hulme PE (2011) Consistent flowering response to global warming by European plants introduced into North America. Funct Ecol 25:1189–1196
117. Dukes JS, Mooney HA (1999) Does global change increase the success of biological invaders? Trends Ecol Evol 14(4):135–139

118. Patterson DT (1995) Weeds in a changing climate. Weed Sci 43:685–701
119. Rosenzweig C, Neofotis P (2013) Detection and attribution of anthropogenic climate change impacts. Wiley Interdiscip Rev Clim Change 4(2):121–150
120. Hussner A, Meyer C, Busch J (2009) The influence of water level and nutrient availability on growth and root system development of *Myriophyllum aquaticum*. Weed Res 49:73–80
121. Gilgen AK, Signarbieux C, Feller U, Buchmann N (2010) Competitive advantage of *Rumex obtusifolius* L. might increase in intensively managed temperate grasslands under drier climate. Agric Ecos Environ 135:15–23
122. Valerio M, Tomecek MB, Lovelli S, Ziska LH (2011) Quantifying the effect of drought on carbon dioxide-induced changes in competition between a C-3 crop (tomato) and a C-4 weed (*Amaranthus retroflexus*). Weed Res 51:591–600
123. Turner NC, Molyneux N, Yang S, Xiong Y-C, Kadambot HMS (2011) Climate change in south-west Australia and north-west China: challenges and opportunities for crop production. Crop Pasture Sci 62(6):445–456
124. Pake CE, Venable DL (1996) Seed banks in desert annuals: implications for persistence and coexistence in variable environments. Ecology 77:1427–1435
125. Schwinning S, Sala OE (2004) Hierarchy of responses to resource pulses in arid and semiarid ecosystems. Oecologia 141:211–220
126. Robinson TMP, Gross KL (2010) The impact of altered precipitation variability of annual weed species. Am J Bot 97:1625–1629
127. Neve P, Vila-Aiub M, Roux F (2009) Evolutionary-thinking in agricultural weed management. New Phytol 184:783–793
128. Franks SJ, Sim S, Weis AE (2007) Rapid evolution of flowering time by an annual plant in response to a climate fluctuation. Proc Natl Acad Sci U S A 104:1278–1282
129. Beaumont LJ, Gallagher RV, Thuiller W, Downey PO, Leishman MR, Hughes L (2009) Different climatic envelopes among invasive populations may lead to underestimations of current and future biological invasions. Divers Distrib 15:409–420
130. Mortensen DA, Bastiaans L, Sattin M (2000) The role of ecology in the development of weed management systems: an outlook. Weed Res 40:49–62
131. GuisanA ZNE (2000) Predictive habitat distribution models in ecology. Ecol Model 135:147–186
132. Guisan A, Thuiller W (2005) Predicting species distribution: offering more than simple habitat models. Ecol Lett 8:993–1009
133. Peterson AT (2003) Predicting the geography of species' invasions via ecological niche modeling. Q Rev Biol 78:419–433
134. Thuiller W, Albert C, Araújo MB, Berry PM, Cabeza M, Guisan A et al (2008) Predicting global change impacts on plant species' distributions: future challenges. Perspect Plant Ecol 9:137–152
135. Estrada-Peña A, Pegram RG, Barré N, Venzal JM (2007) Using invaded range data to model the climate suitability for *Amblyomma variegatum* (Acari: Ixodidae) in the New World. Exp Appl Acarol 41:203–214
136. Petitpierre B, Kueffer C, Broennimann O, Randin C, Daehler C, Guisan A (2012) Climatic niche shifts are rare among terrestrial plant invaders. Science 335:1344
137. Welk E (2004) Constraints in range predictions of invasive plant species due to non-equilibrium distribution patterns: purple loosestrife (*Lythrum salicaria*) in North America. Ecol Model 179:551–567
138. Webber BL, Yates CJ, LeMaitre DC, Scott JK, Kriticos DJ, Ota N et al (2011) Modelling horses for novel climate courses: insights from projecting potential distributions of native and alien Australian acacias with correlative and mechanistic models. Divers Distrib 17:978–1000
139. Mau-Crimmins TM, Schussman HR, Geiger EL (2006) Can the invaded range of a species be predicted sufficiently using only native-range data?: Lehmann lovegrass (*Eragrostis lehmanniana*) in the southwestern United States. Ecol Model 193:736–746
140. Bradley BA, Blumenthal DM, Wilcove DS, Ziska LH (2010) Predicting plant invasions in an era of global change. Trends Ecol Evol 25:310–318

Chapter 3
Role of Allelopathy in Weed Management

Ahmad Nawaz, Muhammad Farooq, Sardar Alam Cheema and Zahid Ata Cheema

Introduction

Weeds cause substantial yield loss of crops and pose a severe threat to food security for future generations. Controlling weeds in field crops is therefore imperative, but this is a hard nut to crack. However, wise management is quite effective in achieving the target weed control. Several methods of weed management, with varying degrees of effectiveness, are practiced according to the climatic conditions, cropping systems and socioeconomic conditions of the region. Manual and mechanical methods of weed control have been practiced for centuries, but these are inefficient methods, labor intensive and weather dependent [1, 2].

Chemical means of weed control are far cheaper, the most prevalent, and quite effective [2]. Nonetheless, continuous and indiscriminate use of herbicides is posing environmental hazards [3], may cause development of herbicide-resistant weed biotypes [4, 5], and is also creating human health concerns [6–8]. For example, babies born to families living near wheat farms, with continuous use of chlorophenoxy herbicides for weed control, may have 65% greater risk of birth defects related to

M. Farooq (✉) · A. Nawaz · S. A. Cheema · Z. A. Cheema
Allelopathy and Eco-physiology Lab, Department of Agronomy,
University of Agriculture, Jail Road, Faisalabad, Punjab 38040, Pakistan
e-mail: farooqcp@gmail.com

A. Nawaz
e-mail: ahmadnawaz2006@gmail.com

S. A. Cheema
e-mail: sardaralam35@gmail.com

Z. A. Cheema
e-mail: cheemaza@gmail.com

M. Farooq
The UWA Institute of Agriculture, The University of Western Australia,
Crawley WA 6009, Australia

College of Food and Agricultural Sciences, King Saud University,
Riyadh 11451, Saudi Arabia

B. S. Chauhan, G. Mahajan (eds.), *Recent Advances in Weed Management*,
DOI 10.1007/978-1-4939-1019-9_3, © Springer Science+Business Media New York 2014

the circulatory/respiratory system [9]. This situation demands to develop environmentally friendly technology for weed control.

Allelopathy, a naturally occurring ecological phenomenon of interference among organisms, involves the synthesis and release of plant bioactive compounds which are known as allelochemicals [10, 11]. These allelochemicals are capable of acting as natural pesticides and can resolve problems of soil and environmental pollution, resistance development in weed biotypes, and health defects caused by the indiscriminate use of synthetic herbicides [11].

Allelopathy may be employed for weed management in field crops through mix cropping intercropping [12], use of surface mulch [13], soil incorporation of plant residue [14], allelopathic aqueous extracts [12, 15], combined application of allelopathic aqueous extracts with lower herbicide doses [16, 17], and crop rotation [11, 18, 19]. In addition, smothering crops, such as rye (*Secale cereale* L.), buck wheat (*Fagopyrum esculentum* Moench), black mustard (*Brassica nigra* L.), and Sorghum–Sudan grass hybrids can also be used for controlling different weeds [20]. Conventional breeding and modern biotechnological approaches can be used to breed the crop cultivars having more weed-suppressive ability through allelopathy.

Most plants with allelopathic properties, including wheat (*Triticum aestivum* L.), rice (*Oryza sativa* L.), maize (*Zea mays* L.), barley (*Hordeum vulgare* L.), sorghum (*Sorghum bicolor* [L.] Moench), oat (*Avena sativa* L.), rye, and pearl millet (*Pennisetum glaucum* [L.] R. Br.), belong to the family Poaceae. However, plants from other families, including *Brassica* spp., alfalfa (*Medicago sativa* L.), eucalyptus (*Eucalyptus* spp.), tobacco (*Nicotiana tabacum* L.), sesame (*Sesamum indicum* L.), sweet potato (*Ipomoea batatas* [L.] Lam.), sunflower (*Helianthus annuus* L.), and mulberry (*Morus alba* L.), also possess allelopathic properties [21–26].

In this chapter, potential application of allelopathy for weed management in field crops is discussed. Furthermore, role of conventional breeding and biotechnology in improving the allelopathic activity of crop genotypes for weed suppression is also included.

Intercropping

Intercropping, growing of two or more crops together at the same time in the same field, can be used as an effective weed management strategy [27]. Recent studies have suggested to use intercropping allelopathic crops as an effective element for integrated weed management, particularly in low-input farming systems [11, 28, 29]. Allelopathic intercrops suppress the weeds by shade effect, weed–crop competition, and by the release of certain allelochemicals [27, 28, 30]. In addition to weed suppression, intercropping may provide several other benefits, including increase in net returns and biological diversity, less chance of complete failure of crop, better use of resources, and suppressive effects on diseases and insect pests [30].

Intercropping maize with fodder legumes like Spanish tick-clover (*Desmodium uncinatum* [Jacq.] DC.) and green leaf desmodium (*Desmodium intortum* [Mill.]

Urb.) significantly reduced giant witchweed (*Striga hermonthica* [Del.] Benth) infestation in maize compared to sole maize crop [31]. In another field study, intercropping sesame, soybean (*Glycine max* [L.] Merr.), and sorghum in cotton (*Gossypium hirsutum* L.) suppressed the density and total dry biomass of purple nutsedge (*Cyperus rotundus* L.) [32]. Intercropping sorghum, sunflower, mungbean (*Vigna radiata* [L.] R. Wilczek; Table 3.1) [33], bean species (Table 3.1) [34], cassava (*Manihot esculenta* [L.]) Crantz) [35], horse gram (*Macrotyloma uniflorum* [Lam.] Verdc.) [36], groundnut (*Arachis hypogaea* L.), sweet potato [37], and legumes [38] with maize reduced the densities and dry biomass of many weed species. Maize–legume intercrop is also effective in reducing weed density and weed biomass compared to sole crops [39]. Bansal found that intercropping of linseed (*Linum usitatissimum* L.) with wheat suppressed corn buttercup (*Ranunculus arvensis* L.; Table 3.1) [40]. Bitter bottle gourd (*Cucurbita pepo* L.) intercropping in maize at lower density also decreased weed biomass (Table 3.1) [41]. In general, crop yield increases with simultaneous decrease in weed growth if the intercrops are more effective than sole crops in usurping resources from weeds [42]. Intercropping sorghum with fodder cowpea (*V. unguiculata* [L.] Walp.) suppressed densities and total biomass of several weeds [43]. Growing leek (*Allium porrum* L.) and celery (*Apium graveolens* L. var dulce [Mill.] Pers.) as intercrop shortened the critical period for weed control in the intercrop compared to pure stand of leek [44]. Likewise, pea (*Pisum sativum* L.) intercropped with barley, instead of sole crop, increased the competitive ability towards weeds [45]. Similarly, intercrops of wheat–canola–pea and wheat–canola provided better weed suppression than each individual crop grown alone [46].

In another study, after first weeding in rice, black gram (*Phaseolus mungo* [L.] Hepper) was seeded as intercrop, which effectively controlled rice weeds (Table 3.1) [47]. Banik et al. found that intercropping wheat and chickpea (*Cicer arietinum* L.) decreased the total weed density and weed biomass compared to monocrop of both crops (Table 3.1) [48]. In a two-year study, intercropping pea with false flax (*Camelina sativa* [L.] Crantz) suppressed the weeds by 52–63% more than sole crop of pea [49]. Similarly, intercrop of finger millet (*Eleusine coracana* [L.] Gaertn.) and green leaf desmodium decreased the density of giant witchweed more than monocrops of these crops [50]. Intercropping wheat with canola (*B. napus* L.) significantly reduced density and fresh/dry weight of littleseed canarygrass (*Phalaris minor* Retz.), broad-leaved duck (*Rumex* obtusifolius L.), swine cress (*Coronopus didymus* [L.] Sm.), and common lambsquarters (*Chenopodium album* L.) than the sole crops of both (Table 3.1) [51]. Similarly, intercropping canola with wheat suppressed annual ryegrass (*Lolium rigidum* Gaud.) and common lambsquarters [52]. In a two-year study, growing one strip of canola between two strips of wheat caused substantial decrease in weed density and dry weight than sole wheat crop [53]. Similarly, weed population was also significantly suppressed when either one strip of lentil or chickpea was planted between two strips of wheat [53].

Although intercrops are able to suppress weeds through the release of allelochemicals, the use of intercropping as a strategy for weed control should be approached carefully.

Table 3.1 Effect of different intercrops on weed suppression

Main crop	Intercrop	Weeds suppressed	Reference
Linseed (*Linum usitatissimum* L.)	Wheat (*Triticum aestivum* L.)	Corn Buttercup (*Ranunculus arvensis* L.)	Bansal [40]
Maize (*Zea mays* L.)	Hyacinth-bean (*Lablab purpureus* (L.) Sweet), Jack-bean (*Canavalia ensiformis* (L.) DC.), Butterfly pea (*Pueraria phaseoloides* (Roxb.) Benth.)	Itchgrass (*Rottboellia cochinchinensis* (Lour.) W.D. Clayton)	Cruz et al. [34]
Rice (*Oryza sativa* L.)	Black gram (*Phaseolus mungo* (L.) Hepper)	Junglerice (*Echinochloa colona* (L.) Link.), large crabgrass (*Digitaria sanguinalis* (L.) Scop.), yellow foxtail (*Setaria glauca* (L.) Beauv.)	Midya et al. [47]
Wheat (*Triticum aestivum* L.)	Chick pea (*Cicer arietinum* L.)	Bermudagrass (*Cynodon dactylon* (L.) Pers.), wild oat (*Avena fatua* L.), purple nutsedge (*Cyperus rotundus* L.), common lambsquarters (*Chenopodium album* L.), sweet clover (*Melilotus indica* (L.) Pall.), honey clover (*Melilotus albus* Medik.), scarlet pimpernel (*Anagallis arvensis* L.), swinecress (*Coronopus didymus* (L.) Sm.)	Banik et al. [48]
Pea (*Pisum sativum* L.)	False flax (*Camelina sativa* (L.) Crantz)	Black bindweed (*Fallopia convolvulus* (L.) Á.Löve), common sowthistle (*Sonchus oleraceus* L.), chamomile (*Matricaria chamomilla*L.)	Saucke and Ackermann [49]
Maize (*Zea mays* L.)	Bitter bottle gourd (*Cucurbita pepo* L.)	Pigweed amaranth (*Amaranthus retroflexus* L.), field bindweed (*Convolvulus arvensis* L.)	Fujiyoshi [41] Fujiyoshi et al. [168]
Cotton (*Gossypium hirsutum* L.)	Sesame (*Sesamum indicum* L.), Soybean (*Glycine max* (L.) Merr.) and Sorghum (*Sorghum bicolor* (L.) Moench)	Purple nutsedge (*Cyperus rotundus* L.)	Iqbal et al. [12]

Table 3.1 (continued)

Main crop	Intercrop	Weeds suppressed	Reference
Finger millet (*Eleusine coracana* (L.) Gaertn.)	Green leaf desmodium (*Desmodium intortum* (Mill.) Urb.)	Giant witchweed (*Striga hermonthica* (Del.) Benth)	Midega et al. [50]
Maize (*Zea mays* L.)	Sorghum (*Sorghum bicolor* (L.) Moench), Sunflower (*Helianthus annuus* L.) and mungbean (*Vigna radiate* (L.) R. Wilczek)	Purple nutsedge (*Cyperus rotundus* L.), field bindweed (*Convolvulus arvensis* L.), horse purslane (*Trianthema portulacastrum* L.)	Khalil et al. [33]
Wheat (*Triticum aestivum* L.)	Canola (*Brassica napus* L.)	Annual ryegrass (*Lolium rigidum* Gaud.), common lambsquarter (*Chenopodium album* L.)	Khorramdel et al. [52]
Wheat (*Triticum aestivum* L.)	Canola (*Brassica napus* L.)	Littleseed canarygrass (*Phalaris minor* Retz.), Broad-leaved dock (*Rumex obtusifolius* L.), Swine cress (*Coronopus didymus* (L.) Sm.), common lambsquarter (*Chenopodium album* L.)	Naeem [51]

Crop Rotation

Accumulation of autotoxins and spread of plant pests are the major limitations of monoculture cropping systems [23, 54, 55]. Crop rotation, growing of different crops in sequence in a particular field over a definite time period, can be helpful in overcoming the autotoxicity and decreasing the pressure of plant pests, including weeds, pathogens and insects [11, 19].

Inclusion of allelopathic crops in crop rotation may be useful to control weeds [27]. In crop rotation, the allelochemicals released in the rhizosphere by plant roots and decomposition of previous crop residues help in weed suppression [56, 57]. For instance, in the crops following sorghum, weed population is significantly reduced due to the release of sorghum allelochemicals [58]. Therefore, in rice–wheat system, growing of allelopathic crops after wheat harvest and prior to rice transplantation may be useful to control weeds in rice.

A 10-year study on different crop rotations, viz. maize–soybean, continuous maize, and soybean–wheat–maize, indicated a significant decrease in giant foxtail (*Setaria faberi* [R.] Hermm.) density in the succeeding crop following wheat [59]. Likewise, in sunflower–wheat rotation, density and dry biomass of wild oat (*Avena fatua* L.) and Canada thistle (*Cirsium arvense* [L.] Scop.) were decreased

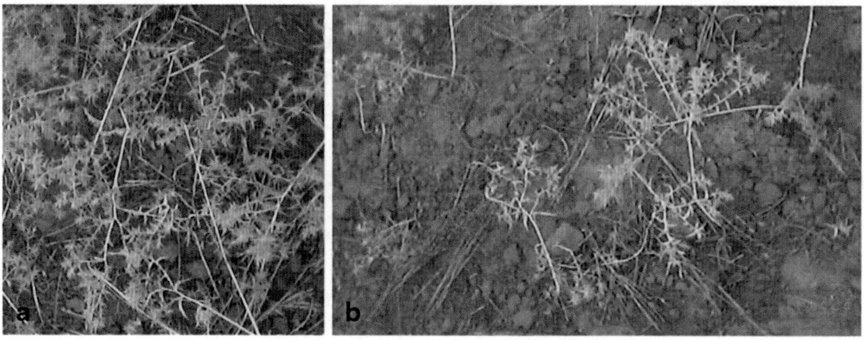

Fig. 3.1 Wild safflower infestation in field previously occupied by wheat and chickpea field. **a** After wheat harvest. **b** After chickpea harvest

significantly in the succeeding wheat crop after sunflower [60]. In a rotation study conducted in Russia, weed suppression of up to 40% was noted in crops raised in rotation with rapeseed [61]. Al-Khatib et al. noted that weed suppression in peas varied between different green manure crops [62]. One month after planting, the highest weed population was in green pea following wheat, whereas the lowest was in green pea following rapeseed. Wild safflower (*Carthamus oxyacantha* [M.] Bieb.) is a noxious weed of the rainfed areas of Pakistan. However, its population in field vacated by wheat is always higher than in the chickpea-vacated fields (Fig. 3.1), owing to release of certain allelochemicals from the chickpea roots. Thus, proper rotation of crops in any cropping system in a specific region can be used as a successful strategy to control weeds without reliance upon chemical, manual, and mechanical methods used for centuries.

Mulching

In mulching, crop residues (or other materials) are applied on soil surface and/or incorporated into the soil. Mulching inhibits the germination and seedling growth of weeds through the release of certain allelochemicals [63, 64], producing microbial phytotoxins during decomposition, and physically obstructing the growth of seedlings [65]. Mulching also increases the soil's water-holding capacity [66].

In 1979, Lockerman and Putnam floated the idea to use allelopathic crop residues as mulch [67]. Afterward, several researchers have evaluated the potential use of allelopathic crop residues as surface-applied or soil-incorporated mulches for weed suppression in field crops [13, 58, 68]. Sorghum is the most-studied crop in this regard. For example, surface-applied sorghum mulch (10–15 t ha^{-1}) in maize at sowing provided weed control of about 26–37% [69], whereas in cotton, surface-applied sorghum mulch (3.5–10.5 t ha^{-1}) reduced the weed density by 23–65% [13]. In aerobic rice, incorporation of sorghum residue (8 t ha^{-1}) reduced the weed density and total dry biomass by 50% [70].

Purple nutsedge is one of the most noxious weeds. Allelopathic mulching has also been very effective in managing this cumbersome weed. For instance, surface-applied and soil-incorporated sorghum mulch (15 t ha^{-1}) reduced the purple nutsedge density by 40–45% [71]. In another study, Ahmad et al. reported that sorghum residues suppressed the broad-leaved dock, littleseed canarygrass, field bind weed, common lambsquarters, purple nutsedge, and scarlet pimpernel (*Anagallis arvensis* L.) [72].

Other than sorghum, several other allelopathic mulches also provide a good weed control. For example, sunflower mulching suppressed the germination and seedling growth of several weeds [73]. Likewise, application of rye mulch and its root residues controlled redroot pigweed (*Amaranthus retroflexus* L.), common lambsquarters, and common ragweed (*Ambrosia artemisiifolia* L.) by 90% in tobacco, sunflower, and soybean in no-tilled system [74]. Mulching of subterranean clover (*Trifolium subterraneum* L.) and rye suppressed different weeds in tobacco, sorghum, sunflower, maize, and soybean [75]. Likewise, application of rice mulch provided a good control of several weeds in wheat [76].

Use of wheat residues as surface mulch suppressed the density and dry weight of several weeds in maize–legume intercropping [64]. Likewise, soil incorporation of wheat straw suppressed the horse purslane (*Trianthema portulacastrum* L.) growth [77]. Soil incorporation of mint marigold (*Tagetes minuta* L.) suppressed purple nutsedge and barnyard grass (*Echinochloa crus-galli* [L.] P. Beauv.), the two most problematic weeds of rice [78], whereas application of root and leaf powder of Malabar catmint (*Anisomeles indica* L.) mulch reduced the density and dry mass of littleseed canarygrass in wheat field [79].

Combined application of more than one allelopathic mulch has been found more effective in weed management than their sole application. For instance, mulching residues of *Brassica,* sunflower, and sorghum suppressed the horse purslane and purple nutsedge; nonetheless, combined application of these residues provided better weed control than sole application of these crop residues [14, 80]. Sunflower mulch applied on soil surface alone or in mixture with legume and buckwheat suppressed weeds; however, the mixed application was more effective in this regard [81]. In another study on wheat, surface application of sorghum, sunflower, or *Brassica* substantially suppressed weeds; however, combined application was more effective [53]. Thus, allelopathic crop mulches, either surface applied or soil incorporated, can be used to control various weed biotypes in different agro-ecological regions of the world.

Use of Cover Crops

Cover crops are widely used for weed management in field crops [82, 83]. Cover crop suppresses weeds by covering the soil surface [84] through competition, release of allelochemicals, stimulation of microbial allelochemicals, shading effect, and through alteration in soil physicochemical properties [85], or weed germination inhibition through physical barriers [86–88]. Most of the crops used as cover crops—including cowpea, sunhemp (*Crotalaria juncea* L.), alfalfa, yellow sweet clover (*Melilotus officinalis* [L.] Pall.), ryegrass, and velvet bean (*Mucuna pruriens* [L.] DC.)—belong to the legume family [89]. Use of leguminous crops as cover crop substantially decreased the population of barnyard grass [90], while use of barley as cover crop suppressed many weed species in soybean [91].

Rye and oat are also considered as potential cover crops. For instance, rye residues reduced the emergence of common ragweed, green foxtail (*Setaria viridis* [L.] P. Beauv.), redroot pigweed, and common purslane (*Portulaca oleracea* L.) by 43, 80, 95, and 100%, respectively [92]. Barnes et al. reported 90% reduction in weed biomass in a cover crop of rye compared to unplanted controls [93]. Similarly, different oat cultivars reduced the germination of common lambsquarters from 10 to 86% [94]. Rye as cover crop inhibited the seedling emergence of yellow foxtail (*Setaria glauca* [L.] Beauv.) [95]. Hoffman et al. reported that due to increase in the density of rye plantation, leaf number, growth, and dry matter production of barnyard grass seedlings were suppressed owing to allelopathy other than weed–crop competition [96].

Sudex hybrid (sorghum × Sudan grass) is often used as summer cover crop due to its rapid growth habit and strong ability to suppress different weed species [97]. Red spiderlily (*Lycoris radiata* [L'Hér.] Herb.) can also be used as ground cover crop to suppress weeds because its dead leaves contain lycorine, an allelochemical with strong suppressive ability against several rice weeds [98]. In Mexico, morning glory (*I. tricolor* Cav.) is used as an important summer cover crop for controlling weeds in sugarcane fields during fallow periods. Peters and Zam opined that tall fescue (*Festuca arundinacea* Schreb.) can be grown as a cover crop for controlling large crabgrass (*Digitaria sanguinalis* [L.] Scop.) weed in multiple crops [99]. In crux, inclusion of cover crops, especially leguminous crops in different cropping systems, can be useful to manage different weed genotypes, depending upon the socioeconomic conditions of the farmers.

Use of Allelopathic Water Extracts

Benefits of using crop allelopathic water extracts have been explored in several studies for their good efficacy to control several weed types. These water-soluble allelochemicals are extracted in water and then are utilized for managing weeds [100]. Application of sorghum water extract (*Sorgaab*) has been very effective in suppressing weeds [19, 101–104]. For instance, *Sorgaab* application suppressed common lambsquarters, broad-leaved dock, swine cress, Indian fumitory (*Fumaria parviflora* Lam.) [101], wild oat, field bindweed, and littleseed canarygrass [103, 104] in wheat. Other than wheat, *Sorgaab* application also suppressed the weeds in rice [105], cotton [106], canola [15, 107], mungbean [102], sunflower [108], soybean [109], and maize [69, 110].

In soybean, *Sorgaab* application at 25 and 50 days after sowing (DAS) reduced the total weed dry weight by 20–42% [109], whereas in maize, *Sorgaab* application reduced the total weed density and total weed dry weight by 34–57 and 13–34%, respectively [110]. In sunflower, *Sorgaab* application 20 DAS decreased the density of purple nutsedge and horse purslane by 10–21% and dry weight of weeds by 18–29%, respectively with yield increase of 25% [108].

Combined application of allelopathic water extracts may be a better option to control weeds than the individual application of these extracts. For example, combined application of sunflower, sorghum, and eucalyptus (*Eucalyptus camaldulensis* Dehnh.) water extracts was more effective for weed suppression in wheat than their sole application [111]. In another study in wheat, mixed application of *Sorgaab* and sunflower water extract was more effective in suppressing the littleseed canarygrass and wild oat than the individual extracts [26]. Mixed application of *Sorgaab* and sunflower and *Brassica* water extracts reduced the total weed dry weight by 55% in wheat [53].

Although complete weed control has not been achieved by the application of allelopathic water extract, there exists a great scope for its use in organic agriculture.

Combined Application of Allelopathic Water Extracts with Reduced Doses of Herbicides

Though weed management through the use of allelopathic water extracts is economical as well as environmentally friendly, the decrease in weed biomass is less than the target. Nonetheless, these allelopathic water extracts may be applied in combination with reduced rates of herbicides for effective weed control [11, 19].

Herbicides applied along with allelopathic compounds could have supportive action, affecting the same or different weed species. A reduced level of herbicide may be feasible to provide weed control when it operates simultaneously with allelopathic compounds [112]. Cheema et al. evaluated the combined effect of concentrated *Sorgaab* with a reduced dose of herbicide in maize crop [113]. Various doses of atrazine (50, 100, and 150 g a.i. ha^{-1}) were combined with *Sorgaab* (12 L ha^{-1}), while atrazine at 300 g a.i. ha^{-1} was sprayed as standard dose. Combined application of atrazine at 150 g a.i. ha^{-1} and *Sorgaab* at 12 L ha^{-1} was as effective as atrazine at 300 g a.i. ha^{-1} alone in controlling weeds such as horse purslane, field bindweed, and purple nutsedge. In another study, combined application of concentrated *Sorgaab* at 12 L ha^{-1} and pendimethalin at 0.5 g a.i. ha^{-1} at sowing decreased the horse purslane density and biomass by 72 and 76%, respectively. Similarly, application of *Sorgaab* at 12 L ha^{-1} + S-metolachlor at 1.0 kg a.i. ha^{-1} enhanced yield of seed cotton by 70% over control [114]. In a similar study, application of *Sorgaab* at 10 L ha^{-1} combined with reduced doses of pendimethalin reduced total weed dry weight by 53–95% [115]. Use of reduced doses of pendimethalin (413 g a.i. ha^{-1}) in combination with sorghum/sunflower water extract (15–18 L ha^{-1} each) was effective in complete suppression of common lambsquarters (Table 3.2) [116].

Iqbal et al. found that application of glyphosate (575–767 g a.i. ha^{-1}) combined with *Sorgaab* + *Brassica* water extracts (15–18 L ha^{-1} each) reduced purple nutsedge dry biomass by 89% (Table 2) [106]. Weeds were controlled successfully with the combined use of allelopathic crop water extract with reduced doses (50–67%) of herbicide in canola crop (Table 3.2) [15, 107]. Similarly, use of reduced doses of S-metolachlor (715–1,075 g a.i. ha^{-1}) combined with sorghum water extract (12–15 L ha^{-1}) reduced purple nutsedge dry biomass by 81% in cotton [16]. Combined application of various crop water extracts and herbicides reduced the dry biomass of many weed species in wheat [17, 117], rice [118, 119], and maize [120, 121].

In another study on mungbean, combined application of S-metolachlor (preemergence) at 1.15 kg a.i. ha^{-1} or pendimethalin at 165 g a.i. ha^{-1} and *Sorgaab* (conc.) at 10 L ha^{1} reduced weed dry weight compared with the control [122]. Cheema et al. reported that combined application of one-third dose of S-metolachlor at 667 g a.i. ha^{-1} or pendimethalin at 333 g a.i. ha^{-1} with concentrated *Sorgaab* at 10 L ha^{-1} provided as good weed control as was achieved by a full dose of these herbicides, that is, S-metolachlor at 2 kg a.i. ha^{-1} and pendimethalin at 1 kg a.i. ha^{-1} [115]. Cheema et al. indicated that *Sorgaab* combined with a lower dose of MCPA (2-methyl-4-chlorophenoxyacetic acid) at 150 g a.i. ha^{-1} and fenoxaprop-p-ethyl at 375 g a.i. ha^{-1} provided effective weed control in wheat crop [123]. Moreover, *Sorgaab* at 12 L ha^{-1} + isoproturon at 500 g a.i. ha^{-1} produced almost equal wheat grain yield as was obtained with a full dose of isoproturon (1,000 g a.i. ha^{-1}), which clearly revealed that the isoproturon dose can be reduced by 50% in combination with *Sorgaab* at 12 L ha^{-1}. Additionally, combined application of *Sorgaab* with a reduced dose of herbicide controlled weeds by 85% than control (Table 3.2) [124].

In conclusion, combined application of allelopathic water extracts with reduced doses of herbicides can control weeds as efficiently as standard dosing of a sole herbicide, thus reducing production costs and protecting the environment.

Improving the Allelopathic Potential of Crops

Conventional Breeding

Interest is increasing among researchers to breed crop cultivars with high weed-suppressive ability because of the development of resistance against herbicides in major weed flora as well as environmental concerns related to herbicide usage [125]. In the current scenario, it is of utmost importance to breed smothering crops with the ability of efficient weed suppression, thus lowering reliance upon herbicide usage. Crop cultivars suppressing weed communities can be used as an alternative to herbicides, often herbicide performance being superior when competitive cultivars are used [126]. Different crop species vary for their capabilities to suppress weeds [127]. Even variability in the genotypes of the same species to suppress weeds has been observed in rice [128], oat [129], *Brassica* [130], and pearl millet [131].

Laboratory and greenhouse bioassays controlling for genotypic variation in competition for light, water, and nutrients should be considered as an initial screening

Table 3.2 Effect of allelopathic water extracts applied in combination with reduced doses of herbicides on weed control

Crop	Allelopathic extracts + herbicides	Percent decrease over control	Weeds suppressed	Reference
Wheat (*Triticum aestivum* L.)	Isoproturon (400–500 g a.i. ha^{-1}) + *Sorgaab* (12 L ha^{-1})	85.5	Littleseed canarygrass (*Phalaris minor* Retz.), yellow sweet clover (*Melilotus parviflora* (L.) Pall.), swine cress (*Cronopus didymus* (L.) Sm.)	Cheema et al. [124]
Canola (*Brassica napus* L.)	Pendimethalin (400–600 g a.i. ha^{-1}) + Sorghum/*Brassica*/Rice water extracts (15 L ha^{-1})	70.76	Purple nutsedge (*Cyperus rotundus* L.), horse purslane (*Trianthema portulacastrum* L.), common lambsquarters (*Chenopodium album* L.), swine cress (*Cronopus didymus* (L.) Sm.)	Jabran et al. [15]
Cotton (*Gossypium hirsutum* L.)	S-metolachlor (715–1,075 g a.i. ha^{-1}) + Sorghum water extract (12–15 L ha^{-1})	81.25	Purple nutsedge (*Cyperus rotundus* L.)	Iqbal and Cheema [16]
Sunflower (*Helianthus annuus* L.)	Pendimethalin (413 mL a.i. ha^{-1}) + Sorghum/Sunflower (15–18 L ha^{-1} each)	72	Common lambsquarters (*Chenopodium album* L.), sweet clover (*Melitotus indica* (L.) Pall.)	Awan et al. [116]
Cotton (*Gossypium hirsutum* L.)	Glyphosate (575–767 g a.i. ha^{-1}) + *Sorgaab* + *Brassica* water extract (15–18 L ha^{-1} each)	89.38	Purple nutsedge (*Cyperus rotundus* L.)	Iqbal et al. [106]
Wheat (*Triticum aestivum* L.)	Metribuzin (52.5 g a.i. ha^{-1})/Isoproturon (315 g a.i. ha^{-1})/Fenoxaprop (57 g a.i. ha^{-1})/Idosulfuron (36 g a.i. ha^{-1})/Idosulfuron (4.32 g a.i. ha^{-1}) + Sorghum/Sunflower water extract (18 L ha^{-1} each)	86.02	Swine cress (*Coronopus didymus* (L.) Sm.), littleseed canarygrass (*Phalaris minor* Retz.)	Razzaq et al. [17]

Table 3.2 (continued)

Crop	Allelopathic extracts + herbicides	Percent decrease over control	Weeds suppressed	Reference
Rice (*Oryza sativa* L.)	Butachlor (1,200 g a.i. ha^{-1})/ Pretilachlor (625 g a.i. ha^{-1})/ Ethoxysulfuronethyl (30 g a.i. ha^{-1}) + Sorghum/Sunflower/Rice water extract (15 L ha^{-1})	53.67	Barnyardgrass (*Echinochloa crus-galli* (L.) P.Beauv., rice flatsedge (*Cyperus iria* L.), crowfootgrass (*Dactyloctenium aegyptium* (L.) Willd.)	Rehman et al. [118]
Rice (*Oryza sativa* L.)	Ryzelan (15 mL ha^{-1}) + Sorghum water extract (7.5 L ha^{-1})	34.76	Barnyardgrass (*Echinocloa crus-galli* (L.) P.Beauv.), rice flatsedge (*Cyperus iria* L.), junglerice (*Echinochloa colona* (L.) Link., purple nutsedge (*Cyperus rotundus* L.), crowfootgrass (*Dactyloctenium aegyptium* (L.) Willd.	Wazir et al. [119]
Maize (*Zea mays* L.)	Furamsulfuron (half dose) + *Sorgaab*	57.33	Field bindweed (*Convolvulus arvensis* L.), redroot pigweed (*Amaranthus retroflexus* L.)	Latifi1 and Jamshidi [120]
Wheat (*Triticum aestivum* L.)	Sorghum + sunflower water extract (18 L ha^{-1} each) + Metribuzin (52.5 g a.i./ha)/Bensulfuron + isoproturon (315 g a.s./ha)/ Metribuzin + phenoxaprop (57 g a.i./ha)/Mesosulfuron + idosulfuron (36 g a.i./ ha)/Mesosulfuron + idosulfuron (4.32 g a.i./ha)	88.24	Swine cress (*Coronopus didymus* L.), littleseed canarygrass (*Phalaris minor* Retz.)	Razzaq et al. [117]
Maize (*Zea mays* L.)	Atrazine (125–250 g a.i. ha^{-1}) + Sorghum + Brassica + Sunflower + Mulberry water extracts (20 L ha^{-1} each)	74.67	Horse purslane (*Trianthema portulacastrum* L.)	Khan et al. [121]

tool for allelopathic research because some lines do not possess high competitiveness but have more allelopathic activity. Variability in traits in major crop genotypes can be used to breed cultivars that possess greater ability to suppress weeds [132, 133]. For example, Haan et al. bred a smother plant by crossing dwarf *B. campestris* with *B. campestris,* and when this plant was intercropped with maize and soybean, it suppressed the weeds for 4–6 weeks without influencing the performance of maize and soybean [134]. In another study, hybrid rice was produced by backcrossing and selfing of two lines, that is, Kouketsumochi (with allelopathic gene) and IR24 (with restoring gene). The specific hybrid rice produced by this method suppressed barnyard grass more effectively [135]. Selection of "STG06L-35-061" developed from crosses between indica (cv. Katy) and commercial tropical japonica (cv. Drew) suppressed the rice weeds, such as barnyard grass, more efficiently [136].

Continuous breeding with barley genotypes has resulted in an increase in allelopathic activity of spring wheat [137] and decrease in barley [138]. Rondo is a line of indica rice developed by mutation breeding that has high weed-suppressive ability and is high yielding [139, 140]. Similarly, present crop cultivars are more allelopathic than older ones [141]. So breeding of old cultivars with modern cultivars is of prime importance to breed crop cultivars having high allelopathic activity.

Environmental variations and environment genotype interactions can obstruct phenotypic selection by obscuring genotypic differences in weed-suppressive ability [142]. For example, Gealy and Yan studied the suppressive ability of different rice genotypes against barnyard grass [140]. Some rice genotypes suppressed barnyard grass 1.3–1.5 times greater than long-grain rice cultivars, but genotypic differences were nonsignificant. These nonsignificant differences among genotypes may be due to environmental variation. Varietal potentials for weed suppression are mostly unpredictable across different study locations [143] and growing seasons [144], indicating strong genotype by environment interactions. Therefore, screening of genotypes for their relative competiveness or allelopathic potential must be carried out in different environments, locations, and years.

Use of Biotechnology

Although less attention has been given to the biotechnological aspect of allelopathy than others, during the last decade, the role of biotechnology in allelopathy has received much attention. Wu et al. tested 453 winter wheat accessions and found a normal distribution of allelopathic activity, indicating a quantitative mode of inheritance [145]. When lines having strong allelopathy activity were crossed with the lines having low allelopathic activity, the allelopathic activity was normally distributed in resulting progenies in rice [146–148] and wheat [149, 150].

Different crop species possess different allelochemicals and each allelochemical suppresses special weed biotype. For example, scopoletin suppresses wild mustard (*B. kaber* [DC] L.; Table 3.3) [129] and hydroxamates suppress wild oat (*Avena fatua* L.; Table 3.3) [151]. Similarly, DIMBOA (2,4-dihydroxy-7-methoxy-1,4-

Table 3.3 Weed-suppressing ability of some allelochemicals

Allelochemicals	Weeds suppressed	Reference
Scopoletin	Wild mustard (*Sinapis arvensis* L. (*Brassica kaber* [DC.]) wheeler var. pinnatifida lStokes] wheeler	Fay and Duke [129]
Hydroxamates	Wild oat (*Avena fatua* L.)	Pérez and Ormemeño-Núñez [151]
DIMBOA	Foxtail amaranth (*Amaranthus caudatus* L.), garden cress (*Lepidium sativum* L.)	Pethó [153]
Gramine/Hordenine	Shepherd's purse (*Capsella bursa-pastoris* (L.) Medik.), white mustard (*Sinapis alba* L.), common chickweed (*Stellaria media* (L.) Vill.)	Overland [152], Liu and Lovett [154]
Hydroxamic acids	Wild oat (*Avena fatua* L.), henbit deadnettle (*Lamium amplexicaule* L.), common lambsquarter (*Chenopodium album* L.), knotgrass (*Polygonum aviculare* L.), black bindweed (*Fallopia convolvulus* (L.) Á. Löve)	Pérez and Ormemeño-Núñez [151], Friebe et al. [155]

benzoxazin-3-one), gramine/hordenine, and hydroxamic acids suppressed various weed biotypes in several studies (Table 3.3) [151–155].

There is a need to identify the genes controlling production of these allelochemicals so that gene expression for production of these allelochemicals may be improved/enhanced, resulting in increased quantity of these allelochemicals production. Some work has been done to map the allelopathic genes found in wheat [149, 156]. Hydroxamic acids are the important allelochemicals found in wheat. Niemeyer and Jerez mapped the position of genes responsible for hydroxamic acid production [156]. The quantitative trait loci (QTLs) responsible for accumulation of hydroxamic acid were identified on chromosomes 4A, 4B, 4D, and 5B. In another study, Wu et al. mapped allelopathic QTLs in a double haploid population, which was obtained from the cross of two cultivars, one being strongly allelopathic and other being less allelopathic [149]. For mapping these QTLs, they used amplified fragment length polymorphism (AFLP), restriction fragment length polymorphism (RFLP), and simple sequence repeat markers (SSRM). Scientists have found two major allelopathic QTLs on wheat chromosome 2B, based on the 189 DH lines and two parents [149].

Extensive work has been carried out for mapping allelopathic QTLs in rice. Ebana et al. mapped seven allelopathic QTLs in rice on chromosomes 1, 3, 5, 6, 7, 11, and 12 by using RFLP markers in an F2 population, which was obtained from the cross of high allelopathic genotype with low allelopathic genotype [157]. Jensen et al. identified four main-effect QTLs on chromosomes 2, 3, and 8, and these QTLs explained the 35% of the total phenotypic variation in the population of rice [158]. In another study, Jensen et al. identified 15 QTLs in a rice population, each explaining 5–11% of phenotypic variation [146]. These QTLs were identified on chromosomes 3, 4, 6, 8, 9, 10, and 12. In a similar study, Zhou et al. identified three main-effect QTLs on chromosomes 5 and 11, which collectively explained phenotypic variation up to 13.6% [147]. These QTLs were identified from different recombinant inbred lines, which were obtained from the cross of two Chinese rice cultivars, one being strongly allelopathic and other being weakly allelopathic. In short, allelopathic QTLs have been identified in multiple rice genomes but still no QTL has been identified for chromosome 2. Discovery of additional fine-resolution QTLs controlling allelopathy in rice and wheat will hopefully result in the development of effective molecular markers that can be used in marker-assisted selection for cultivars with improved allelopathic activity. Marker-assisted selection may be hindered because of the large number of minor-effect QTLs that appear to control allelopathy in various genotypes. Marker-assisted backcrossing can be used as a successful tool for breeding genotypes with high allelopathic activity if major QTLs controlling allelopathy are less than five [141].

Some researchers also suggested transgenic approaches as successful tools to enhance crop allelopathy [159]. However, before moving towards transgenic approaches, it is necessary to have a clear understanding of the genes responsible for the biosynthesis and regulation of allelochemicals and their synthesis pathway. Although QTL mapping facilitates marker-assisted selection, it seldom tells about the gene responsible for allelochemical production. Several candidate genes may be located in an individual QTL spanning 5–10 cM (centimorgans) [160] and knowledge about individual genes is necessary. Genes responsible for regulation and biosynthesis of allelochemicals can be identified through isolation, discovery [161], activation tagging [162], purification of plant enzymes, purification of related bioactive metabolites [161], and through gene knockout libraries [163]. Particular genes responsible for the biosynthesis and regulation of allelochemicals, such as momilactones [164, 165], phenolic compounds [166], and benzoxazinoids [167], have been reported. Antisense knockout techniques and overexpression of genes can be used to change the quantity and quality of secondary metabolites of allelopathic plants. Fortunately, transgenic approaches can be utilized to introduce genes from high allopathic genotypes to low or non-allelopathic genotypes, but the goal is not easy to attain due to complex genetics of allelopathy. According to Bertin et al. expression of multiple genes into crop species and its regulation should be optimized in such a way that the transformed crop will be able to produce the desired allelochemicals successfully [160].

Conclusion

Allelopathy can be used as an environmentally friendly tool to manage weeds in modern agriculture for improving crop yields without reliance on synthetic herbicides, which are posing a severe threat to our environment and human health. Allelopathic strategies, such as intercropping, crop rotation, mulching, use of allelopathic crop water extracts alone or in combination with reduced doses of herbicides, and incorporation of cover crops in cropping systems, may be used as successful tools to manage different weed ecotypes. Conventional breeding of cultivars having more allelopathic activity with cultivars having low allelopathic activity may also be useful to enhance the allelopathic activity of existing crop cultivars. Moreover, Modern biotechnological approaches should be used to identify genes responsible for allelochemical production, and then these genes should be introduced to improve the allelopathic potential of cultivars that are less allelopathic.

References

1. Fahad S, Nie L, Rahman A, Chen C, Wu C, Saud S, Huang J (2013) Comparative efficacy of different herbicides for weed management and yield attributes in wheat. Am J Plant Sci 4:1241–1245
2. Gianessi LP (2013) The increasing importance of herbicides in worldwide crop production. Pest Manage Sci. doi:10.1002/ps.3598.
3. Sodaeizadeh H, Hosseini Z (2012) Allelopathy: an environmentally friendly method for weed control. International conference on applied life sciences (ICALS2012), Turkey, September 10–12, 2012
4. Bhowmik PC, Inderjit J (2003) Challenges and opportunities in implementing allelopathy for natural weed management. Crop Prot 22:661–671
5. Heap I (2008) The international survey of herbicide resistant weeds. http://www.weedscience.com/. Accessed 15 May 2013
6. Kudsk P, Streibig JC (2003) Herbicides—a two-edged sword. Weed Res 43:90–102
7. Juraske R, Antón A, Castells F, Huijbregts MAJ (2007) Pest screen: a screening approach for scoring and ranking pesticides by their environmental and toxicological concern. Environ Int 33:886–893
8. Sethi A, Dilawari VK (2008) Spectrum of insecticide resistance in whitefly from upland cotton in Indian subcontinent. J Entomol 5:138–147
9. Schreinemachers DM (2003) Birth malformations and other adverse perinatal outcomes in four U.S. wheat-producing states. Environ Health Persp 111:1259–1264
10. Rice EL (1984) Allelopathy, 2nd ed. Academic, Orlando
11. Farooq M, Jabran K, Cheema ZA, Wahid A, Siddique KHM (2011) The role of allelopathy in agricultural pest management. Pest Manage Sci 67:494–506
12. Iqbal J, Cheema ZA (2007) Effect of allelopathic crops water extracts on glyphosate dose for weed control in cotton (*Gossypium hirsutum* L.). Allelopathy J 19:403–410
13. Cheema ZA, Asim M, Khaliq A (2000) Sorghum allelopathy for weed control in cotton (*Gossypium arboretum* L.). Int J Agric Biol 2:37–40
14. Matloob A, Khaliq A, Farooq M, Cheema ZA (2010) Quantification of allelopathic potential of different crop residues for the purple nut sedge suppression. Pak J Weed Sci Res 16:1–12
15. Jabran K, Cheema ZA, Farooq M, Basra SMA, Hussain M, Rehman H (2008) Tank mixing of allelopathic crop water extracts with pendimethalin helps in the management of weeds in canola (*Brassica napus*) field. Int J Agri Biol 10:293–296
16. Iqbal J, Cheema ZA (2008) Purple nut sedge (*Cyperus rotundus* L.) management in cotton with combined application of *Sorgaab* and S-Metolachlor. Pak J Bot 40:2383–2391
17. Razzaq A, Cheema ZA, Jabran K, Farooq M, Khaliq A, Haider G, Basra SMA (2010) Weed management in wheat through combination of allelopathic water extracts with reduced doses of herbicides. Pak J Weed Sci Res 16:247–256
18. Cheema ZA, Farooq M, Wahid A (2012a) Allelopathy: current trends and future applications. Springer-Verlag, Heidelberg
19. Cheema ZA, Farooq M, Khaliq A (2012b) Application of allelopathy in crop production: success story from Pakistan. In: Cheema ZA, Farooq M, Wahid A (eds) Allelopathy: current trends and future applications. Springer-Verlag, Heidelberg, pp 113–144

20. Putnam AR, Nair MG, Barnes JB (1990) Allelopathy: a viable weed control strategy. In: Baker RR, Dunn PE (eds) New directions in biological control, alternatives for suppressing agricultural pests and diseases. Proceedings of a UCLA colloquium held at Frisco, Colorado, January 20–27, 1989, pp 317–322
21. Dilday RH, Lin J, Yan W (1994) Identification of allelopathy in the USDA-ARS rice germplasm collection. Aust J Exp Agri 34:907–910
22. Narwal SS (1996) Potentials and prospects of allelopathy mediated weed control for sustainable agriculture. In: Narwal SS, Tauro P (eds) Allelopathy in pest management for sustainable agriculture. Proceedings of the international conference on allelopathy, Scientific Publishers, Jodhpur, pp 23–65
23. Miller DA (1996) Allelopathy in forage crop systems. Agron J 88:854–859
24. Weston LA (1996) Utilization of allelopathy for weed management in agro-ecosystems: allelopathy in cropping systems. Agron J 88:860–866
25. Narwal SS, Sarmah MK, Tamak JC (1998) Allelopathic strategies for weed management in the rice-wheat rotation in northwestern India. In: Olofsdotter M (ed) Allelopathy in rice. Proceedings of the workshop on allelopathy in rice, 25–27 Nov. 1996, Manila (Philippines): International Rice Research Institute, IRRI Press, Manila
26. Jamil M, Cheema ZA, Mushtaq MN, Farooq M, Cheema MA (2009) Alternative control of wild oat and canarygrass in wheat fields by allelopathic plant water extracts. Agron Sustain Dev 29:475–482
27. Liebman M, Dyck E (1993) Crop rotation and intercropping strategies for weed management. Ecol Appl 3:92–122
28. Liebman M, Davis AS (2000) Integration of soil, crop, and weed management in low-external-input farming systems. Weed Res 40:27–47
29. Baumann DT, Bastiaans L, Kropff MJ (2002) Intercropping system optimization for yield, quality, and weed suppression combining mechanistic and descriptive models. Agron J 94:734–742
30. Ali Z, Malik MA, Cheema MA (2000) Studies on determining a suitable canola-wheat intercropping pattern. Int J Agri Biol 2:42–44
31. Khan ZR, Hassanali A, Overholt W, Khamis TM, Hooper AM, Pickett JA, Wadhams LJ, Woodcock CM (2002) Control of Witch weed, Striga hermonthica by intercropping with Desmodium spp, and the mechanism defined as allelopathic. J Chem Ecol 28:1871–1885
32. Iqbal J, Cheema ZA, An M (2007) Intercropping of field crops in cotton for the management of purple nut sedge (*Cyperus rotundus* L.). Plant Soil 300:163–171
33. Khalil SK, Mehmood T, Rehman A, Wahab S, Khan AZ, Zubair M, Mohammad F, Khan NU, Amanullah, Khalil IH (2010) Utilization of allelopathy and planting geometry for weed management and dry matter production of maize. Pak J Bot 42:791–803
34. Cruz RD, Rojas E, Merayo A (1994) Management of Itch grass (*Rottboellia cochinchinensis* L.) in maize crop and in the fallow period with legume crops. Integr Pest Manage 31:29–35
35. Olasantan FO, Lucas EO, Ezumah HC (1994) Effects of intercropping and fertilizer application on weed control and performance of cassava and maize. Field Crops Res 39:63–69
36. Witcombe JR, Billore M, Singhal HC, Patel NB, Tikka SBS, Saini DP, Sharma LK, Sharma R, Yadav SK, Pyadavendra J (2008) Improving the food security of low-resource farmers: introducing horse gram into maize based cropping systems. Exp Agri 43:339–348
37. Steiner KG (1984) Intercropping in tropical smallholder agriculture with special reference to West Africa. GTZ Publication, Eschborn, p 304
38. Gliessman SR, Garcia ER (1979) The use of some tropical legumes in accelerating the recovery of productivity of soils in the low land humid tropics of Mexico. In: Tropical legumes: resources for the future. National Academy of Sciences, Washington, pp 292–298
39. Bilalis D, Papastylianou P, Konstantas A, Patsiali S, Karkanis A, Efthimiadou A (2010) Weed suppressive effects of maize-legume intercropping in organic farming. Int J Pest Manage 56:173–181
40. Bansal GL (1989) Allelopathic potential of linseed on Ranunculus arvensis. In: Plant Science Research in India. Today and Tomorrow Publishers, New Delhi, pp 801–805

41. Fujiyoshi PT (1998) Mechanisms of weed suppression by squash (*Cucurbita spp.*) intercropped in Corn (*Z. mays* L.). PhD Dissertation, University of California, Santa Cruz, p 89
42. Olorunmaiye PM (2010) Weed control potential of five legume cover crops in maize/cassava intercrop in a Southern Guinea savanna ecosystem of Nigeria. Aust J Crop Sci 4:324–329
43. Abraham CT, Singh SP (1984) Weed management in sorghum-legume intercropping systems. J Agri Sci 103:103–115
44. Baumann DT, Krop MJ, Bastiaans L (2000) Intercropping leeks to suppress weeds. Weed Res 40:361–376
45. Hauggaard-Nielsen H, Ambus P, Jensen ES (2001) Interspecific competition, N use and interference with weeds in pea-barley intercropping. Field Crops Res 70:101–109
46. Szumigalski A, Acker RV (2005) Weed suppression and crop production in annual intercrops. Weed Sci 53:813–825
47. Midya A, Bhattacharjee K, Ghose SS, Banik P (2005) Deferred seeding of black gram (*Phaseolus mungo* L.) in rice (*O. sativa* L.) field on yield advantages and smothering of weeds. J Agron Crop Sci 191:195–201
48. Banik P, Midya A, Sarkar BK, Ghose SS (2006) Wheat and chickpea intercropping systems in an additive series experiment: Advantages and weed smothering. Eur J Agron 24:325–332
49. Saucke H, Ackermann K (2006) Weed suppression in mixed cropped grain peas and false flax (*Camelina sativa*). Weed Res 46:453–461
50. Midega CAO, Khan ZR, Amudavi DM, Pittchar J, Pickett JA (2010) Integrated management of *Striga hermonthica* and cereal stem borers in finger millet (*Eleusine coracana* L.) through intercropping with *Desmodium intortum*. Int J Pest Manage 56:145–151
51. Naeem M (2011) Studying weed dynamics in wheat (*Triticum aestivum* L.)-canola (*Brassica napus* L.) intercropping system. M.Sc. thesis, Department of Agronomy, University of Agriculture, Faisalabad, Pakistan
52. Khorramdel S, Rostami L, Koocheki A, Shabahang J (2010) Effects of row intercropping wheat (*Triticum aestivum* L.) with canola (*Brassica napus* L.) on weed number, density and population. Proceedings of 3rd Iranian Weed Science Congress. 17–18 February 2010. Weed biology and ecophysiology, Babolsar, Iran, pp 411–414
53. Arif M (2013) Exploiting crop allelopathy for weed management in wheat (*Triticum aestivum* L.). PhD thesis, Department of Agronomy, University of Agriculture, Faislabad, Pakistan
54. Kimber RWL (1967) Phytotoxicity from plant residues: The influence of rotted wheat straw on seedling growth. Aust J Agri Res 18:361–374
55. Batish DR, Singh HP, Kohli RK, Kaur S (2001) Crop allelopathy and its role in ecological agriculture. In: Kohli RK, Harminder PS, Batish DR (eds) Allelopathy in agroecosystems. Food Products Press, New York, pp 121–162
56. Mamolos AP, Kalburtji KL (2001) Significance of allelopathy in crop rotation. J Crop Prod 4:197–218
57. Voll E, Franchini JC, Tomazon R, Cruz D, Gazziero DL, Brighenti AM (2004) Chemical interactions of *Brachiaria plantaginea* with *Commelina bengalensis* and *Acanthospermum hispidum* in soybean cropping systems. J Chem Ecol 30:1467–1475
58. Einhellig FA, Rasmussen JA (1989) Prior cropping with grain sorghum inhibits weeds. J Chem Ecol 15:951–960
59. Schreiber MM (1992) Influence of tillage, crop rotation and weed management on grain foxtail (*Setaria faberi*) population dynamics and corn yield. Weed Sci 40:645–653
60. Cernusko K, Boreky V (1992) The effect of fore crop, soil tillage and herbicide on weed infestation rate and on the winter wheat yield. Rostlinna Vyroba-UVTIZ 38:603–609
61. Grodzinsky AM (1992) Allelopathic effects of cruciferous plants in crop rotation. In: Rizvi SJH, Rizvi V (eds) Allelopathy: basic and applied aspects. Chapman and Hall, London, pp 77–85.
62. Al-Khatib K, Libbey C, Boydston R (1997) Weed suppression with *Brassica* green manure crops in green pea. Weed Sci 45:439–445
63. Teasdale JR, Mohler CL (2000) The quantitative relationship between weed emergence and the physical properties of mulches. Weed Sci 48:385–392

64. Bilalis D, Sidiras N, Economou G, Vakali C (2003) Effect of different levels of wheat straw soil surface coverage on weed flora in *Vicia faba* crops. J Agron Crop Sci 189:233–241
65. Narwal SS (2005) Role of allelopathy in crop production. J Herbologia 6:31
66. Younis A, Bhatti MZM, Riaz A, Tariq U, Arfan M, Nadeem M, Ahsan M (2012) Effect of different types of mulching on growth and flowering of *Freesia alba* CV. Aurora. Pak J Agri Sci 49:429–433
67. Lockerman RH, Putnam AR (1979) Evaluation of allelopathic cucumbers (*Cucumis sativus*) as an aid for weed control. Weed Sci 27:54–57
68. Weston LA, Harmon R, Mueller S (1989) Allelopathic potential of sorghum sudangrass hybrid (sudex). J Chem Ecol 15:1855–1865
69. Cheema ZA, Khaliq A, Saeed S (2004) Weed control in maize (*Zea mays* L.) through sorghum allelopathy. J Sustain Agric 23:73–86
70. Riaz MY (2010) Non-chemical weed management strategies in dry direct seeded fine grain aerobic rice (*Oryza sativa* L.). M.Sc. (Hons.) Thesis, Department of Agronomy, University of Agriculture, Faisalabad, Pakistan
71. Mahmood A, Cheema ZA (2004) Influence of sorghum mulch on purple nut sedge (*Cyperus rotundus* L.). Int J Agri Biol 6:86–88
72. Ahmad S, Rehman A, Cheema ZA, Tanveer A, Khaliq A (1995) Evaluation of some crop residues for their allelopathic effects on germination and growth of cotton and cotton weeds. In: 4th Pakistan Weed Science Conference, Faisalabad, Pakistan, pp 63–71
73. Wilson RE, Rice EL (1968) Allelopathy as expressed by *Helianthus annuus* and its role in old-field succession. Bull Torrey Bot Club 95:432–448
74. Shilling DG, Liebl RA, Worsham AD (1985) Rye (*Secale cereale* L.) and wheat (*Triticumn aestivum* L.) mulch: The suppression of certain broad-leaves weeds and the isolation and identification of phytotoxins. In: Thompson AC (ed) Chemistry of allelopathy. ACS symposium series, American Chemical Society, Washington, pp 243–271
75. Worsham AD (1991) Allelopathic cover crops to reduce herbicide input. Proc Southern Weed Sci Soc 44:58–64
76. Lee HW, Ghimire SR, Shin DH, Lee IJ, Kim KU (2008) Allelopathic effect of the root exudates of K21, a potent allelopathic rice. Weed Biol Manage 8:85–90
77. Aslam F (2010) Studying wheat allelopathy against horse purslane (*Trianthema portulacastrum*). M.Sc. Thesis, Department of Agronomy, University of Agriculture, Faisalabad, Pakistan
78. Batish DR, Arora K, Singh HP, Kohli RK (2007a) Potential utilization of dried powder of *Tagetes minuta* as a natural herbicide for managing rice weeds. Crop Prot 26:566–571
79. Batish DR, Kaura M, Singh HP, Kohli RK (2007b) Phytotoxicity of a medicinal plant, *Anisomeles indica*, against *Phalaris minor* and its potential use as natural herbicide in wheat fields. Crop Prot 26:948–952
80. Khaliq A, Matloob A, Farooq M, Mushtaq MN, Khan MB (2011) Effect of crop residues applied isolated or in combination on the germination and seedling growth of horse purslane (*Trianthema portulacastrum* L.). Planta Daninha 29:121–128
81. Bernat W, Gawtonska H, Gawtonski SW (2004) Effectiveness of different mulches in weed management in organic winter wheat production. In: Oleszek W, Burda S, Bialy Z, StepienW, Kapusta I, Stepien K (eds) Abstracts, II European allelopathy symposium, allelopathy from understanding to application, 3–5 June 2004, Institute of Soil Science and Plant Cultivation, Czartoryskich 8, 24-100 Pulawy, p 118
82. Ekeleme F, Chikoye D, Akobundu IO (2004) Changes in size and composition of weed communities during planted and natural fallows. Basic Appl Ecol 5:25–33
83. Hiltbrunner J, Liedgens M, Bloch L, Stamp P, Streit B (2007) Legume cover crops as living mulches for winter wheat: components of biomass and the control of weeds. Eur J Agron 26:21–29
84. Qasem JR (2003) Weeds and their control. University of Jordan Publications, Amman, p 628
85. Lehman ME, Blum U (1997) Cover crop debris effects on weed emergence as modified by environmental factors. Allelopathy J 4:69–88

86. Kaspar TC, Radke JK, Laflen JM (2001) Small grain cover crops and wheel traffic effects on infiltration, runoff, and erosion. J Soil Water Cons 56:160–164
87. Sarrantonio M, Gallandt E (2003) The role of cover crops in North American cropping systems. J Crop Prod 8:53–74
88. Price AJ, Stoll ME, Bergtold JS, Arriaga FJ, Balkcom KS, Kornecki TS, Raper RL (2008) Effect of cover crop extracts on cotton and radish radicle elongation. Commun Biomet Crop Sci 3:60–66
89. Fujii Y, Heradata S (2005) A critical survey of allelochemicals in action, the importance of total activity and the weed suppression equation. In: Harper JDI, An M, Wu H, Kent JH (eds) Proceedings of fourth world congress on allelopathy "Establishing the scientific base", 21–26 Aug 2005, Charles Strut University, Wagga Wagga, NSW, pp 73–76
90. Caamal-Maldonado JA, Jimenez-Osorino JI, Barragan AT, Anaya AL (2001) The use of allelopathic legume cover and mulch species for weed control in cropping systems. Agron J 93:27–36
91. Kobayashi H, Miura S, Oyanagi A (2004) Effects of winter barley as a cover crop on the weed vegetation in a no-tillage soybean. Weed Biol Manage 4:195–205
92. Putnam AR, DeFrank J (1983) Use of phytotoxic plant residues for selective weed control. Crop Prot 2:173–181
93. Barnes JP, Putnam AR, Burke BA (1986) Allelopathic activity of rye (*Secale cereal* L.). In: Putnam AR, Tang CS (eds) The science of allelopathy. Willey Interscience, New York, pp 271–286
94. Grimmer OP, Masiunas JB (2005) The weed control potential of oat cultivars. Hort Technol 15:140–144
95. Creamer NG, Bennett MA, Stinner BR, Cardina J, Regnier EE (1996) Mechanisms of weed suppression in cover crop-based production systems. Hort Sci 31:410–413
96. Hoffman ML, Weston LA, Snyder JC, Reigner EE (1996) Allelopathic influence of germinating seeds and seedlings of cover crops on weed spp. Weed Sci 44:579–589
97. Forney DR, Foy CL (1985) Phytotoxicity of products from rhizospheres of a sorghum-sudangrass hybrid (*S. bicolor x S. sudanense*). Weed Sci 33:597–604
98. Iqbal Z, Nasir H, Hiradate S, Fujii Y (2006) Plant growth inhibitory activity of *Lycoris radiate* Herb. and the possible involvement of lycorine as an allelochemical. Weed Biol Manage 6:221–227
99. Peters EJ, Zam AHBM (1981) Allelopathic effects of tall fescue (*Festuca arundinacea*) genotypes. Agron J 73:56–58
100. Bonanomi G, Sicurezza MG, Caporaso S, Esposito A, Mazzoleni S (2006) Phytotoxicity dynamics of decaying plant materials. New Phytol 169:571–578
101. Cheema ZA, Luqman M, Khaliq A (1997) Use of allelopathic extracts of sorghum and sunflower herbage for weed control in wheat. J Anim Plant Sci 7:91–93
102. Cheema ZA, Khaliq A, Akhtar S (2001) Use of *Sorgaab* (sorghum water extract) as a natural weed inhibitor in spring mungbean. Int J Agri Biol 3:515–518
103. Cheema ZA, Iqbal M, Ahmad R (2002a) Response of wheat varieties and some rabi weeds to allelopathic effects of sorghum water extract. Int J Agric Biol 4:52–55
104. Cheema ZA, Khaliq A, Ali K (2002b) Efficacy of *Sorgaab* for weed control in wheat grown at different fertility levels. Pak J Weed Sci Res 8:33–38
105. Irshad A, Cheema ZA (2004) Effect of sorghum extract on management of barnyard grass in rice crop. Allelopathy J 14:205–213
106. Iqbal J, Cheema ZA, Mushtaq MN (2009) Allelopathic crop water extracts reduce the herbicide dose for weed control in cotton (*Gossypium hirsutum*). Int J Agri Biol 11:360–366
107. Jabran K, Cheema ZA, Farooq M, Hussain M (2010) Lower doses of pendimethalin mixed with allelopathic crop water extracts for weed management in canola (*Brassica napus* L.). Int J Agri Biol 12:335–340
108. Nawaz R, Cheema ZA, Mahmood T (2001) Effect of row spacing and sorghum water extract on sunflower and its weeds. Int J Agri Biol 3:360–362
109. Khaliq A, Cheema ZA, Mukhtar MA, Basra SMA (1999) Evaluation of sorghum (*Sorghum bicolor*) water extracts for weed control in soybean. Int J Agri Biol 1:23–26

110. Ahmad A, Cheema ZA, Ahmad R (2000) Evaluation of *Sorgaab* as natural weed inhibitor in maize. J Anim Plant Sci 10:141–146
111. Cheema ZA, Khaliq A, Mubeen M (2003a) Response of wheat and winter weeds to foliar application of different plant water extracts of sorghum (*S. bicolor*). Pak J Weed Sci Res 9:89–97
112. Einhelling FA, Leather GR (1988) Potentials for exploiting allelopathy to enhance crop production. J Chem Ecol 14:1829–1844
113. Cheema ZA, Farid MS, Khaliq A (2003b) Efficacy of concentrated *Sorgaab* with low rates of atrazine for weed control in maize. J Anim Plant Sci 13:48–51
114. Cheema ZA, Khaliq A, Tariq M (2002c) Evaluation of concentrated *Sorgaab* alone and in combination with reduced rates of three pre-emergence herbicides for weed control in cotton (*Gossypium hirsutum* L.). Int J Agri Biol 4:549–552
115. Cheema ZA, Khaliq A, Hussain R (2003c) Reducing herbicide rate in combination with allelopathic *Sorgaab* for weed control in cotton. Int J Agri Biol 5:1–6
116. Awan IU, Khan MA, Zareef M, Khan EA (2009) Weed management in sunflower with allelopathic water extract and reduced doses of a herbicide. Pak J Weed Sci Res 15:19–30
117. Razzaq A, Cheema ZA, Jabran K, Hussain M, Farooq M, Zafar M (2012) Reduced herbicide doses used together with allelopathic sorghum and sunflower water extracts for weed control in wheat. J Plant Prot Res 52:281–285
118. Rehman A, Cheema ZA, Khaliq A, Arshad M, Mohsan S (2010) Application of sorghum, sunflower and rice water extract combinations helps in reducing herbicide dose for weed management in rice. Int J Agri Biol 12:901–906
119. Wazir I, Sadiq M, Baloch MS, Awan IU, Khan EA, Shah IH, Nadim MA, Khakwani AA, Bakhsh I (2011) Application of bio-herbicide alternatives for chemical weed control in rice. Pak J Weed Sci Res 17:245–252
120. Latifi P, Jamshidi S (2011). Management of corn weeds by broomcorn *Sorgaab* and Foramsulfuron reduced doses integration. International conference on biology, environment and chemistry, IACSIT Press, Singapoor
121. Khan MB, Ahmad M, Hussain M, Jabran K, Farooq S, Waqas-Ul-Haq M (2012) Allelopathic plant water extracts tank mixed with reduced doses of atrazine efficiently control *Trianthema portulacastrum* L. in *Zea mays* L. J Anim Plant Sci 22:339–346
122. Khaliq A, Aslam Z, Cheema ZA (2002) Efficacy of different weed management strategies in mungbean (*Vigna radiata* L.). Int J Agri Biol 4:237–239
123. Cheema ZA, Hussain S, Khaliq A (2003d) Efficacy of *Sorgaab* in combination with allelopathic water extracts and reduced rates of pendimethalin for weed control in mungbean (*Vigna radiata*). Indus J Plant Sci 2:21–25
124. Cheema ZA, Iqbal J, Khaliq A (2003e) Reducing isoprotron dose in combination with *Sorgaab* for weed control in wheat. Pak J Weed Sci Res 9:153–160
125. Worthington M, Reberg-Horton SC (2013) Breeding cereal crops for enhanced weed suppression: optimizing allelopathy and competitive ability. J Chem Ecol 39:213–231
126. Lemerle D, Verbeek B, Cousens RD, Coombes N (1996) The potential for selecting wheat varieties strongly competitive against weeds. Weed Res 36:505–513
127. Bertholdsson NO (2005) Early vigour and allelopathy-two useful traits for enhanced barley and wheat competitiveness against weeds. Weed Res 45:94–102
128. Xu GF, Zhang FD, Li TL, Wu D, Zhang YH (2010) Induced effects of exogenous phenolic acids on allelopathy of a wild rice accession (*Oryza longistaminata*, S37). Rice Sci 17:135–140
129. Fay PK, Duke WB (1977) An assessment of allelopathic potential in *Avena* germplasm. Weed Sci 25:224–228
130. Sarmah MK, Narwal SS, Yadava JS (1992) Smothering effect of *Brassica* species on weeds. In: Narwal SS, Tauro P (eds) Proceeding of first national symposium allelopathy in agroecosystems. Haryana Agricultural University, Indian Society of Allelopathy, Hisar, pp 51–55
131. Narwal SS, Sarmah MK, Dahiya DS, Kapoor RL (1992) Smothering effect of pearl millet genotypes on weed species. In: Tauro P, Narwal SS (eds) Proceeding national symposium allelopathy in agro-ecosystems. Indian Society of Allelopathy, Department of Agronomy, Haryana Agricultural University, Hisar, pp 48–50

132. Callaway MB (1990) Crop varietal tolerance to weeds: a compilation. Publication Series No. 1990-1. Cornell University, Ithaca.
133. Shili-Touzi I, Tourdonnet SD, Launay M, Dore T (2010) Does intercropping winter wheat (*Triticum aestivum*) with red fescue (*Festuca rubra*) as a cover crop improve agronomic and environmental performance? A modeling approach. Field Crops Res 116:218–229
134. Haan RL, Wyse DL, Ehike NJ, Maxwell BD, Putnam DH (1994) Simulation of spring seeded smother plant for weed control in corn. Weed Sci 42:35–43
135. Lin W, Kim KU, Liang K, Guo Y (2000) Hybrid rice with allelopathy. In: Kim KU, Shin DH (eds) Rice allelopathy. Proceeding of the international workshop in rice allelopathy, 17–19 August 2000, Kyungpook National University, Taegu, Korea, pp 49–56
136. Gealy DR, Moldenhauer KAK, Jia MH (2013) Field performance of STG06 L-35-061, a new genetic resource developed from crosses between weed-suppressive indica rice and commercial southern U.S. long-grains. Plant Soil 1–17
137. Bertholdsson NO (2007) Varietal variation in allelopathic activity in wheat and barley and possibilities for use in plant breeding. Allelopathy J 19:193–201
138. Bertholdsson NO (2004) Variation in allelopathic activity over 100 years of barley selection and breeding. Weed Res 44:78–86
139. Yan WG, McClung AM (2010) Rondo a long-grain indica rice with resistances to multiple diseases. J Plant Reg 4:131–136
140. Gealy DR, Yan W (2012) Weed suppression potential of 'Rondo' and other indica rice germplasm lines. Weed Technol 26:524–527
141. Courtois B, Olofsdotter M (1998) Incorporating the allelopathy trait in upland rice breeding programs. In: Olofsdotter M (ed) Allelopathy in rice. IRRI Publishing, Los Banos, pp 57–68
142. Coleman RD, Gill GS, Rebetzke GJ (2001) Identification of quantitative trait loci for traits conferring weed competitiveness in wheat (*Triticum aestivum* L.). Aust J Agric Res 52:1235–1246
143. Mokhtari S, Galwey NW, Cousens RD, Thurling N (2002) The genetic basis of variation among wheat F3 lines intolerance to competition by ryegrass (*Lolium rigidum*). Euphytica 124:355–364
144. Seavers GP, Wright KJ (1999) Crop canopy development and structure influence weed suppression. Weed Res 39:319–328
145. Wu HJ, Pratley D, Lemerle, Haig T (2000) Laboratory screening for allelopathic potential of wheat (*Triticum aestivum*) accessions against annual ryegrass (*Lolium rigidum*). Aust J Agri Res 51:259–266
146. Jensen LB, Courtois B, Olofsdotter M (2008) Quantitative trait loci analysis of allelopathy in rice. Crop Sci 48:1459–1469
147. Zhou YJ, Cao CD, Zhuang JY, Zheng KL, Guo YQ, Ye M, Yu LQ (2007) Mapping QTL associated with rice allelopathy using the rice recombinant inbred lines and specific secondary metabolite marking method. Allelopathy J 19:479–485
148. Chen XH, Hu F, Kong CH (2008) Varietal improvement in rice allelopathy. Allelopathy J 22:379–384
149. Wu H, Pratley J, Ma W, Haig T (2003) Quantitative trait loci and molecular markers associated with wheat allelopathy. Theor Appl Genet 107:1477–1481
150. Bertholdsson NO (2010) Breeding spring wheat for improved allelopathic potential. Weed Res 50:49–57
151. Pérez FJ, Ormemeño-Núñez J (1993) Weed growth interference from temperate cereals: the effect of a hydroxamic-acids-exuding rye (*Secale cereale* L.) cultivar. Weed Res 33:115–119
152. Overland L (1966) The role of allelopathic substances in the "smother crop" barley. Am J Bot 53:423–432
153. Pethó M (1992) Occurrence and physiological role of benzoxazinones and their derivates. III. Possible role of 7-methoxybenzoxazinone in the uptake of maize. Acta Agron Hung 41:57–64
154. Liu DL, Lovett JV (1993) Biologically active secondary metabolites of barley. II. Phytotoxicity of barley allelochemicals. J Chem Ecol 19:2231–2244
155. Friebe A, Wieland I, Schulz M (1996) Tolerance of Avena sativa to the allelochemical benzoxazolinone - degradation of BOA by rootcolonizing bacteria. Angew Botanik 70:150–154

156. Niemeyer HM, Jerez JM (1997) Chromosomal location of genes for hydroxamic acid accumulation in *Triticum aestrum* L. (wheat) using wheat aneuploids and wheat substitution lines. Heredity 79:10–14
157. Ebana K, Yan W, Dilday RH, Namai H, Okuno K (2001) Analysis of QTLs associated with the allelopathic effect of rice using water-soluble extracts. Breed Sci 51:47–51
158. Jensen LB, Cortois B, Shen LS, Li ZK, Olofsdotter M, Mauleon RP (2001) Locating genes controlling allelopathic effects against barnyard grass in upland rice. Agron J 93:21–26
159. Duke SO, Bajasa J, Pan Z (2013) Omics method for probing the mode of action of natural and synthetic phytotoxins. J Chem Ecol 39:333–348
160. Bertin C, Weston LA, Kaur H (2008) Allelopathic crop development: Molecular and traditional plant breeding approaches. Plant Breed Rev 30:231–258
161. Yang LT, Mickelson S, See D, Blake TK, Fischer AM (2004) Genetic analysis of the function of major leaf proteases in barley (*Hordeum vulgare* L.) nitrogen remobilization. J Exp Bot 55:2607–2616
162. Hayashi H, Czaja I, Lubenow H, Schell J, Walden R (1992) Activation of a plant gene by T-DNA tagging: auxin-independent growth *in vitro*. Science 258:1350–1353
163. Krysan PJ, Young JC, Sussman MR (1999) T-DNA as an insertional mutagen in Arabidopsis. Plant Cell 11:2283–2290
164. Shimura K, Okada A, Okada K, Jikumaru Y, Ko KW, Toyomasu T, Sassa T, Hasegawa M, Kodama O, Shibuya N, Koga J, Nojiri H, Yamane H (2007) Identification of a biosynthetic gene cluster in rice for momilactones. J Biol Chem 282:34013–34018
165. Kato-Noguchi H, Peters RJ (2013) The role of momilactones in rice allelopathy. J Chem Ecol 39:175–185
166. Fang CX, Xiong J, Qiu L, Wang HB, Song BQ, He HB, Lin RY, Lin WX (2009) Analysis of gene expressions associated with increased allelopathy in rice (*Oryza sativa* L.) induced by exogenous salicylic acid. Plant Growth Regul 57:163–172
167. Frey M, Schullehner K, Dick R, Fiesselmann A, Gierl A (2009) Benzoxazinoid biosynthesis, a model for evolution of secondary metabolic pathways in plants. Phytochem 70:1645–1651
168. Fujiyoshi PT, Gliessman SR, Langenheim JH (2007) Factors in the suppression of weeds by squash inter-planted in corn. Weed Biol Manage 7:105–114

Chapter 4
Weed Management in Organic Farming

Eric Gallandt

Introduction

To minimize loss of crop yield and quality, organic farmers replace herbicides with cultivation. This input substitution maintains focus on the seedling stage of weeds. Efficacy, however, is generally lower for cultivation relative to herbicide application, and as a consequence, weed pressure, that is, weed seedbanks, increase. Albrecht followed a farm in southern Germany through the transition from conventional to organic production [1]. During the first 3 years after conversion, total seed numbers in soil increased from 4,000 to more than 17,000 m^{-2}. These seedbank densities would be considered high on a conventional farm. For example, in the central USA "Corn Belt," a low initial seedbank (i.e., < 100 seeds m^{-2}) resulted in too few seedlings to warrant control. Seedbanks ranging from 100 to 1,000 seeds m^{-2} produced seedling populations that could be controlled by cultivation alone; there were high seedling populations where seedbanks were greater than 1,000 seeds m^{-2} and cultivation alone could not prevent large yield losses [2]. Unfortunately, many organic farmers far exceed this seedbank threshold.

Seedbank data from an organic, diversified vegetable operation in Dixmont, ME, USA, demonstrate the challenging situation facing many such growers: a mean germinable seedbank nearing 25,000 seeds per m^{-2} to 10-cm soil depth, with an average of 14.6 species (Gallandt, unpublished). The top four species included hairy galinsoga (*Galinsoga quadriradiata* Cav.), common lamb's-quarters (*Chenopodium album* L.), redroot pigweed (*Amaranthus retroflexus* L.), and common purslane (*Portulaca oleracea* L.), all considered difficult-to-manage species. This farmer's situation is not unique. In a recent survey of weed seedbanks on 23 organic farms in northern New England, USA, mostly diversified vegetable farms, germinable weed seed densities ranged from 2,500 to 25,000 seeds m^{-2} (Fig. 4.1)

E. Gallandt (✉)
School of Food and Agriculture,
University of Maine, 5722 Deering Hall, Orono, ME 04469-5722, USA
e-mail: gallandt@maine.edu; eric.gallandt@gmail.com

B. S. Chauhan, G. Mahajan (eds.), *Recent Advances in Weed Management*,
DOI 10.1007/978-1-4939-1019-9_4,

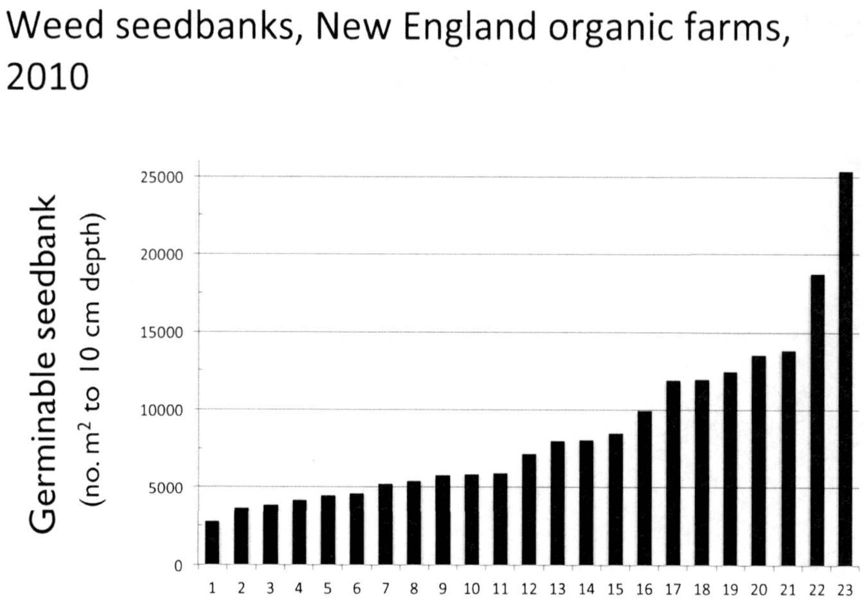

Fig. 4.1 Density of germinable weed seeds on 23 organic farms. Soil samples were collected in the spring of 2010 and subjected to exhaustive germination in the greenhouse. (Source: [30])

[3]. Eight of the 23 farms averaged more than 10,000 seeds m^{-2}, which would be considered heavy weed pressure.

What are the options for a farmer with very high weed densities? One option is to increase cultivation efficacy, but this is easier said than done. Cultivation tools vary in their efficacy based on soil conditions, weed species and growth stage, and operator skill. With practice, trial and error, and perhaps training by an "expert" cultivator, farmers may improve their own skills, and thus efficacy. Alternatively, the simplest solution is to accept the inherent limitations of cultivation, perhaps invest in additional tools and labor, and increase the number of cultivation events. Although cultivating more is an easy solution, it may not be compatible with goals related to fuel economy or soil quality improvement. A second option is to decrease the initial weed seedling density. Cultivation efficacy of 70 %, a typical value for a tine harrow, may be perfectly acceptable for an initial weed density of 10 seedlings m^{-2}, whereas an initial density of 100 seedlings m^{-2} would require two passes to achieve the same result. Reducing the initial weed seedling density, in turn, requires a reduction in the germinable weed seedbank.

Organic farming systems share production-related restrictions, and weed problems are ubiquitous, whether in organic olive in Italy, organic cereals in Australia, or organic vegetables in New England, USA. This chapter does not attempt a global review of weed management in all possible crops; rather, the chapter focuses on several ecologically based weed management principles and sample practices that have broad application.

Knowledge of Weed Biology Required to Guide Management

"Critical-weed-free-period" managers need to focus only on weed seedlings and cultivation tools. Their aim is simply to reduce weed density as much as possible, for as long as practical or affordable, and then rely on crop competition for subsequent weed control. However, as growers shift their management from seedling focus and the short term—toward a longer time horizon and an explicit goal of improving weed management over time—they require increasingly detailed knowledge of weed biology and ecological principles to guide their management.

Three important areas of seedbank ecology are particularly useful in this regard: (1) temporal patterns of weed emergence, (2) temporal patterns of weed seed rain, and (3) seed persistence in the soil, i.e., "half-lives." Emergence periodicity is essential information to direct timing of fallow events deployed to stimulate germination thereby depleting the seedbank (see section "Seedbank management," Maximizing Debits). Seed rain periodicity can guide deployment of short-season cash or cover crops that are terminated prior to weed seed rain. The duration of seed rain remains a priority research topic [4].

While annual weeds are known for extreme longevity in the soil, with *some* seeds persisting for years or even decades, seed densities generally decline exponentially and, thus, *most* seeds are relatively short-lived in the soil. For this reason, half-lives better describe weed survival [5]. In a review of 20 weed species conducted by Roberts and Feast [6], in cultivated soil, only two species had seed half-lives >2 years: black medic (*Medicago lupulina* L.) and annual bluegrass (*Poa annua* L.). Mean half-life for the 18 reported species was 1.38 years. Thus, species will vary in their response to seedbank management; species with short half-lives will be particularly responsive.

Organic farmers are increasingly requesting species-specific information as they work to manage emerging problems or weeds that are escaping other environmental and management stresses. Although most weed species have a considerable presence in the literature, rarely is weed biology information explicitly linked to practical management problems—a critical task for applied weed ecologists [7].

Crop Rotation

Crop rotation resides at the highest level of farm organization and is the foundation on which an ecologically based weed management program can be built. Here, cash and cover crops are chosen thereby defining the temporal sequence of management and disturbance "filters" that will contribute to the control of certain weed species and proliferation of others [8]. In fact, crop rotation is a required practice in the US National Organic Program (§ 205.205): "The producer must implement a crop rotation including but not limited to sod, cover crops, green manure crops, and catch crops that provide the following functions: (a) maintain or improve soil organic matter content; (b) provide pest management in annual and perennial crops; (c) manage deficient or excess plant nutrients; and (d) provide erosion control" [9].

From a weed management perspective, crop sequences should arrange dissimilar species with temporally varying disturbance regimes to challenge weeds at multiple points in their life history, and individual crops managed to preempt resource capture by weeds [10]. Over time, diverse cropping systems deploy a great variety of disturbance factors, including tillage, cultivation, crop competition, termination, and crop harvest. This can be the source of the multiple stresses or "Many Little Hammers" that are fundamental to ecologically based weed management [11]. Cover crops are often an important source of additional crop diversity, especially an opportunity to include sod crops and legumes on cash grain or vegetable farms lacking livestock.

Anderson recently proposed a 9-year crop rotation sequence for organic production in the semiarid Great Plains of the USA, including perennial forages, cool- and warm-season annual crops, and intervals of no-till to stress annual weeds at multiple points in their life history. The sequence included 3 years of alfalfa, two warm-season crops (corn, soybean), two cool-season crops (oat/pea, winter wheat), and again two warm-season crops (soybean, corn) [12]. Notable in this design is the stacking of the warm- and cool-season crops in contrast to an alternating sequence. The 2-year interval exploited the relatively short half-life of downy brome (*Bromus tectorum* L.) seeds in the soil, consistently providing lower weed densities than an alternate year sequence, and matching the weed control achieved in a comparatively less productive winter wheat/fallow system [13]. In a model exploring crop rotation effects on a depth-structured population of spotted lady's thumb (*Polygonum persicaria*), Mertens et al. [14] likewise found important sequence effects in a simple two-crop rotation, with lower populations in a rotation of crop sequence AABB compared to ABAB. Here, the effect was attributed not to seed survival but to tillage effects on the seedbank, establishment probability, and resultant fecundity. Expert organic farmers in the northeast USA frequently use particular short crop sequences or couplets to address multiple rotation goals including weed management; long-term, fixed sequences of crops are comparatively less common [15].

Mohler offered seven principles to guide crop rotation in diversified organic vegetable systems (Table 4.1) [15]. Foremost, and perhaps applicable only in high-value cropping systems, is the inclusion of clean fallow periods to deplete perennial below-ground reserves and to stimulate germination and establishment of annual weeds thereby depleting the seedbank [15]. This strategy, combined with intensive cover cropping, has proven pivotal to the Nordell's weed management program [16] (described later). Together, these seven recommendations reflect the core goals related to managing weed seedbanks: Maximize seed losses through timely fallowing to encourage germination and minimize seed inputs, and use short-season crops or crops planted in different seasons to preempt weed seed rain.

Exploiting benefits of crop rotation in extensive cash-grain cropping systems may require longer sequences, as suggested by Anderson [12]. Expert or rule-based decision support systems may prove useful in this regard. Bachinger and Zander [17] developed and evaluated such a tool for organic farming systems in central

Table 4.1 Principles guiding crop rotation in organic farming systems [15]

Recommendation	Relationship to seedbank management
Include clean fallow periods in the rotation to deplete perennial roots and rhizomes and to flush out and destroy annual weeds	Soil disturbance to stimulate germination losses
Follow weed-prone crops with crops in which weeds can easily be prevented from going to seed	Preempt weed seed rain
Plant crop in which weed seed production can be prevented before crops that are poor competitors	Preempt weed seed rain
Rotate between crops that are planted in different seasons	Avoid particular groups of species; prevent weed seed rain; accumulate 1 year of cumulative seedbank losses
Work cover crops into the rotation between cash crops at times when the soil would otherwise be bare	Disturbance stimulates germination losses; termination of cover crops can preempt weed seed rain; cover crop competition and mowing can reduce weed seed rain
Avoid cover crop species and cover crop; management that promote weeds	Preempt weed seed rain
Rotate between annual crops and perennial sod crops	Avoid particular groups of species; prevent weed seed rain; accumulate 1 or more years of cumulative seedbank losses

Europe. Their static, rule-based model, ROTOR, offers farmers opportunity to evaluate long-term cropping sequence effects on crop yield as well as weed and nitrogen dynamics.

Cover Cropping

A poor cover crop is worse than no cover crop. Eric and Anne Nordell [16]

Cover crops are frequently noted as essential for managing weeds on organic farms, and have been the subject of many weed management research projects. A beneficial effect is assumed and weed biomass is often presented as evidence: e.g., 50 % less weed biomass in a particular cover crop compared to fallow. If cover crops are terminated before weeds set seed, weed biomass is largely irrelevant regarding performance. Relevant parameters include effects on weed seedling recruitment: Did the cover crop contribute to seedbank depletion? Was there weed seed rain within the cover crop, i.e., did the cover crop contribute to or preempt seed rain? In short, cover crops should be considered within the context of weed population dynamics and life history characteristics (Fig. 4.2). Cover crops may contribute less to weed management than often assumed, at least *directly*. Rather, the *indirect* effects of cover cropping, specifically soil disturbance regimes associated with cover crop management may, in fact, drive effects on weed dynamics (e.g., see disturbance regime in Fig. 4.2). Field experiments conducted in Maine and Pennsylvania, USA, examined the seedbank-depleting effects of several full-season cover cropping

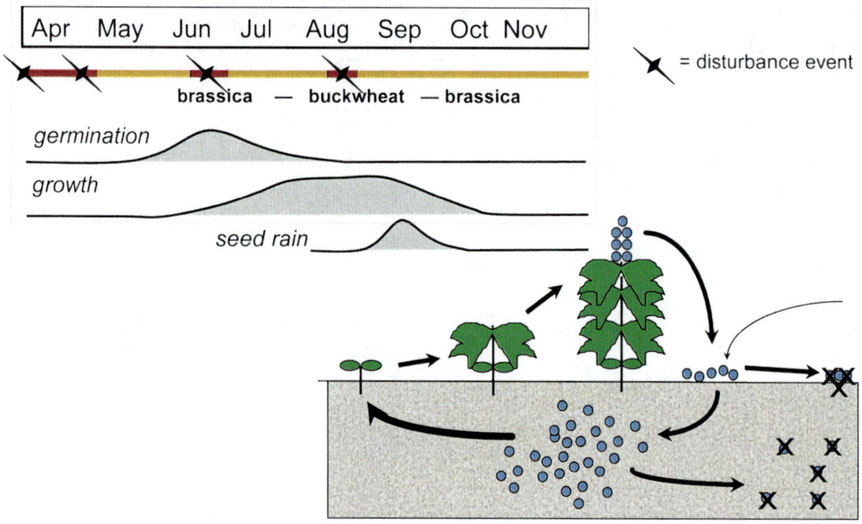

Fig. 4.2 Schematic representation of the life history of an annual weed, showing the weed seedbank and processes affecting population dynamics including establishment, maturation and reproduction, weed seed rain, immigration, seed predation, incorporation into the seedbank, and seed decay or death. Also shown is an example sequence of cover crops, brassica–buckwheat–brassica, deployed to reduce the weed seedbank, including disturbance events strategically timed to promote germination of a hypothetical weed species, and preempt seed rain

treatments [18]. Synthetic weed seedbanks were established in the fall or late winter. The following spring, cover crop treatments were established: oat/red clover; oat/pea, rye/hairy vetch; green bean, rye/hairy vetch; mustard, buckwheat, winter rape; and a summer fallow control. The incorporated green manure crops are well known for their allelopathic, residue-mediated effects on weeds [19–21]. Despite these established mechanisms, the main factor responsible for large and consistent seedbank losses was tillage. The soil disturbance events associated with cover crop management stimulated germination; subsequent disturbance events killed these weeds and prevented weed seed rain. The summer fallow and mustard, buckwheat, winter rape treatments, with four and three tillage events, respectively, eliminated the *Setaria* spp. seedbank, and reduced the *Chenopodium album* seedbank by 85 % and the *Abutilon theophrasti* seedbank by 80 % (Fig. 4.3) [18]. Overall, while cover crops provide many benefits to organic cropping systems [22], their major contribution to managing weed seedbank is a result of associated disturbance regimes.

Organic "No-Till"

Organic systems are frequently criticized for reliance on tillage and cultivation for residue management, seedbed preparation, and weed control. Conservation tillage, no-till in particular, offers well-known improvements to soil quality, but generally

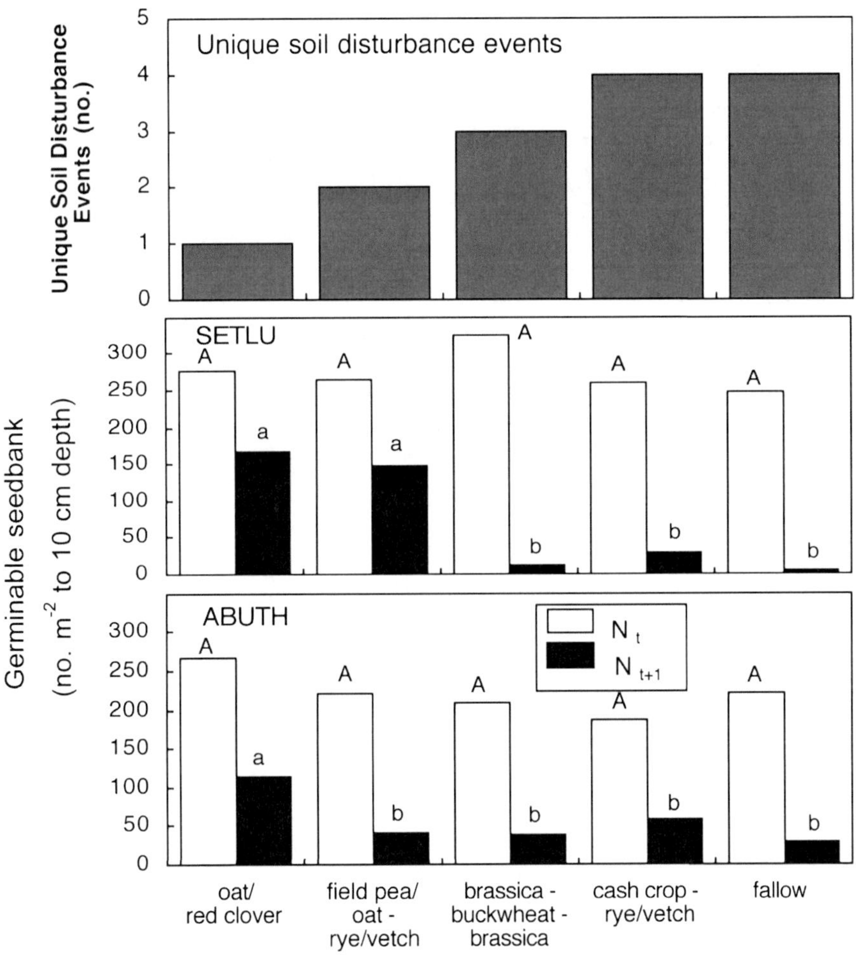

Fig. 4.3 Cover cropping systems evaluated for single-season effects on synthetic seedbanks in Maine and Pennsylvania; data shown are from Maine [18]. Data are arranged in order of increasing number of unique soil disturbance events. Germinable seedbank values are for samples collected at the initiation of the experiment, N_t, *open bars*, and samples collected the following year, N_{t+1}, *solid bars*. Within a sampling period, bars labeled with similar letters are not significantly different ($P>0.05$)

relies extensively on herbicides, both preplanting to burn down weeds and crop volunteers, and postemergence to control weeds within the crop. In recent years, research teams in several regions have challenged the assumption that no-till systems require herbicides, aiming to bring the soil-improving benefits to organic systems [23–25]. Generally, these systems have relied on high-residue cultivation equipment to maintain physical weed control, or the use of mowed or rolled/crimped cover crops in an attempt to establish weed-suppressive mulch. While the ecosystem services of organic grain production are compelling, particularly soil quality benefits of organic no-till [26], there are considerable challenges in managing weeds [27].

In the southeastern USA, rolled/crimped rye has shown some promise in organic no-till soybean, but weed control was highly dependent on the amount of rye biomass, and unfortunately, high levels of rye biomass were related to excessive soybean lodging [24]. In the mid-Atlantic USA, rolled/crimped cover cropping systems have been the focus of considerable attention in the past decade. Fall-sown cereal rye and rye/hairy vetch intercrops, terminated by rolling–crimping the following spring, provided variable weed control largely depending on the timing of cover crop planting, termination, and timing of weed emergence [23]. Again, the challenge is producing sufficient cover crop biomass to establish thick, dense mulch that provides season-long weed suppression. Density of the cover crop may also be an important factor, contributing to greater ground cover in early spring [28].

As common with many ecological approaches to weed management, variability in mulch crops has proven a problem for these systems. One tool to manage this uncertainty is a decision aid that would enable farmers to change to a clean-till system when overwintering cover crop biomass is likely to be insufficient [25]. Routine use of additional ecological weed management tactics may also help make mulch-based systems more reliable. Ryan et al. [29] found that elevated soybean seeding rates could compensate for lower cereal rye biomass levels; however, seeding rate effects were inconsistent over years when tested alone. A common problem, however, with elevated seeding rates in organic systems is the high cost of organic seed.

In the more arid cropping region of the US northern Great Plains, such mulch-based systems have proven to have many problems: excessive water use by cover crops, insufficient nitrogen contribution, termination dates incongruent with cash crop sowing dates, and weed control—perennial weeds in particular [30]. In western Canada, rolling–crimping and mowing were effective alternatives to tillage for the termination of legume green manures, and a rotary hoe designed for minimum tillage systems provided control of small-seeded weeds resulting in wheat yields within 13 % of a herbicide-treated control [31]. While promising, perennial weeds, which are a particular challenge in no-till, are generally unaffected by mulch or rotary hoe treatments.

Case 1: Anne and Eric Nordell, Beech Grove Farm

"Weed the soil, not the crop." This is the philosophy behind the particularly insightful ecologically based weed management program of Anne and Eric Nordell, Beech Grove Farm, Trout Run, PA, USA [16, 32, 33]. The Nordells wanted to farm without outside labor, which would require very low weed pressure. They also wanted to rely upon on-farm resources, which would mean no purchased mulch. They started farming the old hay fields of their current farm in 1983, immediately

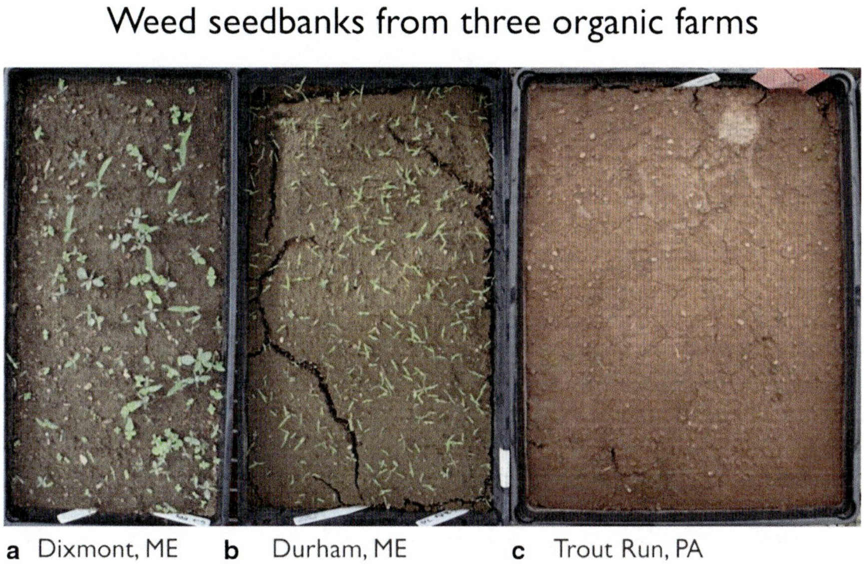

Fig. 4.4 Photograph of soils collected from two organic farms in Maine (**a** and **b**) and the Nordells' organic farm in Trout Run, Pennsylvania (**c**). Soil samples were collected in May, sieved, and spread over medium grade vermiculate in greenhouse flats. Photo was taken 3 weeks after initiating the germination assay. (Source: Gallandt, unpublished)

battling a heavy infestation of quack grass (*Elytrigia repens* L.) by summer fallowing. Early on, they spent a lot of time speaking with older local dairy farmers who told of the time-tested "COWS" (*c*orn, *o*at, *w*heat, and *s*od) rotation that was common to the region in the pre-herbicide era. Including diversity in timing of field operations with the warm- and cool-season grasses, and the soil-improving benefits of a perennial species, the COWS rotation was the model for what the Nordells would develop into their "rotational cover cropping" system for diversified vegetable farms.

Briefly, this approach crops half of the farm each year, the other half dedicated to various cover crops with carefully considered disturbance events, timed to encourage high levels of germination of either winter- or summer-annual weeds. Tillage is generally shallow, maintaining weed seeds closer to the soil surface where germination is more likely. Furthermore, the cover-crop-associated disturbance regime also considers the phenology of weed reproduction, to ensure preemption of seed rain. They now farm with virtually weed-free conditions where crops are cultivated once or twice. Living mulches provide in-season weed suppression and soil improvement. And, occasional hand weeding removes weed escapes before they set seed. Weeds germinating in soil samples collected from two organic farms in Maine, and the Nordell's farm in Trout Run, PA, USA, provide striking evidence of the efficacy of their seedbank management regime (Fig. 4.4).

Case 2: Paul and Sandy Arnold, Pleasant Valley Farm

"Attention to detail covers all aspects of our farm from weed control to preparing produce for markets. We realized early on that if we prevented weeds from going to seed, it would reduce our weed seed banks and labor spent in weeding." This was according to Paul and Sandy Arnold, Pleasant Valley Farm, Argyle, NY, USA. They manage several acres of organic hay for mulch in addition to their vegetable production. A modified forage harvester chops the mulch into a self-unloading wagon that is driven through the vegetable field depositing a windrow of green mulch. This is then spread, using pitchforks and by hand, to place a 15–20-cm deep layer of mulch around and between transplants. No hand weeding is subsequently required and soil organic matter increased from the original of 1–2 % to 3.5–4.0 %.

Harvesting is the largest labor expense on the farm. "Good weed control also increases harvest efficiency, yields, and everyone's morale; we enjoy working on a farm that everyone can be proud of in terms of organization and visual appearance." The intensive mulching system used by the Arnolds is congruent with two guiding rules regarding management of the farm: First is the "US$ 10,000 per acre rule," i.e., each crop is expected to have a minimum gross value of US$ 10,000 per acre. Second is the "US$ 30 per h rule," which means that each employee, while picking and packing for market, must be earning a minimum US$ 30 per h for the farm [34]. In this context, the early-season investment in intensive mulching means that morale remains high and employees are never asked to spend a day "heroic weeding." For the remainder of the season, employees focus on picking, washing, packing, and marketing. The mulch clearly represents considerable early investment in labor, but depreciated over the season, and importantly, the benefit to soil quality has made this strategy central to the success of this farm.

Mechanical Weed Control

> Good weed control leads to better weed control, poor weed control can only get worse.
> Bond et al. [35]

Bond and Turner [36] provided an excellent overview of mechanical weed control, including hand tools, harrows, tractor hoes, mowers, pneumatic tools, and guidance systems. Likewise, the field guide produced by the Wageningen URI Practical Farming group is an invaluable reference [37], as is *Steel in the Field* [38]. There have been notable advances in interrow cultivation in recent years, particularly guidance systems, either GPS or real-time, camera-based systems, which have increased working rates while permitting adjustment of tools very close to the crop row. Autonomous robotic weeding systems are projected for the near future (e.g., [39]), and, if affordable and scalable, could benefit many organic cropping systems. "Many little robots" could be the future of physical weed control on farms of any scale. To date, such systems remain in development attempting to overcome challenges in detection and identification of weeds in the field [40].

The so-called blind cultivation tools, tine weeders, harrows, and rotary hoes, are widely used for a great diversity of organic crops to provide both interrow and intrarow weed control. Selectivity for these tools is rather crude, relying on greater planting depth and initial size advantage of crops over weeds. Some crop damage is expected, by either uprooting or burial, which is affected by crop species, variety, and year [41]. There is an inherent trade-off between yield gains due to reductions in weed density from blind cultivation, and yield loss due to crop damage [42]. As weed seedlings grow beyond the sensitive "white thread" stage, these implements must be adjusted or operated more aggressively to maintain even moderate efficacy, thereby causing even greater crop damage. Organic growers and their advisors would be wise in establishing control strips to evaluate the cost/benefit of blind cultivation as the negative effects on crop yields are often ignored.

Limitations of tine harrows and other blind cultivation tools in cereals has prompted some organic grain farmers in northern Europe to adopt wide-row systems that permit interrow hoeing [42]. This provides opportunity for robust control of weeds at many growth stages, and is particularly effective on perennial weeds, which are the predominant weed problem in northern European organic grains.

Physical weed control research continues to be a priority in Europe, including work on flaming, brush weeding, hoeing, torsion weeding, finger weeding, robotic weeding, and band-steaming the soil [42]. Band-steaming, for example, reduced weed seedlings by 90 % at 61 °C and 99 % at 71 °C [43]. These innovative physical weeding tools are generally evaluated in combination with cultural or ecologically based management tactics, e.g., fertilizer placement, seed vigor, elevated seeding rates, and competitive cultivars [42, 44].

Intrarow weeds have always been the focus of efforts to advance cultivation performance. The justification for this is evident considering the generally linear relationship between the density of intrarow weeds and time spent hand weeding [45]. Finger and torsion weeders offer improved intrarow weed control compared to harrowing, but require accurate steering or precision guidance and thus have relatively low working rates. Nevertheless, these intrarow tools reduced hand weeding by 40–70 % [45]. Other innovative weeding implements from Europe include the Pneumat® weeder, which uses compressed air to blow small weed seedlings out of the crop row, and several nonselective tools that rely on real-time crop sensing to move an implement between crop plants that are widely spaced within the row. Hoes with intrarow crop sensing, as well as propane weeding systems for intrarow weeding are both commercially available in Europe (See http://www.visionweeding.com/); similar technologies are being developed in the USA (See http://bluerivert.com/home).

Despite these exciting technological developments, basic research regarding cultivation efficacy remains lacking. Sources of variability in cultivation efficacy include weed species, weed growth stage, soil moisture, soil quality, and tool design. Future research with cultivation tools should provide accurate estimates of the mean and variance of efficacy under a set of standardized conditions, and should characterize how efficacy changes with aforementioned variables.

Hand weeding is the most common weed management practice on small- to mid-scale diversified organic farms. While exceptionally effective, high labor costs make hand weeding expensive. Moreover, time required for high levels of weed control by hand increases with increasing weed density. A comprehensive weed management plan focused on reducing the weed seedbank will result in both improved weeding outcomes with the use of hand tools and lower hand weeding costs.

Hand weeding may rely on pushed, wheeled tools, long-handled tools, short-handled tools, and/or hand pulling. We conducted eight field experiments, measuring working rate (i.e., row-feet weeded per minute) and efficacy (i.e., proportion of weeds controlled) in a standardized crop/surrogate weed system of corn and condiment mustard (*Sinapis alba,* "Idagold") (Gallandt, unpublished). Wheeled tools generally had highest working rates, but occasionally lower efficacy than other tools or hand pulling. Importantly, working rates for wheeled tools were independent of weed density. Thus, wheeled tools should be used before other hand methods because of their higher working rates. Long-handled tools may offer improved efficacy over wheeled tools, but generally with lower working rates. Short-handled tools and hand pulling offer potentially complete weed control, but with increasing time proportional to weed density. Overall, wheeled tools should be the first step in a hand weeding program, followed by long- and then short-handled tools, with hand pulling a final step where very high efficacy is required.

Qualitative surveys of hand tools indicated a high level of variation in user preference (Gallandt, unpublished). The Glaser® stirrup hoe was top-ranked in aggregate user scores for categories of "feel," "efficacy," and "overall," followed closely by the Glaser® wheel hoe. Contrary to expectations, tool rankings were, with a few minor exceptions, generally unaffected by gender, age, years of experience, or scale of enterprise (Gallandt, unpublished).

While research and development efforts strive to enhance efficacy of future tools, it is possible to overcome rather low efficacy through repeated use of existing tools. This may be necessary due to low efficacy and an abundant seedling population, and/or due to protracted establishment patterns of particular species. An example of this is the recently published work of Peruzzi et al. [46], in the Fucino plateau area of Italy. In organic carrots, their strategy was to reduce the initial weed population with a false seedbed (or stale seedbed); subsequently flame; precision hoe; and hand weed. This example demonstrates the need to reduce the germinable seedbank. Although the authors did not quantify the germinable seedbank, the site clearly required multiple physical weed control passes to bring the population down to an acceptable level. Despite reduction in initial seedling density achieved with false seedbed, subsequent flushes of weeds were generally very high, requiring repeated interventions.

Seedbank Management

Researchers have concluded that relatively low and variable efficacy of alternative weed management practices requires additional efforts to reduce weed seedbanks [47], a conclusion further supported by simulation models [48]. Direct weed control measures (e.g., herbicides) provide density-independent weed seedling control [49]. Thus, the density of surviving weeds is simply a proportion of the initial density. This holds true for cultivation, (Gallandt, unpublished), at least at reasonable weed densities; at exceptionally high densities of selected species, one could imagine inverse density-dependent effects, e.g., a "sod" of large crabgrass (*Digitaria sanguinalis* L.) seedlings may experience reduced mortality as the seedlings are less likely to be dislodged from soil and desiccate. There are at least three solutions to the problem of density-independent efficacy: (1) simply cultivate more, adding additional cultivation events proportional to the initial weed density; (2) cultivate better, by improving the operator's skill in decision making, timing, adjustment, or investment in weeding tools with improved efficacy; or (3) start with fewer weeds, i.e., reduce the germinable weed seedbank. The first two options were described previously; principles and practical recommendations for managing the weed seedbank follow.

> Seedbanks are often thought of as a vault in which past species and genotypes are stored awaiting some future conditions that will break dormancy and initiate germination. The following quotes from our recent interviews of organic farmers reflect this conceptual model:Seeds are buried and lie dormant for up to 50 years until they are stimulated with light or water…7, 8, 9 years, up to 50 year viability…there are even seeds that have lasted 1000 years! Midwestern USA, farmer
>
> I think some of those weeds are historical weeds like the lambsquarter, which can last a long, long time in the soil. So I'm fighting somebody else's weed that they left 40 years ago.…New England, USA, farmer

We recently completed a research project aimed at characterizing organic farmers' beliefs and perceptions regarding weeds and weed management. In comparing farmer interviews with "expert" weed managers, including researchers, extension personnel, and farm advisors, weed seedbanks were a topic where the two groups disagreed [50]. Both groups seemingly shared the conceptual model of seeds entering the seedbank and subsequently declining. However, experts focused on the initial rapid decline of the seedbank, inspired by ecologically based opportunities for management (e.g., [51]), whereas farmers expressed a less optimistic perspective, focusing on the protracted "tail" of the seed decay curve and the few individual seeds with extreme longevity (Fig. 4.5). This incongruity in such a pivotal point for ecological weed management is an opportunity for improved educational programming and applied research. Seedbanks are, in fact, very dynamic, responding to annual seed inputs, germination losses, predation, and decay, the mechanisms related to seedbank management [51].

Minimizing "Credits"

Efforts to minimize credits to the seedbank (i.e., seed rain) start with effective weed seedling control, discussed previously. The next goal is to minimize weed biomass and therefore seed rain by enhancing crop–weed interference in long-season crops,

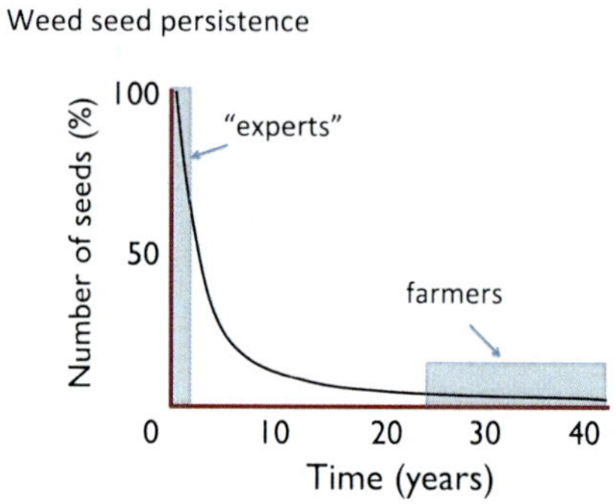

Fig. 4.5 Idealized representation weed seed decay over time noting the incongruity of experts and farmers in the priority of their conceptual models of this process. This diagram represents a relatively long-lived species, with a half-life of 4 years; most problematic annual weeds are far less persistent

or terminating growth and thus preempting weed seed rain in short-season cash or cover crops. Weed ecologists have spent considerable effort characterizing weed–crop competition through experiments of increasing weed density effects on crop yield [52]. While relevant to short-term crop yield and economics, the crop response has few long-term implications. In the context of seedbank management, the interesting effect is the competitive effect of the crop on weeds, specifically the resulting weed fecundity or seed rain [4].

Organic farmers often tolerate elevated weed pressure. A recent analysis of the Rodale Farming Systems Trial indicates that organic crops can tolerate more weeds and still yield well [53]. Corn yields over 27 years were comparable in organic and conventionally managed systems, despite 4.5–6.3 times more weed biomass, on average, in the organic system. This evidence that weed–crop competitive relationships may be fundamentally different in organic and conventional systems is the source of important mechanistic questions regarding crop and weed growth in organic management.

Maximizing "Debits"

Germination, death, and emigration are seedbank debiting mechanisms [54]. Depleting seedbanks by encouraging germination losses is the aim of preparing a "stale seedbed" in which primary and secondary tillage are performed, a seedbed prepared, and weeds are allowed to establish; subsequent shallow tillage kills this first flush of weeds, while preparing a new seedbed for additional weed seed depletion or immediate crop sowing [55]. Tillage is the primary tool used to encourage germination. Working depth should be relatively shallow (< 5 cm), and cultipacking or firming the soil to improve weed seed/soil contact may further improve germination. Soil moisture must be adequate to support germination; irrigation may be used to better control timing of weed establishment [56]. Various implements may be used to control the flush of weeds before crop planting: rototillers, top knives, rotary hoe, and heavy tine weeders. The aim is to cultivate as shallow as possible to avoid bringing up germinable seeds from lower soil strata, thereby minimizing the magnitude of the second flush of weeds. Flaming, which kills small broadleaf weeds but not grasses or perennials, results in no soil disturbance, and thus fewer weeds germinating with the crop [56]. The use of stale seedbeds is generally restricted to high-value or short-season crops and areas with long growing seasons. In northern temperate regions, there generally is not sufficient time, and often not suitable environmental conditions, to delay crop sowing.

Rasmussen [57] introduced the practice of "punch planting," which aimed to further exploit benefits of the stale seedbed by sowing seeds into a hole, created by a dibbler following flaming, thereby minimizing intrarow disturbance caused by planting. In this system, tested with beet (*Beta vulgaris* L.), weeds were controlled by preemergence flaming; intrarow weed density was reduced by 30% in the punch-planted treatment compared to conventional drilling with flame weeding. Subsequent evaluation of the punch-planting system with a prototype dibbler drill reduced intrarow weeds in onion by 37% [58]. However, technical complications with the drill resulted in some unintended effects, e.g., the dibbler wheel engaging the soil surface and functioning as a press wheel causing elevated weed emergence.

Timing of soil disturbance is an important factor affecting potential weed emergence. Fallowing midsummer, for example, will encourage germination of summer annual weeds, but winter annuals will remain dormant. Soil degree-day modeling may be a useful tool to optimize disturbance regimes to target depletion of particular species [59].

Seed Predation

The past decade has seen considerable interest in seed predation as an ecosystem service that could contribute to ecologically based weed management, perhaps even at large landscape scales [60]. Comparisons of organic and conventional seed predation rates are inconclusive. Experiments in New Zealand documented mean seed removal rates of 17% per 48 h in organic fields compared to 10% in conventional fields; video images indicated that birds were the predominant predator [61]. However, similar studies conducted in Germany found similar predation rates in organic and conventionally managed cereal fields [62]. In areas where invertebrate predators predominate, vegetation may provide seed predators cover thereby increasing their activity/density [63]. While measurements of seed removal and predator abundance have provided considerable circumstantial evidence that seed predators are present and active in organic cropping systems, these point estimates offer little insight into their importance to longer-term population dynamics.

Westerman et al. [64] provided compelling evidence using simulation models that seed predators were indeed providing an important and quantifiable ecosystem service in the Midwestern USA. In a more diverse 4-year rotation, with higher rates of seed predation, velvetleaf (*A. theophrasti* L.) density could be maintained over time with 86% efficacy of control in the soybean phase of the rotation. In the 2-year system, with comparably lower rates of seed predation, a higher-level seedling control (93%) was required to prevent an increasing weed population.

Table 4.2 Seed predator exclosure effects on density of total germinable weeds. (Gallandt, unpublished)

Treatment	Germinable weed seeds 2008	2009	2010
	(No. m^{-2} to 10 cm depth)		
Exclosure (−)	36,100	52,500	32,100
Exclosure (+)	62,500	54,000	32,600
P	0.002	0.765	0.989

Exclosures were installed in the fall of 2008–2010 following weed seed rain; soil samples were collected in the following May, and subjected to exhaustive germination in a greenhouse

We recently completed field studies using long-term exclosures, installed in the fall after weed seed rain, and subsequent spring soil sampling within and outside exclosures, to quantify season-long predation effects on the seedbank. We expected to see higher germinable weed seedbank densities within exclosures where seeds were protected from predators. In the first year of the study, there were 42 % fewer germinable weed seeds outside the exclosures, evidence of a large and significant effect of predators over the fall, winter, and spring assay periods (Table 4.2). However, in 2 subsequent years, we did not detect an exclosure effect. Thus, although predation rates may be impressive in certain years, the high level of interannual variation suggests that predation may not be a reliable seedbank debiting mechanism.

We expected that fall tillage used to incorporate residues and establish cover crops could be reducing potential seed predation. Seed burial, even coverage by 1 mm of sand, dramatically reduced seed predation rates [65]. Thus, in a related set of experiments, we examined weed seedbank response to four fall weed management strategies, some designed to retain weed seeds for as long as possible at the soil surface. Surprisingly, seeds do not simply stay at the soil surface, but are buried even in the absence of tillage [66]. Zero seed rain, a control, we included in this study, consistently had the smallest seedbank (Fig. 4.6). Treatments designed to retain weed seeds on the soil surface thereby increasing opportunity for predation losses—i.e., mowing alone, or mowing with a no-till cover crop—offered no advantage over the fall tillage, cover cropped treatment, despite the weed seed burial and protection from predators expected with this treatment (Fig. 4.6).

Seed predation remains a topic of interest among ecological weed managers, inspired at least in part by occasionally large effects and the ubiquitous nature of predators in agroecosystems. More reliable strategies, however, include mechanical weed seed harvesting (e.g., using combines as "predators") and post-dispersal seed flaming, as recent evidence suggests, could be another tool for managing the seedbank.

Mechanical Weed Seed Harvesting

Inspired by increasingly intractable problems with multiple herbicide resistance in annual ryegrass (*Lolium rigidum* Gaudin), Australian farmers and researchers are developing innovative technologies for weed management. The Harrington Seed

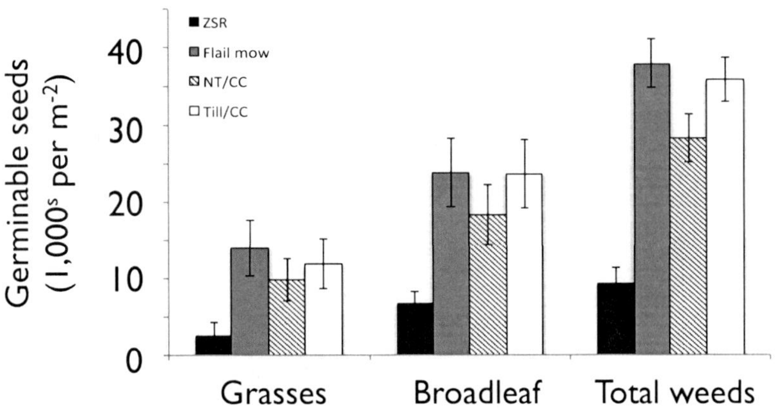

Fig. 4.6 Weed seedbank density, means and standard errors, of grass, broadleaf and total weeds following four seedbank management treatments: (1) zero seed rain, *ZSR*, and *black bars*; (2) fall flail mowing of standing weeds, *flail mow*, and *gray bars*; (3) fall flail mowing followed by no-till sowing a cereal rye cover crop, *NT/CC, textured bars*; and (4) fall tillage and sowing a rye cover crop, *till/CC, open bars*. (Source: Gallandt and Jabbour, unpublished)

Destroyer, for example, is a device that mills grain chaff and weed seeds exiting a combine, resulting in >95 % weed seed mortality [67]. Chaff carts, narrow windrow burning, and bale direct are additional practices being used, or developed, to address the herbicide resistance problem in Australia [68]. These techniques could benefit organic grain systems in locations where the predominant weeds are species that retain their seeds when mature, and species that have relatively short-lived seedbanks, e.g., < 1 year.

Davis [69] conducted a thorough analysis of fall weed seed pools in east-central Illinois, USA, in conventionally managed corn and soybean. The seed pools included seed that was undispersed, recently dispersed, collected by machinery, and previously dispersed. In one or more crops during the 2 experimental years, the ratio of undispersed seeds to seeds in the seedbank was > 1, indicating the complete seedbank replenishment. Although many species dispersed seed prior to harvest, ivyleaf morningglory (*Ipomoea hederacea* L.), giant foxtail (*Setaria faberi* Herrm.), and prickly sida (*Sida spinosa* L.) retained more than 50 % of new seeds on the mother plant, providing an opportunity for mechanical seed capture.

Seed Flaming

Flaming is commonly used on organic and low-external-input farms to control weed seedlings, especially annual dicots in slow-to-emerge crops, such as carrot and beet. Field studies conducted at the Goranson Farm, Dresden, ME, USA, demonstrated that pre-dispersal flaming did not affect viability of common lamb's-quarters or redroot pigweed seeds (Gallandt, unpublished). Flaming, however, showed promise

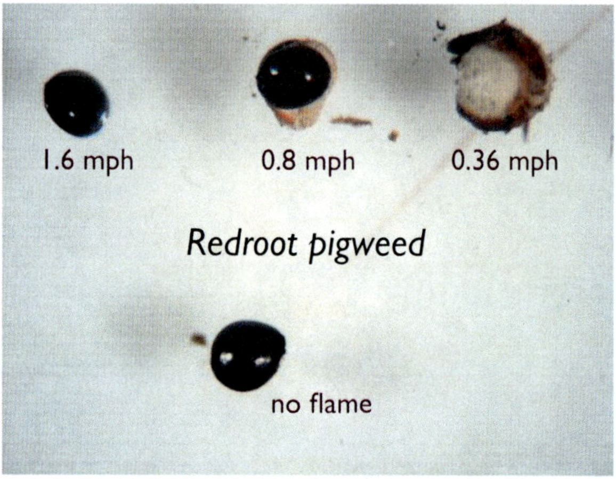

Fig. 4.7 Photograph of redroot pigweed (*Amaranthus retroflexus* L.) seeds following exposure to a farm-scale liquid propane burner (1000°C), with exposure adjusted by the forward tractor speed including 0.6, 1.3, and 2.6 miles km h^{-1}. (Source: Gallandt, unpublished)

as a method to reduce density of weed seeds following dispersal. Greenhouse and field studies demonstrated that flaming could kill weed seeds on the soil surface (Gallandt, unpublished). Typical tractor speeds used for other flaming operations (e.g., 2.6 km h^{-1}) killed about 50% of the most sensitive species (i.e., hairy galinsoga); however, with the flame dosage doubled (i.e., to 1.3 km h^{-1}) flaming reliably killed 75% or more seeds of mustard, large crabgrass, and hairy galinsoga. Flaming effects on redroot pigweed were visually very striking (Fig. 4.7). There was no advantage to further doubling the flaming dosage, as seed mortality was similar to both 1.3 and 0.6 km h^{-1} treatments. With an estimated cost of US$ 375 ha^{-1} for 1.3 km h^{-1} treatment, fall flaming could prevent large weed seedbank credits, especially of relatively sensitive species, including hairy galinsoga.

Soil Quality and Seed Decay

The hypothesis that improved soil quality and concomitant increases in microbial activity could accelerate weed seed decay [70] has been tested, but experimental evidence to date has been inconclusive. In California, USA, organic amendments increased microbial biomass, which was negatively correlated with burning nettle (*Urtica urens* L.) and shepherd's purse (*Capsella bursa-pastoris* L.) [71]. However, experiments conducted with wild oat (*Avena fatua* L.) in long-term conservation-tillage and no-till plots, managed conventionally, indicated that decay and overall seed mortality was similar in the contrasting tillage treatments despite differences in soil quality [72]. Ullrich et al. [73] recently tested this hypothesis in two long-term experiments near Beltsville, MD, USA. Common lamb's-quarters and smooth pigweed (*Amaranthus hybridus* L.) were buried in mesh bags, in both organic and conventional cropping systems. Management system effects on seed mortality were inconsistent, as were relationships between soil microbial biomass and seed mortality. In another study conducted near Presque Isle, ME, USA, a similar mesh bag study characterized fall, overwinter, and spring mortality of six weed species in long-term replicated plots managed for divergent soil quality characteristics; there was no evidence of a soil quality effect on seed mortality (Gallandt, unpublished). There is, in fact, evidence that organic amendments may actually inhibit seed mortality [74].

Conclusion

Ecologically based weed management requires diversity. The most important tools in this regard are cash or cover crops with opportunities for high levels of weed control and varying disturbance regimes, which can preempt weed seed production and encourage seedbank depletion. The disintegration of crop and livestock production characterizes the modern farming system, and limits the potential internal biological controls likely to be essential to the development of whole-farm, systemic resistance to weeds. Animals per se are probably not requisite, the manure, compost, and crop diversification—sod crops, winter and spring cereals, and legumes—may be the elements required to satisfactorily manage weeds on organic farms. Supporting this contention, crop rotation and cover cropping are top-ranked weed management practices noted by organic farmers, practices offering such diversification. However, these practices often do not contribute to weed management goals, and may actually hinder long-term progress in reducing the weed seedbank if they permit abundant weed seed rain. Similarly, cover crops are frequently noted as "competitive" and "weed-suppressive," but they may offer few benefits if associated disturbance events do not prevent weed seed inputs.

Knowledge of weed seedbank ecology and management has grown considerably over the past decade, as has interest and research on weed management in organic farming systems. There is broad consensus in the research community that successful weed management without herbicides requires a systems perspective, a combination of control measures and greater system complexity that aims to compensate for generally moderate reliability and efficacy of any individual practice [75]. There are farmers who embrace such complexity and share researchers' fascination for the ecology of complex systems. However, it is important to note that weeds, while an important problem for organic farmers, are only one of their many problems, and many farmers desire the simplest solution they can find. This, perhaps, explains the continued emphasis on cultivation and managing for the critical weed-free period. Facing an increasing weed problem, the simple solution is to cultivate more. Farmers of varying scales and enterprises manage very successful enterprises using this approach, despite criticism from researchers regarding possible deterioration of soil quality, risk, and labor costs.

There are considerable management costs, investment in education, scouting, and trial and error associated with complex, ecologically based, "Many Little Hammers" approaches to weed management. It is perhaps not surprising that adoption is slow. Bastiaans et al. [75] recommend that, in addition to continued development of additional practices, ecological weed management strategies require clear and quantitative evidence of efficacy, variability, and cost–benefit. Expert farmer testimonials currently provide some of this evidence and simulation models additional support [64].

References

1. Albrecht H (2005) Development of arable weed seedbanks during the 6 years after the change from conventional to organic farming. Weed Res 45:339–350
2. Forcella F, Eradat-Oskoui K, Wagner SW (1993) Application of weed seedbank ecology to low-input crop management. Ecol Appl (JSTOR) 3(1):74–83
3. Jabbour R, Gallandt ER, Zwickle S, Wilson RS, Doohan D (2014) Organic farmer knowledge and perceptions area associated with on-farm weed seedbank densities in northern New England. Weed Sci 62(2):338–349
4. Norris RF (2007) Weed fecundity: current status and future needs. Crop Prot 26(3):182–188
5. Mohler CL (2001) Weed life history: identifying vulnerabilities. In: Liebman M, Mohler CL, Staver CP (eds) Ecological management of agricultural weeds. Cambridge University Press, New York, pp 40–98
6. Roberts HA, Feast PM (1972) Fate of seeds of some annual weeds in different depths of cultivated and undisturbed soil. Weed Res 12(4):316–324
7. Van Acker RC (2009) Weed biology serves practical weed management. Weed Res 49(1):1–5
8. Booth BD, Swanton CJ (2002) Assembly theory applied to weed communities. Weed Sci 50:2–13
9. US National Organic Program (§ 205.205). http://federal.eregulations.us/cfr/section/title7/chapteri/part205/sect205.205?selectdate=11/1/2011. Accessed 27 Nov 2013
10. Liebman M, Staver CP (2001) Crop diversification for weed management. In: Liebman M, Mohler CL, Staver CP (eds) Ecological management of agricultural weeds. Cambridge University Press, New York, pp 322–374
11. Liebman M, Gallandt ER (1997) Many little hammers: ecological approaches for management of crop-weed interactions. In: Jackson LE (ed) Ecology in agriculture. Academic, New York, p 291–343
12. Anderson RL (2010) Rotation design to reduce weed density in organic farming. Renew Agric Food Syst 25(03):189–195
13. Anderson RL (2008) Diversity and no-till: keys for pest management in the U.S. great plains. Weed Sci 56(1):141–145
14. Mertens SK, Bosch F Van Den, Heesterbeek JAPH, Applications E, Aug N (2002) Weed populations and crop rotations: exploring dynamics of a structured periodic system. Ecol Appl 12(4):1125–1141
15. Mohler CLC, Johnson SES, Resource N (2009) Crop rotation on organic farms: a planning manual. In: Mohler CL, Johnson SE (eds) Engineering. NRAES (Natural Resource, Agriculture, and Engineering Service), Ithaca, p 156
16. Grubinger VP (1999) Crop rotation. Sustain vegetable production from start-up to mark. p 69–77
17. Bachinder J, Zander P (2007) ROTOR, a tool for generating and evaluating crop rotations for organic farming systems. Eur J Agron 26(7):130–143
18. Mirsky SB, Gallandt ER, Mortensen DA, Curran WS, Shumway DL (2010) Reducing the germinable weed seedbank with soil disturbance and cover crops. Weed Res 50(4):341–352

19. Conklin AE, Erich MS, Liebman M, Lambert D, Gallandt ER, Halteman WA (2002) Effects of red clover (*Trifolium pratense*) green manure and compost soil amendments on wild mustard (*Brassica kaber*) growth and incidence of disease. Plant Soil 238:245–256
20. Haramoto ER, Gallandt ER (2007) Brassica cover cropping for weed management: a review. Renew Agric Food Syst 19(4):187–198.
21. Kruidhof HM, Gallandt ER, Haramoto ER, Bastiaans L (2011) Selective weed suppression by cover crop residues: effects of seed mass and timing of species' sensitivity. Weed Res 51(2):177–186
22. Sarrantonio M, Gallandt E (2003) The role of cover crops in North American cropping systems. J Crop Prod 8(1/2):53–74
23. Mirsky SB, Curran WS, Mortenseny DM, Ryany MR, Shumway DL (2011) Timing of cover-crop management effects on weed suppression in no-till planted soybean using a roller-crimper. Weed Sci 59(3):380–389
24. Smith AN, Reberg-Horton SC, Place GT, Meijer AD, Arellano C, Mueller JP (2011) Rolled rye mulch for weed suppression in organic no-tillage soybeans. Weed Sci 59(2):224–231
25. Reberg-Horton SC, Grossman JM, Kornecki TS, Meijer AD, Price AJ, Place GT et al (2011) Utilizing cover crop mulches to reduce tillage in organic systems in the southeastern U S A. Renew Agric Food Syst 27(01):41–48
26. Cavigelli MA, Mirsky SB, Teasdale JR, Spargo JT, Doran J (2013) Organic grain cropping systems to enhance ecosystem services. Renew Agric Food Syst 28(2):145–159
27. Carr P, Gramig G, Liebig M (2013) Impacts of organic zero tillage systems on crops, weeds, and soil quality. Sustainability 5(7):3172–3201
28. Ryan MR, Curran WS, Grantham AM, Hunsberger LK, Mirsky SB, Mortensen DA et al (2011a) Effects of seeding rate and poultry litter on weed suppression from a rolled cereal rye cover crop. Weed Sci 59(3):438–444
29. Ryan MR, Mirsky SB, Mortensen DA, Teasdale JR, Curran WS (2011b) Potential synergistic effects of cereal rye biomass and soybean planting density on weed suppression. Weed Sci 59(2):238–246
30. Carr PM, Anderson RL, Lawley YE, Miller PR, Zwinger SF (2011) Organic zero-till in the northern US great plains region: opportunities and obstacles. Renew Agric Food Syst 27(01):12–20
31. Shirtliffe SJ, Johnson EN (2012) Progress towards no-till organic weed control in western Canada. Renew Agric Food Syst 27(01):60–67
32. Barberi P (2002) Weed management in organic agriculture: are we addressing the right issues? Weed Res 42(3):177–193
33. Nordell A, Nordell E (2009) Weed the soil, not the crop. Acres USA 40(6)
34. Arnold P, Arnold S (2003) Putting it all together to have profitability on a small farm. Northeast organic farming association. Northeast Organic Farming Association of New York, Inc, New York. p 10–13
35. Bond W, Burston S, Moore HC, Bevan JR, Lennartsson MEK (1998) Changes in the weed seedbank following different weed control treatments in transplanted bulb onions grown organically and conventionally. Asp Appl Biol (Weed seedbanks: determination, dynamics and manipulation) 51:273–278
36. Bond W, Turner RJ (2005) A review of mechanical weed control. HDRA, Ryton Organic Gardens, Coventry, CV8 3LG, UK, pp 1–16
37. Bleeker P, Molendijk L, Plentinger M, van der Weide RY, Lotz B, Bauermeister R et al (2006) Practical weed control in arable farming and outdoor vegetable cultivation without chemicals. Wageningen UR, Applied Plant Research, Wageningen, p 77
38. Bowman G (2002) Steel in the field: a farmer's guide to weed management tools (Sustainable Agriculture Network). SARE Outreach, Beltsville, p 128
39. Young SL (2012) True integrated weed management. Weed Res 52(2):107–111
40. Slaughter DC, Giles DK, Downey D Autonomous robotic weed control systems: a review. Comput Electron Agric 61(1):63–78.

41. Kolb LN, Gallandt E (2012) Weed management in organic cereals: advances and opportunities. Org Agric 2(1):23–42
42. Melander B, Rasmussen IA, Bàrberi P (2005) Integrating physical and cultural methods of weed control- examples from European research. Weed Sci (BioOne) 53(3):369–381
43. Melander B, Jørgensen MH (2005) Soil steaming to reduce intrarow weed seedling emergence. Weed Res 45(3):202–211
44. Hatcher PE, Melander B (2003) Combining physical, cultural and biological methods: prospects for integrated non-chemical weed management strategies. Weed Res 43(5):303–322
45. Van der Weide RY, Bleeker PO, Achten VTJM, Lotz LAP (2008) Innovation in mechanical weed control in crop rows. Weed Res 48:215–224
46. Peruzzi A, Ginanni M, Raffaelli M, Fontanelli M (2007) Physical weed control in organic fennel cultivated in the Fucino Valley (Italy). EWRS 7th workshop on physical and cultural weed control, Salem, Germany, 11–14 March
47. Jordan N (1996) Weed prevention: priority research for alternative weed management. J Prod Agric 9(4):485–490. (Madison, Wis: American Society of Agronomy, Crop Science Society of America, Soil Science Society of America, 1988–1999)
48. Davis AS, Dixon PM, Liebman M (2004) Using matrix models to determine cropping system effects on annual weed demography. Ecol Appl 14(3):655–668
49. Dieleman JA, Mortensen DA, Martin AR (1999) Influence of velvetleaf (*Abutilon theophrasti*) and common sunflower (*Helianthus annuus*) density variation on weed management outcomes. Weed Sci 47(1):81–89
50. Jabbour R, Zwickle S, Gallandt ER, McPhee KE, Wilson RS, Doohan D (2013) Mental models of organic weed management: comparison of New England U.S. farmer and expert models. Renew Agric Food Syst 1–15. Available on CJO 2013. doi: 10.1017/S1742170513000185
51. Gallandt ER (2006) How can we target the weed seedbank? Weed Sci 54(3):588–596
52. Cousens R (1985) A simple model relating yield loss to weed density. Ann Appl Biol 107(2):239–252
53. Ryan MR, Smith RG, Mortensen DA, Teasdale JR, Curran WS, Seidel R et al (2009) Weed-crop competition relationships differ between organic and conventional cropping systems. Weed Res 49(6):572–580
54. Forcella F (2003) Debiting the seedbank: priorities and predictions. Asp Appl Biol 69:151–162 (1999)
55. Caldwell B, Extension C, Mohler CL (2001) Stale seedbed practices for vegetable production. Crop Prot 36(4):703–705
56. Boyd NS, Brennan EB, Fennimore SA (2006) Stale seedbed techniques for organic vegetable production. Weed Technol 20(4):1052–1057
57. Rasmussen J (2003) Punch planting, flame weeding and stale seedbed for weed control in row crops. Weed Res 43(6):393–403
58. Rasmussen J, Henriksen CB, Griepentrog HW, Nielsen J (2011) Punch planting, flame weeding and delayed sowing to reduce intra-row weeds in row crops. Weed Res 51(5):489–498
59. Myers MW, Curran WS, VanGessel MJ, Calvin DD, Mortensen DA, Majek BA et al (2004) Predicting weed emergence for eight annual species in the northeastern United States. Weed Sci (BioOne) 52(6):913–919
60. Bohan DA, Boursault A, Brooks DR, Petit S (2011) National-scale regulation of the weed seedbank by carabid predators. J Appl Ecol 48(4):888–898
61. Navntoft S, Wratten SD, Kristensen K, Esbjerg P (2009) Weed seed predation in organic and conventional fields. Biol Control 49(1):11–16
62. Daedlow D, Sommer T, Westerman PR (2012) Weed seed predation in organic and conventional cereal fields. 25th German conference on weed biology and weed control. Braunschweig, Germany p 265–271
63. Gallandt ER, Molloy T, Lynch RP, Drummond FA, Drummond A (2005) Effect of cover-cropping systems on invertebrate seed predation. Weed Sci 53(1):69–76

64. Westerman PR, Liebman M, Menalled FD, Heggenstaller AH, Hartzler RG, Dixon PM (2005) Are many little hammers effective? Velvetleaf (*Abutilon theophrasti*) population dynamics in two- and four-year crop rotation systems. Weed Sci 53(3):382–392
65. Harrison S, Gallandt E (2012) Behavioural studies of *Harpalus rufipes* De Geer: an important weed seed predator in Northeastern US agroecosystems. Int J Ecol 2012:1–6
66. Westerman PR, Dixon PM, Liebman M (2009) Burial rates of surrogate seeds in arable fields. Weed Res 49(2):142–152
67. Walsh MJ, Harrington RB, Powles SB (2012) Harrington seed destructor: a new nonchemical weed control tool for global grain crops. Crop Sci 52(3):1343
68. Walsh M, Newman P, Powles S (2013) Targeting weed seeds in-crop: a new weed control paradigm for global agriculture. Weed Technol 27(3):431–436
69. Davis AS (2008) Weed seed pools concurrent with corn and soybean harvest in Illinois. Weed Sci 56(4):503–508
70. Gallandt ER, Liebman M, Huggins DR (1999) Improving soil quality: implications for weed management. J Crop Prod 2(1):95–121
71. Fennimore SA, Jackson LE (2003) Organic amendment and tillage effects on vegetable field weed emergence and seedbanks. Weed Technol 17:42–50
72. Gallandt ER, Fuerst EP, Kennedy AC (2004) Effect of tillage, fungicide seed treatment, and soil fumigation on seed bank dynamics of wild oat (Avena fatua). Weed Sci (BioOne) 52(4):597–604
73. Ullrich SD, Buyer JS, Cavigelli MA, Seidel R, Teasdale JR (2011) Weed seed persistence and microbial abundance in long-term organic and conventional cropping systems. Weed Sci 59(2):202–209
74. Davis AS, Anderson KI, Hallett SG, Renner KA (2006) Weed seed mortality in soils with contrasting agricultural management histories. Weed Sci 54(2):291–297
75. Bastiaans L, Paolini R, Baumannà DT (2008) Focus on ecological weed management: what is hindering adoption? Weed Res 48(6):481–491

Chapter 5
Weed Management in Conservation Agriculture Systems

Seyed Vahid Eslami

Introduction

Feeding the world in the future will need further crop production, intensification, and optimization. Nevertheless, until now, agricultural intensification generally has had a negative impact on the quality of many essential resources, such as the soil, water, land, biodiversity, and the ecosystem services. Another challenge for agriculture is its environmental adverse effects and climate change [1].

Conventional agriculture is facing serious problems due to land degradation and increasingly unreliable climatic conditions. Conventional arable agriculture is normally based on soil tillage as the main operation [2]. Most people understand tillage to be a process of physically manipulating the soil to achieve weed control, fineness of tilth, smoothness, aeration, artificial porosity, friability, and optimum moisture content so as to facilitate the subsequent sowing and covering of the seed. In the process, the undisturbed soil is cut, accelerated, impacted, inverted, squeezed, burst, and thrown in an effort to break the soil physically and bury weeds, expose their roots to drying or to physically destroy them by cutting [3]. Although the excessive tillage of agricultural soils increases soil fertility in the short term, it has resulted in soil degradation in the medium term [2]. Regular tillage breaks down the soil's organic matter through mineralization, more so in warmer climates, thus contributing to deteriorating the soil's physical, chemical, and biological properties. The physical effects of tillage also adversely affect soil structure, with consequences for water infiltration and soil erosion through runoff, and create hardpans below the plow layer [4]. The damage to the environmental resources caused by intensive tillage-based agriculture has forced farmers and scientists to look for alternatives that are ecologically sustainable as well as profitable. The rational strategy to this has been to reduce tillage. The first attempts in this regard were sparked after severe soil erosion in the US Great Plains, known as the "Great Dust Bowl" in the 1930s.

S. V. Eslami (✉)
Agronomy Department, Faculty of Agriculture, University of Birjand,
Birjand, South Khorasan, Iran
e-mail: s_v_eslami@yahoo.com

B. S. Chauhan, G. Mahajan (eds.), *Recent Advances in Weed Management*,
DOI 10.1007/978-1-4939-1019-9_5, © Springer Science+Business Media New York 2014

American farmers started abandoning their traditional practice of plowing. Instead, they left the crop residues on the soil surface, and planted the next crop directly into the stubble. Faced with similar problems, farmers in South America also took up conservation agriculture (CA). They planted cover crops to protect the soil and rotated crops in order to maintain soil fertility. Because of the benefits, knowledge passed quickly from farmer to farmer. While in 1973–1974 the system was used only on 2.8 M ha worldwide, the area had grown in 1999 to 45 M ha, and by 2011, the area had increased to 125 M ha, mainly in North and South America [5].

CA is characterized by three sets of practices that are linked to each other in a mutually reinforcing manner, namely: (1) continuous no or minimal mechanical soil disturbance (i.e., direct sowing or broadcasting of crop seeds, and direct placing of planting material in the soil, minimum soil disturbance from cultivation, harvest operation, or farm traffic); (2) permanent organic matter soil cover, especially by crop residues and cover crops; and (3) diversified crop rotations in the case of annual crops or plant associations in case of perennial crops, including legumes. CA is based on enhancing natural biological processes above and below the ground. Interventions, such as mechanical soil tillage, are reduced to an absolute minimum, and the use of external inputs, such as agrochemicals and nutrients of mineral or organic origin, is applied at an optimum level and in a way and quantity that do not interfere with, or disrupt, the biological processes [1]. A diagram illustrating the comparison of some issues between conventional tillage and CA is given in Fig. 5.1 [6]. CA facilitates good agronomy, such as timely operations, and improves overall land husbandry for rain-fed and irrigated production systems. Complemented by other known good practices—including the use of quality seeds and integrated pest, nutrient, weed, and water management—CA is a base for sustainable agricultural production intensification. The yield levels of CA systems are comparable with and even higher than those under conventional intensive tillage systems, which means that CA does not lead to yield penalties. As a result of the increased system diversity and the stimulation of biological processes in the soil and above the surface, as well as due to reduced erosion and leaching, the use of chemical fertilizers and pesticides, including herbicides, is reduced in the long term. Groundwater resources are replenished through better water infiltration and reduced surface runoff. Water quality is improved due to reduced contamination levels from agrochemicals and soil erosion. It further helps to sequester carbon in soil at a rate ranging from about 0.2 to 1.0 t/ha/year depending on the agroecological location and management practices. Labor requirements are generally reduced by about 50 %, which allows farmers to save on time, fuel, and machinery costs. Fuel savings in the order of around 65 % are in general reported [5]. CA means less work because it is not necessary to plow the soil as many times. It suppresses weeds and reduces erosion.

In conventional farming, tillage (turning the soil over) is a major way to control weeds. Farmers plow repeatedly in order to suppress weeds and have a clean field when they plant their next crop. Plowing buries many weed seeds, but it also brings other seeds back to the soil surface, where they can germinate. Burning crop residues may also stimulate the growth of some types of weeds. CA reduces weed densities in several ways. Adoption of reduced or zero tillage under CA makes

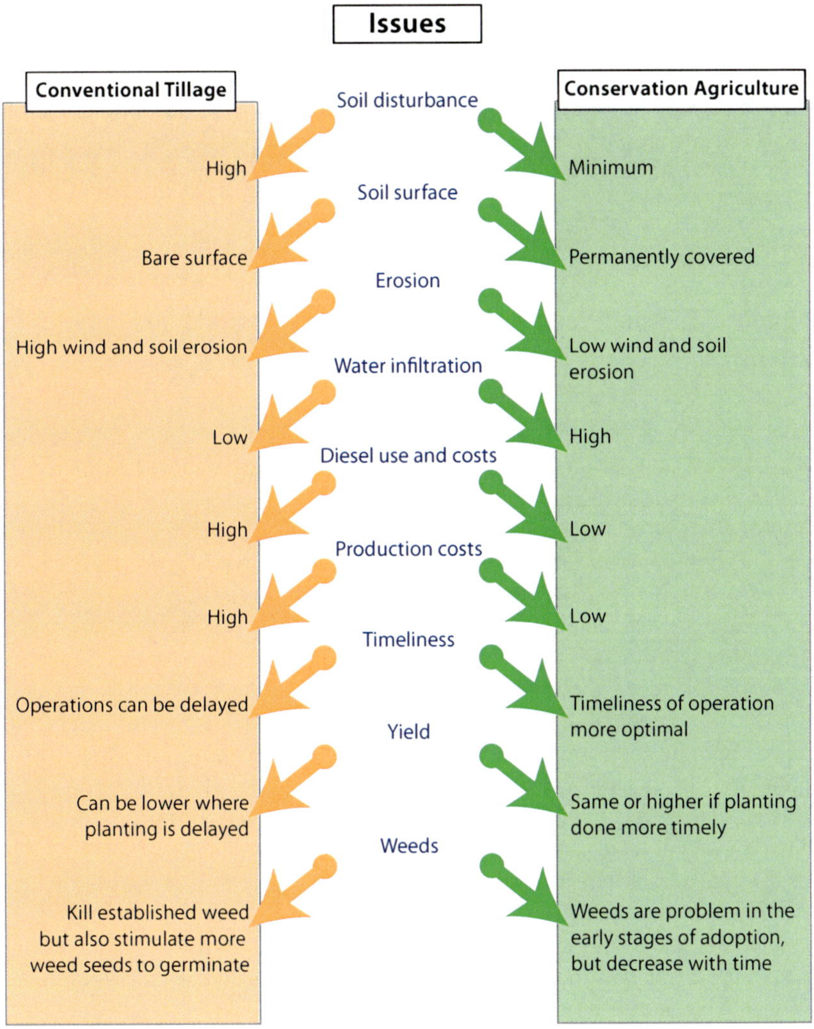

Fig. 5.1 A comparison of some issues between conventional tillage and conservation agriculture. (Modified after [6])

an inappropriate environment for weed seed germination. It disturbs the soil less, thereby bringing fewer buried weed seeds to the surface where they can germinate. Further, reduced or zero tillage improves the soil structure, increases the nutrient recycling, and decreases the pesticides use, greenhouse gas emissions (e.g., nitrous oxide), eutrophication, and cost of the production [7].

The cover on the soil in CA systems (intercrops, cover crops, or mulch) smothers weeds and prevents them from growing. Moreover, crop residues on the soil surface lower the possibility of weed seed germination by creating an obstruction for proper contact to the soil. Weeds under CA may also be controlled when the cover crop is harvested or killed by herbicides [8].

Eventually, rotating crops, as the third main pillar of CA, prevents certain types of weeds from multiplying. Weeds, however, are one of the major limitations to the adoption of CA, especially in the first few years after farmers start practicing CA. Although employing reduced tillage practices has led to increased crop yield in CA systems, weeds are a common problem where these strategies are not successful. In this chapter, the proper and advanced weed management strategies under CA systems will be further discussed.

Impact of CA on Weed Ecology and Weed Population Dynamics

Agroecosystems impart selection pressure on weed communities that inevitably result in weed population shifts [9]. Weed shift refers to a change in the relative abundance or type of weeds as a result of a management practice [10]. Selection pressures usually eliminate susceptible weeds from the existing population and allow surviving species or biotypes to flourish and reproduce [11]. Since communities are influenced by multiple abiotic and biotic factors, weed community shifts cannot be accounted for by a single variable [12]. Weed population shifts have occurred most readily in the presence of a control measure that promotes high selection pressure on the population [11, 13–16]. The control measure could be herbicide, tillage operation, crop rotation, or other agronomic factors, but it must be used continuously throughout the cropping system to cause a shift in the weed population [11]. Generally, weed shifts occur over a relatively long period of time. However, highly effective practices can rapidly cause weed shifts. Variation within and among weedy species—including seed dormancy mechanisms, emergence patterns, growth plasticity, life cycle and overall life duration, shade tolerance, late-season competitive ability, seed dispersal mechanisms, and morphological and physiological variation—can contribute to a population's response to management practices [17]. Changes in weed population dynamics due to the adoption of CA are discussed in separate sections as follows.

Soil Disturbance

Removal of tillage from the crop production systems not only eliminates an important method of weed management but also alters the environment where weeds reside. In fact, the lesser degree of soil disturbance in CA tends to provide safe sites for weed germination and establishment not present in conventional systems [18–22]. The use of reduced or no-till practices in CA systems may lead to shifts in the weed flora and diversity. The amount and type of these variations, however, are inconsistent in the literature and other factors such as region, previous herbicide use, type of weed and crop species, climate (temperature and precipitation patterns), and cultural practices may influence weed community structure [8, 12, 22, 23]. Converting from conventional tillage to no-till systems usually increases weed density and species diversity [18]. Generally, perennial weeds and small-seeded annual weeds that germinate near the soil surface are the dominant weed species under no-till systems. In contrast, annual weeds that germinate from various depths are favored by conventional tillage systems [11]. Froud-Williams found weed communities in reduced tillage systems composed mainly of annual and perennial grasses, wind-disseminated species, and volunteer crops, at the expense of annual dicots [24]. A general consequence of lowering soil disturbance is increasing perennial weed species under CA systems, as perennial weeds prosper in less-disturbed and more stable environments [16, 20, 21, 25–28]. Using moldboard plows in conventional agriculture systems can sever shoots from the roots of perennial weeds, which is partially effective in controlling them. Despite these general trends, perennial weed species show different reactions to tillage practices. Légére and Samson concluded that the relationship between perennials and tillage was highly dependent on the response of perennating structures to soil disturbance [29]. In other words, biennial and perennial weeds are likely to increase under conservation tillage, particularly under no-till—e.g., hemp dogbane (*Apocynum cannabinum* L.) and dandelion (*Taraxacum officinale* [L.] Weber)—because the root systems necessary for perennation are not disrupted. Nonetheless, perennials that reproduce if their underground parts are disrupted—e.g., quackgrass (*Elytrigia repens* L.) and American germander (*Teucrium canadense* L.)—would be expected to be less favored under conservation tillage [16, 29].

In CA systems, weed seeds remain at or near the soil surface, which may lead to changes in weed composition. Chauhan et al. reported that a low-soil-disturbance single-disc system retained more than 75 % of the weed seeds in the top 1-cm soil layer, whereas the high-soil-disturbance seeding system buried more than 75 % of the seeds to a depth of 1–5 cm [30]. While a seed is on the soil surface, it is very likely to suffer one of the two fates: germination or predation. Once it is buried, both of these outcomes become much less likely; predation because most seed predators are surface foragers and germination because, in many seeds, germination is stimulated by light or a light requirement is induced by burial [31]. Weed seeds that remain at or near the soil surface typically have higher mortality rates, probably because of enhanced physiological aging or seed predation. Even in the absence of seed predators, weed seeds from species as diverse as wild proso millet (*Panicum miliaceum* L.), redroot pigweed (*Amaranthus retroflexus* L.), and hairy nightshade (*Solanum sarrachoides* Sendtner) have been shown to lose viability at a greater rate when positioned near the soil surface than when buried below the emergence zone [32, 33]. The impact of seeding depth on weed germination and establishment has been documented in several studies [34–40]. Different soil depths differ in availability of moisture, diurnal temperature fluctuation, light exposure, and activity of predators. Changes in seed depth and corresponding differences in emergence depth may contribute to shifts among weed species under different tillage systems due to differences in temperature and light [41, 42]. Conservation tillage systems reduce soil temperature early in the growing season compared with conventional tillage.

Weed species show different responses to soil temperature and seeding depth, in terms of their germination and establishment, which contribute to population shifts under different tillage systems. For example, giant foxtail (*Setaria faberi* Herrm.) had greater ability than velvetleaf (*Abutilon theophrasti* Medik.) to establish when seeds were at the soil surface [43].

The effect of cultivation on the weed flora and weed seeds and propagules in the soil depends on depth and type of tillage [8, 44], seed emergence [45–47], seedling survival [48], and seed production [49, 50], and also vegetative survival and dispersal of crops in the case of perennial weeds [51, 52].

Although some studies found that there is no significant difference between no-till, minimum tillage, and conventional tillage in terms of diversity indices evenness [53, 54], other studies have shown that the most diverse weed seedbank is observed in no-till systems [51, 55, 56]. Mas and Verdu mentioned several reasons for these inconsistencies, including climate conditions, geographical distribution of particular flora, rotation cycle, herbicide effects, variation in sampling dates, and parameters (density or biomass) used to compute the diversity indices [51]. The relative contributions to the size and diversity of weed flora are likely to be greater by common species under the conventional tillage and by rare species under the reduced- and no-till systems [57]. Cardina et al. studied the weed seedbank size and composition after 35 years of continuous crop rotation and tillage system and concluded that seed density was highest in no-till and generally declined as tillage intensity increased [58]. Seeds accumulated near the surface (0–5 cm) in no-till, but were uniformly distributed with depth in other tillage systems (moldboard plow and chisel plow). Differences in weed seed location in soil may influence interaction between germinating seedlings and herbicide-treated soil. No-till systems act directly by selecting weed species that are able to emerge and develop in soil covered by plant residue, thereby determining the weed flora in those areas [59]. Mas and Verdu [51] pointed out that the no-till system seemed a better means for weed management than the minimum tillage and conventional tillage, as prevention of dominating the weed flora by only a few species is one of the main goals in weed management, and this objective is usually obtained under no-till systems [60].

Cover Crops and Crop Residues

Inclusion of cover crops and their residue as a main component of CA systems may influence weed population dynamics. Crop residues acting as mulches can influence weed seed germination and seedling emergence [61, 62]. Several mechanisms may contribute to reduced weed emergence and growth where surface cover crop residues are present, including reduction in light penetration to the soil [63], physical obstruction resulting in seed-reserve depletion before emergence [64], increased seed predation or decay [65, 66], decreased daily soil temperature fluctuations [63, 67], or the production of allelopathic compounds [63]. Annual broad-leaved weeds often appear in low densities in conservation tillage systems [19, 28]. This happens probably due to the presence of crop residues on the soil surface, which reduces light and temperature on the soil surface. Tuesca et al. attributed the lower densities of common lambsquarters (*Chenopodium album* L.) under no-till systems to the inhibitory effect of crop residues on light interception [28]. In contrast, higher densities of common purslane (*Portulaca oleracea* L.) in tilled systems could be attributed to the light and high temperatures in these environments, which favor the germination and establishment of this summer species. Grass weed infestation is commonly higher in no-till systems than in conventional tillage systems [25, 28, 68]. It appears that providing a rough soil surface covered with residue in reduced or no-till systems that maintain soil moisture could promote the germination and establishment of grass weeds [28]. The higher density of wind-disseminated species in reduced or no-till systems has been documented in several studies [28, 57, 69], which could be attributed to their susceptibility to soil disturbance or to crop residue accumulation on the soil surface under conservation tillage systems, which may catch wind-borne seeds or favors weed establishment.

The degree of weed suppression in cover crops depends largely on the crop species and management system. Cover crop residue can also influence weed populations in no-till cropping systems because of the proximity of the residue to the site of seed germination on the soil surface [70]. Cover crop residue may alter physical conditions in the microsite of seed germination enough to reduce, delay, or even increase weed emergence. Teasdale and Mohler compared the light, temperature, and moisture conditions under desiccated residues of hairy vetch (*Vicia villosa* Roth.) and rye (*Secale cereale* L.) [63]. They found that transmittance through hairy vetch was greater than that through rye because of faster decomposition of hairy vetch residue. It was concluded that reductions in light transmittance and daily soil temperature amplitude by cover crop residue were sufficient to reduce emergence of weeds, whereas maintenance of soil moisture by residues could increase weed emergence. Under drought conditions, residue could maintain soil moisture at levels more favorable than bare soil. In saturated soils, however, residue could limit soil drying, which may lead to inhibition of emergence in weed species intolerant to saturated conditions [63, 67]. This indicates that cover crop residue may have a selective influence on inhibition of weed emergence under different environmental conditions. Gallagher et al. concluded that weed species with a wide germination temperature window might be less affected by cover crops than species that required warm soil temperature or had narrow germination temperature windows [71]. Liebman and Dyck concluded that incorporation of red clover (*Trifolium pratense* L.) residue might reduce weed emergence and growth through changes in nitrogen (N) dynamics or release of allelochemicals [60]. Reddy et al. found higher weed biomass with crimson clover (*Trifolium incarnatum* L.) residue in comparison with rye (*Secale cereale* L.) or no cover crop treatments in soybeans (*Glycine max* [L.] Merr.) [72]. They attributed this to increased N availability to weeds resulting from decomposition of crimson clover root and shoot biomass, as it was evident from the higher NO_3-N level in crimson clover treatment.

Sunlight needed to induce weed seeds to germinate penetrates only a few millimeters into the top soil layer. In addition to reducing the light intensity, the plant residue covering the soil surface affects light quality, by acting as a filter [59]. Far-red-rich light, which is present under green canopy shade, also can inhibit germination [73]. Desiccated cover crop residues will only lower the red to far-red ratio slightly [63] and probably will not influence germination [74].

Maintaining a sufficient level of crop residues on the soil surface, however, is necessary for a successful weed management program. Amuri et al. pointed out that reduced tillage or no-till can decrease weed densities, but without sufficient crop residues covering the soil surface, weed species composition may be increased under no-till [75]. The amount, composition, and stability of the plant residue covering the soil surface are directly correlated to the plant species used to produce the plant residue, the climate, and the management of the crop used to make the soil covering. These factors directly influence the weed flora and herbicide efficacy [76].

Among the chemical soil properties affected by plant residue covering the soil surface are the modification of soil carbon-to-nitrogen ratio (C:N) and the soil nitrate content. It is well known that the lower the C:N of the plant residue the faster is its decomposition, influenced also by the environmental conditions [77]. Nitrate is the only inorganic ion common to the soil solution that affects germination of many weed species, and it is normally present in higher concentrations in the top layer of the soil, due to the decomposition of the organic matter and microbial activity [78]. Research has shown positive correlations between the rate of N fertilizer application and the increase in the weed emergence rates [79]. Therefore, adding plant residue to the soil surface may alter the C:N in the topsoil layer, which may indirectly stimulate weeds in the seedbank to germinate and emerge.

Diversified Crop Rotations

Growing diverse crops in CA systems may affect the weed flora and composition of cropping systems over time. Crop rotations can influence weed species shifts because of diverse cultural practices, competitive ability, and herbicide-use patterns associated with different crops [13, 14, 80]. Weed diversity has been shown to increase under crop rotation compared to monoculture [54, 81]. Greater diversity prevents the domination of a few problem weeds. It has also been reported that weed densities are generally lower in crop rotational systems than in monocultures [60]. Application of the same herbicide each year under monoculture systems increases selection pressures in the plant community for certain weed populations. Crop rotation introduces conditions and practices that are unfavorable for a specific weed species, and thus, growth and reproduction of that species are hampered [81]. The structure of the current species in a seedbank is influenced by the crops that are a part of the rotation [82]. Increasing the diversity of crops in rotations and reduced tillage appears to have long-term benefits in terms of the production of fewer weed seeds, which results in a situation in which fewer weed seeds are incorporated into the seedbank [83]. Murphy et al. observed the highest weed species diversity in no-till fields with a three-crop rotation of corn–soybean–winter wheat [26]. The reduction of soil disturbance and other microenvironmental changes created by no-till and crop rotations probably changes selection pressures, so that the formerly dominant species are no longer at a large selective advantage [65]. Compared with continuous monoculture, diverse rotation may differ in the light transmitted through the crop canopy, the herbicide(s) used, the timing of management operations, and the natural enemies living in the crop; these conditions presumably make it difficult for one weed species to dominate the weed community [18, 26, 55, 60, 84].

The period for which the seeds remain viable in the soil has been shown to influence the composition of the weed population. Thus, rotations that include crops with different life cycles could lead to additional benefits because of their role in restricting seed germination; particularly if the average life in the seedbank is short [85]. Weed species shifts are highly dependent on the species present, differing in susceptibility to the herbicides being used. Population fluctuations rather than shifts have been observed in response to various management strategies, but most often shifts are observed following long-term uninterrupted use of a control measure [11].

Weed species composition would be affected by rotation design, and weed population dynamics are very dependent on the crops included in the rotation [55, 86]. Anderson and Beck found that warm-season weeds were more prevalent in rotations with two warm-season crops in 3 years, whereas these species were rare in rotations that included 2-year intervals of cool-season crops or fallow [86]. The 2-year interval will also result in a rapid decline in weed seed density in the soil seedbank.

Changes in management from one crop to another in a rotation can result in rapid shifts in the composition and abundance of the germinable fraction of the weed seedbank from year to year [87]. Young and Thorne observed a reduction in weed populations in no-till rotations due to controlling downy brome (*Bromus tectorum* L.) and other winter-annual weeds with preplant or prefallow herbicide applications [22].

Inclusion of a fallow period in a rotational could pose a substantial effect on weed populations. Hume et al. observed that green foxtail (*Setaria viridis* [L.] Beauv.) was present at very high densities under a continuously cropped situation, but was almost absent from the rotations, including a fallow year [88]. Anderson and Beck found that weed control during fallow eliminated seed production of both cool-season and warm-season weeds [86].

Explaining the effect of crop rotation on weed communities may be a gross generalization because of existing interactions between crop rotation and management factors. Thus, the effect of crop rotation on weed communities can only be described in terms of interactions between the crop type, crop structure, frequency of occurrence within the rotation, and management variables, such as tillage, planting date, and method of weed control [89].

Herbicides

Reduction or elimination of tillage practices in CA systems has compelled farmers to be more dependent on herbicides for weed control. Herbicides may influence seed densities and species composition of the seedbank. Certain species decrease in the seedbank and others increase, depending on the crop/herbicide systems [90]. Herbicides could cause shifts in species composition in favor of species that are less susceptible to applied herbicides, and there would be a gradual shift to tolerant weed species when practices are continuously used that are not effective against these species [10, 13, 82]. The intense selection pressure from herbicide use will result in the evolution of herbicide-resistant (HR) weed biotypes or shifts in the relative prominence of one weed species in the weed community [91].

The adoption of HR crops in CA systems will also result in greater selection pressure on the weed community due to a limited number of different herbicides used. Increased selection pressure will increase weed population shifts. Selection pressure imparted by herbicide tactics can result in weed shifts attributable to the natural resistance of a particular species to the herbicide or the evolution of herbicide resistance within the weed population. Both of these types of weed shift have occurred in response to grower adoption of crop production systems based on an HR crop and the resultant application of the herbicide [9]. Reddy found that continuous bromoxynil-resistant cotton production resulted in weed species shift toward common purslane, sicklepod (*Senna obtusifolia* [L.] Irwin and Barneby), and yellow nutsedge (*Cyperus esculentus* L.) [92]. The adoption of glyphosate-resistant crops was especially suited to CA systems, and growers rely heavily or completely on glyphosate for weed management, since glyphosate can effectively control many perennial species that appear when tillage practices are reduced [93]. Weeds that are tolerant to glyphosate or emerge after glyphosate applications often escape glyphosate-based weed management programs. The glyphosate-based weed management tactics used in glyphosate-resistant crops impose the selection pressure that supports weed population shifts [91]. Differential tolerance to glyphosate in a glyphosate-based management system would contribute significantly to population growth [17, 94]. Avoidance of the glyphosate application through emergence periodicity may result in species composition changes. Mechanisms that are likely to lead to species shifts in a non-residual herbicide system either allow the plant to escape treatment or tolerate the herbicide [17]. Weed shifts have been observed as the frequency and rate of glyphosate use in glyphosate-resistant crops, which have increased [95–97].

Seed Predation

Weed seeds are lost from the seedbank through mortality because of aging [98], attacks by seed predators [99, 100], seed decay, and germination [101]. As a principle of CA, at least 30% of the soil surface should be covered with a residue [7, 102]. Crop residues provide habitat and invite the diversity of beneficial insects, birds, and a wide range of invertebrates (e.g., earthworms, small rodents, birds, carabid beetles, field crickets, etc.) [103]. Availability of suitable habitats near the crop fields supports early colonization of natural enemies [104]. The field managed under minimum tillage or zero-till holds the weed seeds on the soil surface that increases the predation possibility of weed seeds [105]. Seed predation rates of 32–70 % can be as effective as mechanical weed control and accounted for greater losses to the weed seedbank than aging or microbial activity [99]. Ground-dwelling invertebrates alone can consume 80–90 % of the postdispersal seeds of common lamb's-quarters and barnyardgrass (*Echinochloa crus-galli* [L.] Beauv.) [65]. Rodents, birds, ants, ground beetles, and crickets are important postdispersal weed seed predators, contributing up to 90 % of weed seedbank loss [65, 106, 107]. Carabid beetles in particular have been shown to consume significant amounts of weed seeds in agroecosystems [108]. Decreased levels of disturbance correspond with increases in seed predation, and increased quantities of ground cover may also influence the quantity of seed predation by providing habitat for seed predators [109], thereby increasing seed predator populations and the quantity of seed consumed. This suggests that in a managed agroecosystem, seed predation may be influenced by tillage practices and cropping practices, such as crop residue management. Tillage and mowing, as well as the disturbance from crop harvesting, can drastically decrease insect populations [110]. Tillage affects seed predation potential because: (1) weed seeds may be buried beyond the reach of seed predators, (2) seed predators may be killed during tillage, or (3) critical habitat of seed predators may be destroyed (thus reducing survivorship) [66]. Harrison et al. reported that approximately 90 % of giant ragweed (*Ambrosia trifida* L.) seeds deposited on the soil surface of a no-till cornfield was eliminated by predation when seeds were kept on the soil surface [111]. *Harpalus rufipes* is capable of consuming up to 90 % of the postdispersal seeds of certain weed species [112]. As the density of *H. rufipes* increases, a corresponding increase in seed predation of preferred species, such as common lamb's-quarters and redroot pigweed (*Amaranthus retroflexus* L.), has been observed [112, 113]. Moldboard plowing and rotary tillage reduced *H. rufipes* activity density by 53 and 55 %, respectively [114, 115]. Cover cropping can also indirectly contribute to weed management by promoting populations of beneficial weed seed predators. Gallandt et al. (2005) suggested that one of the multiple benefits of cover cropping may be conservation and enhancement of resident invertebrate weed seed predators, in particular the ground-dwelling carabid *H. rufipes* [113]. Shearin et al. found that pea (*Pisum sativum* L.)/oat (*Avena sativa* L.)—rye/hairy vetch cover crop systems are apparently beneficial for *H. rufipes* during the cover crop year as well as in subsequent crops planted into this cover crop's residues [115]. Crop residues have been shown differential suitability in the different rotations as habitat for seed predators [116]. Westerman et al. reported that loss of velvetleaf seeds to predators was greater in the 4-year rotation (corn–soybean–triticale–alfalfa) than in the 2-year rotation (corn–soybean) [117]. Weed seeds vary in their palatability to granivores, based on seed size, seed coat strength, and secondary metabolite profile, and nutritive value, with significant intraspecific variation in palatability as well [118, 119].

Microorganisms comprise another set of potential allies against the weed seedbank. Many newly shed weed seeds are highly resistant to invasion by bacteria or fungi; however, as these seeds age and weather in the soil, their seed coats eventually become more porous, leaving the seeds more vulnerable to microbial attack. Use of the cover crops promotes the fungal, bacterial, and mycorrhizal communities that may be detrimental to weeds and beneficial for the crops [120]. Davis and Renner reported that soil fungi contributed to mortality of velvetleaf seeds germinating from a 10-cm depth in the soil [101]. Davis found that soil that was enriched in N likely favored greater microbial predation of velvetleaf seeds than soil that was enriched in C, whereas for giant ragweed (*Ambrosia trifida* L.) and wooly cupgrass (*Eriochloa villosa* [Thunb.] Kunth), the effect on seed mortality appeared to be mediated through soil N effects on germination [121]. Mechanisms underlying soil N fertility effects on weed seed mortality appear to be species specific. Conservation tillage creates a favorable condition for microbial population [122]. A study conducted in Brazil showed that conservation tillage increased the soil organic matter by 45 % and soil microbial population by 83 % in a 20-year period in comparison with conventional tillage [123].

Weed Management Strategies Under CA Systems

In CA systems, weed management has a major role in obtaining profitable yields. In such systems, achieving satisfactory weed control requires more intensive management from the farmer [25]. A number of approaches and strategies, including the use of preventive weed control practices, cover crops and crop residue, crop rotation, competitive crops, optimum sowing rate, date and row spacing, stale seedbed, and herbicide resistant crops, have been proven to be effective for weed control under CA systems.

Preventive Weed Management

The prevention of a weed is usually easier and less costly than control or eradication attempts, because weeds are most tenacious and difficult to control after establishment. The major preventive measures include using clean seed; using manure after thorough fermentation; cleaning harvesters and tillage implements; avoiding transportation and use of soil from weed-infested areas; removing weeds that are near irrigation ditches, fence rows, rights-of-way, and other non-crop lands; preventing reproduction of weeds; using weed seed screens to filter irrigation water; and restricting livestock movement into non-weed-infested areas as well as other practices including weed laws, seed laws, and quarantines [8, 124]. In croplands where certified crop seeds are planted, the chance of introducing weed problems into the field is small. However, there are producers who save their seeds for planting next year's crop. The potential is high for this seed to contribute to increased weed problems [125]. The level of crop seed contamination by weed seeds is greatest when weed seeds resemble the shape and size of crop seeds [8]. Even in cleaned seeds, a similarity between certain weed and crop seeds in shape and size makes it very difficult to distinguish between species during the seed-cleaning process [124]. Small grain seeds are often contaminated with seeds of wild oat (*Avena fatua* L.), hairy vetch, wild radish (*Raphanus raphanistrum* L.), and annual ryegrass (*Lolium rigidum* Gaud.). Common weed seed contaminants in soybean seeds include balloon vine (*Cardiospermum halicacabum* L.), annual morning glories (*Ipomoea* spp.), ragweeds (*Ambrosia* spp.), and common cocklebur (*Xanthium strumarium* L.) [125]. Some of other examples include *Echinochloa* spp. in rice (*Oryza sativa* L.), Persian ryegrass (*Lolium persicum* Boiss. and Hohen.) or corn cockle (*Agrostemma githago* L.) in small grains, and common vetch (*Vicia sativa* L.) in lentils (*Lens culinaris* Medik.) [124]. Dastgheib observed that the use of wheat seeds saved from on-farm production contributed 182,000 weed seeds ha^{-1}, representing 11 species [126]. These instances demonstrate the importance of using weed-free crop seeds and innovating improved technologies that allow the separation of weed and crop seeds with the same size and shape with the least errors. An important attempt in this regard would be to persuade farmers not to use saved seeds and encourage them to sow certified seeds, particularly in the developing countries. This should be put into practice through programs run by agricultural extension agents.

To obtain weed-free crop seeds, cultural and mechanical measures need to be adopted. The idea should be to minimize the weed infestation area and decrease the dissemination of weed seeds from one area to another or from one crop to another [8]. Agricultural equipment can disperse weed seeds over large distances [127]. Harvesting, mowing, hay baling, tilling, and earth moving equipment all have the potential to disseminate seeds and/or vegetative reproductive structures of weeds. Froud-Williams reported that of the total seeds shed, 4.5 % were removed by straw baling, 4 % remained on the soil surface, and the vast majority (91.5 %) was removed in the grain [24]. The combine harvester is considered an almost-perfect device for the dispersal of weed seeds. The potential of combine harvesters in dispersing weed seeds to great distances has been reported in several studies [125, 128–132]. Woolcock and Cousens demonstrated that seed dispersal by combines can increase the rate of spread up to 16 times that of natural dispersal [133]. Chaff collection may be an important management tool to reduce the dispersal of weed seeds. Shirtliffe and Entz found that chaff collection consistently reduced the amount and distance that wild oat seeds were dispersed [130]. Chaff collection reduced wild oat seed dispersal past the wild oat patch to less than 10 seeds m^{-2} at 45 m, whereas without chaff collection, there was greater than 10 seeds m^{-2} up to 145 m. Careful cleaning of tractor wheels, parts of implements used in soil preparation and seeding, as well as horizontal surfaces of harvesters is crucial for preventive weed management programs [124, 134]. Boyd and White recommended avoiding dense weed patches, altering harvest timing, and periodic cleaning of harvesting equipment between fields as proper strategies for preventing the spread of weed seeds [127]. Seed cleaning and chaff carts or direct bailing of chaff offer alternative methods to reduce the amount of weed seeds entering the seedbank. Chaff collected by chaff carts is generally burned or used as livestock feed [8]. Walsh and Parker, using a chaff cart attached to the rear of the harvester, achieved up to 85 % efficiency in removing annual ryegrass seeds in Australia [135]. The chaff containing weed seeds could then be destroyed or composted to kill weed seeds. Walsh and Powles recommended the use of baling equipment attached to the harvester to bale the chaff along with weed seeds that could be later fed to the confined livestock [136]. Another strategy is to windrow chaff as it exits the harvester and subsequently burn it; high-temperature conditions during burning can kill weed seeds. Walsh and Newman found that burning windrows was more effective for wild radish, for which trials found 80% of seeds destroyed compared with only 20% in burned standing stubble [132]. The planning of post-infested weed control programs should be done in such a way that the buildup of weed seeds is reduced drastically within a short period of time. Proper care should be taken to restrict the weed seedbank size in the area by using integrated methods of weed control [8]. Walsh et al. introduced harvest weed seed control systems as a new paradigm for global agriculture, which target weed seed during commercial grain harvest operations and act to minimize fresh seed inputs to the seedbank [137]. These systems exploit two key biological weaknesses of targeted annual weed species: seed retention at maturity and a short-lived seedbank. Harvest weed seed control systems, including chaff carts, narrow windrow burning, bale direct, and the Harrington Seed Destructor target the weed-seed-bearing chaff material during commercial grain harvest. The destruction of these weed seeds at or after grain harvest facilitates weed seedbank decline, and when combined with conventional herbicide use can drive weed populations to very low levels.

Cover Crops and Crop Residues

The use of cover crops in conservation tillage offers many advantages, one of which is weed suppression through physical as well as chemical allelopathic effects [68, 138]. Cover crops can influence weeds in the form of either living plants or plant residue remaining after the cover crop is killed. There is often a negative correlation between cover crop and weed biomass [139–141]. Cover crops can be effective in controlling weeds by shading, overcrowding, and competing with them. Some cover crops also control weeds, because they are a source of allelopathic compounds that interfere with weed germination or their later growth. Some cover crops with allelopathic properties capable of suppressing weeds are black oats (*Avena strigosa* Schreb), hairy vetch, oats, rye, sorghum-sudangrass (*Sorghum bicolor* L.) hybrids, subterranean clover, sweet clover, and woolypod vetch. It has also been shown that rice can affect the viability of subsequent broadleaf weeds, with some rice cultivars being more effective than others [142]. Cereal rye (*Secale cereale* L.) and soft red winter wheat (*Triticum aestivum* L.) are winter cover crops recommended for cotton production in the USA [68]. Both of these cover crops also contain allelopathic compounds that inhibit weed growth [139, 143]. Yenish et al. reported an increased short-term weed control using a rye cover crop in no-till corn (*Zea mays* L.) but not season-long control [143]. In southern Brazil, black oat is the predominant cover crop on millions of hectares of conservation-tilled soybean, which is partly due to its weed-suppressive capabilities [68].

Once terminated, residues of cover crops, left as mulch, tend to suppress the emergence of weeds. Plant residues on the soil surface can affect weed seed survival, germination and species composition, and effectiveness of soil-applied herbicides. Chauhan and Abugho in a field study evaluated the effect of rice residue amounts (0, 3, and 6 t ha^{-1}) on seedling emergence of spiny amaranth (*Amaranthus spinosus* L.), southern crabgrass (*Digitaria ciliaris* [Retz.] Koel.), crowfootgrass (*Dactyloctenium aegyptium* [L.] Willd.), junglerice (*Echinochloa colona* [L.] Link.), eclipta (*Eclipta prostrata* [L.] L.), goosegrass (*Eleusine indica* [L.] Gaertn.), and Chinese sprangletop (*Leptochloa chinensis* [L.] Nees.) in zero-till dry-seeded rice [144]. They concluded that increasing residue amounts reduced seedling emergence and biomass of these weeds and also delayed their emergence. Weed control provided by cover crop residues depends on crop and weed species. For example, Saini et al. found that rye provided 81–91 % control of Virginia buttonweed (*Diodia virginiana* L.) and smallflower morning glory (*Jacquemontia tamnifolia* [L.] Griseb.), whereas large crabgrass (*Digitaria sanguinalis* [L.] Scop.) control was only 11 % [145].

Using summer annual cover crops is also a helpful strategy for weed suppression. In many cropping systems, there is a gap between early-harvested summer crops (e.g., peas or snap beans) and winter wheat. Land is often left bare during this period, allowing weeds to grow and reproduce. Inclusion of a summer annual cover crop with strong weed-suppressive ability is useful for suppressing weed growth and improving soil in this late-summer niche. For example, Kumar et al. found that buckwheat (*Fagopyrum esculentum* Moench) residue had no negative effect on wheat yields, but suppressed emergence (22–72 %) and growth (0–95 %) of winter annual weeds [146]. During its growth, buckwheat effectively suppressed many weeds, including quackgrass (*Agropyron repens* [L.] Beauv.).

Some cover crops and their residues contain allelopathic metabolites, which in turn suppress weed germination and growth. Brassica species, such as wild radish, rapeseed (*Brassica napus* L.), black mustard (*Brassica nigra* L.), and white mustard (*Sinapis alba* L.), contain isothiocyanates, derivatives of glucosinolates, that have noted pesticide properties including herbicidal activity [147, 148]. The inhibitory effect of isothiocyanates on germination of some weed species such as redroot pigweed, dandelion, yellow nutsedge (*Cyperus esculentus* L.), sicklepod (*Senna obtusifolia* L.), and Palmer amaranth (*Amaranthus palmeri* [S.] Wats.) has been shown [147, 149, 150]. Evidence from field studies also confirms that residues of brassicas—including canola, rapeseed, and mustards—may contribute to weed management [151].

Cereal cover crops such as rye also have shown effective weed suppression potential in reduced-till row crops [68, 150]. Grain cover crops can provide a level of weed control through both physical and allelopathic means. Rye, in particular, has been shown to be a hardy cover crop that can successfully compete with many weed species while actively growing. Benzoxazinoid compounds produced by some grains, such as wheat and rye, can exhibit allelopathic effects on various weed species [150].

Allelopathy has also been noted for some legume species, such as cowpea (*Vigna unguiculata* [L.] Walp.), sunn hemp (*Crotalaria juncea* L.), and velvetbean

(*Mucuna pruriens* [L.] DC.); however, chemical weed suppression has not been identified for most legumes typically used as cover crops [152, 153]. Successful weed control achieved by legumes is primarily attributed to biomass production, which can shade germinating weeds; however, legume cover crops tend to have a low C:N ratio and decompose very quickly compared to other cover crops, which can reduce their potential for weed control further into the growing season. A number of studies have examined legume cover crop use in reduced-till with results indicating that these covers perform best in a mixture with a cereal grain rather than in a monoculture [143, 154, 155]. Longer-term weed control may also be achieved when planting legumes in a mixture due to rapid decomposition of legume-only residue [156]. Hayden et al. concluded that where winter annual weed control was a primary objective, rye would likely be the most effective and inexpensive cover crop option [157]. However, rye–vetch mixtures can match the level of suppression achieved by rye monoculture, in addition to providing a potential source of fixed N.

The weed-suppressive potential of cover crops may depend on the species (or mixture of species) chosen, the method of cover crop termination, and residue management. Phytotoxin composition differs among and within species, and total production may depend on a variety of biotic and abiotic stresses [158]. Moreover, the allelopathic effects of individual phytotoxic compounds may be weed species specific [159]. Therefore, a diverse mixture of allelopathic cover crop species may be more effective, integrating a broad range of weed species. Mixed species communities also may contribute to improved soil coverage and physical mechanisms of weed suppression [76, 160].

To maximize weed suppression, high-residue cover crop systems that provide at least 4,500 kg ha^{-1} of biomass for ground cover are generally utilized [161]. In these instances, winter cereal grain crops, such as rye or oat, are employed to attain the greatest amounts of residue prior to cash crop planting to maintain ground cover for an extended period into the growing season [162–164].

Incorporated cover crop residues may inhibit weed, but not crop establishment through seed size-dependent effects on germination and emergence. The mass of most weed seeds is one to three orders of magnitude smaller than the seed mass of crops they infest [151]. Liebman and Gallandt compared the responses of common bean (*Phaseolus vulgaris* L.) and wild mustard (*Brassica kaber* [DC.] L. C. Wheeler) to red clover residue and concluded that although wild mustard growth was significantly reduced by red clover residue, bean yield was not inhibited, as it had much heavier seeds (380 mg $seed^{-1}$) than that of wild mustard (2.3 mg $seed^{-1}$) [165].

Manipulating planting date in the fall or termination date in the spring may allow growers to achieve a higher level of weed suppression [166]. The rate of cover accumulation in the spring is influenced by the timing of fall planting. The degree of synchrony between weed species emergence and accumulated cover crop biomass plays an important role in defining the extent of weed suppression. Duiker and Curran showed that average aboveground cereal rye biomass was three times greater when terminated at the late-boot stage (4,200 kg ha^{-1}) compared with the early-boot stage (1,400 kg ha^{-1}) [162]. Mirsky et al. found that delaying cover crop termination reduced weed density, especially for early- and late-emerging summer annual weeds [166]. Implementing an earlier kill date for cover crops, however, can reduce the loss of soil moisture and allow more rapid soil warming, while a later kill date increases the amount of residue produced, which can lead to enhanced weed suppression [161]. In addition to timing, mechanisms for terminating cover crops can vary depending on the type of cover crop. Cover crops can be terminated climatically (e.g., winterkill), chemically, or through various mechanical measures (e.g., plowing, disking, mowing, roller-crimping, or undercutting) [76]. Rolling/crimping of cover crops is a potential means of cover crop termination in low-input or organic systems; however, timing of rolling operations for effective termination without herbicides is species specific [155, 167]. The most appropriate termination method will depend on the farm management objective. When managing for improved weed management, termination methods resulting in maximum surface residue and minimal soil disturbance have the greatest potential to inhibit weed emergence and growth [76]. However, it is possible that allelopathic phytotoxins are most effective when residues are incorporated into the soil [168]; thus, multiple methods of cover crop termination may be effective depending on the targeted mechanism of weed suppression (e.g., physical suppression or allelopathy). When managing cover crops for maximum surface residue and minimal soil disturbance, a sweep plow undercutter may have great potential. Creamer et al. demonstrated that cover crop termination with a sweep plow undercutter created a thick and uniform cover crop mulch, and the subsequent weed suppression was greater than when cover crops were terminated via mowing [169]. While other mechanical termination methods, such as the roller-crimper, have shown great promise for weed control [170], the sweep plow undercutter may be more effective in killing cover crops at younger growth stages [167, 169]. Wortman et al. concluded that terminating cover crops with the undercutter consistently reduced early-season grass weed biomass, whereas termination with the field disk typically stimulated grass weed biomass relative to a no cover crop control [158].

Rotation

Diversified crop rotation, as a main component of CA systems, can increase yield potential by influencing weeds, plant diseases, root distribution, moisture utilization, and nutrient availability. Crop rotation alone can improve weed management regardless of tillage, because of rotations affecting weed populations and composition through altering the weed seedbank and subsequent weed growth [75]. Weeds tend to associate with crops that have similar life cycles. For example, the winter annual weed downy brome proliferates in winter wheat, because seedling emergence and flowering periods coincide. Rotating crops with different life cycles can disrupt the development of weed–crop associations. Different planting and harvest dates of diverse crops as well as different herbicides [82] or other means of weed control practiced in each crop will provide opportunities for producers to prevent either weed establishment or seed production [25].

The time a crop is planted is probably the main factor determining composition of weed flora infesting a crop [46]. Thus, diversifying crops with different growing seasons in a rotation leads to different planting dates, which disrupts the life cycle of predominant species in the seedbank. For example, to manage winter annual grasses in wheat, producers in the winter wheat region of the USA include summer annual crops, such as corn, sorghum, or proso millet in the rotation to lengthen time between wheat crops [171]. Rotations comprised of a diversity of crops lead to a diverse seedbank community, without predominance of one or two species [23]. Leaving the field barren after harvesting of crop, substantial amounts of nitrate are leached out or flushed away from the field as an effect of higher nitrate-N available after harvest. Growing of nonlegume crops in the rotation as a cover crop utilizes the surplus N from the soil that prevents nitrate-N removal and also reduces the available nutrients for weed germination and its growth [7].

Because weeds have a characteristic emergence pattern, weed densities in crop can be altered by a crop's planting date. Crop yields, however, are usually decreased when crops are planted beyond their optimum planting date range. A more favorable strategy is to rotate crops with different optimum planting dates, which can aid weed management without negatively impacting crop yield. For example, normal planting dates vary between crops grown in winter wheat–oilseed crop (sunflower and safflower)–fallow rotation used in the semiarid Great Plains of the USA. In this rotation system, safflower is planted in early April, whereas sunflower is planted in early June. The oilseed crop grown dramatically affects weed densities in the crop, the explanation for this trend being related to weed emergence [172]. Rotating crops that possess definite height advantage over the weeds may increase weed control. Bryson et al. concluded that in cotton production, severe infestations of purple nutsedge (*Cyperus rotundus* L.) can be managed by rotating cotton with soybean [173].

Intercropping

Intercropping refers to growing two or more crops of different growth habits simultaneously on the same piece of land, which offers early canopy cover and seedbed use resulting in reduced weed growth by competition for resources among component crops [8]. The individual crops that constitute an intercrop can differ in their use of resources spatially, temporally, or in form, resulting in overall more complementary and efficient use of resources than when they are grown in monocultures, thus decreasing the amount available for weeds [174]. Intercrops may inhibit weeds by limiting resource capture by weeds or through allelopathic interactions [60]. Liebman and Dyck, in an extensive literature review, reported that weed biomass in intercrops was lower than component crops in 50 % of the studies, intermediate to component crops in 42 % of the studies, and greater than all component crops in 8 % of the studies [60]. Baumann et al. found that intercropping celery with leek can increase light interception by the weakly competitive leek and can, therefore, shorten the critical period for weed control and reduce growth and fecundity of late-emerging weeds [175]. Banik et al. reported 70 % reduction in weed biomass and population under the intercropping of a wheat–chickpea (*Cicer arietinum* L.) system without weeding treatment as compared to unweeded monocrop wheat [176]. Similarly, Szumigalski and Van Acker concluded that annual intercrops can enhance both weed suppression and crop production compared with sole crops [177].

Intercropping of short-duration, quick-growing, and early-maturing legume crops with long-duration and wide-spaced crops leads to covering ground quickly and suppressing emerging weeds effectively [8]. Weed suppression by intercrops is often, but not always, better than that obtained from sole crops [60]. Greater land-use efficiency, yield stability, and weed suppression of intercropping relative to sole cropping all appear to derive from complementary patterns of resource use and facilitative interactions between crop species. Because complementary patterns of resource use and facilitative interactions between intercrop components can lead to greater capture of light, water, and nutrients, intercrops can be more effective than sole crops in preempting resources used by weeds and suppressing weed growth [178].

Seeding Rate and Row Spacing

Higher seeding rate and narrow-row spacing are known strategies for increasing crop tolerance to weeds. Increasing the sowing density improves competitiveness in many crops. The effectiveness of higher seeding rates has been documented in several studies. High sowing rates are commonly used by Latin American farmers to suppress rice weeds [179]. Blackshaw et al. reported that an increase in wheat seed rate from 50 to 300 kg ha^{-1} reduced redstem filaree (*Erodium cicutarium* [L.] L'Her. ex Ait.) biomass and seed production by 53–95 % over the years [180]. Biomass and yield of wild oat were reduced by 20 % when the sowing rate of winter wheat was increased from 175 to 280 plants m^{-2} [181]. Estorninos Jr. et al. reported that increasing seeding rate of rice from 50 to 150 kg ha^{-1} considerably reduced the red/weedy rice (*Oryza sativa* L.) seed yield [182]. Eslami et al. in a no-till system in Australia observed that the growth and seed production of wild radish was adversely affected by increasing the density of wheat from 100 to 400 plants m^{-2} [183]. Paynter and Hills found that increasing barley plant density increased grain yield, and reduced both rigid ryegrass dry matter and tiller number [184].

Planting in narrow rows (19–25 cm) is an option for many crops and can be integrated into most conservation management programs [25]. Soybean planted in 19-cm rows reduced total weed biomass, increased soybean yield, and resulted in similar to higher net return compared to soybean planted in 57-cm and 95-cm rows [185, 186]. Benefits of planting soybean in narrow rather than wide rows are increased crop competitiveness, rapid canopy closure, increased herbicide effectiveness, suppression of late-emerging weeds, suppression of weeds not killed by a postemergence herbicide application, and increased light interception by the crop resulting in improved seed yield [187]. Narrow rows also shorten the weed-free requirement for maximum crop yield. There has been a renewed interest in growing cotton in ultra-narrow rows in recent years, partially because of potential weed control benefits. Ultra-narrow-row cotton production systems have row widths of 19–25 cm, and populations of 210,000–378,000 plants ha^{-1} compared with wide-row systems, which usually have row widths of 76–100 cm, and populations of 80,000–120,000 plants ha^{-1} [188]. The weed control benefits of narrow-row cotton are likely to be derived from more complete and rapid canopy closure compared with wide-row cotton.

Despite all these advantages, there are some limitations to narrow-row crop production in regard to weed management. For example, Reddy concluded that in ultra-narrow row (25-cm row spacing), bromoxynil-resistant cotton—unlike wide-row cotton—banded application of preemergence herbicides, interrow cultivation, postemergence-directed herbicide sprays, and hooded sprayer applications were not possible and late-season weed growth reduced yields where preemergence herbicides were not applied [189].

Crop Type and Cultivar

Using competitive crop types and cultivars can be an important tool for an integrated weed management program, useful in CA cropping systems [190]. Crop choice could be an important tool for the weed control in CA. For example, wild oat infests both spring wheat and barley in Argentina. Nonetheless, wild oat seed production in barley was one-half the production in spring wheat [191]. This difference was attributed to the later harvesting of spring wheat, which allowed more wild oat seed to reach physiological maturity before cutting at harvest. In addition, barley is a more competitive crop than wheat. Asian rice cultivars have suppressed barnyard grass or aquatic weeds without the use of herbicide [192]. Recent reports have indicated that weed-suppressive rice cultivars may lessen reliance on herbicides and facilitate effective weed control at reduced herbicide rates [193–195]. The effects of spurred anoda (*Anoda cristata* [L.] Schlecht.) interference also varied with cotton variety. The yield reductions resulting from spurred anoda interference with "Deltapine 16" (DP 16) were greater than with Stoneville 213 or DES 21326-04. Early-season-spurred anoda interference also reduced the yield of determinate varieties more than that of indeterminate varieties, indicating that early-season weed control is more important in the early-maturing cultivars [188]. Barley cultivars also can vary in their competitiveness with weeds [184, 190]. Watson et al., in a comparison of 29 barley cultivars, found that semidwarf and hull-less cultivars were less competitive than full-height and hulled cultivars, respectively [190]. Yenish and Young concluded that wheat height had a consistent effect on jointed goatgrass (*Aegilops cylindrica* Host.) seed production, with the taller wheat suppressing jointed goatgrass seed production and 1000-kernel weight relative to the shorter wheat [196]. There appears to be a positive relationship between crop height at maturity and weed competitiveness [190]. These differences appear to be related to morphological traits that affect light interception, such as leaf inclination, early vigor, rate of stem elongation, and tiller number [197, 198]. Lemerle et al. indicated that extensive leaf display and shading ability were characteristic of competitive cultivars [197]. Although the height of crop cultivars has shown significant correlation with their competitiveness, there are also other traits that influence the success of a cultivar in competition with weeds. For example, Estorninos Jr. et al. demonstrated that red rice was more competitive when compared with the tropical *japonica* Kaybonnet than the *indica* PI 312777 [199]. Despite its semidwarf stature, PI 312777 tended to suppress red rice more than did Kaybonnet. The ability to produce more tillers and aboveground biomass resulted in sustained competitiveness against red rice by PI 312777 compared with Kaybonnet. Ability to produce allelochemicals may also be an important trait of competitive crop cultivars. Some rice cultivars, including PI 312777, can produce phytotoxins and exhibit allelopathic activity against barnyard grass or other target plants [200, 201].

The maturity date of crop cultivars is also another important issue that affects their impact on weed species, so the desired cultivar for this purpose should be chosen considering factors, such as climatic conditions and phenology of weed species present in the field. Bennett and Shaw found that early-maturing soybean cultivars reduced seed weight, seed production, and seedling growth in pitted morning glory (*Ipomoea lacunosa* L.) in most instances and usually reduced seed weight, germination, emergence, and growth in hemp sesbania (*Sesbania exaltata* [Raf.] Rydb.) by allowing harvest prior to physiological maturity of these weeds [202]. Tall, late-maturing soybean decreased weed seed production and seed weight of both species, presumably through increased competitiveness of soybean. Late-maturing soybean cultivars have also been found to be more competitive with weeds, and they may recover better from early-season weed competition than early-maturing cultivars because of their ability to grow vegetatively for a longer time [203].

Fertilization Strategies

Management practices that increase the competitive ability of crops with weeds can be important components of integrated weed management systems. Fertilizer management is one such practice that can markedly affect crop–weed interference [204]. The greatest competition among plants is usually for N, and it is the major nutrient input that farmers utilize to increase crop yield [205]. It is important to develop fertilization strategies for crop production that enhance the competitive ability of the crop, minimize weed competition, and reduce the risk of nonpoint source pollution from N [206]. N fertilizer is known to break the dormancy of certain weed species and thus may directly affect weed infestation densities [207]. The dormancy of several grass weed species was broken by ammonia, but the gas had no effect on the dormancy of dicotyledonous weed seed [208]. Redroot pigweed seed germination was stimulated by 10–100 ppmv of ammonium nitrate or urea [209]. Some weeds are luxury consumers of N and thus reduce the N available for crop growth [210]. N fertilizer can increase the competitive ability of the weeds more than that of the crop, while crop yield remains unchanged or actually decreases in some cases [205]. Weed growth tends to increase with N application rates. Common lamb's-quarters and redroot pigweed are among the most-responsive weeds, both demonstrating high rates of shoot and root biomass accumulation with increasing N application rate [211]. Ross and Van Aker found that wild oat competitiveness in wheat was significantly greater in the presence of N fertilizer [212]. Weed competition with crops can be affected by the method of placement of N fertilizer. In zero-till crop production systems, weeds tend to germinate at or near the soil surface [44], and it is in this situation that the point-injected or banded fertilizer may have the greatest benefit by physically placing the fertilizer in an area of the soil profile where the crop seed, but not the weed seed, is located [205]. Fertilizer placed in narrow bands below the soil surface compared with being surface broadcast has been found to reduce the competitive ability of wild oat [213], foxtail barley (*Hordeum jubatum* L.) [214], jointed goatgrass [215], and downy brome [216]. Banding N fertilizer with barley seed at planting reduced green foxtail density and interference compared with N-applied broadcast [217]. Barley's access to the banded N favored its early-season growth and competitiveness with weeds.

Weed emergence and growth in the field can be stimulated by N application rate and timing. The timing of N fertilizer application in early-planted crops, such as sugar beet and corn, may influence the germination, emergence, and competitiveness of weeds that might otherwise remain dormant early in the growing season. Sweeney et al. found that if N is broadcast in April at the time of planting, weed germination and emergence may be stimulated [206]. In contrast, N application at the time of planting in May may not influence seed germination and weed emergence, because of greater N availability due to mineralization at this time of the year or because seed germination has been stimulated by other environmental cues. Blackshaw et al. also found that density and biomass of wild oat, green foxtail (*Setaria viridis* [L.] Beauv.), wild mustard (*Brassica kaber* [DC.] L. C. Wheeler), and common lamb's-quarters were sometimes lower with spring-applied N than with fall-applied N [218]. Shoot N concentration and biomass of weeds were often lower with subsurface banded or point-injected N than with surface broadcast N, and concurrent increases in spring wheat yield usually occurred with these N placement treatments. Sweeny et al. found that spring N fertilizer applications increased soil inorganic N and weed growth, but the influence of N on weed emergence was dependent on the weed species, seed source, and environmental conditions [206]. Delaying N applications, applying slow-release N fertilizers, or placing N below the weed seed germination zone could be potential strategies for reducing early-season weed establishment in integrated cropping systems [206].

Some studies indicated that N effects on weed competition might be crop- and weed-specific. Blackshaw and Brandt found that the competitive ability of the low N-responsive species, Persian darnel (*Lolium persicum* Boiss. and Hohen. ex Boiss.) and Russian thistle (*Salsola* spp.), was not influenced by N rate [204]. Conversely, the competitiveness of the high N-responsive species redroot pigweed progressively improved as N rate increased. This suggests that fertilizer management strategies that favor crops over weeds deserve greater attention when weed infestations consist of species known to be highly responsive to higher soil N levels. In these situations, farmers should consider the benefits of specific fertilizer timing and/or placement methods that would minimize weed interference [204].

Although N is often the major nutrient added to increase crop yield, phosphorus (P) is also widely applied to improve plant growth and crop yield. Soil P levels can also have more immediate effects on weeds. Studies have indicated that weed species may inherently be quite different in their level of growth responsiveness to P [219]. Strategic P fertilizer management may potentially reduce growth and competitive ability of weed species having a high P requirement. P fertilizer rate and application method in lettuce (*Lactuca sativa* L.) have been demonstrated to affect the level of weed interference of some weeds, but not others [220, 221]. Blackshaw and Molnar also found that shoot P concentration and biomass of weeds were often lower with seed-placed or subsurface-banded P fertilizer compared with either surface-broadcast application method [219]. This result, however, occurred more frequently with the highly P-responsive weeds. P application method had little effect on weed-free wheat yield, but often had a large effect on weed-infested wheat yield. Seed-placed or midrow-banded P compared with surface-broadcast P fertilizer often resulted in higher yields when wheat was in the presence of competitive weeds. Seedbank determinations at the conclusion of the study indicated that the seed density of five of six weed species was reduced with seed-placed or subsurface-banded P compared with surface-broadcast P.

Chemical Weed Management

Herbicides have an important role in weed control under CA systems, since there is a great reliance on herbicides in such production systems. Herbicides are effective weed control measures and offer diverse benefits, such as saving labor and fuel cost, requiring less human efforts, reducing soil erosion, saving energy, increasing crop production, reducing the cost of farming, allowing flexibility in weed management, and tackling difficult-to-control weeds [8, 222]. The importance of herbicides in modern weed management is underscored by estimates that losses in the agricultural sector would increase about 500 % without the use of herbicides [222]. Despite all these benefits, there are some concerns about herbicides, including injury to succeeding rotation crops due to persistent herbicides, development of resistant weeds, shifts in weed flora, cost, contamination of surface and groundwater, and unknown long-term human health effects [8, 71]. Therefore, herbicide implementation in CA systems should be undertaken with careful consideration and caution. In this regard, several elements such as an appropriate herbicide type, application time, and formulation must be taken into consideration.

Primary tillage operations, practiced in conventional agricultural systems, are usually removed in CA systems. When tillage operations are removed prior to planting, weed species exist on agricultural lands at the time of crop planting.

By eliminating tillage operations before planting, the need for nonselective postemergence herbicides (e.g., glyphosate, paraquat, and glufosinate) to control weeds before planting crops would become inevitable [8, 25]. Nonetheless, application of contact herbicides might not always be an effective tool for weed killing before planting, and glyphosate application might be a better solution. Glyphosate, a nonselective, broad-spectrum herbicide, controls most grass, sedge, and broadleaf weeds [92]. Stale seedbed is a promising approach for killing weeds before planting in CA systems. In this technique, irrigation or a light shower stimulates the germination and emergence of weed species, which mostly remain in the top soil layers (3 cm) in CA systems, and then can be readily killed using nonselective herbicides [223, 224]. Although this technique allows farmers to simply deplete weed seedbank before planting the crop, it usually delays crop planting between 2 and 4 weeks, and this might lead to crop yield loss in some regions. Therefore, this technique should be employed with careful considerations regarding the climatic conditions of the region, and on-farm research must be conducted before the technique is practiced.

Keeping crop residues on the soil surface is one of the main principles of CA systems, and crop residues may intercept an appreciable part of the herbicide applied preemergence for weed control. High levels of cover crop residue can suppress weed emergence and also can intercept preemergence herbicides and potentially reduce their effectiveness. Crop residues can intercept from 15 to 80 % of the applied herbicides, and this may result in reduced efficacy of herbicides in CA systems [8]. Teasdale et al. found that a hairy vetch cover crop both reduced the initial soil solution concentration and increased the rate of decomposition of metolachlor [225]. Consequently, preemergence application of metolachlor with a hairy vetch cover crop provided less grass control and provided a niche for increased smooth pigweed (*Amaranthus hybridus* L.) emergence. However, herbicide application with hairy vetch consistently delayed grass emergence and appeared to reduce the concentration of metolachlor required to delay the initiation of grass emergence. Locke et al. compared the herbicide (fluometuron and norflurazon) dissipation in soil and cover crops under conservation cotton production and found that herbicide dissipation in cover crop residues was often more rapid than in soil [226]. Herbicide retention in cover crop residues and rapid dissipation were attributed to strong herbicide affinity to cover crop residues and herbicide co-metabolism as cover crop residues decomposed. Chauhan and Abugho found that the oxadiazon and pendimethalin applications in the presence of rice residue cover resulted in lower rice flatsedge (*Cyperus iria* L.) control than in the absence of residue [62]. Although crop residues may decrease the preemergence herbicide efficiency, it is also likely that herbicides intercepted by residues become more effective against surface-germinating weed seeds due to retaining herbicides by residues in the vicinity of the germinating seeds. Some researchers believe that nonselective burn-down herbicides should be applied well ahead of the sowing time of crop plants in CA systems [25]. However, Chauhan et al. recommended desiccating the existing weed cover with a nonselective herbicide as close as possible to crop planting, as these herbicides lack residual activity, and their early application might cause further weed emergence prior to crop emergence [8].

Leaving crop residues on the soil surface in CA systems can provide the advantage of reduced herbicide rates. In fact, the contribution of crop residues to weed control may allow implementing the reduced doses of herbicides. Moseley and Hagood observed that a preemergence application of chlorimuron and linuron to a wheat–soybean no-till double crop provided sufficient nonselective activity on winter vegetation so that it was unnecessary to use traditional nonselective herbicides, which might be due to the inhibitory effect of wheat residue on weed growth [227]. Innovative use of low-rate postemergence herbicides may complement the weed control achieved with cover crops and reduce the environmental risk involved with the triazine and chloroacetamide herbicides [71]. The other strategy for an effective and economic use of herbicides in CA is the split application of herbicides. Previous studies suggest that split herbicide applications can, in some cases, improve the efficacy of reduced-rate herbicide programs and control multiple cohorts of weeds [228]. Split applications would provide a grower the option to assess his/her weed control needs after the first application, thereby tailoring the rate and chemistry of any subsequent herbicide to the specific weed infestation that may be present. Such an approach could substantially reduce the amount of herbicide applied and the cost of chemical weed control. It seems that split applications might be useful when competition from early- and late-emerging weeds can be expected [71].

Herbicide formulation also may affect herbicide efficacy under CA systems. It has been indicated that the application of granule formulations of some herbicides (alachlor, cyanazine, metolachlor, and trifluralin) provided a better weed control under no-till cropping systems. This has been attributed to the more effective movement of granules to the soil surface through the stubble compared to the liquid applied herbicide, likely aided by sowing or cultural disturbances [8]. The other effective herbicide formulation for weed control in CA systems is microencapsulated herbicides. As low soil disturbance is one of the major goals of CA systems, using herbicides that enhance the residual activity of herbicides may provide long-term weed control over the growing season. Microencapsulation of herbicides is designed to increase residual activity and decrease volatility compared with the emulsifiable concentrate (EC) formulations [229]. Microencapsulation formulations are those in which herbicides are enclosed by a matrix that slowly releases the herbicide into the environment. Polymeric microcapsules of atrazine showed excellent controlled release activity when compared with a dry flowable formulation [230]. Microencapsulation alachlor increased surface concentrations when compared with the EC formulation up to 70 days after treatment [231].

Interactions of herbicides with crop residues may also affect herbicide fate, transformation, and transport under CA systems. Surface plant residues in CA systems often form dense mats that intercept herbicides, and while at the surface, herbicides are vulnerable to sunlight, higher temperatures, air movement, and evaporation, which favor volatilization and photodegradation and other losses [8]. *S*-Ethyl dipropylthiocarbamate (EPTC), trifluralin, pendimethalin, and other soil-applied herbicides with high vapor pressure are usually prone to vaporization from the soil surface, especially in CA systems where soil disturbances are discouraged [8, 232]. The option of using tillage to incorporate herbicides in soil is more limited with reduced tillage systems. Incorporation can prolong residence time for some herbicides by protecting them from volatilization or photodecomposition [233]. Nevertheless, volatilization may be minimal for herbicides that have a strong affinity for plant residues. For less retentive herbicides that are subject to volatilization, rainfall soon after application may wash herbicides into the soil, where cooler temperatures and less evaporation in reduced tillage conditions may inhibit volatilization. Moreover, as the herbicide diffuses into the soil matrix, sorption to soil may inhibit volatilization [234]. Herbicide dissipation in cover crop residues was often more rapid than in soil. Herbicide retention in cover crop residues and rapid dissipation were attributed to strong herbicide affinity to cover crop residues and herbicide co-metabolism as cover crop residues decomposed [226].

Many other factors can affect herbicide efficacy in crop residues including the composition of herbicide, residue type, and amount and rainfall [124, 235]. Herbicide sorption was consistently greater in surface soil from reduced tillage than from tilled areas, which was primarily attributed to greater quantities of organic carbon in low-disturbed soils [234]. Larsbo et al. observed that adsorption coefficients for both bentazone and isoproturon were larger in the top 5 cm of reduced-till soil compared to the 10–20-cm depth and to conventional till, reflecting the higher organic carbon content [236]. The degradation rate was also generally larger in the top 5 cm of reduced-till soil. As CA creates a favorable condition for microbial populations [122], these systems can also expose herbicides to microbial decomposition.

The effectiveness of postemergence herbicides may also be reduced by the presence of weeds and cover crop residues [8]. Living plants may enhance the degradation of herbicides in soil either by uptake and metabolism or by catalyzing metabolism through rhizosphere interactions [234]. This activity may be important in CA systems where cover crops or standing stubbles are incorporated into the cropping systems. Higher and relatively more abundant microbial populations are found on plant rhizospheres than in adjacent soil due to root exudation of rich substrates (i.e., amino acids and carbohydrates). This enhanced microbial niche can transform herbicides by both metabolic and co-metabolic pathways. The quantity of spray lodged on smooth pigweed was reduced by standing wheat stubble by 38% at a spray travel speed of 8 km h^{-1}, and by 52% at 16 km h^{-1}. Since the timing of weed emergence is less uniform in CA systems than in conventional-tilled systems, it has been suggested that growers wait until weeds become established and then control them with postemergence herbicides [8].

Integrated Weed Management

The commonly accepted best approach to manage weeds is to follow an integrated weed management strategy comprising the combined use of two or more available and effective technologies [237]. For example, corn yield loss due to foxtail millet (*Setaria italica*) interference was 43% when corn was planted at 37,000 plants ha^{-1} in rows 76 cm wide with N fertilizer applied broadcast [238]. Impact of a single cultural practice, such as N banding, narrower-row spacing, or increased crop density, was minimal in reducing crop yield loss. However, where these three practices were combined, yield loss was only 13 %. Therefore, the consistency of weed management can be greatly improved by an intelligent and rational combination of multiple cultural and chemical approaches. Combining good agronomic practices, timeliness of operations, fertilizer and water management, and retaining crop residues on the soil surface improve the weed control efficiency of applied herbicides and competitiveness against weeds [8].

Conclusion

Despite all the benefits, CA faces challenges in regard to weeds and their management. The dynamics of the weed population under CA is completely different from conventional systems, and the ecophysiological responses of weeds and their interactions with crops under CA management practices tend to be more complex [239]. The implementation of the primary principles of CA, including reduction or complete elimination of tillage practices, incorporation of cover crop into the system, and employment of diversified crop rotations, leads to a different environment for germination, emergence, and growth of weed species and causes a variation in the dynamics of the weed populations compared to conventional cropping systems. Although the general variations caused by CA systems might seem identical, inconsistent results have been obtained from research studies. The reasons for these discrepancies are that the climatic conditions, water–soil relations, soil fauna and biota, and the ecophysiological responses of different species and population of weed species in different locations are poles apart. Thus, there is a great need to accomplish site-specific research on the effects of soil disturbance, cover crops and their residues, and different crop rotations on weed population dynamics under contrasting soil and environmental conditions. It is also necessary to find cultivars well adapted to CA conditions with high seed vigor and having the ability to close their canopy early in the season. Early canopy closure can reduce or eliminate light from reaching the soil surface where most of weed seeds accumulate under low-disturbance conditions in CA. Proper leaf architecture also imparts cultivars a higher competitive ability over weeds. Research efforts should be also toward finding high-residue cover crops that provide adequate ground cover for weed suppression. This is a vital requirement, especially in dry and semidry regions, where the production of a sufficient quantity of crop residues appears to be a difficult issue and might discourage the adoption of CA systems in regard to weed problems in these regions.

As mechanical weed control measures have limitations under CA, it is important to develop environmentally safe, preemergence herbicides with high residual activity that control weeds properly over the growing season. Using biotechnology and genetic engineering can help in developing HR crop cultivars that can readily suppress weeds using nonselective herbicides in crop. Nonetheless, there is a strong risk of developing HR weed biotypes due to the gene flow from HR crops to their wild relatives [8]. Chauhan suggested that the use of HR rice cultivars may help in developing resistance in weedy rice through gene flow, making weedy rice control even more difficult [240]. The use of appropriate formulations such as microencapsulated herbicides that gradually release herbicides can be an effective tool for weed control in CA over the growing season. However, more research studies are needed to test the effectiveness of these formulations in different situations under CA.

References

1. Friedrich T, Kienzle J, Kassam A (2009) Conservation agriculture in developing countries: the role of mechanization; Innovation for sustainable agricultural mechanisation. Club of Bologna, FAO, Hannover, Germany, Nov 2009
2. FAO (Food and Agriculture Organization) (2013) Rome: introduction to conservation agriculture (its principles & benefits). http://teca.fao.org/technology/introduction-conservation-agriculture-its-principles-benefits, March 2013
3. Baker CJ, Saxton KE, Ritchie WR, Chamen WCT, Reicosky DC, Ribeiro F, Justice SE, Hobbs PR (2007) No-tillage seeding in conservation agriculture, 2nd edn. FAO and CAB International, Oxford (Baker CJ, Saxton KE (eds))
4. Johansen C, Haque ME, Bell RW, Thierfelder C, Esdaile RJ (2012) Conservation agriculture for small holder rainfed farming: opportunities and constraints of new mechanized seeding systems. Field Crops Res 132:18–32
5. Friedrich T, Derpsch R, Kassam A (2012) Overview of the global spread of conservation agriculture. Field actions science reports (Internet), Nov 2012, Special Issue 6, 7 p. http://factsreports.revues.org/1941, April 2013
6. Hobbs PR, Sayre K, Gupta R (2008) The role of conservation agriculture in sustainable agriculture. Philos Trans R Soc B 363:543–555
7. Baral KR (2012) Weeds management in organic farming through conservation agriculture practices. J Agric Environ 13:60–66
8. Chauhan BS, Singh RG, Mahajan G (2012) Ecology and management of weeds under conservation agriculture: a review. Crop Prot 38:57–65
9. Owen MDK, Zelaya IA (2005) Herbicide-resistant crops and weed resistance to herbicides. Pest Manage Sci 61:301–311
10. Orloff SB, Putnam DH, Canevari M, Lanini WT (2009) Avoiding weed shifts and weed resistance in roundup ready alfalfa systems. ANR publication, Report no. 8362, Division of Agriculture and Natural Resources, University of California, Feb 2009
11. Manley BS, Wilson HP, Hines TE (2002) Management programs and crop rotations influence populations of annual grass weeds and yellow nutsedge. Weed Sci 50:112–119
12. Légére A, Samson DN (1999) Relative influence of crop rotation, tillage, and weed management on weed associations in spring barley cropping systems. Weed Sci 47:112–122
13. Ball DA, Miller SD (1990) Weed seed population response to tillage and herbicide use in three irrigated cropping sequences. Weed Sci 38:511–517
14. Blackshaw RE (1994) Rotation affects downy brome (*Bromus tectorum*) in winter wheat (*Triticum aestivum*). Weed Technol 8:728–732
15. Buhler DD (1992) Population dynamics and control of annual weeds in corn (*Zea mays*) as influenced by tillage systems. Weed Sci 40:241–248
16. Buhler DD, Stoltenberg DE, Becker RL, Gunsolus JL (1994) Perennial weed populations after 14 years of variable tillage and cropping practices. Weed Sci 42:205–209
17. Hilgenfeld KL, Martin AR, Mortensen DA, Mason SC (2004) Weed management in a glyphosate resistant soybean system: weed species shifts. Weed Technol 18:284–291

18. Clements DR, Weise SF, Swanton CJ (1994) Integrated weed management and weed species diversity. Phytoprotection 75:1–18
19. Clements DR, Benoit DL, Murphy SD, Swanton CJ (1996) Tillage effects on weed seed return and seedbank composition. Weed Sci 44:314–322
20. Samarajeewa KBDP, Horiuchi T, Oba S (2005) Weed population dynamics in wheat as affected by *Astragalus sinicus* L. (Chinese milk vetch) under reduced tillage. Crop Prot 24:864–869
21. Thomas AG, Derksen DA, Blackshaw RE, Van Acker RC, Le´ge`re A, Watson PR, Turnbull GC (2004) A multistudy approach to understanding weed population shifts in medium- to long-term tillage systems. Weed Sci 52:874–880
22. Young FL, Thorne ME (2004) Weed-species dynamics and management in no-till and reduced-till fallow cropping systems for the semi-arid agricultural region of the Pacific Northwest, USA. Crop Prot 23:1097–1110
23. Andersson TN, Milberg P (1998) Weed flora and the relative importance of site, crop, crop rotation, and nitrogen. Weed Sci 46:30–38
24. Froud-Williams RJ (1988) Changes in weed flora with different tillage and agronomic management systems. In: Altieri MA, Liebman M (eds) Weed management in agroecosystems: ecological approaches. CRC, Boca Raton, pp 213–236
25. Locke MA, Reddy KN, Zablotowicz RM (2002) Weed management in conservation crop production systems. Weed Biol Manage 2:123–132
26. Murphy SD, Clements DR, Belaoussoff S, Kevan PG, Swanton CJ (2006) Promotion of weed species diversity and reduction of weed seedbanks with conservation tillage and crop rotation. Weed Sci 54:69–77
27. Sosnoskie LM, Herms CP, Cardina J (2006) Weed seedbank community composition in a 35-yr-old tillage and rotation experiment. Weed Sci 54:263–273
28. Tuesca D, Puricelli E, Papa JC (2001) A long-term study of weed flora shifts in different tillage systems. Weed Res 41:369–382
29. Légére A, Samson DN (2004) Tillage and weed management effects on weeds in barley-red clover cropping systems. Weed Sci 52:881–885
30. Chauhan BS, Gill GS, Preston C (2006) Tillage systems affect trifluralin bioavailability in soil. Weed Sci 54:941–947
31. Fenner M, Thompson K (2005) The ecology of seeds. Cambridge University Press, New York
32. Peachey RE, Mallory-Smith C (2007) Influence of winter seed position and recovery date on hairy nightshade (*Solanum sarrachoides*) recruitment and seed germination, dormancy, and mortality. Weed Sci 55:49–59
33. Peachey RE, Mallory-Smith C (2011) Effect of fall tillage and cover crop strategies on wild-proso millet (*Panicum miliaceum*) emergence and interference in snap beans. Weed Technol 25:119–126
34. Chauhan BS, Abugho SB (2012a) Threelobe morning glory (*Ipomoea triloba*) germination and response to herbicides. Weed Sci 60:199–204
35. Chauhan BS, Johnson DE (2008) Influence of environmental factors on seed germination and seedling emergence of eclipta (*Eclipta prostrata*) in a tropical environment. Weed Sci 56:383–388
36. Chauhan BS, Gill GS, Preston C (2006) Factors affecting seed germination of annual sowthistle (*Sonchus oleraceus*) in southern Australia. Weed Sci 54:854–860
37. Chauhan BS, Gill GS, Preston C (2006) Seed germination and seedling emergence of threehorn bedstraw (*Galium tricornutum*). Weed Sci 54:867–872
38. Chauhan BS, Gill GS, Preston C (2006) African mustard (*Brassica tournefortii*) germination in southern Australia. Weed Sci 54:891–897
39. Ebrahimi E, Eslami SV (2012) Effect of environmental factors on seed germination and seedling emergence of invasive *Ceratocarpus arenarius*. Weed Res 52:50–59
40. Eslami SV (2011) Comparative germination and emergence ecology of two populations of common lambsquarters (*Chenopodium album*) from Iran and Denmark. Weed Sci 59:90–97

41. Buhler DD, Owen MDK (1997) Emergence and survival of horsweed (*Conyza canadensis*). Weed Sci 45:98–101
42. Ghosheh H, Al-Hajaj N (2005) Weed seedbank response to tillage and crop rotation in a semi-arid environment. Soil Tillage Res 84:184–191
43. Mester TC, Buhler DD (1991) Effects of soil temperature, seed depth, and cyanazine on giant foxtail (*Setaria faberi*) and velvetleaf (*Abutilon theophrasti*) seedling development. Weed Sci 39:204–209
44. Yenish JP, Doll JD, Buhler DD (1992) Effects of tillage on vertical distribution and viability of weed seed in soil. Weed Sci 40:429–433
45. Anderson RL (1998) Seedling emergence of winter annual grasses as affected by limited tillage and crop canopy. Weed Technol 12:262–267
46. Froud-Williams RJ, Chancellor RJ, Drennan DSH (1984) The effects of seed burial and soil disturbance on emergence and survival of arable weeds in relation to minimal cultivation. J Appl Ecol 21:629–641
47. Mohler CL (1993) A model of the effects of tillage on weed seedlings. Ecol Appl 3:53–73
48. Mohler CL, Callaway MB (1992) Effects of tillage and mulch on the emergence and survival of weeds in sweet corn. J Appl Ecol 29:21–34
49. Froud-Williams RJ (1983) The influence of straw disposal and cultivation regime on the population dynamics of *Bromus sterilis*. Ann Appl Biol 103:139–148
50. Mohler CL, Callaway MB (1995) Effects of tillage and mulch on weed seed production and seed banks in sweet corn. J Appl Ecol 32:627–639
51. Mas MT, Verdú AMC (2003) Tillage system effects on weed communities in a 4-year crop rotation under Mediterranean dryland conditions. Soil Tillage Res 74:15–24
52. Stevenson FC, Légère A, Simard RR, Angers DA, Pangeau D, Lafond J (1998) Manure, tillage, and crop rotation: effects on residual weed interference in spring barley cropping systems. Agron J 90:496–504
53. Derksen DA, Thomas AG, Lafond GP, Loeppky HA, Swanton CJ (1995) Impact of post-emergence herbicides on weed community diversity within conservation-tillage systems. Weed Res 35:311–320
54. Stevenson FC, Légère A, Simard RR, Angers DA, Pangeau D, Lafond J (1997) Weed species diversity in spring barley varies with crop rotation and tillage, but not with nutrient source. Weed Sci 45:798–806
55. Dorado J, del Monte JP, López-Fando C (1999) Weed seedbank response to crop rotation and tillage in semiarid agroecosystems. Weed Sci 47:67–73
56. Bàrberi P, Lo Cascio B (2001) Long-term tillage and crop rotation effects on weed seedbank size and composition. Weed Res 41:325–340
57. Gill KS, Arshad MA (1995) Weed flora in the early growth period of spring crops under conventional, reduced, and zero tillage systems on a clay soil in northern Alberta, Canada. Soil Tillage Res 33:65–79
58. Cardina J, Herms CP, Doohan DJ (2002) Crop rotation and tillage system effects on weed seedbanks. Weed Sci 50:448–460
59. Christoffoleti PJ, de Carvalho SJP, Lo´pez-Ovejero RF, Nicolai M, Hidalgo E, da Silva JE (2007) Conservation of natural resources in Brazilian agriculture: implications on weed biology and management. Crop Prot 26:383–389
60. Liebman M, Dyck E (1993) Crop rotation and intercropping strategies for weed management. Ecol Appl 3:92–122
61. Chauhan BS (2012) Weed ecology and weed management strategies for dry-seeded rice in Asia. Weed Technol 26:1–13
62. Chauhan BS, Abugho SB (2012) Interaction of rice residue and PRE herbicides on emergence and biomass of four weed species. Weed Technol 26:627–632
63. Teasdale JR, Mohler CL (1993) Light transmittance, soil temperature and soil moisture under residue of hairy vetch and rye. Agron J 85:673–680
64. Teasdale JR, Mohler CL (2000) The quantitative relationship between weed emergence and the physical properties of mulches. Weed Sci 48:385–392

65. Cromar HE, Murphy SD, Swanton CJ (1999) Influence of tillage and crop residue on post-dispersal predation of weed seeds. Weed Sci 47:184–194
66. Brainard DC, Peachey RE, Haramoto ER, Luna JM, Rangarajan A (2013) Weed ecology and nonchemical management under strip-tillage: implications for northern U.S. vegetable cropping systems. Weed Technol 27:218–230
67. Mohler CL, Teasdale JR (1993) Response of weed emergence to rate of *Vicia villosa* Roth and *Secale cereale* L. residue. Weed Res 33:487–499
68. Reeves DW, Price AJ, Patterson MG (2005) Evaluation of three winter cereals for weed control in conservation-tillage in nontransgenic cotton. Weed Technol 19:731–736
69. Derksen DA, Thomas AG, Lafond GF, Loeppky HA, Swanton CL (1994) Impact of agronomic practices on weed communities: fallow within tillage systems. Weed Sci 42:184–194
70. Teasdale JR, Beste CE, Potts WE (1991) Response of weeds to tillage and cover crop residue. Weed Sci 39:195–199
71. Gallagher RS, Cardina J, Loux M (2003) Integration of cover crops with postemergence herbicides in no-till corn and soybean. Weed Sci 51:995–1001
72. Reddy KN, Zablotowicz RM, Locke MA, Koger CH (2003) Cover crop, tillage, and herbicide effects on weeds, soil properties, microbial populations, and soybean yield. Weed Sci 51:987–994
73. Smith H (1995) Physiological and ecological function within the phytochrome family. Annu Rev Plant Physiol Plant Mol Biol 46:289–315
74. Gallagher RS, Cardina J (1998) Ecophysiological factors regulating phytochrome-mediated germination in soil seed banks. Asp Appl Biol 51:165–171
75. Amuri N, Brye KR, Gbur EE, Oliver D, Kelley J (2010) Weed populations as affected by residue management practices in a wheat-soybean double-crop production system. Weed Sci 58:234–243
76. Teasdale JR, Brandsæter LO, Calegari A, Neto FS (2007) Cover crops and weed management. In: Upadhyaya MK, Blackshaw RE (eds) Non-chemical weed management: principles, concepts and technology. CABI, Oxfordshire, pp 49–64
77. Vigil MF, Kissel DE (1991) Equations for estimating the amount of nitrogen mineralized from crop residues. Soil Sci Soc Am J 55:757–761
78. Espeby L (1989) Germination of weed seeds and competition in stands of weeds and barley. Influences of mineral nutrients. Crop Prod Sci 6:1–172
79. Agenbag GA, Villiers OT (1989) The effect of nitrogen fertilizers on the germination and seedling emergence of wild oat (*Avena fatua* L.) seed in different soil types. Weed Res 29:239–245
80. Blackshaw RE, Larney FO, Lindwall CW, Kozub GC (1994) Crop rotation and tillage effects on weed populations on the semi-arid Canadian prairies. Weed Technol 8:231–237
81. Karlen DL, Varvel GE, Bullock DG, Cruse RM (1994) Crop rotations for the 21st century. Adv Agron 53:1–45
82. Ball DA (1992) Weed seedbank response to tillage, herbicides and crop rotation sequence. Weed Sci 40:654–659
83. Kegode GO, Forcella F, Clay S (1999) Influence of crop rotation, tillage, and management inputs on weed seed production. Weed Sci 47:175–183
84. Dekker J (1997) Weed diversity and weed management. Weed Sci 45:357–363
85. Martin RJ, Felton WL (1993) Effect of crop rotation, tillage practice, and herbicides on the population dynamics of wild oats in wheat. Aust J Exp Agric 33:159–165
86. Anderson RL, Beck DL (2007) Characterizing weed communities among various rotations in central South Dakota. Weed Technol 21:76–79
87. Smith RG, Gross KL (2006) Rapid change in the germinable fraction of the weed seed bank in crop rotations. Weed Sci 54:1094–1100
88. Hume L, Tessier S, Dyck FB (1991) Tillage and rotation influences on weed community composition in wheat (*Triticum aestivum* L.) in southwestern Saskatchewan. Can J Plant Sci 71:783–789

89. Shrestha A, Knezevic SZ, Roy RC, Ball-Coelho BR, Swanton CJ (2002) Effect of tillage, cover crop and crop rotation on the composition of weed flora in a sandy soil. Weed Res 42:76–87
90. Roberts HA, Neilson JE (1981) Change in the soil seed bank of four long-term crop/herbicide experiments. J Appl Ecol 18:661–668
91. Owen MDK (2008) Weed species shifts in glyphosate-resistant crops. Pest Manag Sci 64:377–387
92. Reddy KN (2004) Weed control and species shift in bromoxynil- and glyphosate-resistant cotton (*Gossypium hirsutum*) rotation systems. Weed Technol 18:131–139
93. Ross MA, Lembi CA (1999) Applied weed science. Prentice-Hall, Upper Saddle River
94. Webster EP, Bryant KJ, Earnest LD (1999) Weed control and economics in nontransgenic and glyphosate-resistant soybean (*Glycine max*). Weed Technol 13:586–593
95. Culpepper AS (2006) Glyphosate-induced weed shifts. Weed Technol 20:277–281
96. Scursoni J, Forcella F, Gunsolus J, Owen M, Oliver R, Smeda R, Vidrine R (2006) Weed diversity and soybean yield with glyphosate management along a north-south transect in the United States. Weed Sci 54:713–719
97. Wilson RG, Miller SD, Westra P, Kniss AR, Stahlman PW, Wicks GW, Kachman SD (2007) Glyphosate-induced weed shifts in glyphosate-resistant corn or a rotation of glyphosate-resistant corn, sugarbeet, and spring wheat. Weed Technol 21:900–909
98. Telewski FW, Zeevaart JAD (2002) The 120-yr period for Dr. Beal's seed viability experiment. Am J Bot 89:1285–1288
99. Westerman PR, Wes JS, Kropff MJ, van der Werf W (2003) Annual weed seed losses due to predation in organic cereal fields. J Appl Ecol 40:824–836
100. Ward MJ, Ryan MR, Curran WS, Barbercheck ME, Mortensen DA (2011) Cover crops and disturbance influence activity-density of weed seed predators *Amara aenea* and *Harpalus pensylvanicus* (coleoptera: carabidae). Weed Sci 59:76–81
101. Davis AS, Renner KA (2007) Influence of seed depth and pathogens on fatal germination of velvetleaf (*Abutilon theophrasti*) and giant foxtail (*Setaria faberi*). Weed Sci 55:30–35
102. Peigné J, Ball BC, Roger-Estrade J, David C (2007) Is conservation tillage suitable for organic farming? A review. Soil Use Manage 23:129–144
103. Andersen A (2003) Long-term experiments with reduced tillage in spring cereals. II. Effects on pests and beneficial insects. Crop Prot 22:147–152
104. Hunter MD (2002) Landscape structure, habitat fragmentation and the ecology of insects. Agric For Entomol 4:159–166
105. Gallagher R, Fernandes E, Mccallie E (1999) Weed management through short-term improved fallows in tropical agroecosystems. Agrofor Syst 47:197–221
106. Carmona DM, Landis DA (1999) Influence of refuge habitats and cover crops on seasonal activity-density of ground beetles (Coleoptera: Carabidae) in field crops. Biol Contr 28:1145–1153
107. Navntoft S, Wratten SD, Kristensen K, Esbjerg P (2009) Weed seed predation inorganic and conventional fields. Biol Control 49:11–16
108. Cardina J, Norquay HM, Stinner BR, McCartney DA (1996) Postdispersal predation of velvetleaf (*Abutilon theophrasti*) seeds. Weed Sci 44:534–539
109. Reader RJ (1991) Control of seedling emergence by ground cover: a potential mechanism involving seed predation. Can J Bot 69:2084–2087
110. Landis DA, Wratten SD, Gurr GM (2000) Habitat management to conserve natural enemies of arthropod pests in agriculture. Annu Rev Entomol 45:175–201
111. Harrison SK, Regnier EE, Schmoll JT (2003) Postdispersal predation of giant ragweed (*Ambrosia trifida*) seed in no-tillage corn. Weed Sci 51:955–964
112. Zhang J (1993) Biology of *Harpalus rufipes* DeGeer (Coleoptera: Carabidae) in Maine and dynamics of seed predation. Dissertation, University of Maine
113. Gallandt ER, Molloy T, Lynch RP, Drummond FA (2005) Effect of cover-cropping system on invertebrate seed predation. Weed Sci 53:69–76

114. Shearin AF, Reberg-Horton SC, Gallandt ER (2007) Direct effects of tillage on the activity-density of ground beetle (Coleoptera: Carabidae) weed seed predators. Environ Entomol 36:1140–1146
115. Shearin AF, Reberg-Horton SC, Gallandt ER (2008) Cover crop effects on the activity-density of the weed seed predator *Harpalus rufipes* (Coleoptera: Carabidae). Weed Sci 56:442–450
116. Davis AS, Liebman M (2003) Cropping system effects on giant foxtail (*Setaria faberi*) demography. I. Green manure and tillage timing. Weed Sci 51:919–929
117. Westerman PR, Liebman M, Menalled FD, Heggenstaller AH, Hartzler RG, Dixon PM (2005) Are many little hammers effective? Velvetleaf (*Abutilon theophrasti*) population dynamics in two- and four-year crop rotation systems. Weed Sci 53:382–392
118. Davis AS, Taylor EC, Haramoto ER, Renner KA (2013) Annual postdispersal weed seed predation in contrasting field environments. Weed Sci 2:296–302
119. Lundgren JG, Rosentrater KA (2007) The strength of seeds and their destruction by granivorous insects. Arthropod Plant Interact 1:93–99
120. Norris RF, Kogan M (2000) Interactions between weeds, arthropod pests and their natural enemies in managed ecosystems. Weed Sci 48:94–158
121. Davis AS (2007) Nitrogen fertilizer and crop residue effects on seed mortality and germination of eight annual weed species. Weed Sci 55:123–128
122. Alvear M, Rosas A, Rouanet JL, Borie F (2005) Effects of three soil tillage systems on some biological activities in an Ultisol from southern Chile. Soil Tillage Res 82:195–202
123. Balota EL, Colozzi Filho A, Andrade DS, Dick RP (2004) Long-term tillage and crop rotation effects on microbial biomass and C and N mineralization in a Brazilian Oxisol. Soil Tillage Res 77:137–145
124. Christoffoleti PJ, de Carvalho SJP, Nicolai M, Doohan D, VanGessel M (2007) Prevention strategies in weed management. In: Upadhyaya MK, Blackshaw RE (eds) Non-chemical weed management: principles, concepts and technology. CABI, Oxfordshire, pp 1–16
125. Walker RH (1995) Preventive weed management. In: Smith AE (ed) Handbook of weed management systems. Marcel Dekker Inc, New York, pp 35–50
126. Dastgheib F (1989) Relative importance of crop seed, manure, and irrigation water as sources of weed infestation. Weed Res 29:113–116
127. Boyd NS, White S (2009) Impact of wild blueberry harvesters on weed seed dispersal within and between fields. Weed Sci 57:541–546
128. Ballaré CL, Scopel AL, Ghersa CM, Sanchez RA (1987) The demography of *Datura ferox* (L.) in soybean crops. Weed Res 27:91–102
129. Ghersa CM, Martinez-Ghersa MA, Satorre EH, VanEsso ML, Chichotky G (1993) Seed dispersal, distribution and recruitment of seedlings of *Sorghum halepense* (L.) Pers. Weed Res 33:79–88
130. Shirtliffe SJ, Entz MH (2005) Chaff collection reduces seed dispersal of wild oat (*Avena fatua*) by a combine harvester. Weed Sci 53:465–470
131. Slagell Gossen RR, Tyrl RJ, Hauhouot M, Peeper TF, Claypool PL, Solie JB (1998) Effects of mechanical damage on cheat (*Bromus secalinus*) caryopsis anatomy and germination. Weed Sci 46:249–257
132. Walsh M, Newman P (2007) Burning narrow wind rows for weed seed destruction. Field Crops Res 104:24–30
133. Woolcock JL, Cousens R (2000) A mathematical analysis of factors affecting the rate of spread of patches of annual weeds in an arable field. Weed Sci 48:27–34
134. Thill DC, Mallory-Smith CA (1997) The nature and consequence of weed spread in cropping systems. Weed Sci 45:337–342
135. Walsh M, Parker W (2002) Wild radish and ryegrass seed collection at harvest: chaff carts and other devices. In: Agribusiness Crop Updates, Western Australian Department of Agriculture Perth, Australia, pp. 37–38
136. Walsh M, Powles S (2004) Herbicide resistance: an imperative for smarter crop weed management. In: Fischer T, Turner N, Angus J, McIntyre L, Robertson M, Borrell A, Lloyd D

(eds) Proceedings of the 4th International Crop Science Congress, The Regional Institute Ltd, Brisbane, Australia, 26 Sep–1 Oct 2004. http://www.cropscience.org.au/icsc2004/symposia/2/5/1401_powles.htm, Nov 2004
137. Walsh M, Newman P, Powles S (2013) Targeting weed seeds in-crop: A new weed control paradigm for global agriculture. Weed Technol 27:431–436
138. Nagabhushana GG, Worsham AD, Yenish JP (2001) Allelopathic cover crops to reduce herbicide use in sustainable agriculture systems. Allelopath J 8:133–146
139. Akemo MC, Regnier EE, Bennett MA (2000) Weed suppression in spring-sown rye-pea cover crop mixes. Weed Technol 14:545–549
140. Ross SM, King JR, Izaurralde RC, O'Donovan JT (2001) Weed suppression by seven clover species. Agron J 93:820–827
141. Sheaffer CC, Gunsolus JL, Grimsbo Jewett J, Lee SH (2002) Annual *Medicago* as a smother crop in soybean. J Agron Crop Sci 188:408–416
142. Wolf B, Snyder GH (2003) Sustainable soils: the place of organic matter in sustaining soils and their productivity, 2nd edn. Food Products, New York
143. Yenish JP, Worsham AD, York AC (1996) Cover crops for herbicide replacement in no-tillage corn (*Zea mays*). Weed Technol 10:815–821
144. Chauhan BS, Abugho SB (2013) Effect of crop residue on seedling emergence and growth of selected weed species in a sprinkler-irrigated zero-till dry-seeded rice system. Weed Sci 61:403–409
145. Saini M, Price MJ, Santen EV (2009) Integration of cover crop residues, conservation tillage and herbicides for weed management in corn, cotton, peanut and tomato Dissertation, Auburn University, Auburn, Alabama
146. Kumar V, Brainard DC, Bellinder RR, Hahn RR (2011) Buckwheat residue effects on emergence and growth of weeds in winter-wheat (*Triticum aestivum*) cropping systems. Weed Sci 59:567–573
147. Norsworthy JK, Meehan IVJT (2005) Use of isothiocyantes for suppression of Palmer amaranth (*Amaranthus palmeri*), pitted morningglory (*Ipomoea lacunosa*), and yellow nutsedge (*Cyperus esculentus*). Weed Sci 53:884–890
148. Malik MS, Norsworthy JK, Culpepper AS, Riley MB, Bridges W Jr (2008) Wild radish (*Raphanus raphanistrum*) and rye cover crops for weed suppression in sweet corn. Weed Sci 56:588–595
149. Norsworthy JK (2003) Allelopathic potential of wild radish (*Raphanus raphanistrum*). Weed Technol 17:307–313
150. Price AJ, Norsworthy JK (2013) Cover crops for weed management in southern reduced-tillage vegetable cropping systems. Weed Technol 27:212–217
151. Haramoto ER, Gallandt ER (2005) Brassica cover cropping: I. Effects on weed and crop establishment. Weed Sci 53:695–701
152. Adler MJ, Chase CA (2007) Comparison of the allelopathic potential of leguminous summer cover crops: cowpea, sunn hemp, and velvetbean. HortScience 42:289–293
153. Price AJ, Stoll ME, Bergtold JS, Arriaga FJ, Balkcom KS, Kornecki TS, Raper RL (2008) Effect of cover crop extracts on cotton and radish radicle elongation. Commun Biometry Crop Sci 3:60–66
154. Brennan EB, Boyd NS, Smith RF, Foster P (2009) Seeding rate and planting arrangement effects on growth and weed suppression of a legume-oat cover crop for organic vegetable systems. Agron J 101:979–988
155. Reberg-Horton SC, Grossman JM, Kornecki TS, Meijer AD, Price AJ, Place GT, Webster TM (2012) Utilizing cover crop mulches to reduce tillage in organic systems in the southeastern USA. Renew Agric Food Syst 27:41–48
156. Burgos NR, Talbert RE (1996) Weed control and sweet corn (*Zea mays* var. rugosa) response in a no-till system with cover crops. Weed Sci 44:355–361
157. Hayden ZD, Brainard DC, Henshaw B, Ngouajio M (2012) Winter annual weed suppression in rye-vetch cover crop mixtures. Weed Technol 26:818–825

158. Wortman SE, Francis CA, Bernards M, Blankenship EE, Lindquist JL (2013) Mechanical termination of diverse cover crop mixtures for improved weed suppression in organic cropping systems. Weed Sci 61:162–170
159. Norsworthy JK, Malik MS, Jha P, Riley MB (2007) Suppression of *Digitaria sanguinalis* and *Amaranthus palmeri* using autumn-sown glucosinolate-producing cover crops in organically grown bell pepper. Weed Res 47:425–432
160. Wortman SE, Francis CA, Lindquist JL (2012) Cover crop mixtures for the western Corn Belt: opportunities for increased productivity and stability. Agron J 104:699–705
161. Balkcom KS, Schomberg H, Reeves DW, Clark A (2007) Managing cover crops in conservation tillage systems. In: Clark A (ed) Managing cover crops profitably. SARE, College Park, pp 44–46
162. Duiker SW, Curran WS (2005) Rye cover crop management for corn production in the northern Mid-Atlantic region. Agron J 97:1413–1418
163. Price AJ, Reeves DW, Patterson MG (2006) Evaluation of weed control provided by three winter cereals in conservation-tillage soybean. Renew Agric Food Sys 21:159–164
164. Price AJ, Balkcom KS, Duzy LM, Kelton JA (2012) Herbicide and cover crop residue integration for *Amaranthus* control in conservation agriculture cotton and implications for resistance management. Weed Technol 26:490–498
165. Liebman M, Gallandt ER (2008) Differential responses to red clover residue and ammonium nitrate by common bean and wild mustard. Weed Sci 50:521–529
166. Mirsky SB, Curran WS, Mortensen DM, Ryan MR, Shumway DL (2011) Timing of cover-crop management effects on weed suppression in no-till planted soybean using a roller-crimper. Weed Sci 59:380–389
167. Mirsky SB, Curran WS, Mortensen DM, Ryan MR, Shumway DL (2009) Control of cereal rye with a roller/crimper as influenced by cover crop phenology. Agron J 101:1589–1596
168. Rice CP, Cai G, Teasdale JR (2012) Concentrations and allelopathic effects of benzoxazinoid compounds in soil treated with rye (*Secale cereale*) cover crop. J Agric Food Chem 60:4471–4479
169. Creamer NG, Plassman B, Bennett MA, Wood RK, Stinner BR, Cardina J (1995) A method for mechanically killing cover crops to optimize weed suppression. Am J Altern Agric 10:157–162
170. Davis AS (2010) Cover-crop roller-crimper contributes to weed management in no-till soybean. Weed Sci 58:300–309
171. Holtzer TO, Anderson RL, McMullen MP, Peairs FB (1996) Integrated pest management for insects, plant pathogens and weeds in dryland cropping systems of the Great Plains. J Prod Agric 9:200–208
172. Anderson RL (1994) Characterizing weed community seedling emergence for a semiarid site in Colorado. Weed Technol 8:245–249
173. Bryson CT, Reddy KN, Molin WT (2003) Purple nutsedge (*Cyperus rotundus*) population dynamics in narrow row transgenic cotton (*Gossypium hirsutum*) and soybean (*Glycine max*) rotation. Weed Technol 17:805–810
174. Mesgaran MB, Mashhadi HR, Khosravi M, Zand E, Mohammad-Alizadeh H (2008) Weed community response to saffron-black zira intercropping. Weed Sci 56:400–407
175. Baumann DT, Kropf MJ, Bastiaans L (2000) Intercropping leeks to suppress weeds. Weed Res 40:361–376
176. Banik P, Midya A, Sarkar BK, Ghose SS (2006) Wheat and chickpea intercropping systems in an additive series experiment: advantages and weed smothering. Eur J Agron 24:325–332
177. Szumigalski A, Van Acker R (2005) Weed suppression and crop production in annual intercrops. Weed Sci 53:813–825
178. Liebman M, Mohler CL, Staver CP (2004) Ecological management of agricultural weeds, 2nd edn. Cambridge University Press, UK
179. Fischer AJ, Ramirez A (1993) Red rice (*Oryza sativa* L.): competition studies for management decisions. Int J Pest Manage 39:133–138

180. Blackshaw RE, Semach GP, O'Donovan JT (2000a) Utilization of wheat seed rate to manage red stem filaree (*Erodium cicutarium*) in a zero-tillage cropping system. Weed Technol 14:389–396
181. Xue Q, Stougaard RN (2002) Spring wheat seed size and seeding rate affect wild oat demographics. Weed Sci 50:312–320
182. Estorninos LE Jr, Gealy DR, Talbert RE, McClelland MR, Gbur EE (2005b) Rice and red rice interference: II. Rice response to population densities of three red rice (*Oryza sativa*) ecotypes. Weed Sci 53:683–689
183. Eslami SV, Gill GS, Bellotti B, McDonald G (2006) Wild radish (*Raphanus raphanistrum*) interference in wheat. Weed Sci 54:749–756
184. Paynter BH, Hills AL (2009) Barley and rigid ryegrass (*Lolium rigidum*) competition is influenced by crop cultivar and density. Weed Technol 23:40–48
185. Reddy KN (2002) Weed control and economic comparisons in soybean planting systems. J Sustain Agric 21:21–35
186. Grey TL, Raymer P (2002) Sicklepod (*Senna obtusifolia*) and red morningglory (*Ipomoea coccinea*) control in glyphosate-resistant soybean with narrow rows and postemergence herbicide mixtures. Weed Technol 16:669–674
187. Norsworthy JK, Oliver LR (2001) Effect of seeding rate of drilled glyphosate-resistant soybean (*Glycine max*) on seed yield and gross profit margin. Weed Technol 15:284–292
188. Molin WT, Boykin D, Hugie JA, Ratnayaka HH, Sterling TM (2006) Spurred anoda (*Anoda cristata*) interference in wide row and ultra narrow row cotton. Weed Sci 54:651–657
189. Reddy KN (2001) Broadleaf weed control in ultra narrow row bromoxynil-resistant cotton (*Gossypium hirsutum*). Weed Technol 15:497–504
190. Watson PR, Derksen DA, Van Acker RC (2006) The ability of 29 barley cultivars to compete and withstand competition. Weed Sci 54:783–792
191. Scursoni J, Benech-Arnold R, Hirchoren H (1999) Demography of wild oat in barley crops: effect of crop, sowing rate, and herbicide treatment. Agron J 91:478–489
192. Dilday RH, Mattice JD, Moldenhauer KA, Yan W (2001) Allelopathic potential in rice germplasm against ducksalad, redstem and barnyardgrass. J Crop Prod 4:287–301
193. Gealy DR, Wailes EJ, Estorninos LE Jr, Chavez RSC (2003) Rice cultivar differences in suppression of barnyardgrass (*Echinochloa crus-galli*) and economics of reduced propanil rates. Weed Sci 51:601–609
194. Mahajan G, Chauhan BS (2011) Effects of planting pattern and cultivar on weed and crop growth in aerobic rice system. Weed Technol 25:521–525
195. Mahajan G, Chauhan BS (2013) The role of cultivars in managing weeds in dry-seeded rice production systems. Crop Prot 49:52–57
196. Yenish JP, Young FL (2004) Winter wheat competition against jointed goatgrass (*Aegilops cylindrica*) as influenced by wheat plant height, seeding rate, and seed size. Weed Sci 52:996–1001
197. Lemerle D, Verbeek B, Cousens RD, Coombes NE (1996) The potential for selecting wheat varieties strongly competitive against weeds. Weed Res 36:505–513
198. Roberts JR, Peeper TF, Solie JB (2001) Wheat (*Triticum aestivum*) row spacing, seeding rate, and cultivar affect interference from rye (*Secale cereale*). Weed Technol 15:19–25
199. Estorninos LE Jr, Gealy DR, Talbert RE, Gbur EE (2005a) Rice and red rice interference. I. Response of red rice (*Oryza sativa*) to sowing rates of tropical *japonica* and *indica* rice cultivars. Weed Sci 53:676 682
200. Ebana K, Yan W, Dilday RH, Namai H, Okuno K (2001) Variation in the allelopathic effect of rice with water soluble extracts. Agron J 93:12–16
201. Olofsdotter M (2001) Rice—a step toward use of allelopathy. Agron J 93:3–8
202. Bennett AC, Shaw DR (2000) Effect of *Glycine max* cultivar and weed control on weed seed characteristics. Weed Sci 48:431–435
203. Nordby DE, Alderks DL, Nafziger ED (2007) Competitiveness with weeds of soybean cultivars with different maturity and canopy width characteristics. Weed Technol 21:1082–1088

204. Blackshaw RE, Brandt RN (2008) Nitrogen fertilizer rate effects on weed competitiveness is species dependent. Weed Sci 56:743–747
205. Blackshaw RE, Semach G, Janzen HH (2002) Fertilizer application method affects nitrogen uptake in weeds and wheat. Weed Sci 50:634–641
206. Sweeney AE, Renner KA, Laboski C, Davis A (2008) Effect of fertilizer nitrogen on weed emergence and growth. Weed Sci 56:714–721
207. DiTomaso JM (1995) Approaches for improving crop competitiveness through the manipulation of fertilization strategies. Weed Sci 43:491–497
208. Cairns ALP, de Villiers OT (1986) Breaking dormancy of *Avena fatua* L. seed by treatment with ammonia. Weed Res 26:191–198
209. Sardi K, Beres I (1996) Effects of fertilizer salts on the germination of corn, winter wheat, and their common weed species. Commun Soil Sci Plant Anal 27:1227–1235
210. Qasem JR (1992) Nutrient accumulation by weeds and their associated vegetable crops. J Hortic Sci 67:189–195
211. Blackshaw RE, Brandt RN, Janzen HH, Entz T, Grant CA, Derksen DA (2003) Differential response of weed species to added nitrogen. Weed Sci 51:532–539
212. Ross DM, Van Acker RC (2005) Effect of nitrogen fertilizer and landscape position on wild oat (*Avena fatua*) interference in spring wheat. Weed Sci 53:869–876
213. Kirkland KJ, Beckie HJ (1998) Contribution of nitrogen fertilizer placement to weed management in spring wheat (*Triticum aestivum*). Weed Technol 12:507–514
214. Blackshaw RE, Semach GP, Li X, O'Donovan JT, Harker KN (2000) Tillage, fertiliser and glyphosate timing effects on foxtail barley (*Hordeum jubatum*) management in wheat. Can J Plant Sci 80:655–660
215. Mesbah AO, Miller SD (1999) Fertilizer placement affects jointed goatgrass (*Aegilops cylindrica*) competition in winter wheat (*Triticum aestivum*). Weed Technol 13:374–377
216. Rasmussen PE (1995) Effects of fertilizer and stubble burning on downy brome competition in winter wheat. Commun Soil Sci Plant Anal 26:951–960
217. O'Donovan JT, McAndrew DW, Thomas AG (1997) Tillage and nitrogen influence weed population dynamics in barley (*Hordeum vulgare*). Weed Technol 11:502–509
218. Blackshaw RE, Molnar LJ, Janzen HH (2004) Nitrogen fertilizer timing and application method affect weed growth and competition with spring wheat. Weed Sci 52:614–622
219. Blackshaw RE, Molnar LJ (2009) Phosphorus fertilizer application method affects weed growth and competition with wheat. Weed Sci 57:311–318
220. Santos BM, Dusky JA, Stall WM, Bewick TA, Shilling DG (2004) Influence of method of phosphorus application on smooth pigweed (*Amaranthus hybridus*) and common purslane (*Portulaca oleracea*) interference in lettuce. Weed Sci 52:797–801
221. Shrefler JW, Dusky JA, Shilling DG, Brecke BJ, Sanchez CA (1994) Effects of phosphorus fertility on competition between lettuce (*Lactuca sativa*) and spiny amaranth (*Amaranthus spinosus*). Weed Sci 42:556–560
222. Pacanoski Z (2007) Herbicide use: benefits for society as a whole- a review. Pak J Weed Sci Res 13:135–147
223. Cloutier DC, van der Weide RY, Peruzzi A, Leblanc ML (2007) Mechanical weed management. In: Upadhyaya MK, Blackshaw RE (eds) Non-chemical weed management: principles, concepts and technology. CABI, Oxfordshire, pp 111–134
224. Mahajan G, Brar LS, Sardana V (1999) Effect of tillage and time of sowing on the efficacy of herbicides against Phalaris minor in wheat. In: The Proceeding of 17th Asian Pacific Weed Science Society Conference, The Organisation of the 17th APWSS Conference, Bangkok, Thailand, 22–27 Nov 1999, pp 193–198
225. Teasdale JR, Shelton DR, Sadeghi AM (2003) Influence of hairy vetch residue on atrazine and metolachlor soil solution concentration and weed emergence. Weed Sci 51:628–634
226. Locke MA, Zablotowicz RM, Bauer PJ, Steinriede RW, Gaston LA (2005) Conservation cotton production in the southern United States: herbicide dissipation in soil and cover crops. Weed Sci 53:717–727

227. Moseley CM, Hagood ES (1991) Decreasing rates of nonselective herbicides in double-crop no-till soybeans (*Glycine max*). Weed Technol 5:198–201
228. Ramsdale BK, Messersmith CG (2002) Low-rate split-applied herbicide treatments for wild oat (*Avena fatua*) control in wheat (*Triticum aestivum*). Weed Technol 16:149–155
229. Mathers HM, Case LT (2010) Microencapsulated herbicide-treated bark mulches for nursery container weed control. Weed Technol 24:529–537
230. Dailey OD Jr, Dowler CC, Mullinix BG Jr (1993) Polymeric microcapsules of the herbicides atrazine and metribuzin: preparation and evaluation of controlled-release properties. J Agric Food Chem 41:1517–1522
231. Fleming GF, Wax LM, Simmons W, Felsot AS (1992) Movement of alachlor and metribuzin from controlled release formulations in a sandy soil. Weed Sci 40:606–613
232. Monaco TJ, Weller SC, Ashton FM (2002) Weed science: principles and practices, 4th edn. Wiley, New York
233. Basham GW, Lavy TL (1987) Microbial and photolytic dissipation of imazaquin in soil. Weed Sci 35:865–870
234. Locke MA, Bryson CT (1997) Herbicide-soil interactions in reduced tillage and plant residue management systems. Weed Sci 45:307–320
235. Flower KC, Cordingley N, Ward PR, Weeks C (2012) Nitrogen, weed management and economics with cover crops in conservation agriculture in a Mediterranean climate. Field Crop Res 132:63–75
236. Larsbo M, Stenström J, Etana A, Börjesson E, Jarvis NJ (2009) Herbicide sorption, degradation, and leaching in three Swedish soils under long-term conventional and reduced tillage. Soil Tillage Res 105:200–208
237. Sanyal D, Bhowmik PC, Anderson RL, Shrestha A (2008) Revisiting the perspective and progress of integrated weed management. Weed Sci 56:161–167
238. Anderson RL (2000) Cultural systems to aid weed management in semiarid corn (*Zea mays*). Weed Technol 14:630–634
239. Chauhan BS, Gill GS, Preston C (2006) Tillage system effects on weed ecology, herbicide activity and persistence: a review. Aust J Exp Agr 46:1557–1570
240. Chauhan BS (2013) Strategies to manage weedy rice in Asia. Crop Prot 48:51–56

Chapter 6
Integrated Weed Management in Rice

Gulshan Mahajan, Bhagirath S. Chauhan and Vivek Kumar

Introduction

Rice (*Oryza sativa* L.) is a staple food for more than 60 % of the world's population and plays a crucial role in the economic and social stability of the world. Meeting rice demand for the burgeoning population will pose a great challenge in the future as the resources for rice production (land, water, nutrients, and labor) are becoming increasingly scarce. Weeds are one of the most important yield-limiting biological constraints in rice production worldwide. Losses caused by weeds vary in different countries as the nature, extent, and intensity of weed problems depend upon the ecology in which the crop is grown and conditions such as hydrology, land topography, establishment methods, and management practices. The idea of dimension of the problem can be realized with the following examples. Annually, ten million tons of rice is lost in China due to weed competition [1]. In Sri Lanka, weeds accounted for 30–40 % of yield losses [2]. In India, about 33 % of rice yield losses are caused by weeds [3].

Globally, about 10 % of the total production of rice is reduced by weeds [4]. Weeds are the universal pests in rice that exceed tolerable levels in all seasons [5]. Therefore, it is necessary to invest in weed management practices to reduce yield losses caused by weed competition. Total loss caused by weeds includes cost on cultural practices pertaining to weed control and land preparation, weed control

B. S. Chauhan (✉)
Queensland Alliance for Agriculture and Food Innovation (QAAFI), The University of Queensland, Toowoomba 4350, Queensland, Australia
e-mail: b.chauhan@uq.edu.au

G. Mahajan
Department of Plant Breeding and Genetics, Punjab Agricultural University, Ludhiana, Punjab 141004, India

V. Kumar
Department of Plant Breeding and Genetics, College of Agriculture, Punjab Agricultural University, Ludhiana, Punjab 141004, India

B. S. Chauhan, G. Mahajan (eds.), *Recent Advances in Weed Management*,
DOI 10.1007/978-1-4939-1019-9_6,

expenses, and reduction in yield quantity and quality. In several rice-producing countries, losses due to weeds have been estimated by many scientists. In 1980, De Datta reported that in India, losses have been estimated at 10% and in the Philippines, losses were estimated at 11% for the dry season and 13% for the wet season [6]. In 1977, Smith et al. estimated the yield and quality losses at 15% in the USA [7]. The average potential reduction in rice yields due to uncontrolled weeds was estimated by Oerke et al. at 55–60% in Europe [8]. In 2003, Ferrero reported that yield loss can be as high as 90% at a yield of 7–8 t ha^{-1}, if weeds are not controlled [9]. Zoschke, in 1990 [10], and Baltazar and De Datta, in 1992 [11], estimated the losses due to weed competition in rice crops worldwide at 10–15% of the potential production. In 2011, Chauhan et al. reported that in dry-seeded rice (DSR), yield losses due to uncontrolled weeds are between 45 and 75% [12].

Rice is cultivated in various ecosystems: irrigated, shallow lowland, mid-deep lands, deep water, and uplands. In Asia, the major method of rice cultivation is transplanting. Weed control in transplanted rice is followed through a combination of hand weeding and water management. For weed control in puddled transplanted rice (PTR), continuous water ponding is required for the first 15 days after transplanting. But, nowadays, PTR is becoming less common with increasing labor and water shortage problems. As a result, farmers now have shown more interest in DSR, as this method is labor and water friendly. However, direct seeding involves intensive use of herbicides for weed control. Competition occurs between rice and weeds for the limiting resources, such as nutrient, moisture, light, space, etc., because both have similar requirements for growth and development. Most of the weeds in rice are C_4 plants, which have high adaptability and faster growth, and therefore, dominate the crop and reduce the yield potential. In general, the weed problem is greater in DSR as compared to PTR. Weed infestation in PTR may be as high as in DSR, if continuous ponding of water cannot be maintained [13]. Both pre- and postemergence herbicides are required for proper weed control in DSR. Lack of knowledge regarding the proper use of herbicides is another challenge, which also brings environmental pollution. Currently, herbicides with acetolactate synthase (ALS) inhibitors are being used in DSR, which have high selection pressure and they may advent the problem of herbicide-resistant weed species. Therefore, it is a worrying problem for farmers, researchers, and policy makers in rice-producing areas of Asia, America, and Latin America. The only way to avoid these types of problems is the implementation of integrated weed management (IWM) practices in rice, which can go a long way in sustainable rice production. In this chapter, an attempt has been made to provide an idea of the work entailed in the implementation of improved weed management practices and suggest areas for future research on IWM for sustainable rice production.

Establishment Methods and Associated Weed Flora

In rice fields, different types of weed species are found and their density varies in different ecological regions. These include grasses, broadleaf, and sedges [14]. Changes in the weed flora occurred with the use of the direct-seeding method of rice [15]. Flo-

Table 6.1 Effect of crop establishment methods on weed flora [15]

Details	Method of establishment (year)		
	Transplanted (1979)	Dry seeded (1987)	Wet seeded (1989)
Number of species	21	50	57
Number of genera	18	38	44
Number of families	13	22	28

ristic diversity was increased by adopting dry and wet direct-sowing methods of rice instead of the transplanting method and a change in the relative dominance of the major weed species occurred (Table 6.1) [15]. Major weed species (ranked by density) in the transplanting method were *Monochoria vaginalis* (Burm. f.) C. Presl. ex Kunth, *Ludwigia hyssopifolia* (G. Don) Exell, *Fimbristylis miliacea* (L.) Vahl, *Cyperus difformis* L., and *Limnocharis flava* (L.) Buch. In the case of dry seeding, *Echinochloa colona* (L.) Link., *E. crus-galli* (L.) Beauv., *Leptochloa chinensis* (L.) Nees., *Scirpus grossus* L.f., and *F. miliacea* dominated, while in wet seeding, *E. crus-galli, L. chinensis, F. miliacea, Marsilea crenata,* and *M. vaginalis* dominated.

Rainfed Upland Rice

Aerobic soil, ideal temperature, and optimum moisture provide favorable environments for diverse weed flora to flourish in rainfed upland rice. As a result, weeds germinate earlier than rice and competition goes in the favor of weeds. The most common weeds that are prevalent in upland rice are as follows:

1. Grasses: *E. colona, E. crus-galli, Echinochloa glabrescens* Munro ex Hook.f., *Eleusine indica* (L.) *Gaertn., Cynodon dactylon* (L.) Pers., *Digitaria sanguinalis* (L.) Scop., *Dactyloctenium aegyptium* (L.) Willd., and *Setaria glauca* (L.) Beauv.
2. Sedges: *Cyperus rotundus* L. and *Cyperus iria* L.
3. Broad-leaved weeds: *Acanthospermum hispidum* DC., *Amaranthus viridis* L., *Ageratum conyzoides* L., *Cleome viscosa* L., *Celosia argentea* L., *Euphorbia hirta* L., *Phyllanthus niruri* L., *Commelina benghalensis* L., *Alternanthera sessilis* (L.) R. Br. ex R.&S., *Caesulia axillaris* Roxb., *Oldenlandia corymbosa* L., and *Physalis minima* L.

Most of these weeds are C_4 types and are highly competitive because of high drought tolerance, water use, and photosynthetic efficiency compared to rice.

Rainfed Lowland Ecology

Rainfed lowlands comprise shallow (0–25 cm), medium deep (0–50 cm), deep (0–100 cm), and very deep water depths (> 100 cm). Since the conditions of rainfed direct-seeded lowlands in the initial stages are similar to that of uplands, weed flora in both the situations are similar. In transplanted conditions, however, there is predominance of sedges and dicots at the initial stages. However, with gradual accumulation of monsoon rains, aquatic weeds (submerged, emerged, and floating types) predominate irrespective of whether the crop is direct seeded or transplanted. The most common weeds that are prevalent in rainfed lowland rice ecology are as follows:

1. Grasses: *Echinochloa crus-galli, Eleusine indica, Cynodon dactylon, Digitaria sanguinalis, Dactyloctenium aegyptium, Coix lachrymal-jobi* L., *Leersia hexandra* Sw., and *Leptochloa chinensis*.
2. Sedges: *C. iria, C. difformis, Scirpus articulatus* L., *and F. miliacea*.
3. Broad-leaved weeds: *Eclipta prostrata* (L.) L., *Sphenoclea zeylanica* Gaertn., and *Ludwigia perennis* L.

Some of the common aquatic weeds are *Chara zeylanica* Willd., *Eichhornia crassipes* (Mart.) Solms., *Monochoria vaginalis, Nelumbo nucifera* Gaertn., *Nitella* spp., *Ipomoea aquatica* Forsk., *Lemna minor* L., *Pistia stratiotes* L., *Vallisneria spiralis* L., *Limnophila heterophylla* (Roxb.) Benth., *Ottelia alismoides* (L.) Vahl., *Marsilea quadrifolia* L., and *Salvinia molesta* D.S. Mitch. Other specific weeds of this ecology are wild rices: *Oryza nivara* S.&S., *O. rufipogon* Griff., *O. officinalis* Walls ex Watt., and *O. sativa* L.f. spontanea Roschev.

Irrigated Ecology

Weeds that can tolerate water are most prominent in irrigated rice:

1. Grasses: *Echinochloa colona, Echinochloa crus-galli, Eleusine indica, Cynodon dactylon, Digitaria sanguinalis*, and *Dactyloctenium aegyptium. E. colona* requires less moisture than *E. crus-galli* so it is predominant in DSR. Similarly, *L. chinensis, Digitaria sanguinalis*, and *Dactyloctenium aegyptium* predominate in DSR. Weedy rice is also emerging as a major problem in DSR
2. Sedges: *C. rotundus* and *C. iria*
3. Broad-leaved weeds: *E. prostrata, Sphenoclea zeylanica* Gaertn., and *L. perennis*

Weeds that predominate in DSR systems are *Amaranthus viridis, Ageratum conyzoides, Celosia argentea, Euphorbia hirta* L., *Phyllanthus niruri* L., *Commelina benghalensis, Alternanthera sessilis, Caesulia axillaris* Roxb, *Oldenlandia corymbosa* (L.)

In direct-seeded rice, the yield is decreased, quality is deteriorated, and cost of production is increased by weeds as a result of competition for various growth factors, such as nutrient, moisture, light, space, etc. The extent of losses may vary from 10% to complete failure and this variation depends upon cultural methods, rice cultivars, weed species associated and their density, and duration of competition. However, in wet-seeded rice, the yield loss from weeds is less than that in DSR (Fig. 6.1) [14].

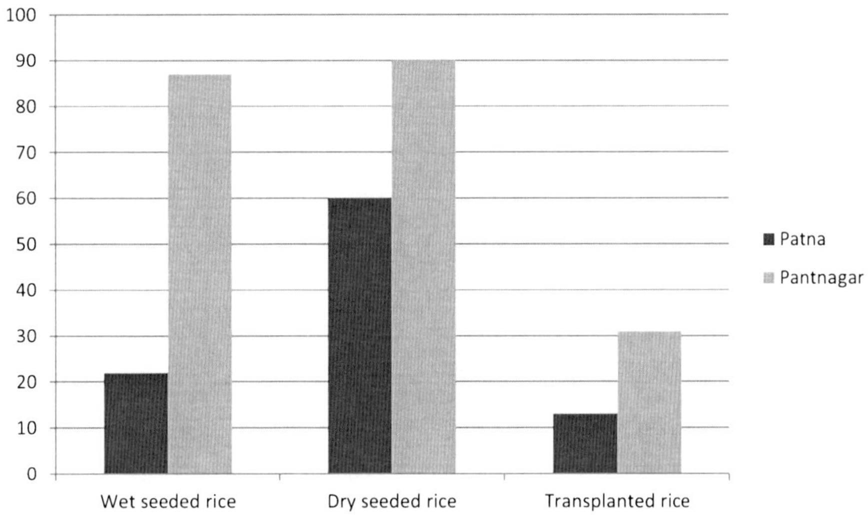

Fig. 6.1 Potential yield loss (%) caused by weeds in different rice cultures at Patna and Pantnagar, India [14]

Weeds reduced rice yield by 12–98 %, depending upon crop establishment methods [16]. Singh et al. reported that yield losses due to weeds were least in transplanted rice but highest in direct-sown rice [17]. Rao et al. reported that yield losses in transplanted rice are less as compared to direct-sown rice when effective weed control options are not applied, because in transplanted rice the initial flush of weeds is controlled by flooding [18]. Besides this, emerging seedlings are less competitive with concurrently emerging weeds in direct-seeded rice [18]. Singh et al. also reported that weed competition is more in direct-sown rice than in transplanted rice [19]. Under the uncontrolled weeding conditions, DSR, wet-seeded rice, and PTR have been reported to reduce the grain yield by 76, 71, and 63 %, respectively [19].

Timsina et al. reported that weed biomass was less in zero-till DSR than conventional till DSR and PTR at 28 days after sowing or transplanting (Table 6.2) [20]. Very limited studies have been conducted on threshold levels of weed species. In PTR, threshold levels of 30 and 20 plants m^{-2} of *C. iria* and *E. crus-galli* caused 7 and 9 % reduction in grain yield of rice, respectively [21].

Factors Influencing Weed Competition and Critical Period

Critical period is the period during which the crop should be kept weed-free after sowing for realizing desired yield levels. The weeds that germinate after this period may not cause much damage. It is necessary to determine the critical periods of weed–crop competition to minimize the labor requirement and maximize economic return. Rice at early stage is most sensitive to competition with weeds and removal

Table 6.2 Effect of crop establishment and weed control methods on weed dry matter at 28 days after sowing/days after transplanting in the aman rice [20]

Crop establishment method	Weed control options	Weed dry matter (g m $^{-2}$)
TPR	W0	8.7
	CW	5.5
	HW	14.9
CTDSR	W0	33.1
	CW	29.9
	HW	22.0
ZTDSR	W0	4.0
	CW	0.7
	HW	5.1
LSD 5 %	10.7	

ZTDSR zero-till dry-seeded rice, *CTDSR* conventional till dry-seeded rice, *TPR* conventional transplanted rice, *W0* no weeding, *CW* one herbicide application + one hand weeding at 28 days after sowing/days after transplanting, *HW* two hand weedings at 28 and 56 days after sowing/days after transplanting

Table 6.3 Estimated yield losses caused by weeds in different methods of rice establishment in India

Methods of rice establishment	Reduction in yield due to weeds (%)	Reference
Upland rice	97	Singh et al. [22]
Upland dry-seeded rice	94	Ladu and Singh [23]
Zero-till dry-seeded rice	98	Singh et al. [17]
Dry-seeded rice	17–26, 10–48, and 34–73	Moorthy and Saha [24]
Wet-seeded rice	85	Singh et al. [17]

of weeds at early stage has direct influence on grain yield (Table 6.3) [17, 22–24]. Anwar et al. reported that weed density and biomass decreased with increasing duration of weed-free periods in rice crop [25]. However, increasing duration of weed interference periods increases the density and biomass of weed (Table 6.4).

Factors, such as crop establishment methods, edaphic factors, crop rotation, type and mode of fertilizer application, rice cultivars, and weed control strategies have great influence on weed abundance in rice fields. These factors are managed to provide competition in the favor of crop. Various studies revealed that the first 30–70 days are critical, depending upon the type of rice cultivar and method of rice establishment. In general, DSR requires an initial weed-free duration of 50–60 days and later crop smothers the weed flora. Losses in grain yield of DSR were more when nitrogen was applied as a basal dose [26]. In PTR, *C. iria* removal at 30–40 days after crop transplanting caused maximum loss in grain yield, indicating the most critical period in PTR for *C. iria*. In another study, Singh et al. recorded a more than 25 % yield loss when *Ischaemum rugosum* Salisb. was allowed to compete

Table 6.4 Density and biomass of weeds in rice as affected by duration of weed competition [25]

Weed competition period	Weed biomass (g m $^{-2}$)	Weed density (no. m $^{-2}$)
Weedy check	432b	521d
Weedy until 14 DAS	188e	950c
Weedy until 28 DAS	239de	1128b
Weedy until 42 DAS	548a	1311a
Weedy until 56 DAS	532a	1033bc
Weed-free until 14 DAS	365c	506d
Weed-free until 28 DAS	286d	421d
Weed-free until 42 DAS	182e	224e
Weed-free until 56 DAS	101f	124e

Data for weedy treatments were taken at the time of weed removal, whereas data for weed-free treatments were taken at the time of rice harvest.
Within a column for each factor, means sharing same alphabets are not significantly different at $p=0.05$ probability level according to least significant difference test.
DAS days after seeding.

for 40 days and concluded that the most critical period of competition was 40–70 days after transplanting for *I. rugosum* [27]. Time of emergence of weed had direct influence on its weed-competitive ability. Weeds emerging at early stages are more competitive than those emerging at later stages [28]. Several scientists in India have conducted studies on the critical period of crop–weed competition in rice (Table 6.5) and they concluded that depending on the method of rice establishment and type of rice cultivar, the critical period may vary from the first 30 to 70 days [23, 27, 29–36].

Principle of Weed Control

Different direct and indirect methods of weed control are used in rice cultivation. Since rice is grown in different types of ecology and weed flora also varies among the ecology, no single method is effective for weed control in rice. Each method has its own pros and cons. Best weed control is achieved with the adoption of multiple weed options simultaneously. Farmers generally select weed control options on the basis of their feasibility and economics. The principle involved in a sustainable weed management approach is to use management options that suit the environment (soil, water, climate, and biota); effective weed management through optimum use of physical, chemical, and biological resources; and reducing weed seed banks on a long-term basis. Therefore, the *mantras* of an integrated weed management system are that it must be effective on a long-term basis, economical, and farmer- and environmentally friendly.

Table 6.5 Critical period of crop-weed competition (CPCWC) for rice under different methods of rice establishment in India

Method of rice establishment	CPCWC	Reference
Transplanted rice	First 20–40 DAT	Mukherjee et al. [29]
Wet-seeded rice	First 15–60 DAS	Mukherjee et al. [29]
Transplanted rice	First 40–70 DAT	Brar et al. [30]
Transplanted rice	Between 4 and 6 weeks after transplanting	Shetty and Gill [31]
Transplanted rice	Between 50 and 70 DAT	Singh et al. [27]
Rainfed direct-seeded rice	0–90 DAS	Arya et al. [32]
Upland direct-seeded rice	First 30 DAS	Tewari and Singh [33], Ladu and Singh [23]
Lowland bunded rice	First 50 days in the monsoon and first 30 days in the summer season	Mohamed Ali and Sankaran [34]
Upland bunded rice	First 60 days in monsoon and 70 days in summer season	Mohamed Ali and Sankaran [34]
Drilled rice	15–45 DAS	Gopal Naidu and Bhan [35]
Dry-seeded rice	15–60 DAS	Singh [36]

DAS days after seeding, *DAT* days after transplanting

Weed Ecology and Biology

For sustainable weed management, knowledge of the behavior of weed species in the region is a must. Sound knowledge—for instance, time of germination of weeds, period of fruit setting, and emission of first vegetative organ—may help in improving weed control. Weed management strategies in the present scenario must be more focused on prevention of the weed seed bank than control of weed species. It is not easy as the weed flora is complex. However, knowledge of appropriate conditions for germination and seed production could prove a suitable guide for farmers for improved weed management. Grass weeds primarily compete for soil water and nutrients, apart from CO_2 and light. Similarly, sedges also pose serious competition for nutrients as their root systems are fibrous. Broadleaved weeds, because of their deep root systems, explore the deeper layer for minerals and exert less competition for nutrients with rice. Rice suffers little competition for light, if any, from *M. vaginalis* (a short-statured weed), whereas *E. crus-galli,* which is tall, poses serious competition to rice for light. Weed distribution in upland rice is highly influenced by management and environmental factors. The emergence pattern of weeds is influenced by soil moisture in the upper layer (0–15 cm). All weeds do not emerge at the same time and emergence may occur in several flushes. Hence, weed control practices adopted at the initial stage of the crop do not control the weeds that emerge late.

Weed Control Methods

Weed control methods can be grouped into cultural, manual, mechanical, and biological techniques.

Preventive Measures

Weed prevention involves all possible measures that restrict the entry and establishment of weeds in an area. The crop seeds used for sowing should be pure, without admixture of weed seeds. For this, it is advisable that certified seeds should be used for sowing. In many countries, weedy or red rice (*Oryza sativa* L.) spread through contamination in rice seeds and now has become a major issue because no selective herbicides work against them. Weed seeds can also be contaminated through tillage and harvest machinery, and therefore, they must be cleaned before operating. Cleaning of bunds and irrigation canal also restricts the movement of weed seeds through irrigation water. Before sowing, mechanical seed separations by dipping the seed in 20 % brine solution help in selection of high-density seeds. Weed seeds that float in brine solution can be separated and removed. Avoid using undecomposable farmyard manure in the field as it contains several weed seeds. In standing crop, where the produce is meant for seeds, there should be thorough rouging.

Stale Seedbed Technique

This technique is very much useful where the weed problem is very severe at the early stages of crop growth, for example, in direct-seeded rice or in rice fields infested with weedy rice. It involves the removal of successive weed flushes before sowing. In this method, weeds are allowed to geminate by providing light irrigation (or after rainfall), and thereafter, chemical (paraquat or glyphosate) or mechanical methods may be used to control the germinated weed seedlings before sowing. Renu et al. concluded that the practice of stale seedbed technique can be used as an efficient tool for the management of weeds in semidry upland rice because this practice can significantly reduce weed density and weed biomass [37]. Significantly higher weed biomass was recorded with normal sowing as compared to stale seedbed with paraquat spray (Table 6.6) [37].

The herbicides should be applied or cultivation should be done when most of the weed seeds in the surface soil have germinated and the weed seedlings have reached the 2–5 leaf stage. Rice crop thus may be sown with minimum soil disturbance after the weeds have been controlled. Stale seedbed technique is very useful in DSR, as it controls the problematic weeds, such as *C. rotundus,*, weedy rice, etc., which are hard to control. The problem of volunteer rice in DSR fields can also be handled with this technique. Although it is very useful for the control of weeds, the practical possibility of this technique depends upon the window period between the subsequent crops

Table 6.6 Weed biomass and grain yield of rice as affected by the stale seedbed method [37]

Treatments	Weed biomass (kg ha^{-1})	Grain yield (kg ha^{-1})
Normal sowing	801a	1745b
Stale seedbed (hoeing)	295b	2531a
Stale seedbed (herbicide)	217b	2567a

Within a column for each factor, means sharing same alphabets are not significantly different at $p = 0.05$

and availability of irrigation/rainfall. Sharma and Pandey showed that stale seedbed technique resulted in significantly higher rice yield than with conventional tillage by better weed management in the rice-wheat cropping system [38].

Tillage Systems and Land Preparation

One deep plowing at the start of the cropping season—for example, during May in north India—helps to bury weed seeds at a depth that prevents germination; but it may stimulate germination of weed seeds in the top soil layer, which can be killed by harrowing before the crop is sown. Land should be properly leveled before sowing for uniform germination and crop stand. Time and level of land preparation greatly influence weed vegetation. Summer plowing brings weed seeds from the subsurface to surface, which are decayed. Some weed seeds from the surface are placed in a deeper layer of soil, which prevents their emergence. The nuts/rhizomes/tubers of perennial weeds are exposed to sun after cutting into pieces, which results in their desiccation. This benefit cannot be taken when field preparation is missed after the onset of monsoon. Poor land preparation enhances weed density and dry matter, making weed menace a challenging task. Well-prepared fields help in reducing weed abundance by providing weed-free seedbed at sowing time in DSR. For uniformity of the crop stand, fields should be laser leveled. Recently, farmers in Southeast Asia (e.g., Cambodia) and South Asia (e.g., India) have shown great interest in laser leveling. Laser leveling not only helps save water and energy but also provides uniform establishment and high weed control and nutrient use efficiency. Rotation of DSR and wet-seeded rice also provides improved weed control in rice.

Tillage can be considered as an important weed management tool, and therefore, any reduction in intensity or frequency of tillage may have an adverse effect on weed management [39]. However, tillage can serve only as a temporary means of weed control because it buries some weed seeds into deep soil layers from where they cannot emerge, but also brings deeply buried seeds to the surface where there is a conducive environment for germination [40]. Auškalnienė and Auškalnis reported that abundance of weed species is affected by different tillage systems and soil depths [41]. In minimum and no-till plots, the effect is more pronounced. The reduced and no-till treatments result in the highest number of weed seeds in the soil layer of 0–5 cm (Fig. 6.2). Besides this, weed seed germination and emergence is affected by seed movement both vertically and horizontally, and by changing the soil environment through tillage [42].

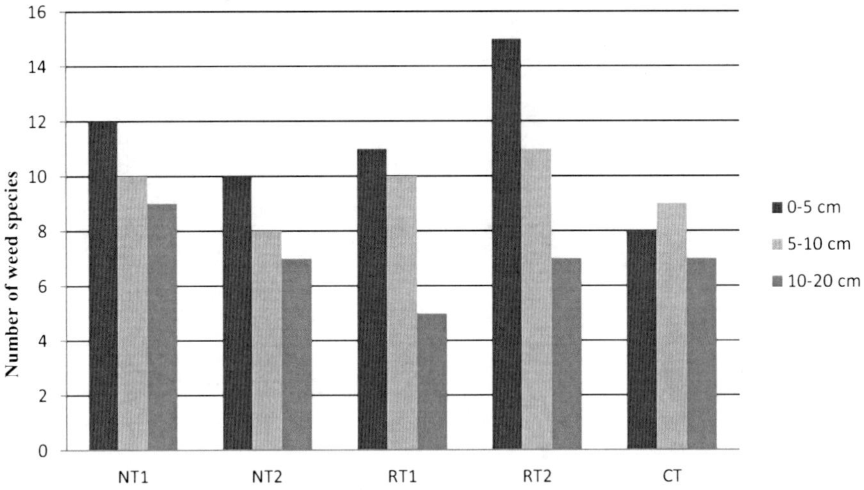

Fig. 6.2 Influence of soil tillage treatments on number of weed species [41] NT1, glyphosate, no-till, rotary drill machine; NT2, glyphosate, no-till, disc drill machine; RT1, reduced tillage 10–12-cm depth, glyphosate treatment; RT2, reduced tillage 10–12-cm depth, disc drill machine; CT, conventional tillage, 20–25-cm depth

Crop Sowing Rate and Geometry

In a crop–weed ecosystem, dense crop plants per unit area help in suppressing weeds by maintaining dominant position over weeds through modification in canopy structure. Narrow row spacing provides a more smothering effect on weeds as less space is available to weeds for flourishing; also a low-light regime is created at ground level by a thick crop canopy. It has been proved that in weed-free conditions, row spacing had no significant effect on grain yield of direct-seeded rice, but under weedy conditions grain yield reduced significantly in the widest spacing [43]. Sunyob et al. reported that plant spacing of rice significantly affected weed biomass at the different stages (Table 6.7) [44]. With the decrease in plant spacing, weed biomass decreased, weed suppression increased, and ultimately, rice yield increased. Phuong et al. reported that higher plant density and narrow row spacing in rice produce favorable conditions for crop to compete with weeds and to produce higher yield [45]. Chauhan and Johnson also reported that total weed biomass is affected by rice row spacing (Table 6.8) [46]. In the row spacing of 30 cm, more weed biomass was produced as compared to the weed biomass produced in 10–20–10-cm paired or 15-cm uniform rows.

Change in plant arrangement with bidirectional sowing also helps in smothering weed flora. Thin crop stand provided favorable environment to weeds, as a result they flourished well. This not only resulted in reduction in grain yield but also caused weed infestation during the following season due to increase in the soil weed seed bank. In Asia, due to small farm size holding, farmers broadcast seeds in dry and wet soil conditions. Although high seeding rate through broadcasting methods favors crop–weed competition toward the crop, sometimes weeds flourish

Table 6.7 Effect of plant spacing on weed biomass (g m^{-2}) in weedy treatment at different growth stages of rice [44]

Plant spacing (cm)	Weed biomass (g m^{-2})		
	25 DAS	50 DAS	At harvest
10 × 10	31c	47c	47d
15 × 15	52b	101b	79c
20 × 20	61ab	115b	101bc
25 × 25	64ab	144a	123ab
30 × 30	75a	157a	128a

Means within a column with the same letter are not significantly different at $p=0.05$ (least significant difference, LSD), *DAS* days after sowing.

Table 6.8 Effect of row spacing on total weed biomass during the wet season of 2009 and dry season of 2010 [46]

Row spacing (cm)	Total weed biomass (g m^{-2})	
	Wet season	Dry season
15	150b	294b
10–20–10	157b	315b
30	203a	390a

Means within a column with the same letter are not significantly different at $p=0.05$ (LSD)

and cause reduction in yield due to patchy crop emergence. It is not possible to do mechanical or manual weeding in that case. Also in broadcast crop, distinction of weedy rice, *E. colona,* and *E. crus-galli* is difficult and therefore such weeds escape manual weeding and cause yield reduction. Therefore, a crop sown in rows has an advantage over broadcast sowing. Row seeding also allows interrow cultivation, which is practically useful for the control of weedy rice. There is no selective weed control herbicide for weedy rice for conventional rice cultivars. In row sowing, weedy rice emerging between the rows can be distinguished and pulled out. The critical period for narrow sown crop is lower than wider sown crop [46]. In some studies, paired-row sowing provided a smothering effect on weeds in DSR; however, the effect depends on the cultivar's traits [47].

In direct-seeded rice, use of seed rate is highly variable at the global level. In countries such as Vietnam, 150–200-kg ha^{-1} seeding rates are being used, while in northwest India, farmers use a rate of 15–25 kg ha^{-1} [48]. Farmers use high seeding rates to compensate for poor quality seeds and to cover the risk of rodent, bird, insect, and nematode attacks [12]. In addition, high seeding rates also suppress weed growth. Moody reported a significant decrease in biomass of broad-leaved, grasses, and sedges as the seeding rate increased from 50 to 250 kg ha^{-1} [49]. Thus, he concluded that high seeding rate can be used as a tool in integrated weed management. Rapid canopy closure occurs with high seeding rates and it helps in reducing weed competition. Thus, grain yield losses due to weeds are reduced by using high seeding rates when other weed management practices are not applied [50]. Different seeding rates affect the weed rating, biomass, and density remarkably and these are decreased gradually with increases in seeding rates [51].

Table 6.9 Effect of different rice varieties on rice–weed competition [57]

Rice variety	Weed-free grain yield (kg ha^{-1})	In competition grain yield (kg ha^{-1})	Reduction (% relative to weed-free plots)	Weed biomass (g m^{-2})	Competitiveness (yield in competition/ yield in weed-free plots
NERICA L 19	3563	3112	13	12.2	0.87
NERICA L20	3611	3176	12	13.1	0.88
NERICA L38	3346	1349	60	36.3	0.40
ROK 10	3175	963	70	44.7	0.30
Butter Cup	2024	1009	50	32.4	0.50
WAS 57-B-B-17-3-3-6-TGR20	3121	2710	13	14.2	0.87
LSD (p=0.05)	542	542	2.2	8.5	0.08

Research evidences revealed that in the presence of weeds, grain yield increased with increases in seed rates. However, high seeding rates may also lead to nitrogen deficiency in crop plants, increase in unproductive tillers, and make the crop prone to lodging and insect and pest attacks [50]. Therefore, an optimum seeding rate should be used, which helps in smothering weeds and provides a suitable environment to the crop.

Weed Competitive Cultivars

Competitiveness is generally increased in tall plants that rapidly establish complete ground covers. But tall cultivars are lower yielding than short cultivars and also prone to lodging. Competitive traits other than height, therefore, are desirable characteristics for breeding weed-competitive cultivars. Quick growing and early canopy-cover-producing cultivars compete better against weeds, such as *Cyperus iria, Cynodon dactylon,* etc. Using competitive varieties to suppress weeds might substantially reduce herbicide use and labor costs, permitting weeds to be controlled with a single herbicide application or hand weeding [52]. Therefore, competitive cultivars can be considered as an important component of integrated weed management strategies [53]. The identification and development of competitive rice varieties can be used as a tool for integrated weed management, because these varieties are quite effective in weed suppression [54] and it is also very easy to adopt this weed control method as compared to others. In West Africa, weed-competitive upland rice varieties New Rice for Africa (NERICA) have been developed to reduce the expenditure on herbicides [55]. Similarly, in Asia, different rice varieties have shown up to 75 % differences in weed suppression [56]. Harding and Jalloh reported that under intense competition, different rice cultivars behave differently for competitiveness against weeds (Table 6.9) [57]. Yield losses due to weeds in rice may range from 12 to 70 %, depending on weed competitiveness provided by the cultivars. Grain yield of rice varieties is reduced due to weed competition and this reduction is positively correlated with weed biomass.

The varieties that are the best competitors have potential in breeding programs to increase their competitiveness without significantly affecting yields since they are both high yielding and weed competitive.

Weed tolerance and weed suppression are the keys for assessing weed competitiveness in different cultivars [58]. Weed-suppressive ability is the ability of plants to suppress the growth of weeds through competition, whereas weed tolerance is the ability to maintain high yields despite weed competition. Both are important, since yield stability and the prevention of weed seed production and subsequent seed bank buildup are desirable in integrated weed management programs [59]. However, weed-suppressive ability is more effective than weed tolerance because weed-suppressive ability reduces weed seed production in current season and also benefits weed management in the future, while weed tolerance benefits only in the current growing season and may result in increased weed pressure from unsuppressed weeds. Therefore, breeding for weed-suppressive ability has been preferred (vs. weed tolerance) [58, 59].

Plant height together with early and rapid growth rates are the important characteristics associated with weed competitiveness [54, 56]. Other characteristics include higher tiller density [60], droopy leaves [61], comparatively high biomass accumulation at the early stage [62], high leaf area index [61] and high specific leaf area [61, 63] during vegetative stage, rapid canopy closure [64], and early vigor [65]. Introducing some of these traits in a variety may result in some yield loss [61, 66–68]. However, the scientific community is of the view that the benefits of having these traits are greater than not having them [53, 56, 62, 65, 69]. Fischer et al. reported that like tall plant-type varieties, semidwarf varieties can also be competitive [60]. Therefore, varieties of intermediate height (between tall traditional and modern semidwarf) may be more desirable, especially for direct seeding [70]. Unlike an initial shock in PTR that causes delay in tillering, this delayed tillering is not a constraint in DSR. Therefore, tillering ability could not be a primary trait for selection [70, 71] in DSR. In fact, Song et al. reported that excessive tillering at an early stage reduced the leaf biomass and photosynthesis at a later stage and caused low yields [71].

Although *Oryza glaberrima,* cultivated rice, has low yield potential, it possesses the trait of droopy leaves with high specific leaf area, and therefore it is effective in weed suppression. Jones et al. suggested that if this trait is restricted to early growth and coupled with the trait of erect leaves with low specific leaf area from *O. sativa,* the cultivars may prove useful for direct seeding [72]. Even though varietal differences in weed competitiveness have been found in rice [65], so far only a limited number of varieties are confirmed to combine superior weed competitiveness with good adaptation to African rice ecosystems. In upland fields in Côte d'Ivoire, *O. glaberrima* varieties IG10 [73], CG14, and CG20 [74] were found to be superior in suppressing weeds but also had low yield potential. On hydromorphic soils in Nigeria, the tall variety OS6 incurred 24 % less yield reductions from weed competition than the semidwarf cultivar ANDNY11 [43]. In Senegal, Haefele et al. reported that lowland rice variety Jaya was weed competitive and high yielding compared to a range of varieties [75]. Jaya incurred lower yield losses due to weeds (<20 %)

compared to popular Sahel 108 (>40%). Superior performance of Jaya under both weedy and weed-free conditions was confirmed in a study carried out in Benin [76]. This study also identified nine superior lowland NERICA varieties as noted previously. Varieties with superior levels of weed competitiveness have been confirmed in other regions, such as Apo and UPLRi-7 in Asia [77], *Oryzica Sabana* 6 in Latin America [53], and M-202 in North America [78]. It was suggested, but not demonstrated, that the weed-suppressive ability of IG10 (*O. glaberrima*) may be, in part, due to allelopathy [73].

A large number of reviews has already been published on crop allelopathy (e.g., [79–83]). Crop allelopathy refers to the process of the release of chemical compounds by living and intact roots of crop plants that affect plants of other species [79, 82]. Allelopathy is suggested by many researchers as one of the potential mechanisms to suppress weeds and as a possible component in IWM (e.g., [59, 73, 79, 80, 82]). Weed suppressiveness and allelopathy may, however, be confounded and coexist in the same variety. Indeed, as Rao et al. suggested, the significance of allelopathy for weed management in rice will remain conjecture until it is clearly demonstrated that differences observed in bioassays also occur in the field [18]. The use of weed-competitive varieties is unlikely to be feasible as a standalone technology but rather it may be a valuable component of integrated measures. In addition to weed competitiveness, suitable varieties should also possess other traits [61], including resistance or tolerance to other biotic and abiotic stresses. Furthermore, a suitable variety needs to be well adapted to the environment and have the specific characteristics desired by farmers and consumers. Crops having strong root systems at early stage compete better with weeds when the leaves are yet to be developed.

Crop Rotation and Green Manuring

Crop rotation can be used as an effective tool in minimizing crop damage from weeds. There are certain weeds that are often associated with specific crops. Rotation procedures recognize these weeds and thus, rotating crops having different life cycle and cultural habits may break the cycle of the weeds. Rotating rice with dryland crops may result in reduced infestation of water-tolerant or water-sustainable weeds. However, the reconversion period greatly influences the composition and growth of weeds. Many studies revealed that response of weeds to intensive cropping may be species specific. With the change in crop, management practices—such as planting time, crop competition, fertility, and herbicide choice—also change and weeds must have to tolerate these practices for their survival. Ultimately, weed management by crop rotation can be successful, depending on the ability to control the weeds in each crop grown in the rotation. Marenco and Santos reported that weed biomass accumulation, weed density, and total weed cover can be reduced by crop rotation (Table 6.10) [84]. Hyacinth bean (*Lablab purpureus* [L.] Sweet) and velvet bean (*Mucuna pruriens* [L.] DC.) rotations reduced weed cover, total weed dry matter accumulation, and weed density by about 70, 80, and 90%, respectively, in comparison to continuous rice [84].

Table 6.10 Effects of rotations on dry matter accumulation, density, and total weed cover (%) at 40 days after crop emergence in the second crop [84]

Treatments	Dry matter accumulation ($g\ m^{-2}$)	Weed density (plants m^{-2})	Total weed cover (%)
Rice–rice	121a	684a	88a
Fallow–rice + N	79a,b	363b	62b
Cowpea–rice	57b	188c	31c
Hyacinth bean–rice	34c	131c	22cd
Velvetbean–rice	17d	30d	13d

Within columns, means followed by the same letters are not significantly different at $p = 0.05$ as determined by the Duncan test

Rotation helps in preventing weed species to become dominant. Crops like maize, pearl millet, and sorghum may drastically reduce weed biomass. The inclusion of these fodder crops in the rice–wheat cropping system could provide satisfactory weed control with fewer amounts of herbicides [85]. Residual weed suppression in the following crop is exhibited by pearl millet. It is also necessary that options of other crops can be successfully implemented in the rice-based system. In a part of shallow lowland of eastern India, jute–rice cropping system is followed. Jute (pre-*kharif*) is considered to be a weed smothering crop and the weeds in the subsequent rice crop (*kharif*) are reduced considerably in this cropping system. In intercropping systems, closure of the crop canopy over weeds decreases sunlight and directly limits weed growth. Practices such as narrow rows that promote early closure of the crop canopy help maximize the effect of crop competition. Weed seed germination is encouraged by a short interval between crop establishment and harvesting in the crop, but it does not allow the weeds to set seed or reproduce vegetatively. Intercropping of upland rice with certain weed-suppressing crops like cowpea and green gram was reported [86]. Once the choice of cropping sequence has been made, then there are opportunities to further improve on the competitive ability of individual crops. In no-till rice systems, Chinese milk vetch (*Astragalus sinicus* L.) in the rice-vetch relay cropping system proved useful [87]. In this method, the vetch provides shading, which reduces the weed stand markedly.

Green manuring can be proved an effective tool to control weeds in rice-based systems. Musthafa and Potty reported that when cowpea is sown as green manure in rice, the dry matter production of weeds is decreased [88]. This reduction in weed growth is due to the successful smothering effect of cowpea on weed flora. The smothering effect on weed plants is caused by the well-developed crop canopy [89]. Weed problem in rice can be effectively controlled by growing *Sesbania* as an intercrop. Compared with control plots, weed density can be decreased by 40 % by intercropping *Sesbania* with rice (Table 6.11) [90].

Table 6.11 Influence of intercropping of *Sesbania* sown at different times on weed density in rice at 60 days after rice sowing (DAS) [90]

Treatments	No. of weeds m^{-2}
Sole crop of rice	68
Sesbania sown at 0 DAS	39
Sesbania sown at 5 DAS	52
Sesbania sown at 10 DAS	65
LSD at p=0.05	11

Mechanical Weed Control

This method is not very common as its scope is limited to favorable physical conditions of the soil and row-seeded crops. In mechanical weeding, different types of weeders or hoes are used either independently or in conjunction with chemical or manual weeding. Mechanical weeding is also done at the same stages recommended for manual weeding. Application of herbicides in rainfed lowlands checks the initial thrust of grassy weeds and sedges and reduces the cost of weed control. Though use of herbicides remarkably reduces the cost of weed control in lowlands, sometimes the farmers face problems regarding their application due to aberrant weather conditions. Continuous rains or dry spells may reduce the effectiveness of the herbicide's activity. Under such situations, use of mechanical methods, such as finger weeder, wheel hoe, or cono weeder, helps to control weeds effectively at early stages in lowland rice fields. Akbar et al. reported that mechanical hoeing resulted in a 72% reduction in total weed density compared with the control [91]. It was also observed in the study that mechanical hoeing increased 25% grain yield over the control, which was even more than grain yield increased by the chemical weed control (Table 6.12) [91].

Herbicides

Herbicides have become an indispensable tool for weed management in rice as they provide superior weed control and their use is more energy and labor efficient than manual or mechanical methods. Farmers consider several factors before choosing a weed management system using herbicides: weed control spectrum, lack of crop injury (or selectivity), cost, and environmental impacts. Farmers rely more on herbicides because cultural and mechanical methods of weed control are time consuming, cumbersome, and laborious. Besides this, weeds have the tendency to regenerate from roots or rhizomes that are left behind. In the present scenario of labor scarcity, herbicide use is more preferable and farmers can easily go for it. Despite the aforementioned advantages of herbicides, several concerns like food safety, ground water and atmospheric contamination, increased weed resistance to herbicides, destruction to beneficial organisms, and concerns about endangered species have also been made with the indiscriminate use of herbicides. Continuous use of the same

Table 6.12 Weed density, weed biomass, and grain yield as influenced by weed control practices in direct sown rice [91]

Treatment	Weed density (no. m^{-2})	Weed biomass (g m^{-2})	Grain yield (t ha^{-1})
Control (weedy check)	34.9a	20.8a	2.5c
Hand pulling (30, 45, and 60 days after sowing)	1.8e	1.0e	3.2a
Mechanical hoeing using kasola (30, 45, and 60 days after sowing)	9.8b	8.8b	3.1a,b
Butachlor (1.8 kg a.i. ha^{-1})	6.8c	5.5c	2.9b
Pendimethalin (1.65 kg a.i. ha^{-1})	6.6c	4.5c,d	2.6c
Pretilachlor (1.25 kg a.i. ha^{-1})	4.5d	2.8d,e	2.9b

Means in column having different letters differ significantly at $p<0.05$

herbicide in the same crop at the same area leads to shift in weed flora. In Korea, Ahn et al. observed that the repeated application of butachlor, thiobencarb, and 2,4-D to rice resulted in predominance of perennial sedges, *Cyperus serotinus,* and *Eleocharis kurogawa* [92]. In the Philippines, application of herbicides over a long time has resulted in *E. crus-galli* and *M. vaginalis* becoming minor weeds and *Scirpus maritimus,* a perennial sedge, becoming increasingly dominant. In India, due to continuous use of butachlor and anilofos in rice, particularly in northwest India, the weed flora is shifting to sedges, such as *Cyperus* sp., *Scirpus* sp., *Fimbristylis* sp., *Eleocharis* sp., etc., and broad-leaved weeds, such as *Caesulia axillaris*.

In puddled transplanted rice, preemergence herbicides (butachlor, thiobencarb, nitrofen, anilofos, oxadiazon, and pendimethalin) are very effective. These preemergence herbicides are applied 4–7 days after transplanting but before weed emergence. Granular formulations are easier to apply but they are more expensive because of transportation costs. Emulsifiable concentrate formulation can be applied by spraying or even by sand mix application. Recently, a number of low-dose sulfonyl herbicides such as metsulfuron, bispyribac, and azimsulfuron have been developed that have a broad spectrum of weed control. In direct-seeded rice, various pre- and postemergence herbicides are recommended in different countries. Some of the preemergence herbicides in direct seeding are pendimethalin, pyrazosulfuron, and oxadiazon. For effective weed control in DSR, sequential application of pre- (2–3 days after sowing) and postemergence herbicides (25–30 days after sowing) proved useful. In some situations, preemergence herbicides, such as oxadiazon, cause phytotoxicity to the crop and the situation becomes worst where

DSR are practiced using low seeding rates [93]. Early postemergence herbicides are preferred in these situations. These herbicides may include 2,4-D, bispyribac-sodium, ethoxysulfuron, azimsulfuron, metsulfuron, penoxsulam, cyhalofop, and their commercial mixtures [50]. Because of complex weed flora in DSR, there is a need to use mixtures of different compatible herbicides. For efficient use of herbicides, the application method should be perfect. Nozzles, spray tips, multiple nozzle booms, pressure regulation, and spray calibration are the essential components of right spray application technology.

Bioherbicides

The rapid pace of environmental changes due to the indiscriminate use of herbicides as witnessed during the past 50 years necessitates the search for an alternative eco-friendly strategy for the management of weeds. Among alternatives, use of microbes and their secondary metabolites are now considered to be the most important, cheaper, and effective eco-friendly means for addressing prevalent weed problems in agriculture as well as other ecosystems. It has reached a point where four strategies, viz. classical, bioherbicidal, phytotoxins, and integrated weed management approaches have been clearly defined. Several microbes and their metabolites have been successfully patented and commercialized in various well-developed countries, including the USA, Canada, and UK. The bioherbicide approach is based on the natural enemies that have an ability to reduce the adverse effects of weeds on crop yield by causing sufficient damage to them. Abundance of the natural enemy in nature is low or there is low abundance at the particular time when it is required to control weeds, and these may be the cause why the potential for damage has not been expressed. In the bioherbicide approach, the abundance of a natural enemy is increased by culturing it in favorable conditions and then these enemies are applied in large amounts to the weed population. Templeton et al. reported that the annual weed species *Aeschynomene virginica* (L.) B.S.P., which is an indigenous weed of the USA, is controlled by the fungus *Colletotrichum gleosporiodes* (Penz.) Sacc. f. sp. *Aeschynomene* (which is also indigenous to the USA) [94]. Many authors have studied the status of bioherbicides and some effective biocontrol agents are listed in Table 6.13 [95–110]. Boyetchko reported that *Cyperus esculentus* can be controlled by *Puccinia canaliculata* (Biosedget) by limiting new tuber formation [111]. Similarly, *Cyperus difformis, C. iria,* and *F. miliacea* are killed by foliar application of conidial suspensions of *Curvularia tuberculata* Jain and *Cyperus oryzae* Bugnicourt [99, 100].

Phatak et al. reported that *C. esculentus* can be controlled by the rust fungus *Puccinia canaliculata* (Schw.) Lagerh [95]. The pathogen is released early in the spring on *C. esculentus,* which reduces the plant population, tuber formation, and flowering. However, the potential of *P. canaliculata* as a mycoherbicide in rice is yet to be determined. Similarly, *Jussiaea decurrens* (Walt.) DC. in rice can be controlled by endemic fungus *Colletotrichum gleosporiodes* f. sp. *jussiaeae* [97].

Table 6.13 List of biocontrol agents reported to be effective in weed management in rice

Weed species	Biocontrol agent	Country	References
Cyperus esculentus L.	*Puccinia canaliculata* (Schw.) Lagerh.	USA	Phatak et al. [95]
Eichhornia crassipes	*Fusarium pallidoroseum*	India	Praveena and Naseema [96]
E. crassipes	*Myrothecium advena*	India	Praveena and Naseema [96]
Ludwigia decurrens Walt.	*Colletotrichum gleosporiodes f. sp. jussiaeae* (C.g.j.)	USA	Boyette et al. [97]
Sphenoclea zeylanica	*Colletotrichum gleosporiodes*	Philippines	Bayot et al. [98]
Cyperus difformis	*Curvularia tuberculata*	Philippines	Luna et al. [99, 100]
C. difformis	*Curvularia oryzae*	Philippines	Luna et al. [99, 100]
Fimbristylis miliacea	*Curvularia tuberculata*	Philippines	Luna et al. [99, 100]
F. miliacea	*Curvularia oryzae*	Philippines	Luna et al. [99, 100]
Aeschynomene virginica (L.) B.S.P.	*Colletotrichum gleosporiodes* (Penz.) Sacc f. sp. *aeschynomene*	USA	Smith [101]
Echinochloa crus-galli	*Cochliobolus lunatus Nelson and Haasis*	Netherlands	Smith [101]
Leptochloa chinensis	*Setosphaeria rostrata*	Vietnam	Chin et al. [102]
Brachiaria platyphylla (Griseb.) Nash	*Bipolaris setariae* (Saw.) Shoem.	North Carolina	Smith [101]
Alismataceae weeds	*Rhynchosporium Alismatis*	Australia	Cother et al. [103]
Alternanthera philoxeroides	*Fusarium* sp.	China	Tan et al. [104]
Cyperus esculentus, C. iria, and C. rotundus	*Dactylaria higginsii*	USA	Kadir and Charudattan [105]
E. crus-galli	*Exserohilum monoceras*	China	Huang et al. [106]
E. crus-galli	*Exserohilum monoceras*	Vietnam	Chin [107]
E. crus-galli	Exserohilum monoceras	Philippines	Zhang and Watson [108]
Hydrilla verticillata	*Plectosporium tabacinum*	USA	Smither-Kopperl et al. [109]
Sagittaria trifolia	*Plectosporium tabacinum*	Korea	Chung et al. [110]

The study reported that the fungus controlled more than 80 % of weed plants in rice after 4 weeks. The fungus *Bipolaris setariae* (Saw.) Shoem. can be used as a tool to control *Brachiaria platyphylla* (Griseb.) Nash, a severe weed of rice that is not controlled effectively by applications of propanil, thiobencarb, or molinate [101]. In the Netherlands, the fungus *Cochliobolus lunatus* Nelson and Haasis kills one to two leaf plants of *E. crus-galli* by inciting leaf necrosis [101].

Integrated Weed Management

Effective and sustainable weed management involves combined use of preventive, cultural, mechanical, chemical, and biological weed control techniques in an effective and economical way. Intensive puddling combined with shallow depth of submergence provided good weed control in transplanted rice [112]. Brar and Walia revealed that high nitrogen rate (180 kg ha^{-1}) along with plant density of 44 plants m^{-2} provided superior weed control [113]. In rainfed upland rice, combined use of good land preparation (two plowings 15 days before sowing and two plowings at sowing) and timely sowing (last week of June) markedly reduced the infestation of all types of weeds [114]. Plowing the land twice in the off-season followed by two hand weedings or growing of *Sesbania* in the off-season provided effective weed control in the rice–rice cropping system [115]. In rice nursery, use of herbicides such as pretilachlor plus safener either alone or in combination with hand weedings results in healthy rice seedlings for transplanting [116].

In wet-seeded rice, *Sesbania* intercropping or azolla dual cropping combined with pretilachlor plus safener (400 g ai ha^{-1}) helped in significant decrease in weed biomass [117]. In rice–green gram intercropping, preemergence spray of pendimethalin 1 kg ai ha^{-1} plus one hand weeding at 25 days after seeding (DAS) provided effective weed control and caused significant improvement in yield in both the crops [118]. High rice yield and superior weed control obtained when rice cultivars "Gautam" (high yielder) and "Prabhat" (better weed competitor) coupled with the application of butachlor at 1.5 kg ai ha^{-1} as preemergence followed by 2,4-D at 0.5 kg ha^{-1} as postemergence [119]. In DSR, combined use of 100 kg ha^{-1} of seed rate and oxyfluorfen 0.25 kg ha^{-1} (3 days after sowing) plus *halod* increased the competitiveness of the crop against weeds [120]. Aulakh and Mehra recorded effective control of *L. chinensis* with increased crop density from 22 to 44 plants m^{-2} coupled with pyrazosulfuron 0.015 kg ha^{-1} [121]. In another study, Singh et al. recorded superior weed control and high yield with combined use of preemergence spray of pendimethalin 1.0 kg ha^{-1} and farm waste as mulch (7.5 t ha^{-1}) plus one hand weeding at 45 days after sowing [122]. In zero-till DSR, preemergence application of pendimethalin supplemented with two hand weedings are needed to reduce weed biomass [19]. Rice in zero- and conventional-tilled conditions provided similar yield when butachlor was supplemented with hand weedings [123]. Sharma and Singh reported that among different weed control treatments, the integrated weed management, including crisscross sowing plus one hand weeding plus herbicide application and one hand weeding plus herbicide application, provided better results than the results obtained by the use of only one weed control method, i.e., two hand weedings (Table 6.14) [124].

Role of Biotechnology in Integrated Weed Management

There are three main ways in which biotechnology can be used in weed management: (1) by using genetically engineering crops with genes conferring resistance to herbicides, (2) by increasing crop competitiveness with weeds using exogenous genes, and (3) by cultivating and modifying biocontrol agents biotechnologically [125]. Competitiveness of crops with weeds can be increased by many different approaches, such as engineering genes into the crops to produce natural allelochemicals, enhanced nutrient uptake, and increased growth rate or habit. Some plant

Table 6.14 Influence of integrated weed management practices on the grain yield and biological yield of the direct sown non-puddled rice [124]

Weed control measures	Grain yield (t ha^{-1})	Biological yield (t ha^{-1})
Weedy	0.7	7.7
Two hand weedings	5.0	12
Herbicide + one hand weeding	5.3	12.6
Crisscross sowing + one hand weeding	3.8	11.5
Crisscross sowing + one hand weeding + herbicide	5.5	13.3
LSD at 5 %	0.1	2.8

species are known to produce allelochemicals that help in competition. Similarly, crop competitiveness is also increased by nutrient uptake. Some weeds are much more efficient at utilizing nitrogen, phosphorus, and other nutrients [126]. If genes of these weeds are introduced into crops, it would facilitate better fertilizer utilization, especially of nitrogen-, phosphorus-, and iron-containing compounds; thus, rice could be a better competitor against weeds. Herbicide-resistant weeds in rice fields have been evolved in some countries by repeated use of the same herbicides year after year. So, biotechnology plays a vital role in producing genetically engineered plants for herbicide tolerance. Herbicide-resistant crops can be used to control weeds that proliferate in conservation (minimum) tillage systems, for example, perennial weeds such as *Cyperus* spp. In addition, there are some weeds that cannot be controlled by herbicides or there are no readily usable selective herbicides for their control. The parasitic broomrapes and witchweeds are examples of such weeds. Therefore, herbicide-resistant crops can be a useful tool to overcome such problems. Besides this, adoption of herbicide-resistant rice may also solve the problem of weed management in DSR [127], and herbicide-resistant crops should be a part of integrated weed management systems for betterment of our agricultural ecosystems.

Three herbicide-resistant systems have been developed for rice: imidazolinone-, glufosinate-, and glyphosate-resistant varieties [128]. Transgenic technologies have been used to develop glufosinate- and glyphosate-resistant rice varieties, having resistance to broad-spectrum, nonselective herbicides. However, imidazolinone-resistant rice was developed through chemically induced seed mutagenesis and conventional breeding. This conveys resistance to the imidazolinone group of herbicides [128]. Herbicide-resistant rice may facilitate adoption of resource conservation technologies by improving weed management options. By introducing herbicide-resistant rice, currently used herbicides can be substituted with new ones that are more efficient and eco-friendly. Besides this, herbicide-resistant rice can be used to control those weeds that have already developed resistance to current herbicides. Thus, herbicide-resistant technology improves the weed management strategies in rice [127]. Gianessi et al. reported that a major benefit of herbicide-tolerant rice in the European Union is a reduction in overall herbicide use [129]. Since glufosinate

Table 6.15 Effect of glyphosate-tolerant rice on herbicide use [9]

Coun-tries	Acreage (000 ha)	Total herbicide use (million kg)			Rate (kg ai/ha)	
		Current	Biotech	Change	Current	Biotech
Greece	20	290	38	−252	14.50	1.92
Italy	218	1853	418	−1435	8.50	1.92
Portugal	25	390	48	−342	15.62	1.92
Spain	113	1765	217	−1548	15.62	1.92
Total	395	4298	721	−3577		

and glyphosate have broad-spectrum activity, there is no need of additional herbicides. Ferrero reported that the use of herbicides in rice production can be decreased by 83 % when two applications of glyphosate are substituted for the current herbicide use (Table 6.15) [9].

Conclusion and Future Research

Weed infestation is a major problem to cultivation of rice, especially direct-seeded rice. Changing from transplanting to direct seeding in rice establishment removes the suppressive advantage of standing water and thus the composition of the weed flora is changed. There are several weed management strategies for direct-seeded rice systems; but the use of any single strategy cannot provide effective, season-long, and sustainable weed control as different weeds have different growth habits [50]. Therefore, based on the available resources and kind of establishment systems, combinations of as many strategies (cultural, chemical, and prevention measures) as possible would control weeds more effectively than with the use of one weed control strategy. In the future, the area under direct-sown rice systems is expected to increase because of shortages in supply of labor and water. However, in direct-seeded rice systems, weeds are the major problem. So, integrated use of different weed management strategies are needed to achieve effective, long-term, and sustainable weed control in direct-seeded systems.

Although the rice-based systems have been benefitted by herbicide-based weed management systems in many ways, the heavy use of herbicides creates an environment favorable for herbicide-resistant weeds, off-site movement of herbicides, and change in weed flora. Rice producers have challenges in using herbicides and other inputs in such a way that prevents adapted species from reaching troublesome proportions. The primary focus of integrated weed management should be on practices that adversely affect the weed propagule production, survival, and the propagule–seedling transition within the agroecosystem. The complex weed problems can be solved by involving weed community analysis; system analysis; weeds' ecophysiology, molecular biology, and genetics; herbicide resistance; assessment of pre- and post-control shifts in weed communities; issues related to transgenic plants; potential benefits of weeds; and environmental issues. To popularize effective and economical options of weed control, that is, integrated weed management strategies to farming community, closer linkage between research and extension is needed. This is a big challenge for weed scientists to develop integrated weed management systems that are innovative, effective, economical, and environmentally safe for current and future cropping systems and which can bring a more diverse and integrated approach to weed management.

References

1. Ze PZ (2001) Weed management in rice in China. Summary presented at FAO workshop on *Echinochloa* spp. control, Beijing, China, 27 May
2. Abeysekera SK (2001) Management of *Echinochloa* spp. in rice in Sri Lanka. Paper presented at the FAO workshop on *Echinochloa* spp. control, Beijing, China. p 3
3. Mukherjee D (2004) Weed management in rice. Agric Today 11:26–27
4. Oerke EC, Dehne HW (2004) Safeguarding production losses in major crops and the role of crop protection. Crop Prot 23:275–285
5. Moody K, Cordova VG (1985) Wet-seeded rice. In: Women in rice farming. International Rice Research Institute. Cower Publishing, England, pp 467–180
6. De Datta SK (1980) Weed control in rice in South and Southeast Asia. Food and Fertilizer Technology Center Ext Bull 156, Taipei City, Taiwan. pp 24
7. Smith RJ Jr, Flinchum WT, Seaman DE (1977) Weed control in US rice production. US Department Agriculture Handbook 497, US Gov Printing Office, Washington, DC, p 78
8. Oerke EC, Dehne HW, Schnbeck F, Weber A (1994) Crop production and crop protection: estimated losses in major food and cash crops. Elsevier, Amsterdam
9. Ferrero A (2003) Weedy rice, biological features and control. In: Labrada R (ed) Weed management for developing countries. Addendum 1. FAO Plant Production and Protection Paper No. 120:89–107
10. Zoschke A (1990) Yield loss in tropical rice as influenced by the competition of weed flora and the timing of its elimination. In: Grayson BT, Green MB, Copping LG (eds) Pest management in rice. Elsevier, London, pp 301–313
11. Baltazar AM, De Datta SK (1992) Weed management in rice. Weed Abstr 41:495–507
12. Chauhan BS, Singh VP, Kumar A, Johnson DE (2011) Relations of rice seeding rates to crop and weed growth in aerobic rice. Field Crops Res 21:105–115
13. Mukherjee D (2006) Weed management strategy in rice—a review. Agric Rev 27:247–257
14. Singh Y, Singh G (2008) Cropping systems and weed flora of rice and wheat in the Indo Gangetic plains. In: Singh Y, Singh VP, Chauhan B, Orr A, Mortimer AM, Johnson DE, Hardy B (eds) Direct seeding of rice and weed management in irrigated rice wheat cropping system of the Indo Gangetic Plains. Los Banos: International Rice Research Institute and Pantnagar: Directorate of Experiment Station, G B Pant University of Agriculture and Technology, pp 33–43
15. Ho NK (1991) Comparative ecological studies of weed flora in irrigated rice fields in the Muda area. Muda Agricultural Development Authority, Telok Chenga, Alor Setar Kedah, Malaysia, p 97
16. Rao AN (2011) Integrated weed management in rice in India. Rice Knowledge Management Portal (RKMP) Directorate of Rice Research, Rajendranagar, Hyderabad 500030, pp 1–35
17. Singh Y, Singh VP, Singh G, Yadav DS, Sinha RKP, Johnson DE, Mortimer AM (2011) The implications of land preparation, crop establishment method and weed management on rice yield variation in the rice-wheat system in the Indo-Gangetic plains. Field Crops Res 121:64–74
18. Rao AN, Johnson DE, Sivaprasad B, Ladha JK, Mortimer AM (2007) Weed management in direct-seeded rice. Adv Agron 93:155–257
19. Singh S, Singh G, Singh VP, Singh AP (2005) Effect of establishment methods and weed management practices on weeds and rice in rice-wheat cropping system. Indian J Weed Sci 37:51–57

20. Timsina J, Haque A, Chauhan BS, Johnson DE (2010) Impact of tillage and rice establishment methods on rice and weed growth in the rice-maize-mungbean rotation in Northern Bangladesh. Presented at the 28th international rice research conference, 8–12 November 2010, Hanoi, Vietnam OP09: Pest, disease and weed management
21. Singh KP, Angiras NN (2003) Ecophysiological studies of *Cyperus iria* L. in transplanted rice under mid hill conditions of Himachal Pradesh, India. Physiol Mol Biol Plants 9:283–285
22. Singh G, Deka J, Singh D (1988) Response of upland rice to seed rate and butachlor. Indian J Weed Sci 20:23–30
23. Ladu M, Singh MK (2006) Crop-weed competition in upland direct seeded rice under foot hill conditions of Nagaland. Indian J Weed Sci 38:131–132
24. Moorthy BTS, Saha S (2001) Study on competitive stress in terms of yield loss caused by pre- and post-submergence weed flora in rainfed lowland direct seeded rice under different seeding densities. Indian J Weed Sci 33:185–188
25. Anwar MP, Juraimi AS, Samedani B, Puteh A, Man A (2012) Critical period of weed control in aerobic rice. Sci World J 2012, Article ID 603043, p 10
26. Sharma AR (1997) Effect of integrated weed management and nitrogen fertilization on the performance of rice under flood-prone lowland conditions. J Agric Sci 129:409–418
27. Singh T, Kolar JS, Sandhu KS (1991) Critical period of competition between wrinkle grass *(Ischaemum rugosum* Salisb.) and transplanted paddy. Indian J Weed Sci 23:1–5
28. Singh G, Singh BB, Agarwal RL, Nayak R (2000) Effect of herbicidal weed management in direct-seeded rice (*Oryza sativa*) and its residual effect on succeeding lentil (*Lens culinaris*). Indian J Agron 45:470–476
29. Mukherjee PK, Sarkar A, Maity SK (2008) Critical period of crop-weed competition in transplanted and wet-seeded *kharif* rice (*Oryza sativa* L.) under *Terai* conditions. Indian J Weed Sci 40:147–152
30. Brar LS, Kolar JS, Brar LS (1995) Critical period of competition between *Caesulia axillaris* Roxb. and transplanted rice. Indian J Weed Sci 27:154–157
31. Shetty SVR, Gill HS (1974) Critical period of crop-weed competition in rice *(Oryza sativa* L.). Indian J Weed Sci 8:101–107
32. Arya MPS, Singha RV, Singh G (1991) Crop weed competition studies in rainfed rice with special reference to *Oxalis latifolia*. Indian J Weed Sci 23:40–45
33. Tewari AN, Singh RD (1991) Crop weed competition in upland direct seeded rainfed rice. Indian J Weed Sci 22:51–52
34. Mohamed Ali A, Sankaran S (1984) Crop-weed competition in direct seeded low land and upland bunded rice. Indian J Weed Sci 16:90–96
35. Gopal Naidu N, Bhan VM (1980) Effect of different groups of weeds and periods of weed free maintenance on the grain yields of drilled rice. Indian J Weed Sci 12:151–157
36. Singh (2008) Integrated weed management in direct seeded rice. In: Singh Y, Singh VP, Chauhan B, Orr A, Mortimer AM, Johnson DE, Hardy B (eds) Direct seeding of rice and weed management in irrigated rice wheat cropping system of the Indo Gangetic Plains. Los Banos (Phillipines): International Rice Research Institute and Pantnagar (India): Directorate of Experiment Station, G B Pant University of Agriculture and Technology, pp 161–175
37. Renu S, George TC, Abraham CT (2007) Stale seedbed: a technique for non-chemical weed control for direct seeded upland rice. In: Proceedings of 19th Kerala Science Congress, 29–31 January, Kannur (Oral presentation)
38. Sharma SK, Pandey DK (2001) Productivity and profitability of nonpuddled direct seeded rice-wheat system as influenced by rice varieties and weed control measures. Annual Report, PDCSR, Modipuram, India, pp 27–28
39. Shrestha A, Lanini T, Mitchell J, Wright S, Vargas R, Tulare UCCE, County M (2005) An update of weed management issues in conservation tillage systems. California Weed Science Society Proceedings, pp 58–63
40. Chauhan BS, Johnson DE (2010) The role of seed ecology in improving weed management strategies in the tropics. Adv Agron 105:221–262

41. Auškalnienė O, Auškalnis A (2009) The influence of tillage system on diversities of soil weed seed bank. Agron Res 7:156–161
42. Clements DR, Benoit DL, Murphy SD, Swanton CJ (1996) Tillage effects on weed seed return and seedbank composition. Weed Sci 44:314–322
43. Akobundu IO, Ahissou A (1985) Effect of interrow spacing and weeding frequency on the performance of selected rice cultivars on hydromorphic soils of West Africa. Crop Prot 4:71–76
44. Sunyob NB, Juraimi AS, Rahman MM, Anwar MP, Man A, Selamat A (2012) Planting geometry and spacing influence weed competitiveness of aerobic rice. J Food Agric Environ 10:330–336
45. Phuong LT, Denich M, Vlek PLG, Balasubramanian V (2005) Suppressing weeds in direct seeded lowland rice: effects of methods and rates of seeding. J Agron Crop Sci 191:185–194
46. Chauhan BS, Johnson DE (2011) Row spacing and weed control timing affect yield of aerobic rice. Field Crops Res 121:226–231
47. Mahajan G, Chauhan BS (2011) Effects of planting pattern and cultivar on weed and crop growth in aerobic rice system. Weed Technol 25:521–525
48. Mahajan G, Gill MS, Singh K (2010) Optimizing seed rate for weed suppression and higher yield in aerobic direct seeded rice in North Western Indo-Gangetic Plains. J New Seeds 11:225–238
49. Moody K (1977) Weed control in rice. Lecture note no. 3. In: 5th Southeast Asian Regional Center for Tropical Biology (BIOTROP). Weed science training course, 14 November-23 December 1977, Rubber Research Institute, Kuala Lumpur, Malaysia, pp 374–424
50. Chauhan BS (2012) Weed ecology and weed management strategies for dry-seeded rice in Asia. Weed Technol 26:1–13
51. Anwar P, Shukor AJ, Puteh A, Selamat A, Man A, Hakim A (2011) Seeding method and rate influence on weed suppression in aerobic rice. African J Biotech 10:15259–15271
52. Mahajan G, Chauhan BS (2013) The role of cultivars in managing weeds in dry-seeded rice production systems. Crop Prot 49:52–57
53. Fischer AJ, Ramirez HV, Gibson KD, Da Silveira PB (2001) Competitiveness of semi dwarf upland rice cultivars against palisadegrass (*Brachiaria brizantha*) and signalgrass (*B. decumbens*). Agron J 93:967–973
54. Caton BP, Cope EA, Mortimer M (2003) Growth traits of diverse rice cultivars under severe competition: implication for screening for competitiveness. Field Crops Res 83:157–172
55. Harding SS, Jalloh AB (2011) Evaluation of the relative weed competitiveness of upland rice varieties in Sierra Leone. African J Plant Sci 5:396–400
56. Garrity DP, Movillon M, Moody K (1992) Differential weed suppression ability in upland rice cultivars. Agron J 84:586–591
57. Harding SS, Jalloh AB (2013) Evaluation of the relative weed competitiveness of some lowland rice varieties in Sierra Leone. Am J Exp Agric 3:252–261
58. Jannink JL, Orf JH, Jordan NR, Shaw RG (2000) Index selection for weed suppressive ability in soybean. Crop Sci 40:1087–1094
59. Jordan N (1993) Prospects for weed control through crop interference. Ecol App 3:84–91
60. Fischer AJ, Ramirez HV, Lozano J (1997) Suppression of junglerice (*Echinochloa colona* (L.) Link) by irrigated rice cultivars in Latin. Am Agron J 89:516–552
61. Dingkuhn M, Johnson DE, Sow A, Audebert AY (1999) Relationships between upland rice canopy characteristics and weed competitiveness. Field Crops Res 61:79–95
62. Ni H, Moody K, Robles RP, Paller EC, Lales JS (2000) *Oryza sativa* plant traits conferring competitive ability against weeds. Weed Sci 48:200–204
63. Audebert A, Dingkuhn M, Jones MP, Johnson DE (1999) Physiological mechanisms for vegetative vigor of interspecific upland rice-implications for weed competitiveness. Proceedings of the international symposium on world food security, pp 300–301, Kyoto
64. Lotz LAP, Wallinga J, Kropff MJ (1995) Crop-weed interactions: quantification and prediction. In: Glen DM, Greaves MP, Anderson HM (eds) Ecology and integrated farming systems. Wiley, Chichester, pp 31–47

65. Zhao DL, Atlin GN, Bastiaans L, Spiertz JHJ (2006) Cultivar weed competitiveness in aerobic rice: heritability, correlated traits and the potential for indirect selection of weed free environments. Crop Sci 46:372–380
66. Jennings PR, Aquino RC (1968) Studies on competition in rice: III. The mechanism of competition among phenotypes. Evolution 22:529–542
67. Kawano K, Gonzalez H, Lucena M (1974) Intraspecific competition with weeds, and spacing response in rice. Crop Sci 14:814–845
68. Pérez de Vida FB, Laca EA, Mackill DJ, Ferna´ndez GM, Fischer AJ (2006) Relating rice traits to weed competitiveness and yield: a path analysis. Weed Sci 54:1122–1131
69. Gibson KD, Fischer AJ, Foin TC, Hill JE (2003) Crop traits related to weed suppression in water-seeded rice (*Oryza sativa* L.). Weed Sci 51:87–93
70. Fukai S (2002) Rice cultivar requirement for direct-seeding in rainfed lowlands. In: Pandey S, Mortimer M, Wade L, Tuong TP, Lopez K, Hardy B (eds) Direct seeding: research strategies and opportunities. Proceedings of the international workshop on direct seeding in Asian rice systems, Strategic research issues and opportunities, pp 15–39
71. Song C, Sheng-guan C, Xin C, Guo-ping Z (2009) Genotypic differences in growth and physiological responses to transplanting and direct seeding cultivation in rice. Rice Sci 16:143–150
72. Jones MP, Mande S, Aluko K (1997) Diversity and potential of *Oryza glaberrima* (Steud.) in upland rice breeding. Breed Sci 47:395–398
73. Fofana B, Rauber R (2000) Weed suppression ability of upland rice under low input conditions in West Africa. Weed Res 40:271–280
74. Jones MP, Johnson DE, Kouper T (1996) Selection for weed competitiveness in upland rice. Int Rice Res Notes 21:32–33
75. Haefele SM, Johnson DE, Mbodj DM, Wopereis MCS, Miezan KM (2004) Field screening of diverse rice genotypes for weed competitiveness in irrigated lowland ecosystems. Field Crops Res 88:39–56
76. Rodenburg J, Johnson DE (2009) Weed management in rice-based cropping systems in Africa. Adv Agron 103:149–218
77. Zhao DL, Bastiaans L, Atlin GN, Spiertz JHJ (2007) Interaction of genotype management on vegetative growth and weed suppression of aerobic rice. Field Crops Res 100:327–340
78. Gibson KD, Hill JE, Foin TC, Caton BP, Fischer AJ (2001) Water-seeded rice cultivars differ in ability to interfere with watergrass. Agron J 93:326–332
79. Belz RG (2007) Allelopathy in crop/weed interactions e an update. Pest Manage Sci 63:308–326
80. Olofsdotter M, Jensen LB, Courtois B (2002) Improving crop competitive ability using allelopathy: an example from rice. Plant Breed 121:1–9
81. Singh HP, Batish DR, Kohli RK (2003) Allelopathic interactions and allelochemicals: new possibilities for sustainable weed management. Crit Rev Plant Sci 22:239–311
82. Weston LA, Duke SO (2003) Weed and crop allelopathy. Crit Rev Plant Sci 22:367–389
83. Xuan TD, Shinkichi T, Khanh TD, Min CI (2005) Biological control of weeds and plant pathogens in paddy rice by exploiting plant allelopathy: an overview. Crop Prot 24:197–206
84. Marenco RA, Santos AMB (1999) Crop rotation reduces weed competition and increases chlorophyll concentration and yield of rice. Pesq agropec bras 34:1881–1887
85. Narwal SS (2000) Weed management in rice: wheat rotation by allelopathy. Critical Rev Plant Sci 19:249–266
86. Hussain A, Gogoi AK (1996) Non-chemical weed management in upland direct seeded rice. Indian J Weed Sci 28:89–90
87. Young Son C, Zhin Ryong C (1999) Vetch effects for the low-input no-till direct-seeding rice-vetch cropping system. Korean J Crop Sci 44:221–224
88. Musthafa K, Potty NN (2001) Effect of *in situ* green manuring on weeds in rice. J Trop Agri 39:172–174
89. Prusty JC, Mishra A, Behera B, Parida AK (1990) Weed competition study in upland rice. Orissa J Agric Res 3:275–276

90. Singh S, Sharma RK, Gupta RK, Singh SS (2008) Changes in rice-wheat production technologies and how rice-wheat became a success story: lessons from zero-tillage wheat. In: Singh Y, Singh VP, Chauhan B, Orr A, Mortimer AM, Johnson DE, Hardy B (eds) Direct seeding of rice and weed management in irrigated rice wheat cropping system of the Indo Gangetic Plains. Los Banos (Philippines): International Rice Research Institute and Pantnagar (India): Directorate of Experiment Station, G B Pant University of Agriculture and Technology, pp 91–106
91. Akbar N, Ehsanullah, Jabran K, Ali MA (2011) Weed management improves yield and quality of direct seeded rice. Aust J Crop Sci 5:688–694
92. Ahn SB, Kim SY, Kim KU (1975) Effect of repeated annual application of preemergence herbicides on paddy field weed population. 287-292 In: Proceedings of 5th Asian-Pacific Weed Science Social Conference Tokyo, Japan
93 .Chauhan BS, Opeña J (2013) Implications of plant geometry and weed control options in designing a low-seeding seed-drill for dry-seeded rice systems. Field Crops Res 144:225–231
94. Templeton GE, TeBeest DO, Smith RJ (1984) Biological weed control in rice with a strain of *Colletotrichum gleosporiodes* (Penz.) Sacc. used as a mycoherbicide. Crop Prot 3:409–422
95. Phatak SC, Callaway MB, Vavrina CS (1987) Biological control and its integration in weed management systems for purple and yellow nutsedge (*Cyprus rotundus* and *C. esculentus*). Weed Technol 1:84–91
96. Praveena R, Naseema A (2003) Effects of two potent biocontrol agents on water hyacinth. Int Rice Res Notes 28:40
97. Boyette CD, Templeton GE, Smith Jr RJ (1979) Control of winged waterprimerose (*Jussiaea decurrens*) and northern jointvetch (*Aeschynomene virginica*) with fungal pathogens. Weed Sci 27:497–501
98. Bayot RG, Watson AK, Moody K (1994) Control of paddy weeds by plant pathogens in the Philippines. In: Shibayama H, Kiritani K, Bay Petersen J (eds) Integrated management of paddy and aquatic weeds in Asia, pp 139–143, FFTC Book Series 45. Food and Fertilizer Technology Center for the Asian and Pacific Region, Taipei
99. Luna LZ, Watson AK, Paulitz TC (2002a) Seedling blights of *Cyperaceae* weeds caused by *Curvularia tuberculata* and *C. oryzae*. Biocontrol Sci Technol 12:165–172
100. Luna LZ, Watson AK, Paulitz TC (2002b) Reaction of rice (*Oryza sativa*) cultivars to penetration and infection of *Curvularia tuberculata* and *C. oryzae*. Plant Dis 86:470–476
101. Smith RJ Jr (1991) Integration of biological control agents with chemical pesticides. In: TeBeest DO (eds) Microbial control of weeds. Chapman and Hall, New York pp 189–208
102. Chin DV, Mai TN, Thi HL (2003) Biological control of *Leptochloa chinensis* (L.) Nees. by using fungus *Setosphaeria rostrata*. In: Annual Workshop of JIRCAS Mekong Delta Project, pp 39–43. Cantho University, Cantho, Vietnam
103. Cother EJ, Jahromi FG, Pitt W, Ash GJ, Lanoiselet V (2002) Development of the mycoherbistat fungus *Rhynchosporium alismatis* for control of *Alismataceae* weeds in rice. In: Hill JE, Hardy B (eds) Proceedings of the second temperate rice conference. International Rice Research Institute, Los Baños, pp 509–513
104. Tan WZ, Li QJ, Qing L (2002) Biological control of alligatorweed (*Alternanthera philoxeroides*) with a *Fusarium* sp. Bio Control 47:463–479
105. Kadir J, Charudattan R (2000) *Dactylaria higginsii*, a fungal bioherbicide agent for purplenutsedge (*Cyperus rotundus*). Bio Control 17:113–124
106. Huang SW, Watson AK, Duan GF, Yu LQ (2001) Preliminary evaluation of potential pathogenic fungi as bioherbicides of barnyardgrass (*Echinochloa crus-galli*) in China. Int Rice Res Notes 26:35–36
107. Chin DV (2001) Biology and management of barnyardgrass, red sprangletop and weedy rice. Weed Biol Manage 1:37–41
108. Zhang WM, Watson AK (1997) Effect of dew period and temperature on the ability of *Exserohilum monoceras* to cause seedling mortality of *Echinochloa* sp. Plant Dis 81:629–634
109. Smither-Kopperl ML, Charudattan R, Berger RD (1998) Plectosporium tabacinum, a pathogen of the invasive aquatic weed *Hydrilla verticillata* in Florida. Plant Dis 83:24–28

110. Chung YR, Ku SJ, Kim HT, Cho KY (1998) Potential of an indigenous fungus, *Plectosporium tabacinum*, as a mycoherbicide for control of arrowhead (*Sagittaria trifolia*). Plant Dis 82:657–660
111. Boyetchko SM (1997) Principles of biological weed control with microorganisms. Hortic Sci 32:201–210
112. Reddy BS, Reddy SR (1999) Effect of soil and water management on weed dynamics in lowland rice. Indian J Weed Sci 31:179–182
113. Brar LS, Walia US (2001) Influence of nitrogen levels and plant densities on the growth and development of weeds in transplanted rice (*Oryza sativa*). Indian J Weed Sci 33:127–131
114. Singh RS, Ghosh DC (1992) Effect of cultural practices on weed management in rainfed upland rice. Trop Pest Manage 38:119–121
115. Gnanavel I, Kathiresan RM (2002) Sustainable weed management in rice-rice cropping system. Indian J Weed Sci 34:192–196
116. Balasubramanian R, Veerabadran V (1998) Herbicidal weed management in lowland rice (*Oryza sativa*) nursery. Indian J Agron 43:437–440
117. Subramanian E, Martin GJ (2006) Effect of chemical, cultural and mechanical methods of weed control on wet seeded rice. Indian J Weed Sci 38:218–220
118. ICAR (2007) Vision 2025. NRCWS Perspective plan. Indian Council of Agriculture Research, New Delhi, India
119. Singh UP, Singh Y, Kumar V (2004) Effect of weed management and cultivars on boro rice (*Oryza sativa* L.) and weeds. Indian J Weed Sci 36:57–59
120. Angiras NN, Sharma V (1998) Integrated weed management studies in direct seeded upland rice (*Oryza sativa*). Indian J Weed Sci 30:32–35
121 .Aulakh CS, Mehra SP (2006) Integrated management of red sprangletop [*Leptochloa chinensis* (L.) Nees] in transplanted rice. Indian J Weed Sci 38:225–229
122. Singh VP, Singh G, Singh RK (2001) Integrated weed management in direct seeded spring sown rice under rainfed low valley situation of Uttaranchal. Indian J Weed Sci 33:63–66
123. Moorthy BTS, Sanjoy S, Poonam A (2002) Effect of different tillage and weed management practices on rainfed lowland rice. Indian J Weed Sci 34:280–281
124. Sharma SK, Singh KK (2008) Production potential of the direct-seeded rice-wheat cropping system. In: Singh Y, Singh VP, Chauhan B, Orr A, Mortimer AM, Johnson DE, Hardy B (eds) Direct seeding of rice and weed management in irrigated rice wheat cropping system of the Indo Gangetic Plains. Los Banos (Philippines): International Rice Research Institute and Pantnagar (India): Directorate of Experiment Station, G B Pant University of Agriculture and Technology, pp 61–73
125. Gressel J (1998) Biotechnology of weed control. In: Altman A (eds) Agricultural biotechnology. Marcel Dekker, New York, pp 295–325
126. Haas H, Streibig JC (1982) Changing patterns of weed distribution as a result of herbicide use and other agronomic factors. In: LeBaron HM, Gressel J (eds) Herbicide resistance in plant. Wiley, New York, pp 57–80
127. Kumar V, Bellinder RR, Gupta RK, Malik RK, Brainard DC (2008) Role of herbicide-resistant rice in promoting resource conservation technologies in rice-wheat cropping systems of India: a review. Crop Prot 27:290–301
128. Gealy DR, Mitten DH, Rutger JN (2003) Gene flow between red rice (*Oryza sativa*) and herbicide-resistant rice (*O. sativa*): implications for weed management. Weed Technol 17:627–645
129. Gianessi L, Sankula S, Reigner N (2003) Plant biotechnology: potential impact for improving pest management in European agriculture. A summary of nine case studies. The natural center for food and agricultural policy, Washington, DC

Chapter 7
Recent Advances in Weed Management in Wheat

Samunder Singh

Introduction

Wheat (*Triticum* sp.) is the second most important cereal crop of the world, grown in several countries in different climatic conditions (sea level to 4000 m above sea level), and meets the nutritional and energy requirements of world's people. It is second only to rice as a vital source of calories in the developing world, has higher protein content than other major cereals, and is the most traded crop of the world. Earlier settlements recorded cultivation of wheat 8000 years ago.

The major breakthrough in wheat production came in the eighteenth century with the introduction of crop rotation and mechanized drill seeding, followed by breeding programs and advances in agronomic practices. In the last century, wheat production was revolutionized with the introduction of Norin10 dwarfing gene identified by Dr. Gonziro Inazuka of Norin Experiment Station, Japan, and incorporated by Dr. Normal E. Borlaug in spring wheat. Another boost came with the modern-day technology of gene pyramiding for resistance to diseases (rust), stress (moisture), and herbicide tolerance (weeds). Demand for wheat will increase by more than 60% by 2050 in the developing countries to meet the dietary requirements of 3.7 billion poor people. By 2030, the world will need 760 m t of wheat, and it is projected that by 2050, the requirement will be 900 m t, whereas the US Department of Agriculture (USDA) production figures for 2022–2023 are expected to be 747 m t with an additional 9-m ha area under wheat compared to 2011–2012 (Table 7.1) [1]. However, in the same period, wheat productivity is expected to decrease by 22% in the former Soviet Union (FSU) countries, and 12% in Australia and also in Morocco, Ukraine, and Turkey. This should be neutralized by increased productivity by 18–22% in North and South African countries, 10–11% in the USA and Mexico, and 8–9% in India and China (Table 7.1) [1]. It is also expected that turbulent weather (climate change) may reduce wheat production by 20–30% in the developing countries. In addition, depletion of natural resources, soil salinization,

S. Singh (✉)
Department of Agronomy, CCS Haryana Agricultural University, Hisar, India
e-mail: Sam4884@gmail.com

B. S. Chauhan, G. Mahajan (eds.), *Recent Advances in Weed Management*,
DOI 10.1007/978-1-4939-1019-9_7,

Table 7.1 Area, production, and productivity in major wheat-producing countries of the world during 2011–2012 and US Department of Agriculture predictions for 2022–2023 [1]

	2011–2012			2022–2023		
	Area (m ha)	Production (m t)	Productivity (t/ha)	Area (m ha)	Production (m t)	Productivity (t/ha)
Argentina	5.170	15.500	2.998	4.102	12.905	3.146
Australia	14.058	29.515	2.100	14.925	27.858	1.867
Brazil	2.170	5.800	2.673	1.852	5.171	2.792
Canada	8.544	25.261	2.957	8.828	26.474	2.999
China	24.270	117.400	4.837	24.327	128.621	5.287
Egypt	1.280	8.400	6.562	1.413	9.544	6.756
EU-27	25.701	137.227	5.339	27.116	149.573	5.516
India	29.400	86.870	2.955	30.985	99.358	3.207
Iran	6.800	13.500	1.985	7.122	15.097	2.120
Iraq	1.587	2.574	1.622	1.773	2.968	1.674
Mexico	0.662	3.628	5.480	0.555	3.433	6.187
Morocco	3.040	5.800	1.908	3.079	5.649	1.835
Other Asia & Ocenia	3.324	5.020	1.510	3.632	6.782	1.867
Other Europe	1.087	4.359	4.010	1.067	4.290	4.021
Other FSU (10)	18.436	36.446	1.977	2.091	33.967	1.625
Other Middle East	1.855	4.300	2.318	1.951	4.614	2.365
Other North Africa	3.022	4.225	1.398	3.129	5.632	1.800
Other S. America	1.545	4.535	2.935	1.630	5.256	3.224
Other Sub-Saharan Africa	1.956	4.136	2.115	2.028	4.543	2.240
Pakistan	8.900	25.000	2.809	9.554	28.240	2.956
Russia	24.900	56.231	2.258	27.345	62.700	2.293
South Africa	0.605	2.005	3.314	0.518	2.093	4.039
Turkey	7.700	18.800	2.442	8.014	19.357	2.415
Ukraine	6.657	22.124	3.323	7.359	23.643	3.213
USA	18.495	54.404	2.942	17.321	56.608	3.268
World	222.11	696.06	3.134	231.36	746.99	3.229

m ha million hectares, *m t* million metric tons, *t/ha* metric tons per hectare

soil fertility, and weed losses due to the evolution of herbicide resistance have the potential to cause food crises of greater proportions.

Weeds not only usurp essential plant nutrients other than moisture but also compete vigorously with wheat to lower its yield by one third; the losses vary from developing to developed countries. To meet the production demand of wheat, weeds need to be effectively controlled. No single approach has been found effective if used repeatedly for a long time, and an integrated approach is required using chemical and nonchemical methods for managing weeds and improving wheat production.

Agronomical Approaches

Planting Methods

A good field preparation enhances wheat seed germination and seedling growth by killing emerged weeds and providing congenial conditions for crop establishment. The conventional tillage (CT) system buries weed seeds in the deeper soil layers, which may take them a longer time to emerge to pose early crop–weed competition. Tillage operations have a significant effect on the vertical distribution of weeds in the wheat fields. Maximum weed seeds emerge from 0.5 to 1 cm, and deeper depths reduce the emergence of most weed species. Singh and Punia [2] reported that increasing the burial depth decreased the emergence of major broadleaf weeds of wheat in north India, viz., *Malva parviflora* and *Rumex dentatus,* but the effect was lower on *Rumex spinosus*. Small-seeded littleseed canarygrass (*Phalaris minor*) emergence was found to be greater at a 0.5–2-cm soil depth [3], and deep placement of seeds by tillage can reduce weed emergence from greater depths.

Weed infestation is generally less under zero tillage (ZT) than CT due to the presence of crop residues on soil surface; it may vary for weed species and growing conditions (soil). ZT in northwest India was found to have significantly more *Rumex* dentatus compared to CT, though the latter favored *P. minor* [4]. On the other hand, in ZT systems, weed seeds present in the top soil zone are forced to germinate with pre-sowing irrigation and can be killed with any contact herbicide; however, in the absence of contact herbicides, the competition from weeds may increase as the weed seeds germinate before/with the crop. Emergence of *P. minor* was found more often in CT than ZT [5]. Mahajan et al. [6] reported that wheat sown with CT resulted in a 25 % reduction in dry matter of *P. minor* as compared to ZT-sown crop. However, the grain yield was statistically similar in both the tillage systems, because the effective tillers were statistically at par in both the tillage systems. Under conservation agriculture (CA) systems, residues maintained on the surface not only help in reducing erosion but also conserve moisture, add organic matter to the soil, and inhibit weed seed germination. Emergence of *P. minor* was reduced from 45 to 75 % (Fig. 7.1) with 6–8 t/ha^{-1} of rice straw [7]. Effect of straw was more significant on *Chenopodium album, Rumex dentatus,* and *Melilotus indica* as 8 t/ha^{-1} rice straw inhibited emergence by 92, 95, and 98 %, respectively, compared to 68 % of *P. minor* (Fig. 7.1). On the other hand, presence of straw on the surface results in decreased herbicide efficacy, particularly for preemergence (PRE) herbicides; however, increased carrier volume has been found to increase efficacy of both water-soluble herbicide (e.g., pyroxasulfone) and non-water-soluble herbicide (e.g., trifluralin) [8].

ZT also has the advantage of early sowing compared to CT, which provides more time for wheat to establish, which smothers late-emerging weeds; but weed management at sowing using burndown or selective herbicides is more important, particularly for emerged broadleaf weeds.

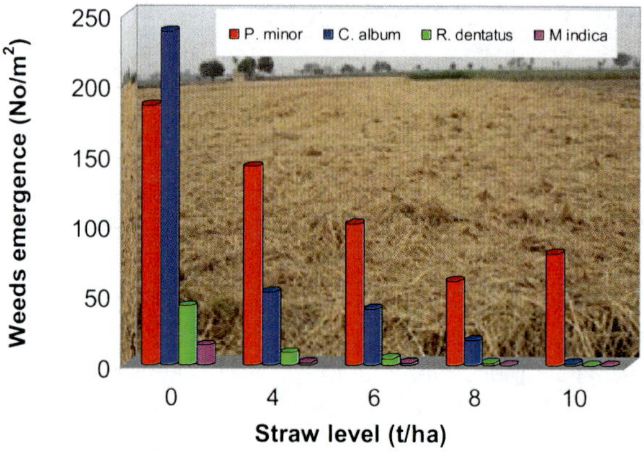

Fig. 7.1 Effect of rice straw on germination of wheat weeds, *Phalaris minor, Chenopodium album, Rumex dentatus*, and *Melilotus indica*. (Data from Kumar et al. [7])

Straw Management

A large number of farmers in the developing countries still have not found an efficient use for crop residues, and a large part of this is burned every year, leading to loss of essential plant nutrients and causing environmental pollution. Crop residues have several uses as mulch, livestock feed, raw material for composting, thatching, energy, biofuel and bio-oil production, biomethanation, gasification, and biochar production. However, a majority of farmers consider crop residues, particularly because of their large amount, as unwanted by-products. India is producing more than 550 m t of crop residues every year; out of this, wheat contributes 22% [9]. Straw burning after rice and wheat harvesting is a big nuisance in mechanized farming in northwest India. Burning 1.41 m^3 of wheat stubbles results in nutrient loss of 17.51, 3.69, and 4.15 kg of N, P, and S, respectively.

Proper straw management will not only improve soil microbial population for increased N mineralization and N fixation but also improve infiltration rate, soil porosity, and water-holding capacity, and will add organic matter to the soil. Since farmers have a short period between the harvest of one crop and planting of the next, lack of heavy machinery for residue management and the notion that straw burning will add nutrients to soil and kill several pests, including weed seeds, should encourage them to burn straw. Stubble burning along narrow windrows has been successfully practiced in several countries for burning weed seeds, until restricted legislatively. Although burning of crop residues chars weed seeds on the surface, it can also break the dormancy of several weeds lying at lower soil depths, which can increase crop–weed competition. Straw burning also has been found to lower the efficacy of several herbicides due to increased adsorption of carbon content (ash), necessitating efficient management of crop residues. ZT drills can successfully be used when stubbles are short and total crop residues of 2–3 t ha^{-1}, for taller plants and higher amount of residues, the Happy and Turbo seeders have been found to work better for wheat planting under Indian conditions. Happy seeder provides an alternative for burning of wheat straw in northwest Indian conditions as it chops and spreads straw, which acts as soil mulch to reduce water losses and adds organic carbon to soil. Harvesting of stubbles by machines and their baling are common practices with heavy machinery in different parts of the world.

Planting Time

Planting time has a significant effect on weed emergence and wheat growth. Winter wheat planted during the second or third week of September in Nebraska (USA) yielded better than during the fourth week of August or the first week of September [10]. This was due to greater weed density when planted before the optimum date and higher vulnerability to crown and root infection. In India, wheat planting in late October had better growth and smothered *P. minor* compared to its late planting at the end of November or early December when low temperatures favored greater *P. minor* emergence [11]. Dhaliwal et al. [12], while analyzing the factors of *P. minor* infestation in Punjab (India), found that 50 % of farmers experienced less population of *P. minor* when they planted wheat in the last week of October compared to mid-November, possibly due to higher temperatures in October—sowing as *P. minor* germinates at a temperature of 17–18°C, which usually prevails in the middle of November.

Manipulation of planting time can provide a significant advantage to wheat, depending upon the competitive weed species as *Avena ludoviciana* has greater emergence in early compared to delayed planted wheat [13]. Emergence time of *Avena ludoviciana* increased from 10 to 16 days when planted on December 30, compared to November 10. Similarly, 82 % lower density of *Avena ludoviciana* was recorded in December 30 sowing compared to November 10; its shoot dry matter accumulation at 45 and 60 days decreased by 65 and 43 %, respectively, when planting was delayed from November 10 to December 30. Under Pakistan conditions, planting of wheat on December 1 provided 38 % higher yields than December 30 planting [14]. Beech and Norman [15] reported that wheat planted in May in the Ord River valley (Australia) had higher plant attributes and yield than late sowing as a result of favorable environmental conditions (24–27 °C). Increased temperature toward maturity (end of July or early August) was the main reason affecting crop yield. Early fall-planted wheat (in eastern Washington, USA) had higher yields than late-planted wheat; in addition to good growth, the early-planted wheat improved erosion control after fallow [16]. Winter wheat planted in Oregon (USA) in the first week of October provided the highest grain yield, whereas delaying planting resulted in a 47 % yield loss. Soil temperature at wheat planting has a significant effect on weed emergence and ensuing competition, whereas temperature at maturity impinges on grain maturity. Wheat planted under ZT in December had higher temperature than CT, which lowers the emergence of *P. minor* (optimum emergence at 15 °C), whereas lower temperature at maturity in April under ZT provided higher translocation of photosynthates from source to sink, resulting in better grain formation and wheat yield (Fig. 7.2).

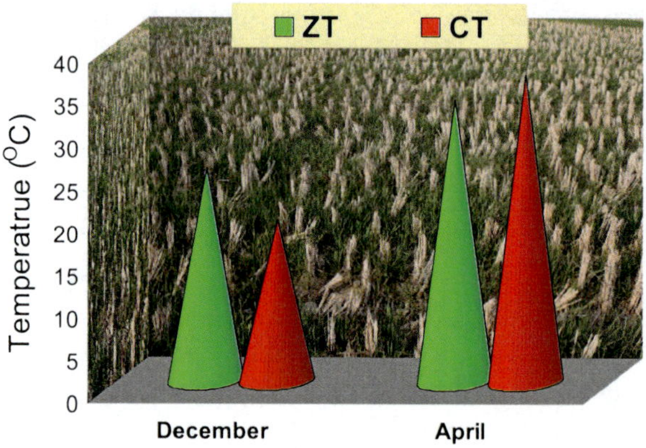

Fig. 7.2 Effect of planting methods on soil temperature at sowing and harvest in wheat

Seed Rate and Planting Methods

Seed rate can be adjusted for improved weed control; higher seed rates are generally used for late planting of wheat to compensate for reduced tillering. In Nebraska (USA), wheat seed rate varies from 45 to 120 kg ha^{-1}, depending upon locations and planting dates. Higher seed rate with narrow spacing has been found to lower weed losses in wheat. Vander Vorst et al. [17] reported that increasing seed rate from 33 to 67 kg ha^{-1} resulted in a 15 % higher wheat yield. The competitive nature of wheat has been found to increase with an increase in seed rate [18]. An increase of the wheat seed rate from 100 to 175 kg ha^{-1} reduced the dry weight of weeds by 29 %. Higher seed rate with cross-row sowing not only reduced weed pressure but also increased herbicide efficacy. Singh and Singh [19] reported that wheat sown by the cross/bidirectional sowing method reduced *P. minor* population by 59.6, 23.4, and 39.0 % and weed dry weight by 59.2, 23.1, and 37.5 % compared to broadcast, closer (15.0 cm) and normal sowing (22.5 cm), respectively. This was due to the smothering effect of crop on weeds in the cross-sowing technique as limited space was available in the cross-sown crop due to a more uniform distribution of crop plants in this technique. Grain yield was significantly higher with cross sowing than other methods. Cross sowing because of lower weed population offered better crop–weed competition in favor of the crop, resulting in 9.8, 17.9, and 25.7 % increases in grain yield over closer (15.0 cm), normal (22.5 cm), and broadcast sowing, respectively. Row orientation is also an important factor influencing light penetration within the crop rows, and hence it affects the growth and development of weeds. Wicks et al. [20] reported that populations of *Eragrostis cilianensis* and *Amaranthus albus* were reduced by 82 % when wheat was planted in a north–south direction rather than an east–west direction due to more crop shading, which reduced weed emergence in the north–south direction. By contrast, Borger et al. [21] reported that east–west-oriented wheat reduced weed biomass by 51 % and increased the wheat grain yield by 24 % compared to north–south plantation. In east–west-planted crops, light interception was 28 % higher in wheat than with a north–south row orientation.

Increased crop density in spring wheat in Denmark had strong and consistent negative effects on weed biomass (>50% reduction) and positive effects on crop biomass and yield [22]. Though a higher seeding rate has the potential to reduce weed competition, it may not be economical until the farmer can compensate for higher seed costs through increased production. Field trials conducted at the university of Maine (USA) by Kolb et al. [23] found that increased wheat density reduced weed density by 64% compared to control and 30% over farmers' practices of 18-cm rows and 400 plants m^{-2}, but also lowered grain protein by 5% compared to standard seeding rates. Doubling the seed rate of wheat varieties resulted in 25% lower dry matter accumulation by *Lolium rigidum* in Australia over the recommended seed rate [24]. Blackshaw et al. [25] showed that under ZT conditions in Canada, increasing the wheat seed rate from 50 to 300 kg ha^{-1} reduced *Erodium cicutarium* (the most competitive weed emerging with wheat) biomass and seed production by 53–95% and the soil seed bank by 79% after 4 consecutive years. Though a higher seed rate (above 50 kg ha^{-1}) had no significant yield benefit under weed-free conditions, under weedy conditions, increasing the seed rate from 50 to 300 kg ha^{-1} resulted in an increased wheat yield from 56 to 498%. This shows that a higher seed rate can also be used in conservation tillage cropping systems.

Fertilizer Application

Application of an optimum amount of fertilizer on soil test basis and at appropriate timing increases the vigor and competitiveness of wheat. Phosphatic fertilizers stimulate broadleaf weeds, whereas nitrogenous fertilizers tend to promote grassy weeds; a balanced amount of fertilizers and their proper application methods and timing can tilt the competition in favor of wheat. Blackshaw et al. [26] found that soil injected with nitrogen resulted in suppressed weed growth, not by reduced uptake by weeds, but due to greater uptake by wheat, which increased its competitiveness. Weed biomass was generally higher with surface broadcast than with either surface pools or soil-injected ammonium nitrate. Kristensen et al. [22] reported that in the presence of weeds, the highest wheat yields were obtained under high-crop density and nitrogen levels.

Under Punjab (India) conditions, application of 175 kg N ha^{-1} reduced the population and dry matter of *P. minor* by 51.3 and 26.5%, respectively, compared to the recommended dose of 125 kg N ha^{-1}, which increased the effective tillers of wheat by 20%, resulting in a 15% higher yield of wheat [27]. Application of 175 kg ha^{-1} of N also resulted in significantly lower leaf area index of *P. minor* compared to 125 kg N ha^{-1}. Placement of fertilizer below the seed at sowing is better than its broadcasting. Soil fertility has a significant effect on weed infestation and crop–weed competition. Singh et al. [28] reported that the occurrence of *Avena ludoviciana* and *P. minor* was reduced from 81 to 30% and 69 to 33%, respectively, from high to low fertility soils of Haryana, India. On the other hand, occurrence of *C. album, Asphodelus tenuifolius,* and *Sonchus arvensis* increased from 72 to 89%,

0 to 52%, and 3 to 22%, respectively, from high- to low-fertility soils. Increasing the application of nitrogen from 30 to 180 kg ha^{-1} decreased the dry matter accumulation of 400 plants m^{-2} of *Lathyrus aphaca* by 45% [29].

Increased nitrogen application is known to increase the ability of cereals to suppress the weeds, but when these are applied uniformly to soil, they may benefit the crops or weeds, depending upon their competitiveness. Kirkland and Beckie [30] studied the effect of methods of fertilizer application in ZT and CT systems in the northern Great Plains (USA) from 1994 to 1996. Tillage system was not found to have a significant effect on weed and crop responses to fertilizer placement; however, broadcast-applied fertilizer over band placement promoted *Avena fatua* and broadleaf weed emergence and growth. Weed densities, biomass, and N uptake averaged 20–40% less, with a 12% higher grain yield of wheat when fertilizer was side-banded compared to broadcast. However, fertilizer application, regardless of method of application, was detrimental to *Setaria viridis* because of enhanced crop competitiveness. Band placement of fertilizer at recommended rates can be an effective cultural practice for managing weeds in ZT and CT wheat-cropping systems in semiarid to subhumid regions of the northern Great Plains, but it is not reliable when used as the sole method of weed management.

Irrigation Management

Irrigation scheduling (frequency and amount) not only decides wheat and weed germination but also sets the competition. Weeds are highly responsive to irrigation and corner a greater amount of nutrients and moisture, depriving them to wheat. Irrigation management has a direct or an indirect effect on weed intensity and crop–weed competition. Frequent irrigation favors grassy weed dominance. Drip and furrow irrigations have lower weed intensities than flooding (check basin) methods. Emergence of second and third flushes of *P. minor* was influenced by irrigation and planting methods—more in CT than ZT methods of planting. Moisture also plays a significant role in herbicide efficacy; lower soil moisture results in a reduced herbicide efficacy of soil-applied herbicides, whereas high moisture content can cause crop injury with acetolactate synthase (ALS)-inhibiting herbicides.

Singh and Yadav [31] found that four irrigations at 22, 65, 85, and 105 days after sowing (DAS) had significantly more populations of *P. minor* as compared to one irrigation at 22 DAS and two irrigations at 22 and 85 DAS. This could be because a higher number of irrigations applied to the crop made more moisture available in the soil, which encouraged more germination and growth of weeds. However, the difference in weed population between one and two irrigations was not significant. Application of four irrigations also increased the grain yield by 28 and 62% over two and one irrigations, respectively.

The depth of irrigation also greatly affects the weed growth and development. Lower panicles and dry matter of *P. minor* were recorded in the crop receiving 7.0 cm depth of water in both first and second irrigations, and these values were significantly less compared to all other irrigation treatments. Crop receiving both heavy irrigations recorded significantly higher panicle number as well as dry matter of *P. minor* as compared to other treatments, indicating thereby that growth and development of *P. minor* are directly related to soil moisture.

Competitive Cultivars

Selecting competitive wheat varieties is important to reduce yield losses from weeds competition. Taller varieties with early canopy cover and higher tiller numbers are more competitive with weeds. Wicks et al. [20] compared several tall-, medium-, and short-statured wheat varieties against weeds and found that medium and tall varieties had 92 and 52 % lower weeds compared to short varieties under North Platte (USA) during 1983 and 1984, respectively. Balyan et al. [32] found that *Avena ludoviciana* reduced wheat yield by 17–62 %, depending upon cultivars. Among the five cultivars, WH 147 and HD 2285 were more competitive. Roberts et al. [33] compared ten cultivars in Oklahoma (USA) for their competitive ability against rye in four experiments at several locations and found that yield losses were less with Jagger and Triumph varieties.

Cultivars with quick initial growth and more leaf area are desired to smother weeds. Fast canopy-forming and tall cultivars generally suffer less from the weed competition than the slow-growing and short-stature Indian wheat varieties (i.e., WH 542, HD 2687, HD 2329, and PBW 343) due to their greater height, dry matter production, and number of tillers, which imposed more suppression of *P. minor,* and were better competitors.

Lolium rigidum is the most troublesome weed of Australian wheat-based cropping systems. A uniform density of *Lolium* rigidum reduced wheat yield up to 80 % in 1993 and 50 % in 1994, depending on wheat genotypes [24]. In order to develop a competitive variety against *Lolium rigidum,* 250 genotypes of *Triticum aestivum* and *T. durum* were screened in 1993, and a subset of 45 genotypes was further examined in 1994, which revealed considerable potential within the wheat genome to breed varieties with greater competitive ability. Durum wheat was found less competitive than *T. aestivum.* The strongly competitive genotypes had high early biomass accumulation, more tillers, and were tall with extensive leaf display.

Korres and Froud-Williams [34] studied the competitive ability of six European winter wheat cultivars for different seed rates in the presence or absence of weeds for traits (crop height and tillers) that confer crop competitiveness and found that Maris Huntsman and Maris Widgeon were the most competitive cultivars, whereas Fresco was the least. Manipulation of seed rate was a more reliable factor than cultivar selection for enhancement of weed suppression, although competitiveness of cultivars Buster, Riband, and Maris Widgeon had no effect of seed rate. Crop densities ranging between 125 and 270 plants m^{-2} were found to offer adequate weed suppression.

Crop Rotation

Rotation of crops will integrate new agronomic practices and more competitive crops to suppress weed species [11, 18]. Crop rotation will exert lower selection pressure due to changes in cultivation practices and herbicides. This is the best nonmonetary technique for weed management, because weeds are associated with certain crops due to their identical ecological requirements. Adoption of rice–potato–wheat, rice–potato–sunflower, and rice-Egyptian clover resulted in significant reduction in dry matter accumulation by *P. minor* as compared to rice–wheat and rice–*Brassica napus*. Rice–fallow–sugarcane–ratoon sugarcane–sunflower–rice–wheat–sugarcane can be adopted as a long-duration (4 year) rotation to take care of *P. minor* and other weeds. This rotation has lower opportunity for grassy weeds to proliferate. Other rotations can include rice–potato–sunflower and rice–mustard–sugarcane, and some labor-intensive rotations of rice–potato–onion can reduce soil seed bank composition of weed species posing problems in rice–wheat rotations.

Anderson et al. [35] reported that crop rotations had a large effect on composition of weed flora. Eight rotations were evaluated for 8 years in the Great Plains (USA) to assess composition of weed communities. Rotations with the least weed seedlings were wheat-fallow and spring wheat-winter wheat–corn–sunflower.

Reducing Weed Seed Input in Soil

Stimulating weed seed germination and destroying them after emergence through cultivation methods or nonselective herbicides have been practiced in many situations around the globe. Double pre-sowing irrigation to stimulate emergence has been commonly used in northwest India for the control of *Avena ludoviciana* in wheat [11]. Decreased soil seed reserve was recorded due to stale seedbeds in Denmark [36]. To lower the seed rain, weed seeds need to be collected before crop harvest. Collecting *P. minor* seeds at the threshing floor was used before the onset of combine harvester in Haryana, India, but not all seeds are collected as 70 % seeds are shed before the wheat harvest, which enriches the soil seed bank. Chemical stimulants to prompt weed seed germination before crop plantation have been tried without such field success. Soil solarization using black polyethylene mulch generally recommended to kill nematodes and harmful pathogens also kills some weed species. Increased soil temperature by 4 to 12 °C at 15 cm depth by using double layer plastic sheets (trapped air in between) has been found effective against *Malva parviflora*. Though soil solarization was found partially or not effective against *Convolvulus arvensis, Conyza canadensis, Cyperus rotundus, Melilotus alba and Eragrostis spp.;* this depends on rise in soil temperature at deeper depths and hard seed coat of weed species. *P. minor* with a hard seed coat was still viable after putting seed in an oven at 60 °C for 16 days.

Herbicide-resistant weeds are a serious threat in Australia due to evolution of multiple resistance mechanism in *Lolium rigidum*. Farmers are striving for a non-chemical strategy to control herbicide-resistant populations. Harvest weed seed control systems were developed for use at wheat harvest to minimize weed seed bank inputs [37]. These machines include chaff carts, narrow windrow burning, bale direct, and the Harrington Seed Destructor (HSD) that collects the weed seeds during wheat harvest resulting in decreased seed bank over the years. Chaff collection using a harvest weed seed control system resulted in 56–86 % weed seed control of *Lolium rigidum* at different locations, 95 % of *Raphanus raphanistrum,* and 74 % of *Avena fatua* compared to 95 % by bale direct for *Lolium rigidum* and 99 % of *Lolium rigidum* and *Raphanus raphanistrum* by narrow windrow burning [37]. Similarly, HSD provided 93–99 % control of *Raphanus raphanistrum*, *Lolium rigidum*, *Avena fatua,* and *Bromus rigidus*. Lower weed seed addition to the soil seed bank will impose lower selection pressure and contribute towards sustainable herbicide control. These (HSD) are heavy machinery for use in large farms, but they have shown a way to get rid of weed seeds effectively and can be replicated for situation-specific use (scale down).

Mechanical Methods of Weed Management

There is an enhanced interest in mechanical methods of weed control, firstly due to organic farming and secondly due to the virulence of herbicide-resistant weeds. Increasing concerns about pesticide use and a steadily increased conversion to organic farming have been major factors in adopting physical and cultural weed control methods in Europe [38]. Herbicides have also been cited for ground and surface water pollution in several European countries [39], and there is a move to lower the use of herbicides by adopting mechanical methods of weed control. Ramsussen [36] conducted several studies in Danish wheat—using different seed rates, planting times and methods, row spacing, and mechanical and chemical weed control—and observed that under heavy weed pressure, intensive mechanical weeding (24 cm spacing) provided weed control similar to herbicides, resulting in a significantly higher yield than control. Melander et al. [38] reviewed several mechanical methods for weed management in different cropping systems in Europe. Planting wheat on raised beds (2–3 rows/bed) in India offers scope for mechanical weeding in case of herbicide failure. Due to narrow row spacing, intercultural operations by tine harrows are cumbersome and less efficient, but on a small scale, a wheel hand hoe is very useful to control weeds in between the rows.

Site-specific Weed Management

Precision agriculture is gaining ground in the USA, Europe, Australia, and other parts of the globe, where remote sensing and mapping tools have reduced input use in agriculture and increased farm profitability. Site-specific weed management (SSWM), a part of precision agriculture, is important in situations where weeds are either in patches or below the threshold level. Mapping the weed infestation through remote sensing and controlled release of herbicides has the additional option of choosing more than one herbicide, depending upon the nature of infestation. Real-time weed detection and patch spraying save on not only time but also herbicide cost. Gerhards and Christensen [40], using online weed detection and digital image analysis through computer-based decision making and global positioning systems (GPS), controlled patch spraying with reduced herbicide use in wheat by 60 % for grassy weeds and 90 % for broadleaf weeds in Denmark without compromising weed control efficiency. As of today, economic and technical limitations for SSWM have not favored widespread adoption. However, as research develops and technology is refined, the opportunities for site-specific control of weeds will greatly increase [41].

Biological Weed Control

Biological control employs the natural enemies (biological control agents) of weeds. Worldwide, 60 % of introduced biological control agents have been successfully established and out of these, 33 % have resulted in at least substantial control of the weeds [42]. These are cost-effective, self-perpetuating, and do not cost the environment. These biocontrol agents (insects, mites, or pathogens) are widely used in the USA, Australia, South Africa, Canada, and New Zealand. Several reviews on the successful use of biocontrol agents have been published ([42–45], and others). Still, their role is insignificant as weed control in major crops is largely dependent on other control measures, but they have a great potential as an effective tool for future weed management programs.

Herbicidal Weed Control

Herbicides have become the dominant technological tool against weeds with several advantages over other methods of weed control since their widespread use from the 1940s. Increased use of herbicides revolutionized agriculture and increased yields by 250 %. Lower use rate and selective herbicides launched in the 1970s along with glyphosate changed the face of weed management practices. Earlier, herbicides were used in an integrated system; however, their sole reliance resulted in the evolution of resistant weed species.

Herbicide Resistance

Herbicide-resistant weeds occurring in major field crops are posing a severe threat to global food security. Worldwide there are approximately 250 weed species (0.1 % of world's flora) which are a potential problem. Out of these 75 weed species have earned notoriety by evolving different mechanisms to defy wheat herbicides. Globally, there are 483 unique cases of resistant weeds from 235 species (138 monocots and 97 monocots) to 155 different herbicides with 22 of 25 known modes of action (MOAs) of herbicides, spread in 82 crops in 65 countries [46]. Currently, wheat has 329 such cases (populations and herbicides MOA which are not controlled by herbicides (Table 7.2). These resistant weeds are spread across 32 countries, including the USA, Australia, and Canada, constituting 52 % of total resistant cases.

Some of these weeds species (*Lolium rigidum, Lolium multiflorum, Alopecurus myosuroides, Phalaris minor*, and *Raphanus raphanistrum*) have evolved multiple resistance to herbicides of different MOAs and pose a challenge to any new herbicide. *Lolium rigidum* is an extreme case that has spread to a large landscape of Australia. *P. minor* (a major weed of wheat in northwest India), which evolved resistance to isoproturon in the 1990s, is the most serious case with a potential to cause complete crop failure [29]. The resistance was characterized by enhanced metabolism by cytochrome P-450 monooxygenase enzyme [47–49]. Enhanced metabolism is the most common mechanism, conferring partial resistance to a wide range of herbicides, though acetyl-CoA carboxylase (ACCase) target-site resistance also occurs widely [50]. After a few applications, *P. minor* populations exhibited multiple resistances to ACCase- and ALS-inhibitor herbicides [51], though resistance was not characterized. Kaundun [52] found that ACCase resistance in *P. minor* was due to target-site mutation. Several weed populations have both metabolic and target-site mutations that offer a major challenge to new herbicide molecules for the long-term control.

Development of New Herbicides

There is greater need than ever to have herbicides with newer MOA for the control of multiple-resistant weed species infesting wheat crops in different parts of the world. However, no new molecules with unique MOA are in the pipeline due to the slowing down of herbicide discovery. High regulatory cost (environment concerns), long discovery-to-commercialization time, smaller market, more generics, glyphosate-/herbicide-resistant crops, and diversion of resources to biotechnology are some of the reasons for trickling new molecule discoveries. Pyroxasulfone, a very long chain fatty acid inhibitor, has been found effective against *Lolium rigidum* in Australia [53], *Lolium multiflorum* in the USA [54], and *P. minor* in India [55]. However, there is greater propensity among these species to evolve quick resistance to this new herbicide if used at lower than recommended rates.

Regulatory removal of some old herbicides from the market due to environmental concerns also limits the choice for herbicides. Thus, continuing herbicide-resistance evolution is a major threat to future crop weed management and a potent driving force in the search for alternate weed control technologies.

Though development of new herbicides has slowed down, there are concerted efforts to develop new herbicide traits in crops for the effective use of several old herbicides. These new traits include metabolic degradation and resistant target site

Table 7.2 List of resistant weeds in wheat to herbicides of different modes of action (MOAs) around the globe (durum, spring and winter wheat)

	Species	Country	MOA
1	*Alopecurus aequalis*	China, Japan	A/1, B/2, K1/3
2	*Alopecurus japonics*	China	A/1, C2/7
3	*Alopecurus myosuroides*	Belgium, Czech Republic, Denmark, France, Germany, Italy, Netherlands, Poland, Spain, Switzerland, Turkey, United Kingdom	A/1, B/2, C1/5, C2/7, K1/3, K3/15, N/8
4	*Amaranthus powellii*	Canada	B/2, C1/5
5	*Amaranthus retroflexus*	Canada	B/2
6	*Anthemis arvensis*	Chile	B/2
7	*Anthemis cotula*	Chile, USA	B/2
8	*Apera spica-venti*	Czech Republic, Denmark, Germany, Poland, Switzerland, Lithuania	A/1, B/2, C2/7
9	*Avena fatua*	Argentina, Australia, Belgium, Canada, Chila, France, Iran, Mexico, Poland, South Africa, Turkey, United Kingdom, USA	A/1, B/2, N/8, Z/8, Z/25
10	*Avena sterilis*	Australia, Greece, Iran, Israel, Italy, Turkey, United Kingdom	A/1, B/2, Z/25
11	*Avena sterilis ssp. ludoviciana*	Australia, France, Iran	A/1, B2, Z/25
12	*Beckmannia syzigachne*	China, Japan	A1, C2/7, K1/3
13	*Brassica tournefortii*	Australia	B/2
14	*Bromus diandrus*	Australia	B/2, G/9
15	*Bromus diandrus ssp. rigidus (=B. rigidus)*	Australia	B/2
16	*Bromus japonicus*	USA	B/2
17	*Bromus secalinus*	USA	B/2
18	*Bromus tectorum*	Spain	C2/7
19	*Buglossoides arvensis (=Lithospermum arvense)*	China	B/2
20	*Camelina microcarpa*	USA	B/2
21	*Capsella bursa-pastoris*	Canada, China	B/2
22	*Centaurea cyanus*	Poland	B/2, O/4
23	*Chenopodium album*	Canada	B/2
24	*Chrysanthemum coronarium*	Israel	B/2
25	*Conyza bonariensis*	Brazil	G/9
26	*Conyza canadensis*	USA	B/2, G/9
27	*Cynosurus echinatus*	Chile	A/1, B/2
28	*Descurainia sophia*	China, USA	B/2, E/14, O/4
29	*Diplotaxis erucoides*	Israel	B/2
30	*Eleusine indica*	Bolivia	A/1
31	*Erucaria hispanica*	Israel	B/2
32	*Erysimum repandum*	USA	B/2
33	*Galeopsis tetrahit*	Canada	B/2, O/4
34	*Galium aparine*	China, Turkey	B/2
35	*Galium spurium*	Canada	B/2, O/4

Table 7.2 (continued)

	Species	Country	MOA
36	*Galium tricornutum*	Australia	B/2
37	*Hordeum glaucum ssp. glaucum*	Australia	B/2
38	*Kochia scoparia*	Canada, USA	B/2, C1/5, G/9, O/4
39	*Lactuca serriola*	Australia, USA	B/2
40	*Lolium perenne*	Argentina, Germany	A/1, B/2, G/9
41	*Lolium perenne ssp. multiflorum*	Argentina, Brazil, Chile, Denmark, France, Italy, UK, USA	A/1, B/2,C2/7,K3/15, G/9
42	*Lolium persicum*	Canada, USA	A/1
43	*Lolium rigidum*	Australia, Chile, France, Greece, Iran, Israel, Saudi Arabia, South Africa, Spain	A/1, B/2, C2/7, F3/13, G/9, K1/3, K2/23, K3/15, N/8
44	*Matricaria recutita (= M. chamomilla)*	Belgium, Germany	B/2
45	*Mesembryanthemum crystallinum*	Australia	B/2
46	*Myosoton aquaticum*	China	B/2
47	*Neslia paniculata*	Canada	B/2
48	*Papaver rhoeas*	Denmark, France, Greece, Italy, Spain, Sweden	B/2, O/4
49	*Phalaris brachystachys*	Italy, Turkey	A/1, B/2
50	*Phalaris minor*	Australia, India, Iran, Israel, Mexico, South Africa	A/1, B/2, C2/7
51	*Phalaris paradoxa*	Australia, Iran, Israel, Italy, Mexico	A/1, B/2
52	*Picris hieracioides*	Russia	B/2
53	*Polygonum convolvulus (=Fallopia convolvulus)*	Australia, Canada	B/2
54	*Polygonum lapathifolium*	Canada	B/2
55	*Polypogon fugax*	China	A/1
56	*Raphanus raphanistrum*	Australia, Brazil, South Africa	B/2, O/4
57	*Raphanus sativus*	Argentina, Brazil, Chile	B/2
58	*Rapistrum rugosum*	Australia, Iran	B/2
59	*Rorippa indica*	China	B/2
60	*Salsola tragus*	Canada, USA	B/2
61	*Sclerochloa kengiana*	China	A/1
62	*Setaria viridis*	Canada, USA	A/1, B/2, K1/3
63	*Silene gallica*	Chile	B/2
64	*Sinapis alba*	Spain	B/2
65	*Sinapis arvensis*	Australia, Canada, Iran, Italy, Turkey	B/2, O/4, C1/5
66	*Sisymbrium orientale*	Australia	B/2
67	*Sisymbrium thellungii*	Australia	B/2
68	*Snowdenia polystachya*	Ethiopia	A/1
69	*Sonchus asper*	USA	B/2
70	*Sonchus oleraceus*	Australia	B/2

Table 7.2 (continued)

	Species	Country	MOA
71	*Stellaria media*	Canada, China, France, Germany, Sweden, United Kingdom, USA	B/2, O/4
72	*Thlaspi arvense*	Canada	B/2
73	*Tripleurospermu m perforatum (=T. inodorum)*	Denmark, Germany	B/2
74	*Urochloa panicoides*	Australia	G/9
75	*Vaccaria hispanica*	Canada	B/2

MOA: A/1= ACCase inhibitors; B/2=ALS inhibitors; C1/5= Photosystem II inhibitors; C2/7 =PSII inhibitors (ureas and amides); E/14=PPO inhibitors; F3/13=Carotenoid biosynthesis (unknown target); G/9=EPSP synthase inhibitors; K1/3=Microtubule inhibitors; K2/23=Mitosis inhibitors; K3/15=Long chain fatty acid inhibitors; N/8=Lipid inhibitors (thiocarbamates); O/4=Synthetic auxins; Z/8=Cell elongation inhibitors; Z/25=Antimicrotubule mitotic disrupter.

for glyphosate, metabolic degradation for dicamba, 2,4-D and ACCase inhibitors, and target-site resistance, overexpression, alternate pathway, and/or pathway flux for p-hydroxyphenylpyruvate dioxygenase (HPPD) inhibitors [56].

Genome sequencing has revealed that P450s constitute the largest family of enzymatic proteins in higher plants. P450s are monooxygenases that insert one atom of oxygen into inert hydrophobic molecules to make them more reactive and hydrosoluble, besides their physiological functions in the biosynthesis of hormones, lipids, and secondary metabolites [57]. These P450s also help plants to cope with xenobiotics, making them less phytotoxic. The recovery of an increasing number of plant P450 genes in recombinant form has enabled their use for engineering herbicide tolerance, biosafening, bioremediation, and green chemistry

Glyphosate-resistant wheat can offer a good solution to manage most of the infested weed species; however, there is a large concern on the use of genetically modified wheat, and the sensitivity varies from country to country. Herbicide-tolerant wheat without genetically modified organisms (GMO) [58], on the other hand, can have large acceptability and such a system can also improve weed control, particularly where traditional wheat herbicides have lost their efficacy against weeds. Imidazolinone-tolerant wheat has been found to perform better in *Lolium multiflorum*-resistant fields [59].

Going by the large numbers of recently evolved resistant weeds to glyphosate, glyphosate-tolerant wheat will not be free from problems in the future, and this technology needs to be used as one of the tools in the management of resistant weeds. Gene pyramiding/stacking with more than one herbicide can lower the risk of quick resistance evolution or shift in weed flora due to the frequent use of herbicides in herbicide-tolerant wheat.

A rapid herbicide-resistance detection through seed or seedling bioassay can easily discriminate herbicide-resistant and herbicide-susceptible weed populations [60, 61], which can be used to issue advisories to the farmers on the selection of herbicides for effective weed management.

Table 7.3 Tank mixture of different herbicides for weed control in wheat

Herbicides mixtures	Application time
1. Pendimethalin + metribuzin	PRE/POE
2. Pendimethalin + flufenacet	PRE
3. Pendimethalin + sulfosulfuron	PRE/POE
4. Pendimethalin + pyroxasulfone	PRE
5. Metribuzin + pyroxasulfone	PRE
6. Metribuzin + flufenacet	PRE
7. Flufenacet + metribuzin	PRE
8. Clodinafop + metribuzin	POE
9. Fenoxaprop + metribuzin	POE
10. Pinoxaden + metsulfuron	POE
11. Pinoxaden + carfentrazone	PRE
12. Pinoxaden + 2,4-D	POE
13. Metsulfuron + carfentrazone	POE

PRE preemergence, *POE* postemergence

Moss et al. [50] suggested that to manage resistant weeds, we should encourage the greater use of cultural control measures, such as plowing, crop rotation, delayed drilling, reduced reliance on high-risk herbicides (ACCase, ALS), and use of mixtures and sequences of herbicides with different MOAs.

Herbicide Rotations, Mixtures, and Sequences

As the synthesis of herbicides with new MOA has slowed down, there is a greater need to use mixtures of existing herbicides in a way to lower the loads on the environment and improve weed control efficacy without any adverse effect on crops. Herbicide mixtures can lower the cost of weed control and also delay the evolution of herbicide-resistant weeds, if used scientifically [62]. Increased efficacy of flufenacet + metribuzin (readymix) was observed by Koepke-Hill et al. [63] against *Lolium multiflorum* in winter wheat. Metsulfuron or carfentrazone was not effective against *Fumaria parviflora* when used alone in wheat, but their mixture was found synergistic [64]. Similarly, higher mortality of *P. minor* was observed with premix of fenoxaprop + metribuzin [65]. Herbicide rotations and mixtures are widely recommended to manage herbicide resistance [66]. Compatibility of the mixture partners is important to have a greater synergy on target weeds; however, ACCase-inhibiting herbicides are not compatible with 2,4-D and need sequential applications. Due to the evolution of resistant weeds to several wheat herbicides, use of PRE herbicides with lower cases of resistance (dinitroanilines and others) is useful as a mixture partner or when used in sequences, depending upon a suitable partner. The following herbicide mixtures have shown promising results against resistant populations of *P. minor* in wheat in northwest India (Table 7.3). Pendimethalin (PRE herbicide) when applied post-emergence as tank mix with metribuzin/sulfosulfuron was found effective in managing subsequent flush of *P. minor*. Similarly surfactants/adjuvants not only increase herbicide uptake but also rainfastness and provides higher weed control efficiency.

Integrated Weed Management

To avoid or delay the development of resistant weeds, a diverse, integrated program of weed management practice is required to minimize reliance on herbicides with the same MOA [67]. An integration of nonchemical methods, knowledge of weed biology, and herbicides will be required for effective weed control [68].

Diverting from chemical-alone methods and diversifying weed control measures are on the rise in many developed countries. Thermal (soil solarization) and mechanical weeding in wide-spaced crops, intra-row weeding, flaming, harrowing, brush weeding, hoeing, torsion weeding, and finger weeding are being considered [38]. Research work is going on for robotic weeding, band-steaming for row crops, and integrating cultural methods such as fertilizer placement, seed vigor, seed rate, competitive varieties, crop rotation, planting time/method, and irrigation scheduling with mechanical and chemical methods needs to be adopted as per field problems to effectively tackle weedy issues. The potential of rhizobacteria which has been found to check germination and growth of weeds can be realized by integrating with herbicides. Reduced population of *Bromus tectorum* through the application of the *Pseudomonas fluorescens* and 18 to 35% yield increase in winter wheat has been reported earlier [70–71]. Greenhouse study by Ehlert et al. [72] reported that inoculation with fungal pathogen *Pyrenophora semeniperda* reduced the emergence of *Bromus tectorum* and subsequent application of imazapic (after emergence) killed the seedlings. Similarly in recent pot studies, seed inoculation with *Pseudomonas* and *Bacillus spp.* reduced the emergence of *Phalaris minor* and higher mortality was recorded with POE application of PSII inhibitors than ACCase and ALS inhibitors [Author's unpublished data, 2014]. Soil and seed inoculation of *Pseudomonas* and *Bacillus* spp. to several susceptible and resistant populations of *P. minor* provided differential effect on emergence and growth and need further verification. Inconsistent performance or low activity by these pathogens, due to poor survival in field soils (environmental conditions) has been observed earlier also and need more detailed investigation of their relationship. However, they provide a niche for sustainable weed management. To increase the effectiveness of these bio-agents soil inoculation can be performed before crop plantation. This integrated approach can lower the selection pressure and increase herbicide efficacy.

Programs for herbicide-resistance management must consider the use of all cultural, mechanical, and herbicidal options available for effective weed control in each situation and employ the following best management practices [69]:

1. Understand the biology of the weeds present.
2. Use a diversified approach toward weed management focused on preventing weed seed production and reducing the number of weed seed in the soil seed bank.
3. Plant into weed-free fields and then keep fields as weed free as possible.
4. Plant weed-free crop seed.
5. Scout fields routinely.
6. Use multiple herbicide MOAs that are effective against the most troublesome weeds or those most prone to herbicide resistance.

7. Apply the labeled herbicide rate at recommended weed sizes.
8. Emphasize cultural practices that suppress weeds by using crop competitiveness.
9. Use mechanical and biological management practices where appropriate.
10. Prevent field-to-field and within-field movement of weed seed or vegetative propagules.
11. Manage weed seeds at harvest and after harvest to prevent a buildup of the weed seed bank.
12. Prevent an influx of weeds into the field by managing field borders.

Conclusion

Weeds were in abundance when humans started selective crop cultivation; they still exist even after putting all our efforts into eradicating them, and they will exist in the future also. The only wise thing will be to lower their losses by reducing their preponderance and increasing farm productivity and profitability in a sustainable manner.

References

1. US Department of Agriculture (2013) http://www.ers.usda.gov/publications/whs-wheat-outlook/whs-13i.aspx#.UkIrq9I0WuJ. Accessed 21 Jan 2014
2. Singh S, Punia SS (2008) Effect of Seeding depth and flooding on emergence of *Malva parviflora, Rumex dentatus* and *R. spinosus*. Ind J Weed Sci 40:178–186
3. Chhokar RS, Malik RK, Balyan RS (1999) Effect of moisture stress and seedling depth on germination of littleseed canarygrass (*Phalaris minor* Retz). Indian J Weed Sci 31:78–79
4. Chhokar RS, Sharma RK, Jat GR, Pundir AK, Gathala MK (2007) Effect of tillage and herbicides on weeds and productivity of wheat under rice-wheat growing system. Crop Prot 26:1689–1696
5. Franke AC, Singh S, McRoberts N, Nehra AS, Godara S, Malik RK, Marshall G (2007) *Phalaris minor* seedbank studies: longevity, seedling emergence and seed production as affected by tillage regime. Weed Res 47:73–83
6. Mahajan G, Brar LS, Walia US (2002) *Phalaris minor* response in wheat in relation to planting dates, tillage and herbicides. Indian J Weed Sci 34:114–115
7. Kumar V, Singh S, Chhokar RS, Malik RK, Brainard DC, Ladha JK (2013) Weed management strategies to reduce herbicide use in zero-till rice-wheat cropping systems of the Indo-Gangetic plains. Weed Technol 27(1):241–254
8. Borger CPD, Riethmuller GP, Michael A, David M, Abul H, Powles SB (2013) Increased carrier volume improves preemergence control of rigid ryegrass (*Lolium rigidum*) in zero-tillage seeding systems. Weed Technol 27:649–655
9. IARI (2012) Crop residues management with conservation agriculture: Potential, constraints and policy needs. Indian Agricultural Research Institute, New Delhi, vii + 32 pp
10. Wicks GA, Martin DA, Mahnken GW (1995) Cultural practices in wheat (Triticum aestivum) on weeds in subsequent fallow and sorghum (*Sorghum bicolor*). Weed Sci 43:434–444
11. Singh S (2007) Role of management practices on control of isoproturon-resistant littleseed canarygrass (*Phalaris minor*) in India. Weed Technol 21:339–346

12. Dhaliwal HS, Singh R, Brar LS (2007) Impact analysis of factors affecting *Phalaris minor* infestation in wheat in Punjab. Indian J Weed Sci 39:66, 73
13. Singh S, Malik RK, Panwar RS, Balyan RS (1995a) Influence of sowing time on winter wild oat (*Avena ludoviciana*) control in wheat (*Triticum aestivum*) with isoproturon. Weed Sci 43:370–374
14. Tahir M, Ali A, Nadeem MA, Hussain A, Khalid F (2009) Effect of different sowing dates on growth and yield of wheat (*Triticum aestivum* L.) varieties in district Jhang, Pakistan. Pak J Life Soc Sci 7(1):66–69
15. Beech DF, Norman MJT (1966) The effect of time of planting on yield attributes of wheat varieties in the Ord River valley. Aust J Exp Agric Anim Husb 6:183–192
16. Thill DC, Witters RE, Papendick RL (1978) Interactions of early and late planted winter wheat with their environment. Agron J 70:1041–1047
17. Vander Vorst PB, Wicks GA, Burnside OC (1983) Weed control in a winter wheat-corn-ecofarming rotation. Agron J 75:507–511
18. Singh S, Kirkwood RC, Marshall G (1999) A review of the biology and control of *Phalaris minor* Retz. (littleseed canarygrass) in cereals. Crop Prot 18:1–16
19. Singh G, Singh OP (1996) Response of late sown wheat (*Triticum aestivum*) to seeding methods and weed control measures in flood prone areas. Indian J Agron 41:237–242
20. Wicks GA, Popken DH, Mahnken GW, Hanson GE, Lyon DJ (2003) Survey of winter wheat (*Triticum aestivum*) stubble fields sprayed with herbicides in 1998: cultural practices. Weed Technol 17:467–474
21. Borger CPD, Hashem A, Pathan S (2010) Manipulating crop row orientation to suppress weeds and increase crop yield. Weed Sci 58:174–178
22. Kristensen L, Olsen J, Weiner J (2008) Crop density, sowing pattern, and nitrogen fertilization effects on weed suppression and yield in spring wheat. Weed Sci 56:97–102
23. Kolb LN, Gallandt ER, Mallory EB (2012) Impact of spring wheat planting density, row spacing, and mechanical weed control on yield, grain protein, and economic return in Maine. Weed Sci 60:244–253
24. Lemerle D, Verbeek B, Cousens RD, Coombes NE (1996) The potential for selecting wheat varieties strongly competitive against weeds. Weed Res 36:505–513
25. Blackshaw RE, Semach GP, O'Donovan JT (2000) Utilization of wheat seed rate to manage redstem filaree (*Erodium cicutarium*) in a zero-tillage cropping system. Weed Technol 14:389–396
26. Blackshaw RE, Semach G, Janzen HH (2002) Fertilizer application method affects nitrogen uptake in weeds and wheat. Weed Sci 50:634–641
27. Bhullar MS, Walia US (2004) Studies on integration of nitrogen and clodinafop for controlling isoproturon resistant *Phalaris minor* in wheat. Fertilizer News 49:41–48
28. Singh S, Malik RK, Balyan RS, Singh S (1995b) Distribution of weed flora of wheat in Haryana. Indian J Weed Sci 27:114–121
29. Malik RK, Singh S (1993) Evolving strategies for herbicide use in wheat: resistance and integrated weed management. In: Proceedings of the Indian Society of Weed Science International Symposium on Integrated Weed Management for Sustainable Agriculture, Nov 18–20, Hisar, India, 1:225–238
30. Kirkland KJ, Beckie HJ (1998) Contribution of nitrogen fertilizer placement to weed management in spring wheat (*Triticum aestivum*). Weed Technol 12:507–514
31. Singh T, Yadav SK (1998) Effect of irrigation and herbicides on weed control and yield of *durum* wheat. Indian J Weed Sci 30:136–140
32. Balyan RS, Malik RK, Panwar RS, Singh S (1991) Competitive ability of winter wheat cultivars with wild oat (*Avena ludoviciana*). Weed Sci 39:154–158
33. Roberts JR, Peeper TF, Solie JB (2001) Wheat (*Triticum aestivum*) row spacing, seeding rate, and cultivar affect interference from rye (*Secale cereale*). Weed Technol 15:19–25
34. Korres NE, Froud-Williams RJ (2002) Effects of winter wheat cultivars and seed rate on the biological characteristics of naturally occurring weed flora. Weed Res 42:417–428
35. Anderson RL, Stymiest CE, Swan BA, Rickertsen JR (2007) Weed community response to crop rotations in Western South Dakota. Weed Technol 21:131–135

36. Rasmussen IA (2004) The effect of sowing date, stale seedbed, row width and mechanical weed control on weeds and yields of organic winter wheat. Weed Res 44:12–20
37. Walsh M, Newman P, Powles S (2013) Targeting weed seeds in-crop: a new weed control paradigm for global agriculture. Weed Technol 27:431–436
38. Melander B, Rasmussen IA, Bàrberi P (2005) Integrating physical and cultural methods of weed control-examples from European research. Weed Sci 53:369–381
39. Mogensen BB, Spliid NH (1995) Pesticides in Danish watercourses: occurrence and effects. Chemosphere 31:3977–3990
40. Gerhards R, Christensen S (2003) Real-time weed detection, decision making and patch spraying in maize, sugarbeet, winter wheat and winter barley. Weed Res 43:385–392
41. Shaw DR (2005) Remote sensing and site specific weed management. Front Ecol Environ 3:526–532
42. McFadyen REC (1998) Biological control of weeds. Annu Rev Entomol 43:369–393
43. Templeton GE, TeBeest DO, Smith Jr RJ (1979) Biological weed control with mycoherbicides. Annu Rev Phytopathol 17:301–310
44. Hoagland RE (2001) Microbial allelochemicals and pathogens as bioherbicidal agents. Weed Technol 15:835–857
45. Sheppard AW, Shaw RH, Sforza R (2006) Top 20 environmental weeds for classical biological control in Europe: a review of opportunities, regulations and other barriers to adoption. Weed Res 46:93–117
46. Heap I (2014) International Survey of Herbicide Resistant Weeds. http://weedscience.com/summary/home.aspx. Accessed 25 May 2014
47. Singh S, Kirkwood RC, Marshall G (1997) Effects of isoproturon on photosynthesis in susceptible and resistant biotypes of *Phalaris minor* and wheat. Weed Res 37:315–324
48. Singh S, Kirkwood RC, Marshall G (1998a) Effect of the monooxygenase inhibitor piperonyl butoxide on the herbicidal activity and metabolism of isoproturon in herbicide resistant and susceptible biotypes of *Phalaris minor* and wheat. Pestic Biochem Physiol 59:143–153
49. Singh S, Kirkwood RC, Marshall G (1998b) Effect of ABT on the activity and rate of degradation of isoproturon in susceptible and resistant biotypes of *Phalaris minor* and in wheat. Pestic Sci 53:123–132
50. Moss SR, Perryman SAM, Tatnell LV (2007) Managing herbicide-resistant blackgrass (*Alopecurus myosuroides*): theory and practice. Weed Technol 21:300–309
51. Chhokar RS, Sharma RK (2008) Multiple herbicide resistance in littleseed canarygrass (*Phalaris minor*): a threat to wheat production in India. Weed Biol Manage 8:112–123
52. Kaundun SS (2010) An aspartate to glycine change in the carboxyl transferase domain of acetyl CoA carboxylase and non-target-site mechanism(s) confer resistance to ACCase inhibitor herbicides in a *Lolium multiflorum* population. Pest Manage Sci 66:1249–1256
53. Walsh MJ, Fowler TM, Bronwyn C, Ambe T, Powles SB (2011) The potential for pyroxasulfone to selectively control resistant and susceptible rigid ryegrass (*Lolium rigidum*) biotypes in Australian grain crop production systems. Weed Technol 25:30–37
54. Hulting AG, Dauer JT, Barbara H-C, Daniel C, Koepke-Hill RM, Mallory-Smith C (2012) Management of Italian ryegrass (*Lolium perenne* ssp. *multiflorum*) in Western Oregon with preemergence applications of pyroxasulfone in winter wheat. Weed Technol 26:230–235
55. Singh S. (2014). Pyroxasulfone efficacy against *Phalaris minor* in wheat in India. Joint 2014 Meeting Weed Science Society of America and Canadian Weed Science Society. Vancouver, BC, Canada, Abstract 239
56. Green JM (2013) State of herbicides and herbicide traits at the start of 2013. Global Herbicide Resistance Challenge, Fremantle (Perth), Western Australia, Feb 18–23, Abst P 27
57. Werck-Reichhart D, Hehn A, Didierjean L (2000) Cytochromes P450 for engineering herbicide tolerance. Trends Plant Sci 5:116–123
58. Anderson JA, Matthiesen L, Hegstad J (2004) Resistance to an imidazolinone herbicide is conferred by a gene on chromosome 6DL in the wheat line cv. 9804. Weed Sci 52:83–90
59. Grey TL, Cutts III GS, Johnson J (2012) Imidazolinone-resistant soft red winter wheat weed control and crop response to ALS-inhibiting herbicides. Weed Technol 26:405–409

60. Tal A, Kotoula-Syka E, Rubin B (2000) Seed-bioassay to detect grass weeds resistant to acetyl coenzyme A carboxylase inhibiting herbicides. Crop Prot 19:467–472
61. Singh S, Singh K, Yadav A, Dhawan RS, Punia SS (2013) Detection of ACCAse herbicide resistance in *Phalaris minor* and its management in India. Global Herbicide Resistance Challenge, Fremantle (Perth), Western Australia, Feb 18–23, Abst P 104
62. Wrubel RP, Gressel J (1994) Are herbicide mixtures useful for delaying the rapid evolution of resistance?—A case study. Weed Technol 8:635–648
63. Koepke-Hill RM, Armel GR, Bradley KW, Bailey WA, Wilson HP, Hines TE (2011) Evaluation of flufenacet plus metribuzin mixtures for control of Italian ryegrass in winter wheat. Weed Technol 25:563–567
64. Singh S, Punia SS, Yadav A, Hooda VS (2011a) Evaluation of carfentrazone-ethyl + metsulfuron-methyl against broadleaf weeds of wheat. Indian J Weed Sci 43:12–22
65. Singh S, Singh K, Punia SS, Yadav A, Dhawan RS (2011b) Effect of stage of *Phalaris minor* on efficacy of Accord Plus (Fenoxaprop + Metribuzin, Readymix). Indian J Weed Sci 43:23–31
66. Beckie HJ, Reboud X (2009) Selecting for weed resistance: herbicide rotation and mixture. Weed Technol 23:363–370
67. Vencill WK, Nichols RL, Webster TM, Soteres JK, Mallory-Smith C, Burgos NR, Johnson WG, McClelland MR (2012) Herbicide resistance: toward an understanding of resistance development and the impact of herbicide-resistant Crops. Weed Sci 60(Special Issue):2–30
68. Singh S, Chhokar RS (2011) Integrated weed management strategies for sustainable wheat production. In: Singh SS, Hanchinal RR, Singh G, Sharma RK, Tyagi BS, Saharan MS, Sharma I (eds) Wheat productivity enhancement under changing climate. Narosa Publishing House, New Delhi, pp 197–205
69. Norsworthy JK, Ward SM, Shaw DR, Llewellyn RS, Nichols RL, Webster TM, Bradley KW, George F, Powles SB, Burgos NR, Witt W, Barrett M (2012) Reducing the risks of herbicide resistance: best management practices and recommendations. Weed Sci. 60(Special Issue):31–62
70. Kennedy AC, Elliot LF, Young FL, Douglas CL (1991) Rhizobacteria suppressive to the weed downy brome. Soil Sci Soc Am J 55:722–727
71. Johnson BN, Kennedy AC, Ogg AG (1993) Suppression of downy brome growth by a rhizobacterium, in controlled environments. Soil Sci Soc Am J 57:73–77
72. Ehlert KA, Mangold JM, Engel RE (2014) Integrating the herbicide imazapic and the fungal pathogen Pyrenophora semeniperda to control Bromus tectorum. Weed Res Published online 16 May 2014

Chapter 8
Integrated Weed Management in Maize

Amit J. Jhala, Stevan Z. Knezevic, Zahoor A. Ganie and Megh Singh

Introduction

Maize (*Zea mays* L.), also known as corn in the Americas, is one of the most important cereal crops. Corn belongs to the grass family Poaceae and tribe Maydeae. Among other cereals, corn has the highest genetic yield potential; therefore, it is known as "queen of cereals." Flint, dent, floury, sweet or sugary, popcorn, multicolored, and other types of corn are grown throughout the world, with color, size, kernel shape, and other attributes varying significantly. The production of yellow corn predominates in the USA, Brazil, and China. However, white corn is preferred in Africa, Central America, and the northern part of South America [1]. Using climatic data where corn is most productive, Harshberger reported that corn originated in Mexico and had once been a wild plant in central Mexico [2]. A closely related species of corn, teosinte, and the landrace diversity of corn have been found on

A. J. Jhala (✉)
Department of Agronomy and Horticulture, University of Nebraska–Lincoln,
279 Plant Science Hall, Lincoln, NE 68583, USA
e-mail: amit.jhala@unl.edu

S. Z. Knezevic
Northeast Research and Extension Center, Department of Agronomy and Horticulture,
Haskell Agricultural Laboratory, University of Nebraska–Lincoln,
57905 866 Road, Concord, NE 68728, USA
e-mail: sknezevic2@unl.edu

Z. A. Ganie
Department of Agronomy and Horticulture, University of Nebraska–Lincoln,
279 Plant Science Hall, East Campus, Lincoln, NE 68583, USA
e-mail: zahoorganie11@huskers.unl.edu

M. Singh
Citrus Research and Education Center, University of Florida,
700 Experiment Station Road, Lake Alfred, FL 33850, USA
e-mail: msingh2500@gmail.com

B. S. Chauhan, G. Mahajan (eds.), *Recent Advances in Weed Management*,
DOI 10.1007/978-1-4939-1019-9_8,

the central plateau and western escarpment of Mexico–Guatemala that supports the theory of origin of corn in this region.

Corn is the most versatile crop with wider adaptability to varied agroecological regions and diverse growing seasons. Besides serving as human food and animal feed, the importance of this crop also lies in its wide industrial applications. For example, corn oil is used in margarine, corn syrup sweeteners in marmalade, and corn syrup solids in instant non-dairy coffee creamer. In addition, corn is fed to cows, chickens, and pigs, which produce milk, eggs, and bacon, respectively. Furthermore, corn finds application in a candy bar, a beer or bourbon whisky, a hamburger, industrial chemicals, ethanol in gasoline, plastics, and in the paper sizing of a glossy magazine [3]. Responding to its multiple uses, the demand for corn is constantly increasing in the global market. New production technologies, such as improved hybrid cultivars, precision agriculture, herbicide-resistant traits, and biotechnological innovations, such as drought-tolerant corn, offer great promise for increasing corn productivity to meet the growing demand.

Globally, corn is grown on more than 175 million ha across 166 countries with a production of around 880 million t [4]. The global output of corn in 2013 was forecast at about 963 million t, 10 % up from 2012 [4]. The top six corn-producing countries are the USA, China, Brazil, India, Mexico, and Argentina. The USA is producing about 30 % of the total corn produced in the world. In addition, the USA is the largest exporter of corn to several destinations in the world. In 2013, it was expected that corn production in the USA would reach about 340 million t [4].

Excluding environmental variables, yield losses in corn are caused mainly by competition with weeds. Weed interference is a severe problem in corn, especially in the early part of the growing season, due to slow early growth rate and wide row spacing. Weeds compete with the corn plants for resources such as light, nutrients, space, and moisture that influence the morphology and phenology of crop, reduce the yield, make harvesting difficult, and mar the quality of grains. Furthermore, high weed infestation increases the cost of cultivation, lowers value of land, and reduces the returns of corn producers. In order to realize the yield potential of corn, weed management becomes indispensable. Weed species infesting the corn crop are functions of a complex interaction among soil characteristics, climate, and cultural practices. These factors vary across regions and influence the composition and number of predominant weeds of economic importance to corn production [5].

The critical period of crop–weed competition and weed threshold are two important aspects in a weed management program in any crop. The critical period may be defined as the time period after crop emergence during which crop must be kept weed-free to prevent yield losses, described as losses greater than 5 % in earlier studies [6, 7]. Likewise, weed threshold, defined as the weed density above an acceptable count, provides an opportunity to decide the right time to take appropriate control measures to avoid yield loss [8, 9]. Weeds that emerge at the time of crop germination or within a few days of crop emergence cause greater yield loss than weeds emerging later in the growing season [8, 10, 11]. The critical period is useful in defining the crop growth stages most vulnerable to weed competition. The critical period of weed control in corn ranges from 1 to 8 weeks after the crop emergence [12–14]; however, to avoid limitations associated with critical period for weed control (CPWC) like weed-species specificity and inconsistency across climate and locations, the onset of critical period for crop weed control is reported to occur on an average between the first (V1) and the third (V3) leaf stages of development [15], while the end of critical period typically coincides with the V8–V10 stages, which is the time of canopy closure in 76-cm row spacing [16–20].

A number of weed species compete with corn plant (Table 8.1) and have been observed to reduce yield as much as 65 % with delay in weed control [15]. Some of the weeds in corn are difficult to control, known as problem weeds, because they have similar life cycle and growth habits as those of the corn plant. Weed species, densities, and their interactions influence corn yield loss [21, 22]. Massing et al. reported yield reduction in corn as much as 91 % by competition with eight Palmer amaranth (*Amaranthus palmeri* S. Wats) plants per meter row length [23].

Corn-based cropping systems in the USA are heavily dependent on herbicide-resistant corn hybrids (e.g., glyphosate-, glufosinate-, or imidazolinone-resistant corn). These crop production systems rely heavily on the use of postemergence herbicides, such as glyphosate, as glyphosate-resistant hybrids dominate the market. Repeated use and solely relying on glyphosate for weed control resulted in an increasing number of herbicide-resistant weeds, shifts in weed species population, higher cost of chemical control measures, and leaching of herbicide into groundwater and surface water as well as herbicide residues in drinking water and food, which have sparked public awareness and restrictions on herbicide use [24–26]. Herbicides have often been cited as one of the main factors responsible for causing a general impoverishment of the flora and fauna in the agricultural landscape [27, 28]. To address these challenges, many countries have developed policies that mandate the reduction of herbicide use and provide incentives to producers for reducing overall chemical use [29–31].

Integrated Weed Management in Corn

Integrated weed management (IWM) has been defined as a multidisciplinary approach to weed control, utilizing the application of numerous alternative control measures [32]. The IWM involves a combination of cultural, mechanical, biological, genetic, and chemical methods for an effective and economical weed control that reduces weed interference with the crop while maintaining acceptable crop yields [18, 33]. None of the individual control measures can provide complete weed control. However, if various components of IWM are implemented in a systematic manner, significant advances in weed control technology can be achieved [32].

The IWM approach advocates the use of all available weed control options that include:

1. Selection of a well-adapted crop variety or hybrid with good early-season vigor and appropriate disease and pest resistance

Table 8.1 Major weeds of corn in the USA listed by family name, scientific name, and life cycle. (Reprinted with permission from Kremer [5])

Family/common name	Scientific name	Life cycle
Monocots		
Cyperaceae		
Purple nutsedge	*Cyperus rotundus* L.	Perennial
Yellow nutsedge	*C. esculentus* L.	Perennial
Poaceae		
Barnyardgrass	*Echinochloa crus-galli* (L.) Beauv.	Annual
Bermudagrass	*Cynodon dactylon* (L.) Pers.	Perennial
Broadleaf signalgrass	*Brachiaria platyphylla* (Griseb.) Nash.	Annual
Crabgrass	*Digitaria* spp.	Annual
Fall panicum	*Panicum dichotomiflorum* Michaux.	Annual
Field sandbur	*Cenchurus incertus* M.A. Curtis	Annual
Foxtails	*Setaria* spp.	Annual
Goose grass	*Eleusine indica* (L.) Gaertn.	Annual
Johnsongrass	*Sorghum halepense* (L.) Pers.	Perennial
Quack grass	*Elytrigia repens (*L.) Nevski	Perennial
Red/weedy rice	*Oryza sativa* L.	Annual
Shattercane	*Sorghum bicolor* (L.) Moench	Annual
Wild proso millet	*Panicum miliaceum*	Annual
Wooly cupgrass	*Eriochloa villosa* (Thunb)Kunth	Annual
Dicots		
Amaranthaceae		
Common waterhemp	*Amaranthus rudis* Sauer	Annual
Palmer amaranth	*A. palmeri* S. Wats.	Annual
Powell amaranth	*A. powellii* S. Wats.	Annual
Redroot pigweed	*A. retroflexus* L.	Annual
Smooth pigweed	*A. hybridus* L.	Annual
Spiny pigweed	*A. spinosus* L.	Annual
Tall waterhemp	*A. tuberculatus* (Moq.) Sauer	Annual
Apocynaceae		
Hemp dogbane	*Apocynum cannabinum* L.	Perennial
Asclepiadaceae		
Common milkweed	*Asclepias syriaca* L.	Perennial
Honeyvine milkweed	*A. albidus* (Nutt.) Britt.	Perennial
Asteraceae		
Canada thistle	*Cirsium arvense* (L.) Scop.	Perennial
Common cocklebur	*Xanthium strumarium* L.	Annual
Common ragweed	*Ambrosia artemisiifolia* L.	Annual
Giant ragweed	*A. trifida* L.	Annual
Horseweed	*Conyza canadensis* (L.) Cronq.	Annual
Jerusalem artichoke	*Helianthus tuberosus* L.	Annual/biennial
Wild lettuce	*Lactuca* spp.	Annual
Wild sunflower	*Helianthus annuus* L.	Annual
Brassicaceae		
Wild mustard	*Brassica* spp.	Annual

Table 8.1 (continued)

Family/common name	Scientific name	Life cycle
Chenopodiaceae		
Common lambsquarters	*Chenopodium album* L.	Annual
Kochia	*Kochia scoparia* (L.) Roth.	Annual
Russian thistle	*Salsola iberica* Sennen and Pau	Annual
Convolvulaceae		
Field bindweed	*Convolvulus arvensis* L.	Perennial
Morning glories	*Ipomoea* spp.	Annual
Cucurbitaceae		
Burcucumber	*Sicyos angulatus* L.	Annual
Malvaceae		
Prickly sida	*Sida spinosa* L.	Annual
Spurred anoda	*Anoda cristata* (L.) Schlecht.	Annual
Velvetleaf	*Abutilon theophrasti* Medik.	Annual
Polygonaceae		
Curly dock	*Rumex crispus* L.	Perennial
Pennsylvania smartweed	*Polygonum pensylvanicum* L.	Annual
Wild buckwheat	*P. convolvulus* L.	Annual
Portulacaceae		
Common purslane	*Portulaca oleracea* L.	Annual
Solanaceae		
Eastern black nightshade	*Solanum ptycanthum* Dun.	Annual
Horsenettle	*S. carolinense* L.	Perennial
Groundcherry	*Physalis* spp.	Perennial
Jimsonweed	*Datura stramonium* L.	Annual

2. Appropriate planting patterns/spacing and optimal plant density, improved timing, placement, and amount of nutrient application
3. Appropriate crop rotation, tillage practices, and cover crops
4. Suitable choice of mechanical, biological, and chemical weed control methods
5. Alternative weed control tools (flaming, steaming, infrared radiation, sand blasting, etc.)

Cultural Control

Cultural practices play an important role in weed management program in corn. Corn is a very competitive crop; so if managed properly, it provides considerable competition against weeds. Research has shown that weeds that emerge after 4 weeks of corn establishment have less impact on corn yield [8, 18]; therefore, early-season weed control is extremely important to get a competitive corn yield. It is important to establish a uniform plant stand at desired density. Soil tilth, fertility, pH, and drainage must be suitable for the crop to be competitive with weeds. As much as possible, the crop must be managed to minimize stresses on the crop from insect and disease damages and environmental stresses (frost, flooding, drought, etc.). Row spacing is an important cultural practice affecting weed control because corn in narrow rows will shade soil surface earlier than corn in wider rows. Once the canopy has closed, very little light reaches the soil surface or weeds beneath the canopy. The value of early canopy closure for weed control is especially evident when weed control program in corn is dependent on postemergence herbicides only.

Historically, crop rotation has been one of the most common methods of managing weeds. The more diverse the crop in rotation in planting time, growth habit, and life cycle, the more effective the rotation will be in controlling weeds. Thus, the selection of a crop in rotation that includes small grains, forages, and legumes is significant; however, such crops are no longer widely grown in the North Central USA. While modern rotations tend to include shorter cycles and fewer crops, a 2-year corn–soybean (*Glycine max* [L.] Merr.) rotation, especially if it includes a different tillage system for each crop, can help to manage some weeds. As in any rotation used over many years on the same field, certain weeds will often adapt to the rotation and become problem weeds or evolve resistance to herbicides over time.

Use of cover crops is another example of cultural control of weeds. Cover crops can be used for a variety of purposes including protecting the soil against erosion, improving soil structure, fixing nitrogen, feeding the soil biological life, and managing soil moisture [34]. A key soil health concept is that there should be something green and growing during as much of the year as possible. Grasses provide the long-lasting residue cover because they have a higher carbon to nitrogen ratio in their biomass compared to non-grass species. In addition, they improve snow catch in the winter and reduce wind erosion in the spring compared to the bare soil. Taller brassicas with broad leaves like rape, mustards, and canola will also effectively reduce wind erosion and catch snowfall, but they provide less residue. In conclusion, a healthy, vigorous corn crop with a high yield potential will be very competitive with weeds; however, competition from the crop alone is not sufficient to provide a season-long weed control. Other methods of control must be used in conjunction with cultural control measures.

Mechanical Weed Control

Tillage is the most common method of mechanical weed control and it can be divided into two categories: (1) preplant tillage and (2) in-row cultivation. The purpose of preplant tillage is to kill all the weeds present before planting corn to give the crop a better start to compete with weeds during the initial stage. Field cultivators and discs are commonly used by growers, and they are highly effective for controlling weed seedlings if used properly. The in-row cultivation is used to remove weeds after the crop has been planted, usually using rotary hoe or an inter-row cultivator. Rotary hoes are most effective on small-seeded broad-leaved weeds and grasses, but they are less effective on large-seeded broad-leaved weeds, such as giant ragweed, velvetleaf (*Abutilon theophrasti* Medik.), cocklebur (*Xanthium strumarium* L.), etc. Rotary hoes are usually operated at the speed of 13–19 km/h and should be used after planting the crop but before weeds have emerged or after weed germination. Another advantage of in-row cultivation is that they are useful when soil-applied herbicides fail to control weeds due to lack of rainfall. Several types of in-row cultivators are available in the market, but it is important to adjust the equipment to effectively kill as many weeds as possible in the interrow area while minimizing the disturbance of the crop plants.

Flame Weeding

Flaming controls weeds primarily by rupturing the cell membranes that leads to subsequent tissue desiccation [35]. Propane burners can generate combustion temperatures up to 1900 °C, which raises the temperature of the exposed plant tissues rapidly [36]. An increase of temperature above 50 °C inside the plant cells can result in the coagulation (denaturation and aggregation) of membrane proteins leading to loss of the membrane integrity [35, 37, 38]. Consequently, flamed weeds would die or their competitive ability against the crop would be severely reduced. The susceptibility of plants to flame largely depends on their heat avoidance, heat tolerance, or both [39]. The extent to which heat from the flames penetrates plants depends on the flaming technique and leaf surface moisture [37]. The effects of flaming on plants are influenced by several factors including temperature, exposure time, and energy input [40]. Depending on the exposure time, protein denaturation may start at 45 °C [40]. Temperatures in the range of 95–100 °C at least for 0.1 s have been reported to be lethal for leaves and stems [40].

Heat from the flames has a direct effect on the cell membranes and an indirect effect on the subsequent tissue desiccation. Cellular death after flame treatment is primarily due to the initial thermal disruption of cellular membranes rapidly followed by dehydration of the affected tissue. Tissue dehydration occurs mainly due to expansion of the cell contents (made of up to 95 % water), subsequent bursting of the cell membranes, and coagulation of membrane proteins [41, 42].

The efficacy of flame weeding was reported to be influenced by several factors, including the presence of protective layers of hair or wax and lignification [39, 40], the physical location of the growing point at the time of flaming [39,43, 44], plant growth stages [39, 45–51], the regrowth potential of plant species [39, 40], the technique of flaming [37], and the relative leaf water content of plant species [52]. Ulloa et al. conducted a series of studies where the authors intentionally flamed several agronomic crops such as field corn, popcorn (*Z. mays* L. var. *everta*), and sweet corn (*Z. mays* L. var. *rugosa*) [48–51, 53, 54]. Response to broadcast flaming varied among corn types, their growth stages, and propane dose. Popcorn was the least tolerant while field corn was the most tolerant to broadcast flaming based on the maximum yield reduction obtained with the highest dose of propane (85 kg ha^{-1}).

Field corn flamed broadcast at the five-leaf stage (V5) was the most tolerant while the two-leaf stage (V2) was the most susceptible, which had the highest visual crop injury and the largest loss of yield and yield components [53]. Visual crop injury symptoms included initial whitening and then browning of leaves. Stunting of growth was especially evident when the plants were flamed with higher propane doses (44 and 85 kg ha^{-1}). Most visual crop injuries, however, were transient as corn plants appeared to be visually recovered within a few weeks [46, 47, 53].

Popcorn flamed at the V5 stage was the most tolerant while the V2 was the most susceptible stage for broadcast flaming [50]. Plants flamed at the V2 stage had the highest yield loss and the lowest yield components. This might be explained by the fact that the ear and tassel tissues are not differentiated at the V2 stage [55]; thus, exposing the plants to the stress from heat can result in potentially shorter cobs. In comparison, flaming popcorn plants at later growth stages (e.g., V5 or V7) had less effects on cob size as the ear and tassel tissues start to differentiate at the V5 stage, and by the V7 stage, cob and tassel sizes are already predetermined [55]. A propane dose of 60 kg ha^{-1} resulted in 8, 9, and 21 % yield reductions at the V5, V7, and V2 stages, respectively, which would not be acceptable by organic farmers. These yield reductions were the result of the intentional flaming where torches were positioned directly over the crop rows. However, positioning flames below the popcorn canopy would reduce the exposure time to the heat and, therefore, should reduce popcorn yield losses.

Sweet corn flamed at the V7 stage was the most tolerant while the V2 was the least tolerant stage for broadcast flaming [49]. Sweet corn flamed at the V7 stage had the least yield loss and the least affected yield components compared to plants flamed at the V5 and V2 stages. The V2 was the most sensitive stage for broadcast flaming, resulting in the highest yield loss and the largest effects on yield components. Among the yield components, number of plants per square meter and seeds per cob were the most affected parameters when flaming was conducted at the V2 and V5 stages. Sweet corn generally starts to accelerate its growth around the V6–V7 stages (growing point reaches soil surface). This growth acceleration in sweet corn is also coupled with increasing concentration of sugars in cell and stem tissues, which requires more energy to boil water in the cell [56]. A propane dose of 60 kg ha^{-1} caused yield losses of 6, 11, and 20 % for the V7, V5, and V2 stages, respectively. From a practical standpoint, the 6 % yield reduction of sweet corn flamed broadcast at the V7 stage may not be acceptable by organic growers. However, yield reductions were the result of the intentional flaming directly over the crop. An alternative might be to direct the flame below the crop canopy in order to spare foliage from the heat, which could result in lower yield losses (e.g., <5 %).

It is important to understand that propane flaming should not be the only method for nonchemical weed control; it should be a part of an IWM program. Other measures are still needed to control weeds that emerge later during the growing season. More research is needed to perhaps develop new flaming equipment and methods, or to examine different positioning of the burners to avoid any significant crop damage and yield reductions. Information from such research would expand flaming options as part of an IWM program for both organic and conventional crop production systems.

Biological Control

The biological control approach makes use of the weed's naturally occurring enemies to help reduce the weed's impact on agriculture and the environment. It simply aims to reunite weeds with their natural enemies and achieve sustainable weed control. These natural enemies of weeds are often referred to as biological control agents. For example, a commercial bio-herbicide Colego, a fungal herbicide, has been used to control northern jointvetch (*Aeschynomene americana* L.) in rice (*Oryza sativa* L.) in the southern USA [57]. It is critical that the biological control agents do not become pests themselves. Considerable host-specificity testing is mandatory as per many government rules and regulations prior to the release of biological control agents to ensure that they will not pose a threat to nontarget species, such as native and agricultural plants. Not all weeds are suitable for biological control. Developing a biological control project requires a substantial investment, sometimes costing millions of dollars. Currently, there are no commercial products for biological weed control in corn, though this area offers great potential for new weed control options in the future.

Chemical Weed Control

Application of herbicides is the most important method of weed control in corn. Herbicides have been adopted by a majority of corn growers in the USA and many other parts of the world because they are effective and economical. Herbicides can be applied at different time intervals, such as before the crop is planted (preplant), after the crop is planted but before emergence (preemergence), and after crop emergence (postemergence). The choice of herbicide application timing depends on many factors and varies from grower to grower and field to field. Many corn growers use more than one herbicide applications that may provide a season-long weed control.

Preplant Herbicides

For control of winter annuals and early-spring annual weeds, herbicides applied on emerged weeds are known as "burndown herbicide treatment." Foliar active herbicides, such as glyphosate, 2,4-dichlorophenoxyacetic acid (2,4-D), or dicamba, are the most common herbicides used as burndown before planting corn. Many farmers include residual herbicides with early-spring burndown treatments. While this may provide a clean seedbed at planting and crop emergence, the longevity of weed control is likely to be shortened significantly. The magnitude of this reduction will depend on the time period and weather encountered between application and planting and the herbicide rate [58]. The rates of many residual products have been reduced due to the reliance on postemergence products, primarily glyphosate in glyphosate-resistant corn. If applications are to be made a few weeks earlier than normal, the product rates should be evaluated carefully in order to maximize the contribution of the residual weed control after crop emergence.

If the residual herbicide is applied before planting corn and is incorporated in the soil with light tillage, it is known as the preplant incorporated method of herbicide application. With this application method, the herbicide is applied to the soil surface and mechanically incorporated into the top 5–8 cm of soil with tillage. Preplant incorporation is a preferred method in corn production where spring rainfall is limited and, therefore, the likelihood of adequate rainfall to incorporate herbicides is low. In addition, it also reduces the chance of herbicide loss through volatilization. For example, in Kansas and Missouri, herbicide incorporation is proposed as one of the best management practices to reduce herbicide runoff from soils with poor internal drainage [59]. Buttle observed that soil incorporation led to a significant reduction in the total metolachlor loss in runoff water relative to application as preemergence [60]. However, in recent years, preplant incorporation has declined in part due to increases in no-tillage and reduced-tillage production systems.

Preemergence Herbicides

Herbicides applied after corn planting, but before emergence and having soil residual activity, are known as preemergence herbicides. Soil-applied preemergence herbicides may either be broadcast on the field or be applied in bands over the planted crop rows. Preemergence herbicides require irrigation or rainfall within 7–10 days of application to activate herbicides and enter the weed germination zone by water infiltration [58]. If there is no rainfall or source of irrigation, mechanical incorporation by a rotary hoe can move some of the herbicide into the weed germination zone. The preemergence herbicides will have little or no foliar activity, so they will not be effective for the control of already emerged weeds at the time of application. If weeds are emerged at the time of application, preemergence herbicide can be tank-mixed with foliar active herbicides to expand weed control spectrum. Excess rainfall can reduce weed control efficacy of preemergence herbicides and increase the risk of corn injury. Several preemergence herbicides have been registered for weed control in corn (Table 8.2). Due to wet soil conditions or other factors, it is quite often that many corn growers are not in a position to apply preemergence herbicides prior to corn emergence. Several residual preemergence herbicides can be applied after corn emergence (Table 8.3). For example, herbicides (e.g., atrazine and mesotrione) have foliar activity on small, emerged weeds.

Metolachlor, alachlor, and dimethenamid are acid amide herbicides, also known as chloroacetamide herbicides. The acid amide herbicides have much more activity on grass weeds, such as crabgrass (*Digitaria sanguinalis* [L.] Scop.), barnyardgrass (*Echinochloa crus-galli* [L.] Beauv.), and broadleaf signalgrass (*Urochloa platyphylla* [Munroex C. Wright]). Tank-mixing these herbicides with

Table 8.2 List of preemergence herbicides registered for weed control in corn [64]

Herbicide	Commercial product kg per hectare		
	Sandy loam	Silt loam	Silty-clay Loam
	< 1 % OM	1–2 % OM	> 2 % OM
Atrazine[a]	[b]Do not use	1.12–2.46	1.12–2.46
Isoxaflutole[b]	0.21	0.21–0.35	0.21–0.42
Isoxaflutole[b] +	0.07–0.21	0.21–0.35	0.21–0.42
Atrazine[a]	1.12	1.45	1.70
S-metolachlor + atrazine[a]	4.06	4.06–4.74	4.74
Mesotrione alone or with	0.42	0.42	0.42
S-metolachlor + atrazine	1.13	1.46	1.46
Thiencarbazone + isoxaflutole[b]	0.23	0.23–0.40	0.23–0.40
Thiencarbazone + isoxaflutole + trazine[a]	2.26	2.26	2.26
Acetochlor	2.60–3.61	3.61–4.52	3.61–4.52
Acetochlor + atrazine[a]	6.10	7.91	7.91
S-metolachlor + benoxacor	1.13	1.46	1.46
S-metolachlor + glyphosate + atrazine[a]	5.65	6.78	8.47
Flumioxazin + pyroxasulfone	0.21	0.21	0.21
Encapsulated acetochlor + atrazine[a]	5.65–6.10	6.10–7.45	6.78–7.91
Dimethenamid-P + atrazine[a]	2.26	2.82	3.95
Dimethenamid-P + atrazine[a]	2.71–3.16	3.16–3.84	3.84–4.52
Acetochlor + MON 13900 safener	1.41–1.97	1.97–2.54	1.97–2.54
Acetochlor + atrazine + MON 13900 safener	4.06	4.06–5.19	4.52–5.19
Flumetsulam + clopyralid	0.28	0.28	0.28–0.35
Acetochlor + atrazine + dichlormid	4.97–5.42	5.42–6.32	5.87–6.78
S-metolachlor + mesotrione + atrazine	6.78	6.78	6.78
S-metolachlor + mesotrione + atrazine	5.65	5.65	5.65–6.78
Dimethenamide-P	0.70–0.98	0.98–1.12	1.12–1.26
Rimsulfuron + isoxaflutole	Do not use	0.11–0.17	0.11–0.18
Pendimethalin + atrazine	Do not use	4.06	4.06
Rimsulfuron + atrazine	Do not use	0.07–0.10	0.07–0.10
Saflufenacil	0.14	0.17	0.21
Acetochlor + dichlormidsafener + flumetsulam + clopyralid	1.70	1.70–1.97	2.26
Acetochlor + dichlormidsafener	1.70–2.82	1.70–2.82	1.70–3.10
Acetochlor + dichlormidsafner alone or with	2.26	4.52–5.65	5.08–6.78
atrazine	1.23	1.45	1.68
Clopyralid + flumetsulam + acetochlor	1.70	1.97	2.26
Saflufenacil + dimethenamid-P	0.7	0.91	1.12
S-metalochlor + mesotrione + benoxacor (safener)	4.52	4.52	4.52

OM organic matter

[a] Do not apply atrazine within 20 m of where water runoff from a field will enter a stream, river, or standpipe. The total amount of atrazine (active ingredient per hectare) applied cannot exceed 2.8 kg ai/ha per calendar year. Use no more than 1.80 kg ai/ha on highly erodible land with less than 30 % crop residue. Using atrazine on soils with less than 1 % organic matter increases carryover injury risk to susceptible crops, especially high pH soils. Do not use on sandy soils if water table is less than 30 ft

[b] Do not use isoxaflutoleon coarse-textured soils of less than 2 % organic matter if the water table is less than 7.6 m. Do not use on fields prone to runoff or flooding. Crop response is most likely to occur where soils are coarse, organic matter content is less than 1.55 %, and the pH is greater than 7.4. Corn seed must be covered with 3–5 cm inches of soil. Avoid planting when soil surface is wet

Table 8.3 List of preemergence herbicides also registered for postemergence (in-crop) application in corn [64]

Herbicide	Crop stage	Maximum weed stage
Atrazine	0–30 cm	4 cm
Isoxaflutole[a]	V2	4 cm
S-metolachlor + atrazine	0–30 cm	Two-leaf
Acetochlor + atrazine + dichlormid	0–28 cm	Unemerged
Mesotrione[b]	0–76 cm	13 cm
Thiencarbazone + isoxaflutole	V2	4 cm
Acetochlor	0–28 cm	Unemerged
Acetochlor + atrazine	0–28 cm	Two-leaf
S-metolachlor + benoxacor	0–101 cm	Unemerged
S-metolachlor + glyphosate + atrazine	0–30 cm	15 cm
Encapsulated acetochlor + atrazine	0–28 cm	Unemerged
Dimethenamid-P + atrazine	0–30 cm	4 cm
Acetochlor + atrazine + MON 13900 safener	0–28 cm	Two-leaf
Flumetsulam + clopyralid	0–5 cm	20 cm
Acetochlor + atrazine + dichlormid	0–28 cm	Unemerged
S-metolachlor + mesotrione + atrazine	0–30 cm	7 cm
Dimethenamide-p	0–30 cm	Unemerged
Pendimethalin	0–76 cm	3 cm
Flumetsulam	0–51 cm	15 cm
Rimsulfuron	0–30 cm	7 cm
Atrazine + metolachlor	0–13 cm	Two-leaf
Acetochlor + dichlormidsafener + flumetsulam + clopyralid	0–28 cm	5-cm broad leaves
Acetochlor + dichlormidsafener	0–28 cm	Unemerged
Atrazine + metolachlor	0–13 cm	Two-leaf
Actochlor + atrazine + safener	0–28 cm	Unemerged

[a] If isoxaflutole is applied after the corn has emerged, do not add oil concentrate

[b] Severe injury may occur if mesotrione is applied postemergence to corn that has been treated with Counter. Do not tank-mix with any organophosphate or carbamate insecticide. Do not cultivate within 7 days of application

atrazine-applied preemergence can provide effective broad-spectrum weed control for about 3 weeks after application. Soil texture, pH, and organic matter content are the soil properties most commonly used to determine the application rates of preemergence herbicides. For example, isoxaflutole, a preemergence herbicide of corn, showed a considerable crop injury [61, 62]. It was concluded that isoxaflutole rates should be carefully selected for soils with low organic matter and high pH [63]. In the past few years, several preemergence herbicides have been tank-mixed with postemergence herbicides and are now available as a prepackaged mixture that expands weed control spectrum and provides more flexibility with application timing (Tables 8.2 and 8.3).

Postemergence Herbicides

Herbicides applied after corn and weed emergence are known as postemergence herbicides. They usually have foliar activity on emerged weeds with a good crop safety if applied as directed on the label. Postemergence herbicides can be broadcast-applied on crop and weeds or with the equipment that directs the herbicide to weeds and minimizes exposure of the crop [64]. Foliar-applied postemergence herbicides do have a requirement for rainfall after application. In fact, a certain time is required after application of postemergence herbicides that should be free from rainfall or overhead irrigation to avoid washing the chemicals of the plant and leaf surface. For example, time until herbicides are rainfast for 2,4-D is 1 h, glyphosate 1–4 h depending on glyphosate formulation, and glufosinate 4 h. Several postemergence herbicides have been registered for weed control in corn (Table 8.4).

Wide-scale adoption of glyphosate-resistant corn has resulted in heavy reliance on glyphosate for weed control for many years in Midwestern United States. Multiple glyphosate applications are relied upon for weed management in glyphosate-resistant corn, which comprise approximately 60 % of the corn hectares in the USA. In addition, more than 90 % of the soybean hectares are planted with glyphosate-resistant cultivars, placing extreme selection pressure for glyphosate resistance in weeds. Although corn and soybean are commonly rotated in North Central and Midwestern USA, corn for seed production is continually grown on the same land without rotation with other crops. The hydroxyphenylpyruvate dioxygenase (HPPD)-inhibiting herbicides, such as mesotrione, tembotrione, topramezone, and isoxaflutole, are important herbicides for control of broadleaf weeds in grain and seed corn.

Atrazine has been in use since 1958 and is applied on several million hectares in the USA and several other counties. Atrazine is the base for the weed control program in corn in the USA. It is widely used because of its low cost, control of a broad spectrum of broadleaf weeds, flexible application timing, such as preemergence or postemergence, and compatibility to mix with several other herbicides. However, a long-term and continuous use of atrazine resulted in accumulation of atrazine and its breakdown products in the environment, groundwater, and aquatic systems [65, 66]. In the USA, a recent national survey of leopard frogs (*Rana pipiens*), a species sensitive to atrazine, has shown that defects linked to atrazine exposure tended to be greater in areas of high atrazine use [67]. Therefore, the use of atrazine for crop production has been banned in several European countries, including France, Germany, Italy, and Sweden. A 3-year study conducted in Canada reported that the addition of atrazine to preemergence herbicides increased weed control (25 %), improved herbicide performance consistency, increased corn yields (8 %), increased adjusted gross return, and reduced risk over sites and years [68]. More research is required to explore the potential of reducing atrazine-use rates while maintaining effective weed control in corn and environmental quality.

Although atrazine effectively controls many broadleaf and some grass weeds, it has been inconsistent for the control of velvetleaf, common cocklebur, and *Ipomoea* spp. Because most corn growers have a number of broadleaf and grass

Table 8.4 List of postemergence herbicides registered for weed control in corn [64]

Herbicide	Rate kg per hectare	Application time
2,4-D ester or 2,4-D amine	0.56–1.12	Spike to 91 cm corn; broadleaf weeds 2–6 leaves
Nicosulfuron	0.05	Corn 10–91 cm (V10); if greater than 50 cm, use drop nozzle
Nicosulfuron + atrazine[a]	0.06 1.23	With atrazine, corn less than 30 cm
Carfentrazone-ethyl	0.03	Corn less than V14, but if greater than V8, use drop nozzles; broadleaves 2–10 cm; velvetleaf up to 90 cm
Atrazine[a]	1.60–2.5	Corn less than 30 cm; broadleaves 5–15 cm; grass weeds 2 cm or less
Atrazine[a] + dicamba[b]	0.62–1.12 + 0.63–1.12	Corn less than V5
Rimsulfuron (50 %) + thifensulfuron (25 %)	0.02	Corn spike to V2; grasses 2–5 cm; broadleaves 2–8 cm
Primisulfuron 75 %	0.03–0.05	Corn 10–50 cm; shattercane 10–30 cm; broadleaves 2–10 cm; grasses 2–8 cm
Bromoxynil + atrazine[a]	1.13–1.70 + 0.61–1.23	Corn three-leaf to 30 cm Broadleaves 5–15.24 cm
Bromoxynil + dicamba[b]	1.13–1.70 + 0.60	
Fluthiacet-methyl	0.04–0.06	Corn emergence to 120 cm
Mesotrione	0.21	Corn to 75 cm or V8; broadleaves less than 12 cm
Mesotrione + atrazine[a]	0.17 + 0.56	Corn less than 30 cm
Thiencarbazone-methyl tembotrione	0.21	V1–V6
Clopyralid + MCPA	2.26	Spike to V4
Dicamba[b]	0.56–1.12	Spike to 90 cm; if greater than 20 cm, use drops
Dicamba[b] + 2,4-D ester or amine	0.56 + 0.3 or0.30	Broadleaves 2–6 leaves
Diflufenzopyr + dicamba	0.42 + 0.30	Corn 10–25 cm; corn 25–60 cm; if 60–90 cm, use drops
S-metolachlor + glyphosate[c] + atrazine (glyphosate-resistant corn only)	6.21–8.50	Corn 0–30 cm
Glyphosate[c](glyphosate-resistant corn only)	Up to 3.40	Corn to 122 cm (V12); if over 60 cm, use drops
S-metolachlor + glyphosate[c] + mesotrione (glyphosate-resistant corn only)	4.10–4.52	Corn to 76 cm (V8); before weeds exceed 10 cm
Flumetsulam + clopyralid	0.14–0.35	Spike to 50 cm, if greater than 50 cm use drops; broadleaf weeds less than 20 cm
Topramezone	0.05–0.07	Broadleaf weeds 5–15 cm; for corn 60–71 cm, apply with drop nozzles
Topramezone + atrazine[a]	0.05 + 0.33–1.80	Corn less than 30 cm

Table 8.4 (continued)

Herbicide	Rate kg per hectare	Application time
Tembotrione	0.21	Corn emergence V8; broadleaf weeds less than 15 cm; grass weeds less than 8 cm
Tembotrione + atrazine[a]	+ 0.56	Corn less than 30 cm; weeds 2–8 cm
Tembotrione + bromoxynil	0.14 + 0.42	Corn less than 30 cm; weeds 2–15 cm
Glufosinate (Liberty Link hybrid required)	1.54–2.03	Corn at 60 cm (V7); for corn 60–90 cm, use drop nozzles to avoid spraying in whorl
Glufosinate (Liberty Link hybrid required) + tembotrione	1.54–1.70 0.10–0.21	Up to 60 cm (broadcast) or 75 cm (drops); weeds 2–10 cm
Imazethapyr + imazapyr alone (Clearfield hybrid required) or with dicamba[b]	0.08 0.56–1.12	Corn up to 50 cm (V6); weeds up to 10 cm; weeds to 10 cm
Foramsulfuron	0.09–0.12	Corn 0–90 cm, if greater than 40 cm, use drops; weeds less than 10 cm
Halosulfuron	0.05–0.09	Corn spike lay-by greater than 50 cm use drops; broadleaf weed 5–15 cm
Rimsulfuron + mesotrione	0.28	Corn up to 50 cm or V7
Rimsulfuron + dicamba[b]	0.28	V2–V7 corn; 2–8 cm weeds
Rimsulfuron + thifensulfuron	0.08	Corn up to 50 cm or V7
Flumiclorac	0.28–0.42	Corn V2–V10; broadleaf weeds less than 10 cm
Rimsulfuron	0.07	Corn up to 30 cm or V6, whichever is most restrictive
Prosulfuron + primsulfuron	0.07	Corn 10–60 cm; if greater than 50 cm (V6), use drops; weeds 5–20 cm
Fluroxypry + bromoxynil	0.45	VE–V5 corn; sweet corn up to V4; weeds less than 20 cm
Diflufenzo-pyr + dicamba[b] + isoxadifen	0.17	Corn 10–90 cm
Nicosulfuron + rimsulfuron	0.05–0.10	Corn to 50 cm or V6; weeds 5–10 cm
Nicosulfuron + acifluorfen	0.05–0.10	Corn to 40 cm or V5
Fluroxypyr + clopyralid	1.50	Corn spike to V5; weeds less than 20 cm
Halosulfuron + dicamba[b]	0.28–0.56	Spike to 90 cm; weeds 2–15 cm

OM organic matter

[a] Do not apply atrazine within 66 ft of where water runoff from a field will enter a stream, river, or standpipe. The total amount of atrazine applied (active ingredient per hectare) cannot exceed 2.8 kg ai/ha per calendar year. Use no more than 1.8 kg ai/ha on highly erodible land with less than 30 % crop residue. Using atrazine on soils with less than 1 % organic matter increases carryover injury risk to susceptible crops, especially high pH soils. Do not use on sandy soils if water table is shallower than 30 ft

[b] Dicamba rates are based on a 4.5–3.4 kg ae/ha formulation

[c] Glyphosate rates are based on a 4.5–3.4 kg ae/ha formulation

weed species in their fields, tank-mixing atrazine with other herbicides—such as mesotrione, isoxaflutole, or acetochlor—might be desirable to broaden the weed control spectrum. Mixtures of two or more herbicides may provide more consistent control of certain weeds, reduce the risk of evolving weed resistance, and may reduce the amount of total active ingredient applied [69, 70]. Synergistic interactions have been observed between mesotrione- and atrazine-applied postemergence for the control of velvetleaf, sunflower (*Helianthus annuus* L.), and Palmer amaranth [71]. Furthermore, tank-mixing atrazine with mesotrione-applied preemergence in corn increased the control of common ragweed (*Ambrosia artemisiifolia*), common lambsquarters (*Chenopodium album* L.), and *Ipomoea* spp. [72]. Several new herbicides have been registered for weed control in corn in the past few years that are tank mixtures of existing herbicides (Table 8.3) [64].

Herbicide Injury

Corn plants are occasionally injured by herbicides. To minimize crop injury, herbicides must be applied uniformly at the stage of crop growth specified on the label. Unfavorable conditions, such as cool, wet weather, delayed crop emergence, deep or shallow planting, seedling diseases, soil in poor physical condition, and poor quality seeds, may contribute to crop stress and herbicide injury. Corn hybrids and cultivars may vary in their tolerance to herbicides and environmental stress. Crop planting options for next season also must be considered when selecting the herbicide program. Corn herbicides may have restrictive cropping intervals for some agronomic and many vegetable crops.

Multiple Herbicide-Resistant Corn

Since 1998, genetically modified herbicide-resistant corn, primarily glyphosate-resistant, has helped to revolutionize weed management and has become an important tool in corn production practices in the Americas [73]. Glyphosate has performed long and well, but due to its widespread and repeated use, 13 weed species in the USA have evolved resistance to glyphosate and 23 species worldwide by 2012 [74]. Unfortunately, most companies are not developing any new selective herbicides with new modes of action that can be effective for the control of glyphosate-resistant weeds [75]. However, they are developing new multiple herbicide-resistant corn traits through genetic engineering to combine with glyphosate resistance and expand the utility of existing herbicides [76]. For example, glyphosate plus glufosinate-resistant corn is already available in the market.

Despite the fact that auxin group herbicides, such as 2,4-D and dicamba, have been used for many years, only a few weed species have evolved resistance to this group of herbicides. Efforts are under way to commercialize 2,4-D plus glyphosate-resistant corn and soybeans, known as Enlist™ System as well as dicamba + glyphosate-resistant corn and soybeans, known as Roundup Ready™ 2 Xtend System. There are several controversies prevailing about multiple herbicide-resistant corn and soybeans that are currently pending regulatory approval. Many groups and individuals are concerned, and they argue that multiple herbicide-resistant crop varieties will make growers more dependent on the intellectual property held by large corporations, will injure nontarget crops due to drift and volatility of 2,4-D and dicamba, and will accelerate the evolution of multiple herbicide-resistant weeds [77, 78]. Others argue that multiple herbicide-resistant crop cultivars will help growers controlling herbicide-resistant weeds. The message is clear that no weed management technology used alone is sustainable since weeds will adapt to any single tactic used repeatedly for many years. Therefore, an IWM approach is required for sustainable corn production to meet the growing demand.

References

1. Rooney LW, McDonough CM, Waniska RD (2004) The corn kernel. In: Smith CW, Betran J, Runge ECA (ed) Corn: origin, history, technology, and production. Wiley, Hoboken, pp 273–304
2. Harshberger JW (1896) Fertile crosses of teosinte and maize. Garden Forest 9:522–523
3. Wilkes G (2004) Corn, strange and marvelous: but is a definitive origin known? In: Smith CW, Betran J, Runge ECA (eds) Corn: origin, history, technology and production. Wiley, Hoboken, pp 3–64
4. Anonymous (2013a) Food outlook-biannual report on global food markets (June 2013). http://www.fao.org/docrep/018/al999e/al999e.pdf. Accessed 7 Nov 2013
5. Kremer RJ (2004) Weed control. In: Smith CW, Betran J, Runge ECA (eds) Corn: origin, history, technology and production. Wiley, Hoboken, pp 717–752
6. Hall MR, Swanton CJ, Anderson GW (1992) The critical period of weed control in grain corn. Weed Sci 40:441–447
7. Van Acker RC, Swanton CJ, Weise SF (1993) The critical period of weed control in soyabean [*Glycine max* (L.) Merr.]. Weed Sci 41:194–200
8. Knezevic SZ, Weise SF, Swanton CJ (1994) Interference of redroot pigweed (*Amaranthus retroflexus*) in corn (*Zea mays*). Weed Sci 42:568–573
9. Oliver LR (1988) Principles of weed threshold research. Weed Technol 2:398–403
10. O'Donovan JT, de St. Remy EA, O'Sullivan PA, Dew DA, Sharma AK (1985) Influence of wild oat (*Avena fatua*) on yield loss of barley (*Hordeum vulgare*) and wheat (*Triticum aestivum*). Weed Sci 33:498–503
11. Swanton CJ, Weaver S, Cowan P, Van Acker R, Deen W, Shreshta A (1999) Weed thresholds: theory and applicability. J Crop Prod 2:9–29
12. Perry K, Evans MR, Jeffery LS (1983) Competition between Johnsongrass (*Sorghum halepense*) and corn (*Zea mays*). Proc South Weed Sci Soc 36:345
13. Vernon R, Parker JMH (1983) Maize/weed competition experiments: implications for tropical small-farm weed control research. Exp Agric 19:341–347
14. Ghosheh HZ, Holshouser DL, Chandler JM (1996) The critical period of Johnsongrass (*Sorghum halepense*) control in field corn (*Zea mays*). Weed Sci 44:944–947
15. Page ER, Cerrudo D, Westra P, Loux M, Smith K, Foresman C, Wright H, Swanton CJ (2012) Why early season weed control is important in maize. Weed Sci 60:423–430
16. Evans SP, Knezevic SZ, Shapiro C, Lindquist JL (2003a) Nitrogen level affects critical period for weed control in corn. Weed Sci 51:408–417

17. Evans S, Knezevic SZ, Shapiro C, Lindquist JL (2003b) Influence of nitrogen level and duration of weed interference on corn growth and development. Weed Sci 51:546–556
18. Knezevic SZ, Evans SP, Blankenship EE, Van Acker RC, Lindquist JL (2002) Critical period of weed control: the concept and data analysis. Weed Sci 50:773–786
19. Knezevic SZ, Evans SP, Mainz M (2003a) Yield penalty due to delayed weed control in corn and soybean. Crop Manag J. http://www.plantmanagementnetwork.org/pub/cm/research/2003/delay/ Accessed 20 Oct 2013
20. Knezevic ZS, Evans SP, Mainz M (2003b) Row spacing influences critical time of weed removal in soybean. Weed Technol 17:666–673
21. Fausey JC, Kells JJ, Swinton SM, Renner KA (1997) Giant foxtail (*Setaria faberi*) interference in non-irrigated corn (*Zea mays*). Weed Sci 45:256–260
22. Scholes C, Clay SA, Brix-Davis K (1995) Velvetleaf (*Abutilon theophrasti*) effect on corn (*Zea mays*) growth and yield in South Dakota. Weed Technol 9:665–668
23. Massing RA, Currie RS, Trooien TP (2003) Water use and light interception under Palmer amaranth and corn competition. Weed Sci 51:523–531
24. Rifai MN, Astatkie T, Lacko-Bartosova M, Gadus J (2002) Effect of two different thermal units and three types of mulch on weeds in apple orchards. J Environ Eng Sci 1:331–338
25. Geier PW, Stahlman PW, Frihauf JC (2006) KIH-485 and *S*-metolachlor efficacy comparisons in conventional and no-tillage corn. Weed Technol 20:622–626
26. Knezevic SZ, Datta A, Scott J, Klein RN, Golus J (2009) Problem weed control in glyphosate-resistant soybean with glyphosate tank mixes and soil-applied herbicides. Weed Technol 23:507–512
27. Marshall EJP, Brown VK, Boatman ND, Lutman PJW, Squire GR, War LK (2003) The role of weeds in supporting biological diversity within crop fields. Weed Res 43:77–89
28. Melander B, Rasmussen IA, Barberi P (2005) Integrating physical and cultural methods of weed control—examples from European research. Weed Sci 53:369–381
29. Kristoffersen P, Rask AM, Grundy AC et al (2008) A review of pesticide policies and regulations for urban amenity areas in seven European countries. Weed Res 48:201–214.
30. Rask AM, Andreasen C, Kristoffersen P (2012a) Response of *Lolium perenne* to repeated flame treatments with various doses of propane. Weed Res 52:131–139
31. Rask AM, Kristoffersen P, Andreasen C (2012b) Controlling grass weeds on hard surfaces: effect of time intervals between flame treatments. Weed Technol 26:83–88
32. Swanton CJ, Weise SF (1991) Integrated weed management: the rationale and approach. Weed Technol 5:657–663
33. Swanton CJ, Murphy SD (1996) Weed science beyond the weeds: the role of integrated weed management (IWM) in agroecosystem health. Weed Sci 44:437–445
34. Teasdale J (1996) Contribution of cover crops to weed management in sustainable agricultural systems. J Prod Agric 9:475–479
35. Lague C, Gill J, Peloquin G (2001) Thermal control in plant protection. In: Vincent C, Panneton B, Fleurat-Lessard F (eds) Physical control methods in plant protection. Springer, Berlin, pp 35–46
36. Ascard J (1998) Comparison of flaming and infrared radiation techniques for thermal weed control. Weed Res 38:69–76
37. Parish S (1990) A review of non-chemical weed control techniques. Biol Agric Hort 7:117–137
38. Pelletier Y, McLeod CD, Bernard G (1995) Description of sub-lethal injuries caused to the Colorado potato beetle by propane flamer treatment. J Econ Entomol 88:1203–1205
39. Ascard J (1995) Effects of flame weeding on weed species at different developmental stages. Weed Res 35:397–411
40. Ascard J, Hatcher PE, Melander B, Upadhyaya MK (2007) Thermal weed control. In: Upadhyaya MK, Blackshaw RE (eds) Non-chemical weed management (Chapter 10). CAB International, Wallingford, pp 155–175
41. Bond W, Grundy AC (2001) Non-chemical weed management in organic farming systems. Weed Res 41:383–405

42. Leroux GD, Douheret J, Lanouette M (2001) Flame weeding in corn. In: Vincent C, Panneton B, Fleurat-Lessard F (eds) Physical control methods in plant protection. Springer, Berlin, pp 47–60
43. Hansson D, Ascard J (2002) Influence of developmental stage and time of assessment on hot water weed control. Weed Res 42:307–316
44. Knezevic SZ, Ulloa SM (2007) Flaming: potential new tool for weed control in organically grown agronomic crops. J Agric Sci 52:95–104
45. Ascard J (1994) Dose–response models for flame weeding in relation to plant size and density. Weed Res 34:377–385
46. Ulloa SM, Datta A, Knezevic SZ (2010a) Growth stage influenced differential response of foxtail and pigweed species to broadcast flaming. Weed Technol 24:319–325
47. Ulloa SM, Datta A, Knezevic SZ (2010b) Tolerance of selected weed species to broadcast flaming at different growth stages. Crop Prot 29:1381–1388
48. Ulloa SM, Datta A, Knezevic SZ (2010c) Growth stage impacts tolerance of winter wheat (*Triticum aestivum* L.) to broadcast flaming. Crop Prot 29:1130–1135
49. Ulloa SM, Datta A, Malidza G, Leskovsek R, Knezevic SZ (2010d) Timing and propane dose of broadcast flaming to control weed population influenced yield of sweet maize (*Zea mays* L. var. *rugosa*). Field Crops Res 118:282–288
50. Ulloa SM, Datta A, Cavalieri SD, Lesnik M, Knezevic SZ (2010e) Popcorn (*Zea mays* L. var. *everta*) yield and yield components as influenced by the timing of broadcast flaming. Crop Prot 29:1496–1501
51. Ulloa SM, Datta A, Malidza G, Leskovsek R, Knezevic SZ (2010f) Yield and yield components of soybean [*Glycine max* (L.) Merr.] are influenced by the timing of broadcast flaming. Field Crops Res 119:348–354
52. Ulloa SM, Datta A, Bruening C, Gogos G, Arkebauer TJ, Knezevic SZ (2012) Weed control and crop tolerance to propane flaming as influenced by the time of day. Crop Prot 31:1–7
53. Ulloa SM, Datta A, Bruening C, Neilson B, Miller J, Gogos G, Knezevic SZ (2011a) Maize response to broadcast flaming at different growth stages: effects on growth, yield and yield components. Eur J Agron 34:10–19
54. Ulloa SM, Datta A, Knezevic SZ (2011b) Growth stage influenced sorghum response to broadcast flaming: effects on yield and its components. Agron J 103:7–12
55. McWilliams DA, Berglund DR, Endres GJ (1999) Corn growth and management quick guide. http://www.ag.ndsu.edu/publications/landing-pages/crops/corn-growth-and-management-quick-guide-a-1173. Accessed 16 March 2013
56. Taiz L, Zeiger E (2002) Plant physiology, 3rd edn. Sinauer Associates, Sunderland
57. Smith RJ Jr (1986) Biological control of northern jointvetch (*Aeschynomene virginica*) in rice (*Oryza sativa*) and soybeans (*Glycine max*)—a researcher's view. Weed Sci 34(Suppl 1):17–23
58. Hoeft RG, Nafziger ED, Johnson RR, Aldrich SR (eds) (2000) Modern corn and soybean production. MCSP Publications, Savoy
59. Baker JL, Mickelson SK (1994) Application technology and best management practices for minimizing herbicide runoff. Weed Technol 8:862–869
60. Buttle JM (1990) Metolachlor transport in surface runoff. J Environ Qual 19:531–538
61. Knezevic SZ, Sikkema PH, Tardif F, Hamill AS, Chandler K, Swanton CJ (1998) Biologically effective dose and selectivity of RPA 201772 (isoxaflutole) for preemergence weed control in corn. Weed Technol 12:670–676
62. Simmons TJ, Kells JJ (2003) Variation and inheritance of isoxaflutole tolerance in corn (*Zea mays* L.). Weed Technol 17:177–180
63. Wicks GA, Knezevic SZ, Bernards M, Wilson RG, Klein RN, Martin AR (2007) Effect of planting depth and isoxaflutole rate on crop injury in Nebraska. Weed Technol 21:642–646
64. Anonymous (2013b) Guide for weed management in Nebraska with insecticide and fungicide information. University of Nebraska-Lincoln Extension, Lincoln, NE. http://www.ianrpubs.unl.edu/epublic/pages/index.jsp?what=publicationD&publicationId=941. Accessed 7 Nov 2013

65. Ahel M, Evans KM, Fileman TW, Mantoura RFC (1992) Determination of atrazine and simazine in estuarine samples by high-resolution gas chromatography and nitrogen selective detection. Anal Chim Acta 268:195–204
66. Ying GG, Kookana RS, Mallavarpu M (2005) Release and behavior of triazine residues in stabilised contaminated soils. Environ Pollut 134:71–77
67. Hayes TB, Haston K, Tsui M, Hoang A, Haeffele C, Vonk A (2003) Atrazine-induced hermaphrodisom at 0.1 ppb in American leopard frogs (*Rana pipiens*): laboratory and field evidence. Environ Health Perspect 111:568–575
68. Swanton CJ, Gulden RH, Chandler K (2007) A rationale for atrazine stewardship in corn. Weed Sci 55(1):75–81
69. Harker KN, O'Sullivan PA (1991) Synergistic mixtures of sethoxydim and fluazifop on annual grass weeds. Weed Technol 5:310–316
70. Zhang J, Hamill AS, Weaver SE (1995) Antagonism and synergism between herbicides: trends from previous studies. Weed Technol 9:86–90
71. Abendroth JA, Martin AR, Roeth FW (2006) Plant response to combinations of mesotrione and photosystem II inhibitors. Weed Technol 20:267–274
72. Armel GR, Wilson HR, Richardson RJ, Hines TE (2003) Mesotrione, acetochlor, and atrazine for weed management in corn. Weed Technol 17:284–290
73. Green JM (2012) The benefits of herbicide-resistant crops. Pest Manag Sci 68:1323–1331
74. Heap I (2012) International survey of herbicide resistant weeds. www.weedscience.org/In.asp. Accessed 10 June 2012
75. Duke SO (2012) Why have no new herbicide modes of action appeared in recent years? Pest Manag Sci 68:505–512
76. Green JM, Hazel CB, Forney DR, Pugh LM (2008) New multiple-herbicide crop resistance and formulation technology to augment the utility of glyphosate. Pest Manag Sci 64:332–339
77. Knezevic ZS, Cassman K (2003) Use of herbicide tolerant crops as a component of an IWM. Crop Manag J. http://www.plantmanagementnetwork.org/pub/cm/management/2003/htc. Accessed 20 Oct 2013
78. Knezevic SZ (2007) Herbicide tolerant crops:10 years later. Maydica 52:245–250

Chapter 9
Integrated Weed Management in Cotton

Mehmet Nedim Doğan, Khawar Jabran and Aydin Unay

Introduction

The name of cotton (*Gossypium* spp.) is derived from the Arabic language, and it is the most important fiber crop of the world—providing enormous utility for our daily lives. Cotton plant has a perennial growth habit; thus, plants get the shape of a bush if it is not cut [1]. Currently, it is grown on nearly 2.5% of the world's arable lands, with an approximate annual production of 2.5 million tons. Towels, socks, tissue papers, bedsheets, curtains, clothes, jeans, and shoestrings are among the hundreds of cotton lint products [1]. "Linters" is the small fuzz that remains adhered to the cottonseed after ginning. Several products of linters importantly include paper, plastic, explosives, cushions, and mattresses [2]. Hulls, oil, and meal are obtained after crushing the seeds of cotton. Hull and meal are used as animal feed along with several other uses. The use of oil from cotton is not restricted to human food only. The stalks and leaves of cotton are a good source of organic fertilizer for the soil. The modern cotton industry has a nearly 300-year-old history, which peaked with the invention of most advanced ginners and other mechanization, taking it from the field to final products. Currently, it is among the biggest industries of the world, providing livelihood to millions of people [1].

Cotton probably originated in what is now Pakistan. Most probably, it was grown for the first time 7000 years ago in the city of Mehrgarh, and later it spread to other parts of Asia and across the world [3]. Cotton fabric approximately dating to 5000 BCE has been excavated from the Kachi Plain of Pakistan (Indus Valley

M. N. Doğan (✉) · K. Jabran
Department of Plant Protection, Adnan Menderes University, Aydin, 09100, Turkey
e-mail: doganmn@hotmail.com; mndogan@adu.edu.tr

K. Jabran
e-mail: khawarjabran@gmail.com

A. Unay
Department of Field Crops, Agricultural Faculty, Adnan Menderes University,
Aydin, 09100, Turkey
e-mail: aunay@adu.edu.tr

B. S. Chauhan, G. Mahajan (eds.), *Recent Advances in Weed Management*,
DOI 10.1007/978-1-4939-1019-9_9,

Civilization) [3, 4]. Four species of cotton, including *G. hirsutum* L., *G. barbadense* L., *G. arboreum* L., and *G. herbaceum* L., are grown in various parts of the world. However, nearly 90 % area is occupied by *G. hirsutum*.

Recently, the development of genetically modified cotton varieties for resistance against insect pests and herbicides has pertinently improved the world cotton production with a reduced input of pesticides [5]. Further, the development of herbicide-resistant cultivars has improved the cotton yield by facilitating the weed management [5].

Weeds are the second most important pests after insects for the cotton crop [6]. The annual cotton yield losses due to weeds range between 10 and 90 %. Integrated weed management (IWM) possesses particular importance for cotton crop in wake of the troublesome weed flora, slow growth rate at start, long crop duration, and less availability of some suitable postemergence herbicides [7]. Long crop duration of cotton requires a long-term weed control for better yield and enhanced fiber quality. Further, the growth of the cotton plant is rather slow in its initial stages; hence, there is a likelihood that improperly controlled weeds can overcome the cotton crop. Therefore, IWM is the solid solution for effective weed control, improved fiber quality, and increased lint yield of cotton.

Currently, a document comprehensively addressing the IWM in cotton is rarely available in the literature. This book chapter is an effort to provide the researchers, academicians, and the extension community with comprehensive information regarding the practical IWM in cotton. We have reviewed the cotton weed flora, the influence of cropping systems on cotton weed flora, losses caused by weeds in cotton, critical periods for weed control in cotton, decision making, and the strategies for IWM in cotton.

Weed Flora

Diverse and complex weed flora have been recorded in cotton fields throughout the world [8–10]. All kinds of weeds—including grasses, broad-leaved weeds, sedges, and annual and perennial weeds—have been reported in cotton [11]. In addition to annual weeds, perennial weeds and sedges also pose a serious threat to cotton productivity in many countries of the world [12–14]. Application of pre-sowing or preemergence herbicides is the most practiced strategy for controlling weeds in cotton crop. However, many of the weeds, especially the perennial ones, can escape from this control strategy [10, 14, 15]. There is greater likelihood for the escape of a few weeds from the complex of weeds if a single management practice is followed. However, IWM would be more effective in order to tackle such issues [16]. Information regarding the weed flora in cotton grown throughout the world has been complied and presented in Table 9.1.

Table 9.1 A comprehensive list of weeds reported in cotton throughout the world

Botanical name	Common name	References
Abutilon theophrasti Medic.	Velvetleaf	Kalivas et al. [10]
Acalypha ostryifolia Riddell	Hophornbeam copperleaf	Norsworthy et al. [9]
Aeschynomene virginica (L.) B.S.P.	Northern jointvetch	Norsworthy et al. [9]
Amaranthus albus L.	Tumble pigweed	Rushing et al. (1985b) [88]
Amaranthus blitoides S.Wats.	Matweed	Kalivas et al. [10]
Amaranthus palmeri S. Wats.	Palmer amaranth	Sosnoskie et al. [89]
Amaranthus retroflexus L.	Redroot pigweed	Scroggs et al. [14]; Kalivas et al. [10]
Ambrosia artemisiifolia L.	Common ragweed	Everman et al. [5]
Anoda cristata (L.) Schlecht.	Spurred anoda	Molin et al. [90]
Brachiaria platyphylla (Griseb.) Nash.	Broadleaf signalgrass	Clewis et al. [13]
Brunnichia ovata (Walt.) Shinners	Redvine	Norsworthy et al. [9]
Capsella bursa-pastoris (L.) Medik.	Shepherd's-purse	Norsworthy et al. [9]
Cassia occidentalis L.	Coffee senna	Higgins et al. [91]
Chamaesyce maculata (L.) Small	Spotted spurge	Norsworthy et al. [9]
Chenopodium album L.	Common lambsquarters	Everman et al. [5]
Chrozophora tinctoria (L.) A.Juss.	Dyer's croton	Kalivas et al. [10]
Commelina benghalensis L.	Benghal dayflower	Webster and Sosnoskie [19]
Commelina diffusa Burm. f.	Spreading dayflower	Norsworthy et al. [9]
Convolvulus arvensis L.	Field bindweed	Ali et al. [8]; Kalivas et al. [11]
Conyza canadensis (L.) Cronquist	Horseweed	Everitt and Keeling [18]; Steckel and Gwathmey [92]
Corchorus tridens L.	Wild jute	Ali et al. [8]
Cucumis melo L.	Smellmelon	Scroggs et al. [14]
Cynodon dactylon (L.) Pers.	Bermudagrass	Ali et al. [8]; Kalivas et al. [11]
Cyperus esculentus L.	Yellow nutsedge	Norsworthy et al. [9]
Cyperus rotundus L.	Purple nutsedge	Ali et al. [8]; Kalivas et al. [11]
Dactyloctenium aegyptium (L.) Willd.	Crowfootgrass	Jarwar et al. [12]
Datura stramonium L.	Jimsonweed	Scott et al. [27]; Everman et al. [5]; Kalivas et al. [10]
Digera muricata (L.) Mart.	False amaranth	Ali et al. [8]
Digitaria sanguinalis (L.) Scop.	Large crabgrass	Clewis et al. [13]
Echinochloa colona (L.) Link	Junglerice	Ali et al. [8]; Jarwar et al. [12]
Echinochloa crus-galli (L.) Beauv.	Barnyardgrass	Scroggs et al. [14]; Kalivas et al. [10]
Eclipta prostrata (L.) L.	Eclipta	Norsworthy et al. [9]
Eleusine indica (L.) Gaertn.	Goosegrass	Scroggs et al. [14]
Euphorbia helioscopia L.	Sun spurge	Ali et al. [8]
Euphorbia prostrata L.	Prostrate sandmat	Ali et al. [8]
Flaveria bidentis L.	Yellowtops	Yong et al. [47]

Table 9.1 (continued)

Botanical name	Common name	References
Galium aparine L.	Cleavers	Clewis et al. [13]
Geranium carolinianum L.	Carolina geranium	Norsworthy et al. [9]
Glycine max (L.) Merr.	Volunteer soybean (herbicide-resistant)	Norsworthy et al. [9]
Hibiscus trionum L.	Bladder weed	Kalivas et al. [10]
Ipomoea hederacea Jacq.	Ivyleaf morningglory	Everman et al. [5]
Ipomoea lacunosa L.	Pitted morningglory	Everman et al. [93]
Ipomoea spp.	Morningglory	Norsworthy et al. [9]
Lolium spp.	Ryegrass	Norsworthy et al. [9]
Oenothera laciniata Hill	Cutleaf evening-primrose	Norsworthy et al. [9]
Oryza sativa L.	Red / weedy rice	Norsworthy et al. [9]
Panicum dichotomiflorum Michx.	Fall panicum	Norsworthy et al. [9]
Persicaria maculosa L.	Ladysthumb	Askew and Wilcut [94]
Physalis spp.	Groundcherries	Norsworthy et al. [9]
Poa annua L.	Annual bluegrass	Norsworthy et al. [9]
Polygonum spp.	Smartweed	Norsworthy et al. [9]
Portulaca oleracea L.	Common purslane	Ali et al. [8]; Kalivas et al. [10]
Rumex crispus L.	Curly dock	Norsworthy et al. [9]
Salsola iberica Sennen & Pau	Russian thistle	Everitt and Keeling [18]
Senna obtusifolia (L.) H.S.Irwin & Barnaby	Sicklepod	Scroggs et al. [14]; Everman et al. [5]
Sesbania herbacea (P. Mil.) McVaugh	Hemp sesbania	Scroggs et al. [14]
Setaria pumila (Poir.) Roem. & Schult.	Yellow foxtail	Clewis et al. [13]
Setaria spp.	Foxtails	Norsworthy et al. [9]
Setaria viridis (L.) P.Beauv.	Green foxtail	Ali et al. [8]
Sida spinosa L.	Prickly sida	Norsworthy et al. [9]
Solanum nigrum L.	Black nightshade	Kalivas et al. [10]
Solanum rostratum Dunal.	Buffalobur	Rushing et al. [95]
Sorghum halepense (L.) Pers.	Johnsongrass	Scroggs et al. [14]; Kalivas et al. [11]
Trianthema monogyna L.	Horsepurslane	Ali et al. [8]
Tribulus terrestris L.	Puncturevine	Ali et al. [8]
Urochloa ramosa L.	Browntop millet	Scroggs et al. [14]
Xanthium strumarium L.	Common cocklebur	Kalivas et al. [10]

Cropping System and Weed Flora

Weed flora of an agroecosystem potentially changes with the changing cropping systems or with the shift in the agronomic aspects of crop production. Variation in the composition and the structure of the weed community occurs with varying management practices. Understanding the relationship between the cropping systems and the weed flora is critical for formulating strategies for properly managing the weeds.

Conventional tillage includes excessive soil manipulations using high-energy and resource inputs [17]. Conversely, conservation tillage focuses on minimum soil disturbance. Many cotton growers are adopting conservation cropping systems, including conservation tillage as a component in order to reduce the cost of production and hence increase the net income with additional benefits of soil and environmental conservation [18]. The rate of organic matter deterioration is far less in conservation tillage due to less soil stirring compared to conventional tillage. Similarly, the microbial activities are also accelerated under the conservation tillage system. This ultimately results in healthier soil and higher water-holding capacity and a changed weed flora [17]. The variable soil disturbance in conventional and conservation tillage leads to a variable seed bank and hence the variable weed flora in an agroecosystem. Nevertheless, the adoption of conservation cropping systems can potentially increase the weed seed bank, perennial weed intensity, and the intensity of some small-seeded weeds. Conversely, an opposite opinion depicts a reduction in weed intensity through adoption of conservation agriculture practices. According to this view, the minimum or no soil disturbance would not allow the access of weed seeds to the deep soil weed seed bank. Further, the current crop rotation systems also influence structuring of weed flora in cotton and other crops.

The use of the same herbicide(s) for a longer time also results in a persistent weed flora in some particular cropping systems. The new persistent weed flora is resistant to the continuously applied herbicide. For example, the weed flora of many cotton fields has been drastically changed by the use of glyphosate for managing weeds in glyphosateresistant cotton [19]. The technology was immediately adopted in some countries (e.g., USA) due to several of its benefits, such as wider application timing window, fast translocation, and an effective as well as easy weed control. However, weed control through application of glyphosate on glyphosate-resistant cotton has resulted in changed weed communities. For instance, the order of top ten troublesome weeds of Georgia (state of USA) [20, 21] changed from 1995 levels:

1. Nutsedges (*Cyperus* spp.)
2. Sicklepod (*Senna obtusifolia* [L.] H. S. Irwin and Barnaby)
3. Coffee senna (*Senna alexandrina* Mill.)
4. Texas millet (*Urochloa texana* [Buckley] R.D. Webster)
5. Pigweeds (*Amaranthus* spp.)
6. Common cocklebur (*Xanthium strumarium* L.)
7. Morningglories (*Ipomoea* spp.)
8. Wild poinsettia (*Euphorbia pulcherrima* Willd. ex Klotzsch)
9. Bristly starbur (*Acanthospermum hispidum* D.C.)
10. Bermudagrass (*Cynodon dactylon* [L.] Pers.)

to 2005 levels [20, 21]:

1. Benghal dayflower (*Commelina benghalensis* L.)
2. Palmer amaranth (*Amaranthus palmeri* S.Wats.)
3. *Ipomoea* spp.
4. Florida pusley (*Richardia scabra* L.)

5. Nutsedges (*Cyperus* spp.)
6. Asiatic dayflower (*Commelina communis* L.)
7. Smallflower morningglory (*Jacquemontia tamnifolia* [L.] Griseb.)
8. *U. texana*
9. *E. pulcherrima*
10. *C. dactylon*

A continuation of weed management practices like the use of single active ingredient would result in a changed weed flora and development of herbicide-resistant weed biotypes. Hence, an induced change in weed flora as a result of certain changes in the cropping system as well as constant weed management practice offers a challenging task of controlling complex weed flora. Undoubtedly, adoption of IWM is the sole solution to control the complex weed flora under the complicated set of management practices.

Effects of Weeds on Cotton Production (Cotton–Weed Interference)

Direct and indirect losses are posed by weeds to the cotton crop. Direct losses include the competition for resources such as nutrients, water, light, and space as well as the damages caused by allelopathic interactions. The indirect losses mostly include the quality deterioration, harvesting obstructions, and provision of growth media for insect pests and disease pathogens. Although the profitability of cotton production is influenced by all these factors collectively, the yield losses due to direct effects can be measured more accurately in most cases while the quantification of indirect effects is a difficult task.

Many investigations show that cotton yield is greatly reduced by direct weed competition. Oerke estimated the potential and actual yield losses of cotton due to weeds from different regions of the world (North Africa, West Africa, East Africa, southern Africa; North America, Central America, northern part of South America, southern part of South America; Near East, South Asia, Southeast Asia, East Asia, northwest Europe, southern Europe, northeast Europe, southeast Europe, Oceania) [6]. The overall yield losses in cotton crop were 36 % if no weed management was implemented and 9 % if weed control measures were adopted. Although these yield losses are considered average worldwide estimations, the losses can be variable depending upon weed species, severity, and duration of concurrence as well as the location [22]. A study from Turkey [23] showed that whole-season weed competition with cotton caused seed yield reduction by 70–85 %, where jimsonweed (*Datura stramonium* L.) dominated on the field accompanied by purple nutsedge (*Cyperus rotundus* L.), common lambsquarter (*Chenopodium album* L.), and common purslane (*Portulaca oleracea* L.). The yield losses of naturally colored cotton due to weed competition were investigated in Brazil. There were 21 different weed species on the field with higher relative importance and dominance of monocotyledonous species. The whole-season weed competition caused nearly 83 % yield losses in naturally colored cotton cultivar [24].

Although yield losses due to mixed weed populations are more realistic, some studies report single-weed-species-based cotton yield losses. Crowley and Buchanan reported that tall morningglory (*Ipomoea purpurea* [L.] Roth.) reduced cotton yield by 88 % at a density of 32 weed plants per 15 m of row [25]. Similarly, the prevalence of Johnsongrass (*Sorghum halepense* [L.] Pers.) reduced the cotton yield by 65–90 % [22, 26]. Scott et al. conducted field experiments to evaluate density-dependent effects of *D. stramonium* on cotton yield and found that this weed reduced cotton yield by 10–25 % depending on the densities [27]. The effect of *C. dactylon* competition periods on cotton yield was investigated by Keeley and Thullen, who found that cotton yield was reduced by 16 and 26 % with *C. dactylon* competition periods of 12 and 25 weeks, respectively [28]. Keeley and Thullen investigated the efficacy of prometryn on cotton–black nightshade (*Solanum nigrum* L.) competition and found that it reduced cotton yield by 22 and 100 % under dry and moist conditions, respectively [29].

In addition to the yield losses through direct competition, the weeds also harbor cotton disease pathogens and insect pests, which not only damage the cotton crop but also deteriorate the fiber quality. For example, the weeds serve as off-season host plants for verticillium wilt disease of cotton caused by *Verticillium dahliae* Kleb. [30]. Thus, weeds can be responsible for spreading the fungus in nearby and/or virgin fields. In a survey study on fields severely infested with verticillium wilt disease in the western part of Turkey, Yildiz et al. found that these fields were heavily infested with *S. nigrum, X. strumarium,* redroot pigweed (*Amaranthus retroflexus* L.), *P. oleracea,* and *D. stramonium* [31]. In laboratory studies, *V. dahliae* was isolated from *X. strumarium* and *A. retroflexus,* and their pathogenicities on cotton were quite high. Weeds also harbor the viruses that cause diseases in cotton. For instance, *X. strumarium* serves as the source of cotton leaf curl Burewala virus while some other weeds act as sources for begomoviruses to cause diseases in cotton [32, 33].

Weeds not only harbor dangerous cotton insect pests but also act as their alternative hosts. Sow thistle (*Sonchus oleraceus* L.) and morningglory (*Ipomoea* spp.) are considered as harboring sites and alternate hosts for *Helicoverpa armigera* and *Spodoptera eridania* [34, 35]. A broad range of plant species were investigated as the alternate hosts for the cotton insect pests from the *Adelphocoris* spp. insects (Hemiptera: Miridae). The results indicated that the weeds hosted these insects for living, egg laying, and overwintering [36].

In conclusion, the weeds not only reduce the cotton yield through competition for resources but also harbor cotton diseases, pathogens, and insect pests. Hence, the implementation of a sound weed management plan would help in eliminating the weeds as harboring sites for pests as well as the resource-competing element for cotton crop.

Critical Period for Weed Control

Critical period for weed control (CPWC) is the concept referring to the period in which weed competition reduces the crop yield unacceptably. This concept indicates that weed control should be applied for a certain period to avoid unnecessary applications. CPWC provides crop species-based information about the starting date, duration, and end period of weed control that are expressed as the days or growing-degree days after crop emergence as well as the growth stages of crops. Previous studies focused on the critical period for weed control in cotton on individual weeds and the total populations. Bryson conducted experiments for critical period for *Sesbania exaltata* (Raf.) in cotton and found that the critical period for this weed's removal is less than or equal to 62 days after planting [37]. Blanco et al. found that *C. rotundus* emerging 6 weeks after cotton planting did not cause damage to the cotton plant [38].

In addition to weed-specific defined critical periods, there are some other studies in which CPWC was determined for total weeds. Deazevedo et al. investigated the CPWC in cotton in Brazil and found that weed presence up to 20 days after cotton planting did not affect the crop yield adversely [39]. Weed-free periods of 40–80 days after emergence were enough to obtain maximum yield. Another study from Greece showed that postemergence weed control should be started within 2 weeks after crop emergence, and fields should be kept weed-free for at least 11 weeks, so that the critical period was defined as the period between the 2nd and 11th week after emergence [40]. Similar results were obtained by Bukun for south-eastern Turkey [41]. A 4-year study showed that CPWC in cotton for 5 % yield loss starts at 100–159 growing-degree days and ends at 1006–1174 growing-degree days, depending on weed flora and year. It means weed control measures should be effectively applied between the periods of 1–2 and 11–12 weeks after cotton emergence. Mavunganidze et al. investigated CPWC for cotton under dry and wet season conditions and found that CPWC is variable depending on water conditions [42]. Weeds were more damaging to crops in the dry season as compared to the wet season. CPWC for 90 % of the yield was defined as the period between 3 and 8 weeks after crop emergence in the wet season, while it was between 2 and 11 weeks in the dry season.

The discussion indicated that the weeds compete with cotton at early stages after emergence. In conclusion, the weed control should be started within 1–2 weeks after crop emergence, and fields should be maintained weed-free for 8–9 weeks. However, this period can be variable depending on the cultural measures prior to planting, prevailing weed species, their initial densities, agronomic factors such as sowing date and row spacing, and climatic conditions.

Importantly, the timing for weed control in cotton can be optimized by considering critical period concepts together with the economic damage thresholds. Thus, the number of weed control measures and their cost, and environmental concerns can be limited within the frame of IWM.

Decision Making for Weed Control

Decision making is an important step of IWM strategies that provides the information regarding the "need for" and "optimum timing of" weed control. This step is mainly considered after preventive measures, and thus associated more with the postemergence weed control. Postemergence weed control is preferably applied in IWM, because all measures can be adjusted to the factors in a given time on the field [43]. Dominating weed species and their distribution, densities, growth stages, and competitive abilities are important factors in decision making for herbicide application. Since all these factors can be immediately assessed after emergence of weeds and the crop, the decision-making process is more reliable when planned for postemergence measures. The decision-making process contains two different concepts that serve jointly for the optimization of weed control within the frame of IWM practices.

One of these concepts is "economical damage threshold" (EDT), which refers to the population densities of weeds at which they cause considerable yield losses and hence the weed control activity becomes economical. In the other words, the EDT is the weed population at which the costs of control are equal to or lower than the increase in crop value from control. There are limited studies concerning the EDT in cotton. Charles et al. investigated the EDT for noogoora burr (*X. occidentale* Bertol.) and fierce thornapple (*Datura ferox* L.) in irrigated cotton and found that thresholds for mechanical control of average-sized weeds were one *X. occidentale* in 195 m and one *Datura ferox* in 73 m of cotton row [44]. Bailey et al. found EDT of velvetleaf (*Abutilon theophrasti* Medic.) in cotton for 5 and 10% yield losses as 0.2 and 0.4 plants per m of row (or 1930 and 4110 plants ha^{-1}), respectively, in the first year and 0.03 and 0.08 plants per m of row (or 360 and 850 plants ha^{-1}), respectively, in the second year [45]. In another study with *Abutilon theophrasti,* Cortes et al. determined the EDT as 0.1–0.5 plants per m^2, depending on the weed control measures used [46]. Yong et al. evaluated the EDT of *Flaveria bidentis* (L.) Kuntze in the cotton fields, which is a type of aggressive alien plant that has invaded farmland in Hebei Province, China [47]. The EDT of this weed was found as 0.69–0.77 plants per m^2.

Although EDT provides information for the need for weed control, it has some concerns for practical use, because it is calculated on a single-weed-species basis, as also given in the aforementioned examples. However, more than one weed species occur in fields and they compete simultaneously with each other as well as with crop plants. Therefore, it is difficult to distinguish the competitive effect of one weed species from another. Additionally, the EDT only addresses the weed population level at which the control should be done, but the timing for weed control remains unknown while weed emergence is a dynamic process and some weeds emerge several times in a season. Therefore, there is probability that the treatments would be repeated unnecessarily when only EDT levels are considered as criteria for weed control.

Integrated Weed Management

Proper weed control possesses an inevitable contribution for the successful cotton outputs. IWM possesses several benefits over using a single weed control strategy. Cost-effective and environment friendly weed control techniques are integrated for the sake of keeping weeds below economic threshold levels. According to Buhler, IWM focuses on avoiding the causes of weed prevalence with basis from scientific knowledge and available management options rather than the mere reaction to the present weed flora [48]. One of the most important considerations entails limiting the weed population through implementation of all the best available options so that the surviving weeds are least damaging for the crop [16]. IWM strategy mainly consists of three different steps. The first step is the cultural measures that are mostly taken prior to planting. The second step is the decision-making process at which some thresholds are considered. Choice of appropriate weed control method is the last step of IWM [43]. Complete information regarding the existing weed flora is compulsory for making the decision regarding the implementation of the IWM plans.

Cultural Weed Management Options

Cultural weed management is a broader term used to describe a set of crop management practices that affect the weed flora in a crop. These crop management practices are manipulated in such a way that the harmful crop weeds are prevented and suppressed in terms of their seed bank, germination, growth, seed production, and seed dispersal [16, 49].

Cultural weed management may include the clean cultivation, stale seedbed preparation, crop rotation, use of appropriate planting method and cultivar, planting the crop at proper distance between the rows using a suitable seed rate, and crop rotation for controlling weeds. Although these are the most common, effective, and popular cultural weed management practices, the options are not limited to the mentioned techniques [16]. Cultural weed management practices can be implemented with little or no expense. Manipulation of the available environment results in a reasonable weed suppression without any additional costs. Environmental gains are the other major harvest of implementing cultural weed management. Cultural strategies do not harm, but rather benefit the environment under certain cases. Importantly, the options under the umbrella of cultural weed management can be practiced in an integrated way to harvest the additive effects of the individual techniques [49]. Cultural practices are most important from an IWM perspective. A successful IWM plan is incomplete without cultural weed management practices [50]. The facts are especially true for the cotton crop having a long crop duration and abundance of weeds and other pests.

Use of cottonseed free from weed seeds along with clean cultivation is among the most important preventive cultural measures for managing weeds in cotton crops [51]. Field banks, paths, and water channels can be kept free from weeds to avoid seed dispersal to the crop. Irrigation water should be free from weed seeds. This is especially true for the water from canals, which contains seeds of several weed species compared with tube well water, which is generally free from weed seeds. Sieves are available in the market that can be fixed at the point where the water enters the field from a channel. This will inhibit the entry of weed seeds into the cotton field through canal irrigation. Importantly, the fertilizer distribution and spray equipment should also be free from weed seeds. Such equipment is used in all fields and can carry weed seeds from one to other fields. Such weed seed dispersal can be avoided with a little attention given to cleaning the equipment before using them in a new field [51].

Continuously planting the same field with the same crop for years makes certain weeds adapted to the growing conditions. Similarly, repeatedly practicing the same weed control practices in each season reduces the populations of susceptible weeds, while the less or non-susceptible weeds escape and develop resistance against this management practice. The weeds escaping from herbicide treatment or mechanical hoeing survive and develop their population in the cotton fields. Crop rotation is therefore an important strategy for managing weeds because different weed control practices are applied in different crop species [52]. The varying growing practices can include irrigation, fertilization, seeding rate or row spacing of the crop in rotation, herbicide mode and sites of action, and morphological and allelopathic properties of plants. Rotating the crops having wide row spacing (e.g., cotton) with narrow spaced crops (e.g., cereals) can be considered as an appropriate strategy to reduce the number of weeds.

The weed populations are reduced when cotton is rotated with maize (*Zea mays* L.), winter cereals, alfalfa (*Medicago sativa* L.), and dry bean (*Phaseolus vulgaris* L.). In southeast USA, some winter cereals are preferably grown as cover crops, then the cotton is sown after the harvest of these cereals, and the weed number is reported to be reduced [53, 54]. Cover crops can also be a suitable rotation partner in terms of weed control in cotton because they exhibit the mulch effect as well as allelopathic properties [55, 56]. In a study by Eiszner et al., it was found that *Cyperus rotundus* population was reduced significantly by rotation of cotton with sesame (*Sesamum indicum* L.) and soybean (*Glycine max* [L.] Merr.), while the population of annual weeds was increased [57]. After 5 years of rotation, lowest weed biomass was achieved by cotton–soybean rotation. Similarly, studies by Johnson and Mullinix showed that yellow nutsedge (*Cyperus esculentus* L.) population is reduced by the rotation of peanut (*Arachis hypogaea* L.) with cotton [58].

Stale seedbed technique is an important preventive weed control strategy that is made prior to planting and helps reduce weed populations at the beginning of the growing period when weed concurrence damages the cotton growth the most. In conventional cotton-growing systems, the first soil tillage is done by means of plow and seedbed preparation is done subsequently within a couple of days prior to planting. However, in the stale seedbed technique, the period between first tillage and seedbed preparation is prolonged to allow weed emergence and development. The weeds emerging after the first soil tillage are then killed mechanically during seedbed preparation or by means of nonselective herbicides such as glyphosate and/or paraquat [59–62]. The stale seedbed preparation system has been widely accepted and practiced by cotton and other row crop growers in which the tillage and seedbed preparation are done in the fall or early spring, and fields are then left weedy until planting, ultimately doing an effective weed control before planting [63]. Interestingly, the stale seedbed technique for weed control can also be practiced in fields under conservation tillage. A light irrigation or shower allows the weed seeds to germinate and later the germinated weeds can be killed by a nonselective herbicide [64]. Using the technique of stale seedbed enables the provision of a weed-free environment at the time of crop emergence. Thus, the crop plants have a competitive advantage over the later emerging weeds. Cotton plots were plowed 4–6 weeks prior to planting and weeds were allowed to germinate. The germinated weeds included *Amaranthus* spp., *P. oleracea, S. halepense,* and *C. rotundus,* which were then desiccated using glyphosate before the seedbed preparation. Hence, the cotton fields with stale seedbed had weed intensities 90 % lower than conventionally prepared seedbeds [62]. In conclusion, cotton crop plants can be provided with a weed-free environment at the initial growth stage using the technique of stale seedbed.

Planting method can potentially impact the occurrence and growth of weeds in a crop. For example, sowing the cotton on a flat field and then "earthing up" 30–45 days after sowing to make ridges, kills the weeds through burial. The method is gaining a quick popularity due to weed suppression and water saving. For instance, the weed intensity was decreased when the flat cotton field was modified to ridges after earthing up with a subsequent increase in cotton yield [65]. A planting method accommodating higher plant density can be effective for suppressing weeds through earlier covering of the space by crop plants rather than weeds. For example, Reddy et al. compared the weed occurrence and weed control efficacy of glyphosate in twin- and single-row systems [66]. The twin-row system included two cotton rows of 38 cm distance on a 102-cm bed, while the single row system included only one cotton row on a 102-cm bed. The cotton canopy closure was 2 weeks earlier on the rows having twin rows and, therefore, the total weed biomass (containing nine predominant weed species) was 35 % lower in twin rows as compared to single rows.

Weed prevalence in cotton can be influenced by the plant-to-plant spacing. Deazevedo et al. pointed out that row spacing is an important factor influencing the starting time and duration of the critical period [39]. Cotton is generally planted using wide row spacing ranging from 70 to 102 cm in various parts of the world. Wide row spacing together with slow growth of cotton in the earlier growth stages makes the crop quite sensitive to weed occurrence. Webster evaluated the effect of cotton row spacing on the growth of *S. obtusifolia* and found that cotton grown on 25-cm row spacing (called an ultra-narrow row) suppressed the weed biomass by 80 % as compared to conventional row spacing (91 cm) [67]. This information and example show that narrow row spacing in cotton provides opportunities to suppress weed growth. Therefore, CPWC in cotton is also shortened and, generally, an effective weed control for 3–4 weeks is enough to eliminate the yield losses due to weeds.

The physiological and morphological characters, such as germination speed, emergence, height, leaf shape, and growth of cotton cultivars, can influence their ability to compete with the weeds. The taller cultivars with a higher number of leaves may be more competitive against weeds. However, the immediate weed control is compulsory regardless of the plant cultivars' characters. Hence, the cotton cultivars with better competitive ability against weeds should be prioritized for a better crop.

Seeding rate is another important factor influencing weed–crop competition in cotton. Increasing the number of crop plants per unit area results in effective suppression of weeds. This is because most spaces are occupied by crops, leaving less space to weeds. Webster investigated the effect of seeding rate on the growth and seed production of *Senna obtusifolia* and seed production of *J. tamnifolia* [67]. Cotton was seeded at the rates of 49,000; 99,000; 118,000; and 148,000 plants ha$-^{1}$. Results showed that *Senna obtusifolia* was reduced by 70% at the maximum seeding rate. Seed production of *Senna obtusifolia* and *J. tamnifolia* was reduced by 72 and 82%, respectively.

Intercropping is mainly done with the aim of mixing different crops or planting them in close sequence to maximize the land use and reduce the risk of crop failure [68]. Intercropping has benefits of maintaining soil fertility or reducing erosion and the weeds in cropping systems. According to Farooq et al., effective weed control is among the major gains of intercropping [52]. Although there is not much published literature on the quantitative effect of intercropping on weeds in cotton, some of the studies showed that weeds in cotton could be reduced by this method. Chatterjee and Mandol [69] and Thakur [70] mentioned that weed competition was reduced when legume crops are grown as intercrops. In cotton–black gram (*Vigna mungo* [L.] Hepper) and cotton–cluster bean (*Cyamopsis tetragonoloba* [L.] Taub.) systems, the weed density was reduced and cotton dry weight was increased.

Allelopathic Weed Control

The phenomenon of allelopathy can be employed for controlling weeds in the cotton crop [52]. Several studies indicate the utilization of allelopathic potential of plants for suppressing weeds in cotton crop [71, 72].

The mulching of residues from allelopathic crops has been documented as an important strategy for managing weeds in various crops [73]. Cheema et al. evaluated the allelopathic sorghum mulch for weed control in cotton (*G. arboreum*) [74]. The sorghum (*Sorghum bicolor* [L.] Moench) mulch was incorporated at 3.5, 7.0, and 10.5 t ha^{-1}, which resulted in improved weed control and increased leaf area per plant, plant height, number of bolls per plant, and cotton yield compared with the control treatment. All of the three mulch rates not only controlled the weeds including *C. rotundus,* field bindweed (*Convolvulus arvensis* L.), *C. dactylon,* and horsepurslane (*Trianthema portulacastrum* L.) but also increased the cotton yield by more than 50% compared with the control treatment.

Iqbal and Cheema intercropped some allelopathic crops in cotton for controlling purple *C. rotundus* [72]. The intercrops, including *S. bicolor, G. max,* and *S. indicum,* were effective in suppressing *C. rotundus* in terms of weed density and weed dry weight in a 2-year field experiment. The cotton yield was decreased after the introduction of intercrops. However, this decrease in cotton yield was compensated by the yield of intercrops. Overall, the plots with intercrops had a better weed control and higher net benefits compared with the sole crop and untreated control.

An important way of using the phenomenon of allelopathy for managing weeds in cotton is the spray of water extracts obtained from allelopathic crops [75, 76]. These solutions containing allelochemicals for weed suppression in cotton either are sprayed alone or can be sprayed after mixing with reduced doses of herbicides. For example, the *S. bicolor* allelopathic water solution (12 L ha^{-1}) was evaluated for controlling the *C. rotundus* in cotton. Treatment with *S. bicolor* allelochemicals decreased the dry weight of *C. rotundus* (41 %) over the nontreated control. Application of *S. bicolor* allelopathic solution also increased the leaf area index, leaf area duration, crop growth rate, plant height, boll weight, seed index, ginning out turn, seed oil content, and seed cotton yield of cotton. The improved cotton growth and yield-contributing parameters might be the result of effective *C. rotundus* control as a result of application of *S. bicolor* allelopathic solution. The yield of cotton was increased by 15 % over the control treatment [77]. Similarly, two sprays of *S. bicolor* allelopathic water extracts were effective in suppressing the cotton (*G. arboreum*) weeds such as *Cynodon dactylon, Cyperus rotundus, T. portulacastrum,* and *Convolvulus arvensis*. Total weed dry weight was decreased by 35 % while seed cotton yield was increased by more than 50 % over the control treatment [74].

The combination of allelopathic solutions with reduced herbicide rates helps in not only effective weed management but also decreasing the herbicide inputs for protecting the environment and economizing weed control expenses. Application of pendimethalin at half of the recommended rate (625 g a.i. ha^{-1}) after mixing with an allelopathic solution of *S. bicolor,* mulberry (*Morus alba* L.), and canola (*Brassica campestris* L.; 15 L ha^{-1}) was effective for controlling weeds and increasing the yield of cotton over the control treatment. The total weed dry weight was reduced by nearly 76 % compared with the control while yield was increased by 57 % [71]. In another study, the *S. bicolor* allelopathic solution at 12 L ha^{-1} was combined with half of the recommended rate of S-metolachlor (1075 g a.i. ha^{-1}) and sprayed for managing *C. rotundus* in cotton crop. The *C. rotundus* dry weight was decreased by 77 % as a result of application of this treatment compared with the control. The seed cotton yield was improved (32 %) with an improvement in leaf area index, leaf area duration, crop growth rate, plant height, sympodial branches per plant, number of bolls per plant, boll weight, seed index, ginning out turn, and seed oil content [77].

In conclusion, the weeds in cotton can be controlled by the use of allelopathy in the form of mulch, intercropping, allelopathic water extracts, and allelopathic water extracts combined with reduced herbicide rates. Further, the allelopathic weed control can be included as an important component of IWM in cotton.

Mechanical Weed Control

Mechanical weed control includes the efforts to stamp down the weeds. The activities like tillage and mowing can be importantly included in mechanical weed control [78]. Probably these weed control methods cannot be implemented as a full-fledged weed control strategy due to some of their shortfalls [51]. For example, tillage deteriorates the soil structure, results in soil moisture loss from the soil, and mixes the weed seeds in the soil for further weed propagation [17]. The mechanical weed control techniques also carry certain benefits, especially in the context of integrated weed control. For example, if the weeds are present in a specific patch of the field, these can be cut using a tool and fed to animals, especially before the seed setting so that the weed propagation is avoided. The cottonweeds such as barnyardgrass (*Echinochloa crus-galli* [L.] Beauv.) escaping other weed control strategies can be controlled with such kind of control. The mechanical weed control does not involve the use of any chemical and, therefore, it may be considered as environment friendly [50]. The mechanical weed control can be chosen for integration in the comprehensive, economical, and environment-friendly weed management plans instead of implementing it as a single strategy [50, 78].

Tillage is an important growing practice for crops, which is done for seedbed preparation, weed control, trash burial, soil aeration, and water infiltration. In a conventional growing system, the soil is first deep plowed (20–40 cm) to cut and invert soil and bury some residues in the soil. The upper layer of soil is then cultivated by disc harrow, cultivator, or rototiller for seedbed preparation. Tillage is also practiced between the rows of crops like cotton after the crop and weeds have emerged. Tillage is an effective weed control method in which the weeds are buried, shoots are separated from roots, and their seeds or vegetative buds are stimulated for germination that would be controlled by using the subsequent control measures. The shoots are desiccated and carbohydrate reserves of perennial weeds are exhausted. However, the weed control through tillage also carries some disadvantages, for example, the soil compaction, soil erosion, moisture loss, breaking seed and vegetative bud dormancy, and high costs needed for carrying out the tillage [17]. However, when an efficient weed control strategy in reduced or no-till system is available, soil tillage may not be necessary for weed control.

In conclusion, mechanical weed control can be used to supplement the other weed control strategies when practicing IWM in cotton. The mechanical weed control of weeds present in patches carries special benefits of avoiding seed dispersal. Tillage can be practiced to control weeds escaping from other strategies under the auspices of integrated weed control.

Chemical Weed Control

The most prevalent weed control method in cotton crop includes the use of residual herbicides (Table 9.2) along with mechanical control done after the crop emergence. Postemergence herbicides are used successfully for the control of monocotyledonous weeds, while a selective control of annual broad-leaved weeds in cotton with herbicides is limited, except from herbicide-resistant cultivars. Although current weed control methods are successful in suppressing some weeds, they provide opportunities for spread of some non-susceptible weed species. In Turkey, for example, most farmers use trifluralin as preplant application, mechanical hoeing between cotton rows, and hand hoeing within the rows. As a result of long-term application of these control methods, some weeds' occurrence and populations have increased. This can be observed especially in the case of perennial weeds, such as *Cyperus rotundus,* which has nearly 100 % frequency in cotton fields in the western part of Turkey. As an alternative to this system, herbicide-resistant cotton was introduced in the early 1990s. This system has been accepted by many cotton farmers, because it offers a selective broad-spectrum, postemergence weed control with an herbicide glyphosate, even in reduced- or no-till systems. Because of this, areas under herbicide-resistant crops have increased recently. However, an intensive use of glyphosate leads to herbicide resistance problems.

Chemical weed control in cotton is among the most important components of IWM. The prevailing weed flora, the kind and nature of weeds, and the weed intensity are desired to be comprehensively assessed before the application of the most suitable herbicide. However, mostly herbicides are likely to be used in cotton fields. Hence, the assessment of weed flora during the field history and selection of a suitable herbicide with its accordance are most important for a successful weed control. Crop injury is likely to result from the application of an incorrect dose, application method, or application timing [15, 79, 80].

Preemergence Herbicides

Inclusion of preemergence herbicides in the IWM plan is inevitable for sustainable weed management in cotton (Table 9.2) [7]. Use of preemergence herbicides along with other strategies is helpful in providing a weed-free environment to cotton plants, especially during the critical weed–crop competition [7]. Hence, effective weed control early in the season results in faster crop growth, aggressive plant vigor, and healthier cotton growth. Everitt and Keeling evaluated preemergence herbicides, including 2,4-dichlorophenoxy acetic acid (2,4-D; 560 and 1120 g a.i. ha^{-1}), dicamba (140 and 280 g a.i. ha^{-1}), and diflufenzopyr plus dicamba (100 and 200 g a.i. ha^{-1}) for controlling horseweed (*Conyza canadensis* [L.] Cronquist) in cotton [18]. They also assessed the crop injury resulting from the application of these herbicides. Diflufenzopyr plus dicamba and dicamba posed some serious injuries to cotton crop with a non-consistent efficacy on weeds. However, 2,4-D application 2 or 4 weeks before cotton planting was not only safe against the cotton crop but also effective in suppressing *C. canadensis* and improving cotton lint yield. Pendimethalin is an effective herbicide, which has been applied for successful weed control in cereals, oilseeds, vegetables, and several other crops for a long time [12, 81–83]. This herbicide is also effective for weed control in the cotton crop. For example,

Table 9.2 Herbicides for weed control in cotton

Herbicide	Dose (a.i. g ha^{-1})	Weed flora	Weed suppression (%)	Yield increase (%)	Reference
Preemergence herbicides					
2,4-dichlorophenoxy acetic acid (2,4-D)	560–1120	*Conyza canadensis* (L.) Cronquist			Everitt and Keeling [18]
Pendimethalin	825	Broad-leaved and narrow-leaved	69–76	100	Ali et al. [8]
Pendimethalin	825	*Trianthema monogyna* L.	83	15	Jarwar et al. [12]
		Digera arvensis Forsk.	87		
		Echinochloa colona (L.) Link	89		
		Cyperus rotundus L.	77		
		Tribulus terrestris L.	77		
		Dactyloctenium aegyptium (L.) Willd.	78		
		Portulaca oleracea L.	87		
Pendimethalin	690	Annual weeds			Richardson et al. [15]
Pendimethalin	1205	Broad-spectrum weeds	78	63	Khaliq et al. [71]
Pendimethalin	1000	Broad-spectrum weeds	53	16	Cheema et al. [84]
S-metolachlor	1920	*Portulaca oleracea* L.	94	40	Jarwar et al. [12]
		Cyperus rotundus L.	93		
		Trianthema monogyna L.	90		
		Tribulus terrestris L.	90		
		Digera arvensis Forsk.	93		
		Echinochloa colona (L.) Link	95		
		Dactyloctenium aegyptium (L.) Willd.	92		

Table 9.2 (continued)

Herbicide	Dose (a.i. g ha^{-1})	Weed flora	Weed suppression (%)	Yield increase (%)	Reference
Preemergence herbicides					
S-metolachlor	2150	*Cyperus rotundus* L.	86	34	Iqbal and Cheema [77]
S-metolachlor	2000	*Cyperus rotundus* L., *Trianthema portulacastrum* L.	51–68	18	Cheema et al. [84]
Postemergence herbicides					
Trifloxysulfuron	4–6	*Cyperus rotundus* L., *Cyperus esculentus* L.	56		Burke et al. [85]
Directed postemergence herbicides					
Glyphosate	2300	Narrow-leaved weeds	42	74	Ali et al. [8]
		Broad-leaved weeds	56		
Flumioxazin	70	*Sida spinosa* L., *Amaranthus palmeri* S.Wats., *Chenopodium album* L., *Ambrosia artemisiifolia* L., *Ipomea* spp., *Senna obtusifolia* (L.) H.S.Irwin & Barnaby and *Amaranthus hybridus* L.			Askew et al. [79]
Paraquat	480	*Convolvulus arvensis* L. and *Trianthema portulacastrum* L.			Cheema et al. [74]

the application of pendimethalin at 825 g a.i. ha^{-1} was effective in suppressing the broad-leaved and narrow-leaved weeds as well as improving the lint yield of cotton [8]. The weed biomass reduction in the treated fields was 76 and 69 % for grasses and broad-leaved weeds, respectively, with an approximate 100 % increase in cotton yield. In another study, the application of pendimethalin at 825 g a.i. ha^{-1} effectively suppressed weeds, including *Trianthema monogyna* L. (83 %), *Digera arvensis* Forsk. (87 %), *Echinochloa colona* (L.) Link (89 %), *C. rotundus* (77 %), *Tribulus terrestris* L. (77 %), *Dactyloctenium aegyptium* (L.) Willd. (78 %), and *P. oleracea* (87 %), over the nontreated control [12]. The average decrease in weed density over the nontreated control in the 2-year field experiment was 72 % with a 15 % increase in cotton yield. Similarly, the preemergence application of pendimethalin at 690 g a.i. ha^{-1} was effective in controlling annual weeds; however, it resulted in ineffective control of perennial weeds [15]. The weed control was improved when trifloxysulfuron was applied as postemergence after the application of pendimethalin. In another study, the application of pendimethalin at 1205 g a.i. ha^{-1} reduced the total weed biomass by 78 % compared with the control treatment and increased the cotton yield by 63 % [71].

Preemergence application of S-metolachlor at 1920 g a.i. ha^{-1} was effective on weeds in the cotton crop. Cotton yield was increased by almost 40 % compared to the untreated crop, with approximately 93 % decrease in total weed density. Compared with the control, the decrease in weed density was: *P. oleracea* by 94 %, *C. rotundus* by 93 %, *T. monogyna* by 90 %, *T. terrestris* by 90 %, *Digera arvensis* by 93 %, *E. colona* by 95 %, and *Dactyloctenium aegyptium* by 92 % [12]. Iqbal and Cheema evaluated the preemergence application of S-metolachlor at 2150 g a.i. ha^{-1} for controlling the *C. rotundus* in cotton [77]. The results indicated that the herbicide was effective in reducing the *C. rotundus* biomass (86 %) and increasing the cotton yield (34 %) in comparison with the untreated control. Similarly, in another study, the applications of pendimethalin (1000 g a.i. ha^{-1}) and S-metolachlor (2000 g a.i. $ha-^{1}$) were effective in suppressing *C. rotundus* and *T. portulacastrum* in cotton. Pendimethalin and S-metolachlor effectively reduced the total weed density by 53 and 51 %, respectively; total weed biomass by 53 and 68 %, respectively; and increased the cottonseed yield by 16 and 18 %, respectively, over the control treatment [84].

In conclusion, preemergence herbicides are effective on cottonweeds during the critical periods for weed–crop competition. Nevertheless, some weeds escape from the application of preemergence herbicides. The situation necessitates the use of preemergence herbicides as a component of IWM.

Postemergence Herbicides

Under the umbrella of IWM, the weeds that are successful in escaping the other control strategies can likely be controlled through either the use of the interrow tillage or postemergence herbicide applications [10, 85]. Tillage, however, can be expensive and may skip weeds close to the crop plant trunk. Nevertheless, it can be employed after assessing the nature and intensity of weeds and crop growth stage. On the other hand, very few broad-spectrum, postemergence herbicides are available for use in cotton crop (Table 9.2).

Burke et al. evaluated the effectiveness of trifloxysulfuron (4 and 6 g a.i. ha^{-1}) on *C. rotundus* and *C. esculentus* at either of 10–15-cm or 20–30-cm height in cotton [85]. More than 56 % reduction in root and shoot biomass was recorded and the higher herbicide dose was more effective in suppressing the weeds, especially at lower plant height.

Directed postemergence application of herbicides is an attractive option for the postemergence weed control in cotton. In this herbicide application, the sprayers are fitted with a protective shield near the sprayer nozzle. Thus, the herbicide droplets during the spray are allowed only to fall on the weeds directly and avoid the cotton crop. Although some minute herbicide particles may reach the surface of the cotton trunk as a result of drift, their intensity may not be so strong to injure the cotton crop [80]. The direct herbicide spray, especially of nonselective herbicides (e.g., glyphosate) requires care to avoid any of the likely crop damage [80]. Several studies indicated the effectiveness of directed herbicide application on cottonweeds (Table 9.2). For example, the directed application of glyphosate (2300 g a.i. ha^{-1}) 40 days after crop planting reduced the intensity of grass (42 %) and broad-leaved weeds (56 %), and improved the cotton yield (74 %) over the control treatment [8]. Similarly, directed application of flumioxazin at 70 g a.i. ha^{-1} in cotton effectively suppressed weeds, such as prickly sida (*Sida spinosa* L.), *Amaranthus palmeri, C. album,* common ragweed (*Ambrosia artemisiifolia* L.), *Ipomoea* spp., *Senna obtusifolia,* and smooth pigweed (*Amaranthus hybridus* L.) [79]. Directed application of glyphosate at 1120 g a.i. ha^{-1} also decreased the weed intensity; however, it was less effective than that of flumioxazin [79]. Cheema et al. directed spray paraquat (480 g a.i. ha^{-1}) at 20 days after planting with a sprayer fitted with a protective shield [74]. The herbicide was not effective on *Cyperus rotundus* and *Cynodon dactylon*; however, it effectively reduced biomass of *Convolvulus arvensis* and *T. portulacastrum,* and increased the plant height, leaf area, number of bolls, and yield of cotton compared with the control treatment.

In conclusion, selective and nonselective postemergence (as directed spray) herbicides can be applied to control weeds in cotton. This is particularly important for the weeds that escape the preemergence weed control methods.

Herbicide-Resistant Cotton and Weed Control

Herbicide-resistant cotton cultivars were developed for attaining better weed control, better lint quality, and increased yield. The introduction of herbicide-resistant cultivars has been a breakthrough development for better weed control and increased farm income. The time window for herbicide application is increased in case of herbicide-resistant cotton. Such cultivars can be included as a component in the IWM plan. Relying merely on herbicides for weed control can lead to the development of herbicide resistance in weeds.

Glyphosate-resistant cotton has attained special attention of the farming community in strengthening IWM plans. In glyphosate-resistant cotton, glyphosate was effective in controlling (62–99 %) weeds, including *A. theophrasti,* pitted morningglory (*Ipomoea lacunosa* L.), *E. crus-galli, Senna obtusifolia, Sorghum halepense, Sida spinosa,* and browntop millet (*Urochloa ramosa* L.) [86]. Cotton crop injury was less than 13 and 5 % at 2 and 3 weeks after treatment, respectively. The number of nodes in cotton or bolls opening was not affected by glyphosate application. Glyphosate application was also effective in suppressing the notorious summer weed *E. crus-galli* when applied in glyphosate-resistant cotton with a subsequent reduction in *E. crus-galli* seed bank [7].

The enzyme glutamine synthase is responsible for catalyzing the conversion of glutamic acid and ammonia into glutamine. The nonselective, contact herbicide glufosinate inhibits the activity of glutamine synthase resulting in deposition of ammonia ions in plant body. This leads to the fracturing of chloroplast and severe disturbances in the process of photosynthesis. Hence, the necrosis of the tissues occurs. Cotton plants have been genetically modified to confer resistance against glufosinate through incorporation of the *pat* or *bar* gene. The glufosinate-resistant cotton cultivars offer an excellent opportunity to control weeds effectively. However, the use of glufosinate in rotation with other herbicides or weed control techniques would be effective to tackle the issue of resistance development in weeds. The weed control through glufosinate-resistant cultivars can be included as a component of IWM in cotton. The early postemergence application of glufosinate (468 g a.i. ha^{-1}) at the two-leaf stage of glufosinate-resistant cotton was effective in controlling the weeds, including broadleaf signalgrass (*Brachiaria platyphylla* [Griseb.] Nash.), large crabgrass (*Digitaria sanguinalis* [L.] Scop.), fall panicum (*Panicum dichotomiflorum* Michx.), *A. palmeri, A. hybridus,* and goosegrass (*Eleusine indica* L.). Weed control was improved when the early postemergence application of glufosinate was combined with preemergence application of pendimethalin (1110 g a.i. ha^{-1}). The weed suppression was greater than 95 % in the case of the sequential application of preemergence pendimethalin followed by early postemergence glufosinate application [87].

In conclusion, the introduction of herbicide-resistant cotton is a breakthrough for effective weed control in cotton. However, the resistance development in weed biotypes due to continuous use of the same herbicide is a critical issue. Nevertheless, the measures to deal with the issue of herbicide resistance development, particularly for glyphosate, are inevitable. Rotational use of herbicides and IWM in cotton is the important strategy to tackle the resistance development issue in cottonweeds.

Conclusion

Weeds seriously threaten the cotton productivity and the quality of produce, while a diverse weed flora competes with the cotton plants. The presence of hardy weed flora in cotton crop necessitates a comprehensive and wise weed management plan. Such a weed management plan should be based on the correct information on the prevalence status of weeds in the cotton field, gained through systematic observation in the field. This weed management plan should also consider the environmental concerns as well as the costs of weed control. Several effective weed management options are available for controlling weeds in cotton. However, practicing a single weed control strategy may not provide an effective weed control due to hardy weeds and the perennial growth habit of cotton crop. IWM truly provides the opportunity to combine several attractive weed control techniques systematically for effective, economical, season-long, and environment-friendly weed control in cotton crop.

Acknowledgment Thanks to the Scientific and Technological Research Council of Turkey (TUBITAK) for supporting the stay of Khawar Jabran in Turkey to complete this book chapter.

References

1. Roche J (1994) The international cotton trade. Woodhead Publishing Limited, Sawston
2. Sczostak A (2009) Cotton linters: an alternative cellulosic raw material. Macromol Symp 280:45–53
3. Moulherat C, Tengberg M, Haquet J-F, Mille B (2002) First evidence of cotton at Neolithic Mehrgarh, Pakistan: analysis of mineralized fibres from a copper bead. J Arch Sci 29:1393–1401
4. Iqbal MJ, Reddy OUK, El-Zik KM, Pepper AE (2001) A genetic bottleneck in the 'evolution under domestication' of upland cotton *Gossypium hirsutum* L. examined using DNA fingerprinting. Theor Appl Genet 3:547–554
5. Everman WJ, Burke IC, Allen JR, Collins J, Wilcut JW (2007) Weed control and yield with glufosinate-resistant cotton weed management systems. Weed Technol 21:695–701
6. Oerke EC (2006) Crop losses to pests. J Agric Sci 144:31–43
7. Werth JA, Preston C, Roberts GN, Taylor IN (2008) Weed management impacts on the population dynamics of barnyardgrass (*Echinochloa crus-galli*) in glyphosate-resistant cotton in Australia. Weed Technol 22:190–194
8. Ali H, Muhammad D, Abid SA (2005) Weed control practices in cotton (*Gossypium hirsutum* L.) planted on bed and furrow. Pak J Weed Sci Res 11:43–48
9. Norsworthy JK, Smith KL, Scott RC, Gbur EE (2007) Consultant perspectives on weed management needs in Arkansas cotton. Weed Technol 21:825–831
10. Kalivas DP, Economou G, Vlachos CE (2010) Using geographic information systems to map the prevalent weeds at an early stage of the cotton crop in relation to abiotic factors. Phytoparasitica 38:299–312
11. Kalivas DP, Vlachos CE, Economou G, Dimou P (2012) Regional mapping of perennial weeds in cotton with the use of geostatistics. Weed Sci 60:233–243
12. Jarwar AD, Baloch GM, Memon MA, Rajput LS (2005) Efficacy of pre and post-emergence herbicides in cotton. Pak J Weed Sci Res 11:51–55
13. Clewis SB, Wilcut JW, Porterfield D (2006) Weed management with S-metolachlor and glyphosate mixtures in glyphosate-resistant strip- and conventional-tillage cotton (*Gossypium hirsutum* L). Weed Technol 20:232–241
14. Scroggs DM, Miller DK, Griffin JL, Wilcut JW, Blouin DC, Stewart AM et al (2007) Effectiveness of preemergence herbicide and postemergence glyphosate programs in second-generation glyphosate-resistant cotton. Weed Technol 21:877–881
15. Richardson RJ, Wilson HP, Hines TE (2007) Preemergence herbicides followed by trifloxysulfuron postemergence in cotton. Weed Technol 21:1–6

16. Vitelli JS, Pitt JL (2006) Assessment of current weed control methods relevant to the management of the biodiversity of Australian rangelands. Range J 28:37–46
17. Farooq M, Flower KC, Jabran K, Wahid A, Siddique KHM (2011a) Crop yield and weed management in rainfed conservation agriculture. Soil Till Res 117:172–183
18. Everitt JD, Keeling JW (2007) Weed control and cotton (*Gossypium hirsutum*) response to preplant applications of dicamba, 2,4-D, and diflufenzopyr plus dicamba. Weed Technol 21:506–510
19. Webster TM, Sosnoskie LM (2010) Loss of glyphosate efficacy: a changing weed spectrum in Georgia cotton. Weed Sci 58:73–79
20. Dowler CC (2005) Weed survey—southern states—broadleaf crops subsection. In Street JE (ed) Proceedings of the Southern Weed Science Society, 2005 January 1618. Southern Weed Science Society, Memphis, pp 290–305
21. Webster TM (2005) Weed survey—southern states: broadleaf crops subsection. In Vencill WK (ed) Proceedings of the Southern Weed Science Society. Southern Weed Science Society, Charlotte, pp 291–306
22. Vargas RN, Fischer WB, Kempen HM, Wright SD (1996) Cotton weed management. In: Hake SJ, Kerby TA, Hake KD (eds) Cotton production manual. University of California, Division of Agriculture and Natural Resources. Publication 3352:187–202
23. Doğan MN, Boz O, Unay A (2008) Wirkung von glyphosat auf unkraeuter in baumwolle - bedeutung der glyphosat resistenz unter den anbaubedingungen in der Türkei. J Plant Dis Prot (Sp. Iss. XXI):45–50. (In German)
24. Cardoso GD, Alves PLCA, Severino LS, Vale LSD (2011) Critical periods of weed control in naturally green colored cotton BRS Verde. Ind Crop Prod 34:1198–1202
25. Crowley RH, Buchanan GA (1978) Competition of four morningglory (*Ipomoea spp.*) species with cotton (*Gossypium hirsutum*). Weed Sci 26:484–488
26. Doğan MN, Boz O (2005)The concept of reduced herbicide rates for the control of johnsongrass (*Sorghum halepense* (L.) Pers.) in cotton during the critical period for weed control. J Plant Dis Prot 112:71–79
27. Scott GH, Askew SD, Wilcut JW, Brownie C (2000) *Datura stramonium* interference and seed rain in *Gossypium hirsutum*. Weed Sci 48:613–617
28. Keeley PE, Thullen RJ (1991a) Growth and interaction of bermudagrass (*Cynodon dactylon*) with cotton (*Gossypium hirsutum*). Weed Sci 39:570–574
29. Keeley PE, Thullen RJ (1991b) Biology and control of black nightshade (*Solanum nigrum*) in cotton (*Gossypium hirsutum*). Weed Technol 5:713–722
30. Ligoxigakis EK, Vakalounakis DJ, Thanassoulopoulos CC (2002) Weed hosts of *Verticillium dahliae* in Crete: susceptibility, symptomatology and significance. Phytoparasitica 30:141–146
31. Yildiz A, Doğan MN, Boz Ö, Benlioğlu S (2009) Weed hosts of *Verticillium dahliae* in cotton fields in Turkey and characterization of *V. dahliae* isolates from weeds. Phytoparasitica 37:171–178
32. Mubin M, Akthar S, Amin I, Briddon RW, Mansoor S (2012) *Xanthium strumarium*: a weed host of components of begomovirus-betasatellite complexes affecting crops. Virus Genes 44:112–119
33. Paul S, Ghosh R, Roy A, Ghosh S (2012) Analysis of coat protein gene sequences of begomoviruses associated with different weed species in India. Phytoparasitica 40:95–100
34. Gu H, Walter GH (1999) Is the common sowthistle (*Sonchus oleraceus*) a primary host plant of the cotton bollworm, *Helicoverpa armigera* (Lep., Noctuidae)? Oviposition and larval performance. J App Entmol 123:99–105
35. Dos Santos KB, Menegium AM, Neves PMOJ (2005) Biology and consumption of *Spodoptera eridania* (Cramer) (Lepidoptera: Noctuidae) in different hosts. Neotrpical Entomol 34:903–910
36. Lu YH, Jiao ZB, Li GP, Wyckhuys KAG, Wu KM (2011) Comparative overwintering host range of three *Adelphocoris* species (Hemiptera: Miridae) in northern China. Crop Prot 30:1455–1460

37. Bryson CT (1990) Interference and critical time of removal of hemp sesbania (*Sesbania exaltata*) in cotton (*Gossypium hirsutum*). Weed Technol 4:833–837
38. Blanco HG, Arevalo RA, Chiba S (1991) Effect of *Cyperus rotundus* L. on herbaceous cotton plants. Pesqui Agropecu Bras 26:169–176
39. Deazevedo DMP, Beltrao NED, Danobrega LB, Dossantos JW, Vieira DJ (1994) Critical period of weed competition on irrigated annual cotton. Pesqui Agropecu Bras 29:1417–1425
40. Papamichail D, Elefterohorinos I, Froud-Williams R, Gravanis F (2002) Critical periods of weed competition in cotton in Greece. Phytoparasitica 30:105–111
41. Bukun B (2004) Critical periods for weed control in cotton in Turkey. Weed Res 44:404–412
42. Mavunganidze Z, Mashingaidze AB, Chivinge OA, Riches C, Ellis-Jones J, Mutenje M (2006) Critical period of cotton weed control in the Zambezi valley. Fifteenth Australian Weeds Conference, pp 387–390
43. Kudsk P, Streibig JC (2003) Herbicides—a two-edged sword. Weed Res 43:90–102
44. Charles GW, Murison RD, Harden S (1998) Competition of noogoora burr (*Xanthium occidentale*) and fierce thornapple (*Datura ferox*) with cotton (*Gossypium hirsutum*). Weed Sci 46:442–446
45. Bailey WA, Shawn AD, Rai SD, Wilcut JW (2003) Velvetleaf (*Abutilon theophrastii*) interference and seed production dynamics in cotton. Weed Sci 51:94–101
46. Cortes JA, Mendiola MA, Castejon M (2010) Competition of velvetleaf (*Abutilon theophrasti* M.) weed with cotton (*Gossypium hirsutum* L.). Economic damage threshold. Spanish J Agric Res 8:391–399
47. Yong FJ, MinHao P, YingChao L, JingAo D (2009) Economic threshold and critical period of competition of *Flaveria bidentis* (L.) Kuntze in the cotton fields. Acta Phytophylacica Sin 36:561–566
48. Buhler DD (2002) 50th anniversary-invited article: Challenges and opportunities for integrated weed management. Weed Sci 50:273–280
49. Bond W, Grundy AC (2001) Non-chemical weed management in organic farming systems. Weed Res 41:383–405
50. Hatcher PE, Melander B (2003) Combining physical, cultural and biological methods: prospects for integrated non-chemical weed management strategies. Weed Res 43:303–322
51. Labrada R (2006) Weed management: a basic component of modern crop production. In Singh HP, Batish DR, Kohli RK (eds) Handbook of sustainable weed management. Haworth Press, New York, pp 2150
52. Farooq M, Jabran K, Cheema ZA, Wahid A, Siddique KHM (2011b) The role of allelopathy in agricultural pest management. Pest Manage Sci 67:493506
53. Price AJ, Reeves DW, Lamm DA (2009) Glyphosate resistant Palmer amaranth-a threat to conservation tillage. Proceedings of the Beltwide Cotton Conference, San Antonio, National Cotton Council, pp 1335–1336
54. Reiter MS, Reeves DW, Burmester CH, Torbert HA (2008) Cotton nitrogen management in a high-residue conservation system: cover crop fertilization. Soil Sci Soc Am J 72:1321–1329
55. Price AJ, Reeves DW, Patterson MG (2006) Evaluation of weed control provided by three winter cereals in conservation-tillage soybean. Renew Agric Food Syst 21:159–164
56. Norsworthy JK, McClelland M, Griffith G, Bangarwa SK, Still J (2001) Evaluation of cereal and Brassicaceae cover crops in conservation-tillage, enhanced, glyphosate resistant cotton. Weed Technol 25:6–13
57. Eiszner H, Blandon V, Pohlan J (1996) Agronomic and ecological impacts on cotton with changing crop rotation. Tropenlandwirt 97:7583
58. Johnson WC, Mullinix BG (1996) Population dynamics of yellow nutsedge (*Cyperus esculentus*) in cropping systems in the southeastern coastal plain. Weed Sci 45:166171
59. Barberi P (2003) Preventive and cultural methods for weed management. In Labrada R (ed) Weed management for developing countries, FAO Plant Production and Protection Paper 120, Add 1, p 179–194
60. Veeramani A, Prema P, Guru G (2006) Effect of pre- and post-sowing weed management on weeds and summer irrigated cotton. Asian J Plant Sci 5:174–178

61. Veeramani A, Prema P, Ganesaraja V (2008) Pre and post-sowing control of weeds, their influence on nutrient uptake in summer irrigated cotton (*Gossypium hirsutum* L.). Res J Agric Biol Sci 4:643–646
62. Doğan MN, Ünay A, Boz Ö, Öğüt D (2009) Effect of pre-sowing and pre-emergence glyphosate applications on weeds in stale seedbed cotton. Crop Prot 28:503–507
63. Fairbanks DE, Reynolds DB, Griffin JL, Jordan DL, Corkern CB, Vidrine PR, Crawford SH (2001) Cotton tolerance and weed control with preplant applications of thifensulfuron plus tribenuron. J Cotton Sci 5:259–267
64. Chauhan BS, Singh RG, Mahajan G (2012) Ecology and management of weeds under conservation agriculture: a review. Crop Prot 38:57–65
65. Cheema MS, Nasrullah M, Akhtar M, Ali L (2008) Comparative efficacy of different planting methods and weed management practices on seed cotton yield. Pak J Weed Sci Res 14:153–159
66. Reddy KN, Boykin JC, Klif J (2010) Weed control and yield comparisons of twin- and single-row glyphosate-resistant cotton production systems. Weed Technol 24:95–101
67. Webster TM (2007) Cotton row spacing and plant population affect weed seed production. World Cotton Research Conference-4, 2007 September 10–14, Texas, USA
68. Khan MB, Khan M, Hussain M, Farooq M, Jabran and K, Lee D-J (2012) Bio-economic assessment of different wheat-canola intercropping systems. Int J Agric Biol 14:769–774
69. Chatterjee BN, Mandal RK (1992) Present trends in research on intercropping. Indian J Agric Sci 62:507–518
70. Thakur DR (1994) Weed management in maize based intercropping systems under rainfed mid hill condition. Ind J Agron 39:203–206
71. Khaliq A, Jabran K, Mushtaq MN, Razzaq A, Cheema ZA (2007) Reduction of herbicide dose using allelopathic crop/plant water extracts with lower rates of pendimethlin in cotton. (*Gossypium hirsutum* L.) 8th National Weed Science Conference, 2007 June 2527, G.C. University, Lahore
72. Iqbal J, Cheema ZA (2007) Intercropping of field crops in cotton for management of purple nutsedge (*Cyperus rotundus* L.). Plant Soil 300:163171
73. Jabran K, Farooq M (2012) Implications of potential allelopathic crops in agricultural systems. In: Cheema ZA et al (eds) Allelopathy: current trends and future applications. Springer, Dordrecht, p 349385. doi: 10.1007/978-3-642-30595-5_15
74. Cheema ZA, Asim M, Khaliq A (2000) Sorghum alleopathy for weed control in cotton (*Gossypium arboreum* L.). Int J Agri Biol 2:37–41
75. Jabran K, Farooq M, Hussain M, Rehman H, Ali MA (2010a) Wild oat (*Avena fatua* L.) and canary grass (*Phalaris minor* Ritz.) management through allelopathy. J Plant Prot Res 50:32–35
76. Jabran K, Cheema ZA, Farooq M, Hussain M (2010b) Lower doses of pendimethalin mixed with allelopathic crop water extracts for weed management in canola (*Brassica napus* L.). Int J Agric Biol 12:335–340
77. Iqbal J, Cheema ZA (2008) Purple nutsedge (*Cyperus rotundus* L.) management in cotton with combined application of sorgaab and s-metolachlor. Pak J Bot 40:2383–2391
78. Melander B, Rasmussen G (2001) Effects of cultural methods and physical weed control on intrarow weed numbers, manual weeding and marketable yield in direct-sown leek and bulb onion. Weed Res 41:491–508
79. Askew SD, Wilcut JW, Cranmer JR (2002) Cotton (*Gossypium hirsutum* L.) and weed response to flumioxazin applied preplant and post-emergence directed. Weed Technol 16:184–190
80. Ferrell JA, Faircloth WH, Brecke BJ, Macdonald GE (2007) Influence of cotton height on injury from flumioxazin and glyphosate applied post-directed. Weed Technol 21:709–713
81. Jabran K, Cheema ZA, Farooq M, Basra SMA, Hussain M, Rehman H (2008) Tank mixing of allelopathic crop water extracts with pendimethalin helps in the management of weeds in canola (*Brassica napus*) field. Int J Agric Biol 10:293–296

82. Jabran K, Farooq M, Hussain M, Ehsanullah, Khan MB, Shahid M, Lee DJ (2012a) Efficient weeds control with penoxsulam application ensures higher productivity and economic returns of direct seeded rice. Int J Agric Biol 14:901–907
83. Jabran K, Ehsanullah, Hussain M, Farooq M, Babar M, Doğan MN, Lee D-J (2012b) Application of bispyribac-sodium provides effective weed control in direct-planted rice on a sandy loam soil. Weed Biol Manage 12:136–145
84. Cheema ZA, Khaliq A, Hussain R (2003) Reducing herbicide rate in combination with allelopathic sorgaab for weed control in cotton. Int J Agri Biol 5:4–6
85. Burke IC, Troxler SC, Wilcut JW, Smith WD (2008) Purple and yellow nutsedge (*Cyperus rotundus* and *C. esculentus*) response to postemergence herbicides in cotton. Weed Technol 22:615–621
86. Koger CH, Price AJ, Reddy KN (2005) Weed control and cotton response to combinations of glyphosate and trifloxysulfuron. Weed Technol 19:113–121
87. Wilson DG Jr, York AC, Jordan DL (2007) Effect of row spacing on weed management in glufosinate-resistant cotton. Weed Technol 21:489–495
88. Rushing DW, Murrary DS, Verhalen LM (1985b) Weed interference with cotton (*Gossypium hirsutum*). II Tumble pigweed (*Amaranthus albus*). Weed Sci 33:815–818
89. Sosnoskie LM, Kichler JM, Wallace RD, Culpepper AS (2011) Multiple resistance in palmer amaranth to glyphosate and pyrithiobac confirmed in Georgia. Weed Sci 59:321–325
90. Molin WT, Boykin D, Hugie JA, Ratnayaka HH, Sterling JM (2006) Spurred anoda (*Anoda cristata*) interference in wide row and ultra narrow row cotton. Weed Sci 54:651–657
91. Higgins JM, Walker RH, Whitwell T (1985) Coffee senna (*Cassia occidentalis*) competition with cotton (*Gossypium hirsutum*). Weed Sci 34:52–56
92. Steckel LE, Gwathmey CO (2009) Glyphosate-resistant horseweed (*Conyza canadensis*) growth, seed production, and interference in cotton. Weed Sci 57:346–350
93. Everman WJ, Thomas WE, Burton JD, York AC, Wilcut JW (2009) Absorption, translocation and metabolism of glufosinate in transgenic and nontransgenic cotton, palmer amaranth (*Amaranthus palmeri*), and pitted morningglory (*Ipomoea lacunosa*). Weed Sci 57:357–361
94. Askew SD, Wilcut JW (2002) Ladysthumb interference and seed production in cotton. Weed Sci 50:326–332
95. Rushing DW, Murrary DS, Verhalen LM (1985a) Weed interference with cotton (*Gossypium hirsutum*). I Buffalobur (*Solanum rostratum*). Weed Sci 33:810–814

Chapter 10
Integrated Weed Management in Soybean

Stevan Z. Knezevic

Introduction

Integrated weed management (IWM) is a component of integrated pest management (IPM), which is an interdisciplinary practice that may involve agronomy, entomology, ecology, economics, horticulture, nematology, plant pathology, and weed science [1]. IWM became a commonly used term in the early 1970s [2], and since then it has been defined in many different ways [1, 3, 4]. Buchanan described IWM as a combination of mutually supportive technologies in order to control weeds [5], while Swanton and Weise described it as a multidisciplinary approach to weed control utilizing the application of numerous alternative control measures [4]. The development of an IWM system is essential in order to efficiently utilize herbicides in the environment [4]. Regardless of the definition, in practical terms, it means developing a weed management program using a combination of preventive, cultural, mechanical, and chemical practices. It does not mean abandoning chemical weed control, but relying on it less.

Chemical weed control has been a primary means of weed management in the developed world for the past six decades. This continues to be true even with the introduction of herbicide-tolerant crops (HTCs), which represent relatively new weed control technology. Examples of HTCs include soybean (*Glycine max* L.), corn (*Zea mays* L.), and canola (*Brassica napus* L.) tolerant to glyphosate and glufosinate [6, 7]. Growers have readily integrated HTCs into crop production practices in the USA and Canada. For example, currently more than 95 % of 25 million hectares of soybean grown in the USA are glyphosate-tolerant cultivars [8]. The introduction of HTCs into the North American market enhanced the availability of weed control options and greatly expanded the demand for these herbicides. Glyphosate or glufosinate can also be used as an alternative tool against weeds, thus playing an important role in the development of IWM systems [9].

S. Z. Knezevic (✉)
Northeast Research and Extension Center, Department of Agronomy and Horticulture, Haskell Agricultural Laboratory,
University of Nebraska–Lincoln, Concord, NE 68728, USA
e-mail: sknezevic2@unl.edu

B. S. Chauhan, G. Mahajan (eds.), *Recent Advances in Weed Management*,
DOI 10.1007/978-1-4939-1019-9_10, © Springer Science+Business Media New York 2014

There are many kinds of weeds with different life cycles; thus, a single control method is not effective [3]. In addition, controlling weeds with one or two methods provides the weeds a chance to adapt to those practices. Therefore, instead of using a particular weed control method, IWM suggests the use of a mixture of control methods that minimize the economic impact of weeds. Applying the principles of IWM can reduce the use of herbicides applied to the environment, and at the same time provide optimum economic returns to the producers.

Many authors have described various components of an IWM system [3, 4], including tillage, critical period of weed interference [10], thresholds [11–14], alternative methods of weed control (cover crops, biocontrol, interrow cultivation, and thermal control), enhancement of crop competitiveness, crop rotation, seed bank dynamics, modeling crop–weed interference, education, and extension [1]. Models also have been developed to describe weed–crop competition [15, 16], weed and crop seed germination [17], weed developmental rate [18], and the threshold levels for common weeds [11–13, 19, 20]. Furthermore, IWM systems must remain flexible to adjust to changing technological, economic, environmental, and social factors and agroecosystem health [4].

There has been much written about IWM in various manuscripts and books (e.g., Hartfield et al. [21]); thus, the objective of this chapter is not necessarily to provide an overview of the literature of various aspects of IWM, but rather to synthetize them into a package that can be utilized by those who work (or do research) in the applied world of weed science (e.g., practitioners). In essence, the development of an IWM program should be based on a few general principles that can be used in any farm operation:

1. Use agronomic practices that limit the introduction and spread of weeds (preventing weed problems before they start)
2. Help the crop compete with weeds (help "choke out" weeds)
3. Use practices that keep weeds "off balance" (do not allow weeds to adapt)
4. Make a weed control decision
5. Documentation and record keeping

Combining agronomic practices based on these principles will allow designing an IWM program for any farming operation. Also, it is important to understand that an IWM program is not a "recipe"; it needs to be changed and adjusted to the particular farming operation, and from year to year. The goal is to manage and not eradicate weeds, as weed eradication is not possible due to environmental and economic reasons.

Preventing Weed Problems Before They Start

Many basic textbooks about pest management and weed control cover the topic of preventive pest and weed management in much detail [22]. It is important to understand that the best start of any weed management program is to reduce the potential for introducing weed seeds to the field. Preventive practices may include field sanitation, planting certified crop seeds, controlling volunteer weeds and patches of new species or herbicide-resistant weeds, cleaning equipment, tarping grain trucks, and using well-composted manure. Field sanitation involves practices that prevent weeds from entering or spreading across the field. Cleaning equipment, especially combines, before moving from field to field can further reduce the spread of weeds. Tarping grain trucks prevents introduction of weeds on roadsides, which can then invade neighboring fields. Planting certified crop seeds produces vigorous crop seedlings and improves crop emergence and establishment, which is important for improving soybean competition against weeds and overall yields. Control of volunteer weeds along edges of the field, fence lines, and ditches is useful in preventing the spread of weeds. Patches of newly invading weeds or herbicide-resistant weeds should be controlled to prevent their spreading. Manure can be a problem by increasing weed numbers and introducing new weed species to a field, especially if either the animals or livestock feed was imported from a different region. Therefore, aging or composting manure for at least a year before spreading in the field will reduce viability of weed seeds. In general, preventive weed control techniques are usually the least expensive but routinely the most overlooked.

Make the Crop Better Compete Against Weeds

A few simple practices can be adopted to provide soybean the advantage over weeds, including adjusting row spacing, increasing planting density, or selecting the most competitive variety. Narrowing soybean row spacing has been known to improve crop competitiveness [2]. For example, soybean planted at 18 and 38 cm is more competitive against weeds than the one planted at 76 or 96 cm [23]. It was suggested that narrower rows and higher soybean density significantly reduced the biomass of late-emerging weeds. Similarly, Mulugeta and Boerboom indicated that soybean planted in 18-cm rows was more competitive against weeds than in 76-cm-wide rows [24]. Several studies indicated that soybean planted in 18-cm rows was more competitive against weeds than in 75-cm-wide rows [24, 25]. Knezevic et al. also suggested that planting soybean in narrow rows improved early-season crop tolerance to weeds, delayed the critical time for weed removal (CTWR), and required less intensive weed management programs than in wide rows [26].

Certain soybean varieties can be more competitive than others. For example, taller soybean varieties close their canopy more completely than shorter types, which helps to shade out weeds. Taller varieties still need to be sprayed, but weed control could be better due to added crop competition [23].

Keep Weeds "Off Balance": Do Not Let Them Adapt

Developing a cropping system based on a variety of practices that reduce and minimize weed establishment is the key in not allowing weeds to adapt to the production system; thus, the goal is to develop a system that "keeps weeds off balance." Such a goal can be achieved with a variety of practices, of which the most important ones are crop and herbicide rotations, while adjusting planting date or the use of cover crops can also be helpful. In addition, utilizing various mechanical and alternative methods of weed control, such as flaming, can also help the process.

Crop rotation has multiple benefits to the cropping systems [27] and should be the first step for keeping weeds "off balance." Diversified crop rotation will help manage weeds in many different ways and at many different times over the growing season. For example, including forage crops in multiyear rotation allows cutting weeds before they set seeds, which is an important form of weed removal. Soybean can be rotated with a variety of crops. In Midwestern cropping systems of the USA, soybean is rotated primarily with corn, while in other parts of the world (e.g., Europe), soybean might be rotated with corn, wheat (*Triticum aestivum* L.), sunflower (*Helianthus annuus* L. [= *H. aridus* Rydb.]), or sugar beets (*Beta vulgaris*).

Crops also differ in their competitive ability. For example, cereals are generally better competitors than corn and soybean. Rotating crops with different life cycles will also help in preventing weeds from adapting. Annual weeds are more common in annual crops while biennial and perennial weeds are mostly found in perennial crops. Rotating crops with different life cycles will prevent weeds with specific life cycles from adapting and establishing. Rotating crops will also allow rotating herbicide choices. Rotating herbicides with different modes of action and application times will help delay weed adaption and reduce a chance for weed resistance. Furthermore, selecting herbicides for a particular application window (e.g., preplant incorporated, preemergence, postemergence) will help keep weeds off balance. Widespread use of postemergence herbicides, for example, may shift weed population towards late-emerging weeds [8].

The current cropping system in the USA is based on herbicide-tolerant crops (e.g., Roundup-Ready and Liberty-Link), which also should be rotated [8]. Technology of herbicide tolerance should be viewed as just another tool for weed control [8].

Crops can also be selected to vary their planting date. Planting date may be chosen to aid in managing the particular weed species. Planting "early" may provide a crop a better start against late-emerging weeds, such as common water hemp (*Amaranthus rudis* Sauer [= *A. tamariscinus auct. non* Nutt.]), ivy leaf morning glories (*Ipomoea hederacea* Jacq. [= *I. barbigera* Sweet, *I. desertorum* House]), and fall panicum (*Panicum dichotomiflorum* Michx). Planting "late" may allow the use of a preplanting (burndown) herbicide or a tillage operation to control early-emerging weeds, such as winter annuals: field pennycress (*Thlaspi arvense* L.), shepherd's purse (*Capsella bursa-pastoris* [L.] Medik. [= *Bursa b.* (L.) Britt.]), tansy mustards (*Descurainia pinnata* [Walt.] Britt.), henbit (*Lamium amplexicaule* L.), velvetleaf (*Abutilon theophrasti* Medik; [= *A. abutilon* (L.) Rusby]), lamb's-quarters (*Chenopodium album* L.), and green foxtail (*Setaria viridis* [L.] Beauv. [= *Panicum viride* L.]). Changing planting dates from year to year will not allow specific weeds to adapt.

The use of cover crops and their residues also has the potential to keep weeds "off balance" through competition, physical suppression, and allelopathic effects [28]. Cultural practices, such as the use of cover crops, high-density planting, and crop rotation, are instead used to keep the weed populations under check [29].

Mechanical and alternative weed control methods also can be valuable tools for keeping weeds "off balance." They might include a variety of mechanical implements to do various types of disking and cultivation. Tillage and cultivation have proven to be effective tools for controlling emerged weeds, or burying their seeds below optimal emergence zones. Many studies have shown that the weed seed bank in the surface soil layer increases with reduced tillage [30–32]. The primary goal of cultivation is to uproot weed seedlings; however, the use of cultivators to control weeds in conventionally produced soybeans in the USA has decreased over the years due to widespread use of postemergent herbicides, especially in HTCs. Most growers prefer not to cultivate because the cultivation process is slow. Examples of mechanical implements used in soybean might include rotary hoes, flex-tine weeders, and spike-tooth harrows. These implements are designed for use early in the cropping cycle on very small weeds. Interrow cultivation equipment includes sweep shovels, knives for low residue fields, and undercutting sweeps for high-residue fields. Traditional operator-guided interrow cultivators require constant attention and skills to avoid damage to the crop. There are also recent developments in the machine vision guidance systems, which can speed up the cultivation process and reduce the need for highly skilled cultivator operators [33]. However, due to the dominance of herbicide-based systems and reduced tillage systems, there is little use of vision guidance systems for cultivation of agronomic crops in the USA.

Alternative weed control methods, such as flame weeding, can be very useful, especially in those fields where weed resistance to herbicides has developed, or during wet spring when mechanical means are not possible. Propane-fueled flame weeding is a process of exposing plant tissues to the heat coming from a propane burner. Flaming should not be confused with burning, as plant tissue does not ignite, but heats rapidly to the point of rupturing cell membranes and tissues, which results in weed death [34]. The heat that comes from the burners can have temperatures of up to 1900 °C, which raises the temperature of the exposed plant tissues rapidly [35]. An increase of temperature above 50 °C inside plant cells can result in the coagulation (denaturation and aggregation) of membrane proteins leading to loss of membrane integrity [34, 36–38]. Consequently, flamed weeds die or their competitive ability against the crop is severely reduced.

The efficacy of flame weeding was reported to be influenced by several factors, including the growth stages of the plant [39–47], the physical location of the growing point at the time of flaming [40, 48], the presence of protective layers of hair or wax and lignification [40, 49], the technique of flaming [36], the regrowth potential of plant species [40, 49], and the leaf relative water content of plant species [50].

Knezevic et al. determined that the propane doses of 60–80 kg/ha were highly effective in controlling many broadleaf weeds at early growth stages (up to 25 cm tall) [51]. Such doses provided more than 90 % control of major broadleaf species:

- Velvetleaf
- Ivyleaf morning glory (*Ipomoea hederacea* Jacq. [= *I. barbigera* Sweet, *I. desertorum* House])
- Redroot pigweed (*Amaranthus retroflexus* L.)
- Common water hemp

- Lamb's-quarters
- Field bindweed (*Convolvulus arvensis* L. [= *C. ambigens* House, *C. incanus* auct. Non Vahl])
- Kochia (*Kochia scoparia* [L.] Schrad. [= *K. alta* Bates, *K. sieversiana* (Pallas) C.E. Mey.])
- Venice mallow (*Hibiscus trionum* L. [= *Trionum t.* (L.) [= *Trionum t.* (L.) Woot. & Standl.])

And the same doses also provided 80 % control of several grass species:

- Barnyard grass (*Echinochloa crus-galli* [L.] Beauv. [= *E. pungens* (Poir.), Rydb., *Panicum c.* L.])
- Green foxtail
- Yellow foxtail (*Setaria glauca* [L.] Beauv. [= *S. lutescens* (Wreigle), F.T. Hubbard, *S. pumila* (Poir.) Roemer and J.A. Schultes])

Control of the aforementioned weeds can be done prior to soybean planting as well as before and after crop emergence [51]. The costs of a single flaming operation applied broadcast below crop canopy could be US$ 30–40 per hectare, without taking into account the costs of the equipment and labor—current price of propane (US$ 0.5/kg × 60–80 kg) banded application over the crop row of flaming can cost US$ 12–20/ha due to lower propane use rates (30–40 kg/ha). Soybean is tolerant to flaming only at the VE–VC stages (emergence-unfolded cotyledon) and after the V4 (four trifoliate leaves) growth stages [51].

Flame weeding has some major advantages over herbicides and repeated tillage. Unlike herbicide application, flame weeding has no negative impact on the quality of surface or underground water. Flame weeding does not disturb soil structure as repeated mechanical weeding does. It has been reported that repeated cultivation promotes loss of organic matter through the dust particles and soil erosion induced by wind and heavy rains [52]. Flame weeding is also significantly less expensive than hand weeding and organic herbicides (e.g., vinegar, aromatic acids) [53], and it is unlikely for weeds to develop resistance to the high level of instant heat produced by flaming torches. In comparison with cultivation, flame weeding can be carried out on wet or stony soils, does not disrupt the soil surface, and does not bring buried weed seeds to the soil surface [49, 52]. Flame weeding can significantly reduce the need for hand weeding in organic systems.

Making a Weed Control Decision

One of the most common dilemmas that farmers and practitioners face is how to make a decision on the timing of weed control operation, or simply said "when to spray an herbicide." Before initiating weed control procedures, the following are some general guidelines to consider: field scouting and mapping weed patches and utilizing the concepts of (1) critical period of weed control (CPWC), (2) weed threshold, and (3) decision support computer models.

Field scouting typically involves assessing the type and number of weeds to determine if a spray operation is necessary. The entire field should be walked in a "W" pattern and weed density assessed, on at least 20 randomly selected spots, using a 1 × 1-m quadrat. Weed count should be done in each quadrat, and then averaged to determine weed density for the field. Some weeds are not distributed uniformly and can be found in patches, or in low spots of the field. These areas should be sprayed separately, as field-wide spraying may not be required. Mapping and monitoring weed patches over time will also help assess the effectiveness of the control program.

Studies of crop–weed competition showed that yield loss is sensitive to small differences in the period between crop and weed emergence [10]. It brings to light the importance of the concepts of CPWC and economic thresholds.

Critical Period of Weed Control

The CPWC has been defined in several ways. Zimdahl defined it as a "span of time between that period after seeding or emergence, when weed competition does not reduce crop yield, and the time after which weed competition will no longer reduce crop yield" [54]. Swanton and Weise defined the CPWC as the time interval when it is essential to maintain a weed-free environment to prevent crop yield loss [4]. In recent years, university extension weed specialists [10] have described the CPWC as a "window" in the crop growth cycle during which weeds must be controlled to prevent unacceptable yield losses.

Examples of historical motivations for studying this concept include: (1) potential for reduction in the amount of herbicides used by achieving efficient timing of their application [55–57], (2) potential for reduction in environmental and ecological degradation associated with the prophylactic use of herbicides [4, 58, 59], and (3) providing a test to determine whether methods of weed control are based on biological necessity [58]. Most importantly, there is a need for the economic optimization of weed control tactics in herbicide-tolerant soybean through timely application of postemergence herbicides. It is believed that knowing the CPWC can aid in making decisions on the need for and timing of weed removal in cropping systems that use both HTCs and conventional soybean cultivars [10]. The popularity of crops resistant to glyphosate has generated many studies across the corn and soybean production areas of the USA [24, 60–63]. The common objective of those studies was to determine the optimum timing of weed control in glyphosate-resistant crops. The timing of weed removal has been reported based on weed height [60, 62, 63], weeks after crop emergence [64], and crop growth stage [24, 61].

Soybeans are grown in row spacing ranging from 19 to 97 cm, with 19-cm and 76-cm row spacing being the most common. The general belief is that reducing row spacing affects the competitiveness of both crops and weeds. Knezevic et al. confirmed that the soybean row spacing significantly influenced crop–weed interference relationships, suggesting that weeds may be allowed to compete with soybeans for longer periods when crop is planted in narrow rows [25]. For example, the CTWR in narrow row beans coincided with the V3 stage compared to V1 stage in 76-cm-wide rows [26]. The differences in the CTWR documented in this study highlight the importance of integrating decisions regarding crop row spacing and the timing of weed control. A practical implication of this study is that planting soybean in wide rows (i.e., 76 cm) reduces early-season crop tolerance to weeds, thus requiring earlier weed management practices than in narrower rows. In contrast, a reduction in soybean row spacing increases soybean tolerance to weeds and may require less intensive weed management programs.

With the growing popularity of HTCs, dependence upon total postemergence herbicide programs became prevalent in US cropping systems. Such a shift in cropping practices highlights the importance of appropriately timed weed control measures. Since glyphosate-tolerant soybeans have been widely adopted, the concept of CTWR is an important part of IWM in answering some fundamental questions such as "if" and "when" to apply postemergence herbicides. A possible strategy for weed control, for example in glyphosate-tolerant soybean, would be to apply glyphosate tank-mixed with a short residual herbicide at the CTWR, which may provide adequate weed control during the entire critical period [10].

In addition, with the widespread use of glyphosate-resistant soybeans, another common question that can help optimize the weed control program is: "What is the cost of a delayed weed control operation?" Knezevic et al. reported that the soybean row spacing can significantly influence crop–weed interference relationships [25]. The timing for weed removal was affected by row spacing in soybean, resulting in about 2% yield penalty for every leaf stage of delayed weed control [26]. Therefore, an average of 2% yield loss per leaf stage of delay past the CTWR (5% yield loss) was determined as the cost of not controlling weeds on time in both corn and soybean. For example, the CTWR in 20-cm-row soybean was the V3 (third trifoliate) stage [26]. If weed control was delayed to the V4 (fourth trifoliate), the yield loss was about 7%, costing about 2% in yield losses due to prolonged competition from weeds. The same was true if weed control was delayed after the recommended critical time in other row spacing in soybean [26].

It is also interesting to note that insect and weeds can have additively detrimental effects on soybean yield loss. Insect and weed management practices exist typically as separate entities to those who develop pest management strategies. The need for an integrated approach is particularly evident in soybeans, where early-season defoliation by bean leaf beetle (*Certoma trifurcata* Förster) is a common occurrence [65, 66] along with weed interference. Typically, early-season defoliation by bean leaf beetles is not considered economically damaging to soybean [67–69] because soybean has enough time in the growing season to recover from bean leaf beetle injury. However, early-season defoliation can delay canopy development and reduces plant height [68], predisposing soybean to economic damage by subsequent stresses including weed interference. Delay in canopy development allows more light transmittance, which can favor weed growth and competition directly affecting weed management programs. No research has specifically addressed the impact of early-season insect defoliation on the need for timing of weed control. Gustafson et al. confirmed that the insect-induced defoliation of soybean canopy significantly influenced crop–weed interference relationships [70]. These results suggest that weeds should not be allowed to compete with soybean under any defoliation stress. Gustafson et al. also suggested that the practical implication of their study is that soybean canopy defoliation reduces early-season crop tolerance to weeds, thus requiring earlier weed management practices than in undefoliated crop [71]. There is a need to monitor bean leaf beetle populations early in the season and weed density in order to design appropriate pest management practices to protect the crop. For example, early-season control of insects can enhance crop tolerance to weed presence and vice versa, or simultaneous control of both pests may be needed. In practical terms, it means that tank-mixing an insecticide and herbicide in a single application may be necessary (consult product labels for the compatibility of the pesticide mixes).

Weed Threshold

The concept of "threshold" was used first in the field of entomology and plant pathology [72]. In the discipline of weed science, the weed threshold—"a point at which weed density causes important crop losses"—is an integral component of an IWM system. Knowledge of thresholds can help agriculturists make decisions on the timing of herbicide applications, in deciding whether remedial weed control efforts are necessary or economically justified [20].

Several types of thresholds have been described in literature, including economic thresholds [72–75]; damage thresholds, action thresholds, and period thresholds [76]; competition thresholds and statistical thresholds [74]; and predictive thresholds, safety thresholds, and visual thresholds [74]. Only economic threshold will be described in this chapter, but the rest can be found in the cited papers.

Cousens defined economic thresholds as "the weed density at which the cost of weed control equals the increased return on yield in the current year" [74]. Because they account for crop losses only in the current cropping season, economic thresholds are single-year measures of weed effects [77]. In addition, economic thresholds are based on factors such as the price of the crop at harvest, herbicide, application cost [75], anticipated crop yield, and the yield loss–weed density relationships, which are a function of environmental factors (e.g., soil types and climate). Since a major cause of yield reductions by weeds is through competition for growth-limiting resources (light, water, and nutrients), the economic threshold is not therefore constant for particular weed–crop combinations and can differ within the same geographic region [74].

Outside of academic circles there is no need for so many definitions of threshold [78], but there is a need to understand that many factors must be considered in calculating thresholds and not only present-year economics. For the practical use of thresholds, the farmer does not need to know a definition but rather would like to know what factors are considered in the calculations and what the risks associated with using thresholds in managing weeds are [74].

Managing weeds according to threshold concepts may mean leaving some weeds in the agricultural system, which may have some advantages [79], such as reduced cost of herbicides required to obtain a high weed control efficacy in the current year, and the long-term maintenance of genotypes susceptible to different weed control tools [79]. If weeds are allowed to remain in the field to maintain susceptibility to weed control tools, then we must balance their genetic diversity qualities with the influence they will have on yield [79]. Also, to reduce spread of herbicide-resistant weeds by balancing genetic diversity, the knowledge of how gene flow and fitness processes interact is needed, which establishes a new rationale for a threshold [79].

Disadvantages associated with the use of weed thresholds were also reported by some authors [74, 80–82], which are summarized as follows:

1. The applicability of single-year economic thresholds for weed management has been discussed by several workers [74, 80, 81]. These theoretical discussions concluded that the use of single-year thresholds could lead to the buildup of weed seeds in the soil. Also, the single-year threshold is much higher than multiple-year thresholds, which take weed seed production into account, and which are based on the multiyear economics.
2. Most fields have multiple weed species. Even if each species were below its threshold level, their combined impact on the crop might warrant treatment [82].
3. Calculation of average weed density in a field is based on the sample size; however, defining the appropriate sample size is difficult since weeds usually emerge in patches [82].
4. Since thresholds are associated with the postemergence application of herbicides, a farmer must be able to identify weeds at the seedling stage before control measures are to be taken. However, if he/she does not know how to identify weed seedlings, or has no time to do it, he/she must be prepared to hire a scout or consultant, thus adding to production cost [82].

To argue in favor of threshold, the experimentation and collection of data for determining thresholds helped us understand at least a small part of the very complex nature of crop–weed relationships. The influence of competition on crop yield and the collection of herbicide performance data that are used to calculate financial information for spray decisions are critical for implementation of a threshold-based weed management plan. Also, an individual grower may adjust the threshold values based on his/her own experience and attitudes about risk. Therefore, the use of herbicides will always be an individual decision because the farmer is the one who will make the decision about spraying [74].

Computer-Based Models and Decision Support Systems

Computer-based models and decision support systems have also been developed to aid in making spray decisions. Modeling as a research tool is used much less in weed science than in other scientific fields [83]. The reasons for this are not obvious but perhaps some are the relatively small number of researchers in that area and the straightforward use of herbicides marketed by large chemical corporations.

WeedSOFT is an example of a practical decision support software developed by the University of Nebraska. This software can aid in selecting the most economical herbicide based on weed species, density, weed emergence time, herbicide efficacy, cost effectiveness, environmental safety, yield loss estimates, and corresponding weed management recommendations. A list of herbicide programs are combined and evaluated based on weed control efficacy [84], site conditions, yield loss prevention, and net gain [85]. About 15 years ago, WeedSOFT was used in the north-central USA, including Illinois, Indiana, Kansas, Michigan, Missouri, Nebraska, Ohio, Pennsylvania, and Wisconsin [84]. Regionalization of WeedSOFT in the north-central USA has allowed specialists to evaluate predictions made by this program in their state, and to further customize it for the specific set of local weed species and environmental conditions. An additional goal of the regionalization effort was also to develop an educational decision support module that aids in choosing weed management tactics that minimize weed seed production [85]. However, practitioners stopped utilizing WeedSOFT around 2005 due to the overwhelming use of HTCs, and widespread use of glyphosate herbicide, which is nonselective and controls many weed species, thus reducing the need for appropriate herbicide selection.

Documentation and Record Keeping

Documentation and record keeping is an essential part of an IWM program. Information on cropping practices and history of each field will help evaluate the weed control program over time. Information can be recorded on paper forms or directly on the computerized versions of those forms, which can be developed as a database application. Data forms should have basic information including site description; weed species composition; evaluation of herbicide performance; weed response to weed control methods; records of the amount, type, and methods of herbicide applications; and other methods of weed control. Record keeping pays off because knowing the weeds on the farm, taking notes, and watching for possible shifts in weed species may prevent costly surprises.

Conclusion

IWM: Making It Work on Any Farm

Since there are many kinds of weeds with highly differing life cycles, they obviously cannot be managed by a single control method. However, if they are implemented in a systematic manner, significant advances in weed management can be achieved. Obviously, no one can use all of the previously described techniques at once. There are a number of ways to start developing an IWM program. The easiest start will be to try one or two techniques and then add more practices as time goes on. After a few years, there will be a program of different techniques working together in an integrated approach. The use of a variety of weed control tools reduces the reliance on any specific tool, which means that those tools will still be effective in the years to come. The use of various weed control methods keeps weeds "off balance" and prevents them from adapting to a particular IWM strategy. And remember, there is no such thing as a "silver bullet" when it comes to weed control.

References

1. Thill DC, Lish JM, Callihan RH, Bechinski EJ (1991) Integrated weed management—a component of integrated pest management: a critical review. Weed Technol 5:648–656
2. Walker RH, Buchanan GA (1982) Crop manipulation in integrated weed management systems. Weed Sci Suppl 30:17–24
3. Shaw WC (1982) Integrated weed management systems technology for pest management. Weed Sci 30 Suppl:2–12
4. Swanton CJ, Weise SF (1991) Integrated weed management: the rationale and approach. Weed Technol 5:648–656
5. Buchanan GA (1976) Weeds and weed management in cotton. In: Proceedings of beltwide cotton production resource conference, pp 166–168
6. Moll S (1997) Commercial experience and benefits from glyphosate tolerant crops. In: Proceedings of Brighton crop protection conference—weeds, vol 3, pp 931–940
7. Rasche E, Gadsby M (1997) Glufosinate amonium tolerant crops—international commercial developments and experiences. In: Proceedings of the Brighton crop protection conference—weeds, vol 3, pp 941–946
8. Knezevic S (2007) Herbicide tolerant crops: 10 years later. Maydica 52:245–250
9. Hamill A, Knezevic SZ, Chandler K, Sikkema P, Tardif F, Swanton CJ (1998) Weed control in glufosinate-tolerant corn (*Zea mays* L.). Weed Technol 14:578–585
10. Knezevic SZ, Evans SP, Blankenship EE, Van Acker RC, Lindquist JL (2002) Critical period for weed control: the concept and data analysis. Weed Sci 50:773–786
11. Knezevic SZ, Weise SF, Swanton CJ (1994) Interference of redroot pigweed (*Amaranthus retroflexus* L.) in corn (*Zea mays* L.). Weed Sci 42:568–573
12. Knezevic SZ, Weise SF, Swanton CJ (1995) Comparison of empirical models depicting density of *Amaranthus retroflexus* L. and relative leaf area as predictors of yield loss in maize (*Zea mays* L.). Weed Res 35:207–214
13. Knezevic SZ, Horak MJ, Vanderlip RL (1997) Relative time of redroot pigweed (*Amaranthus retroflexus*) emergence is critical in pigweed—sorghum (*Sorghum bicolor*) competition. Weed Sci 45:502–508
14. Bosnic CA, Swanton CJ (1997) Influence of barnyardgrass (*Echinochloa crus-galli*) time of emergence and density on corn (*Zea mays*). Weed Sci 45:276–282
15. Rejmanek M, Robinson GR, Rejmankova E (1989) Weed crop competition: experimental designs and models for an analysis. Weed Sci 37:276–284
16. Roush ML, Radosevich SR, Wagner RW, Maxwell BD, Peterson TD (1989) A comparison of methods for measuring effects of density and proportion in plant competition experiments. Weed Sci 37:268–275
17. Bridges DC, Wu H, Sharpe PJH, Chandler JM (1989) Modelling distributions of crop and weed seed germination time. Weed Sci 37:724–729
18. Cudney DW, Jordan LS, Corbett CJ, Bendixen WE (1989) Developmental rates of wild oats (*Avena fatua*) and wheat (*Triticum aestivum*). Weed Sci 37:512–524
19. Zanin G, Sattin M (1988) Threshold level and seed production of velvet leaf (*Abutilon theophrasti* Medicus) in maize. Weed Res 28:347–352
20. Weaver SE (1986) Factors affecting threshold levels and seed production of Jimsonweed (*Datura stramonium*) in soybeans (*Glycine max*). Weed Res 26:215–223

21. Hartfield JL, Buhler DD, Stewart BA (1998) Integrated weed and soil management. Ann Arbor Press, Chelsea
22. Ross, L (1985) Applied weed science. Macmillan, New York. ISBN 0-02-403911–X
23. Teasdale JR (1995) Influence of narrow row/high population corn (*Zea mays*) on weed control and light transmittance. Weed Technol 9:113–118
24. Mulugeta D, Boerboom CM (2000) Critical time of weed removal in glyphosate-resistant *Glycine max*. Weed Sci 48:35–42
25. Knezevic S, Evans S, Mainz M (2003a) Yield penalty due to delayed weed control in corn and soybean. Crop Manag J. Available from: http://www.plantmanagementnetwork.org/pub/cm/research/2003/delay/
26. Knezevic SZ, Evans S, Mainz M (2003b) Row spacing influences critical time of weed removal in soybean. Weed Technol 17:666–673
27. Liebman M, Ohno T (1998) Crop rotation and legume residue effects on weed emergence and growth: application for weed management. In: Hartfield JL, Buhler DD, Stewart BA (eds) Integrated weed and soil management. Ann Arbor Press, Chelsea, pp 181–221
28. Teasdale J (1998) Cover crop, smother plants, and weed management. In: Hartfield JL, Buhler DD, Stewart BA (eds) Integrated weed and soil management. Ann Arbor Press, Chelsea, pp 247–266
29. Price AJ, Balkcom KS, Culpepper SA, Kelton JA, Nichols RL, Schomberg H (2011) Glyphosate-resistant Palmer amaranth: a threat to conservation tillage. J Soil Water Conserv 66:265–275
30. Anderson RL, Tanaka DL, Black AL, Schweizer EE (1998) Weed community and species response to crop rotation, tillage and nitrogen fertility. Weed Technol 12:531–536
31. Barberi P, LoCascio B (2001) Long-term tillage and crop rotation effects on weed seedbank size and composition. Weed Res 41:325–340
32. Feldman SR, Alzugaray C, Torres PS, Lewis P (1997) The effect of different tillage systems on the composition of the seedbank. Weed Res 37:71–76
33. Slaughter DC, Chen P, Curley RG (1999) Vision guided precision cultivation. Precis Agric 1:199–216
34. Lague C, Gill J, Peloquin G (2000) Thermal control in plant protection. In: Vincent C, Panneton B, Fleurat-Lessard F (eds) Physical control methods in plant protection/La lute physique en phytoprotection. Springer, Berlin, pp. 35–46
35. Ascard J (1998) Comparison of flaming and infrared radiation techniques for thermal weed control. Weed Res 38:69–76
36. Parish S (1990) A review of non-chemical weed control techniques. Biol Agric Hortic 7:117–137
37. Pelletier Y, McLeod CD, Bernard G (1995) Description of sub-lethal injuries caused to the Colorado potato beetle by propane flamer treatment. J Econ Entomol 88:1203–1205
38. Rifai MN, Lacko-Bartosova M, Puskarova V (1996) Weed control for organic vegetable farming. Rostl Vyroba 42:463–466
39. Ascard J (1994) Dose-response models for flame weeding in relation to plant size and density Weed Res 34:377–385
40. Ascard J (1995) Effects of flame weeding on weed species at different developmental stages. Weed Res 35:397–411
41. Ulloa SM, Datta A, Knezevic SZ (2010a) Growth stage influenced differential response of foxtail and pigweed species to broadcast flaming. Weed Technol 24:319–325
42. Ulloa SM, Datta A, Knezevic SZ (2010b) Tolerance of selected weed species to broadcast flaming at different growth stages. Crop Prot 29:1381–1388
43. Ulloa SM, Datta A, Knezevic SZ (2010c) Growth stage impacts tolerance of winter wheat (*Triticum aestivum* L.) to broadcast flaming. Crop Prot 29:1130–1135
44. Ulloa SM, Datta A, Malidza G, Leskovsek R, Knezevic SZ (2010d) Timing and propane dose of broadcast flaming to control weed population influenced yield of sweet maize (*Zea mays* L. var. *rugosa*). Field Crops Res 118:282–288
45. Ulloa SM, Datta A, Cavalieri SD, Lesnik M, Knezevic SZ (2010e) Popcorn (*Zea mays* L. var. *everta*) yield and yield components as influenced by the timing of broadcast flaming. Crop Prot 29:1496–1501

46. Ulloa SM, Datta A, Cavalieri SD, Lesnik M, Knezevic SZ (2010f) Popcorn (*Zea mays* L. *var. everta*) yield and yield components as influenced by the timing of broadcast flaming. Crop Prot 29:1496–1501
47. Ulloa SM, Datta A, Malidza G, Leskovsek R, Knezevic SZ (2010g) Yield and yield components of soybean [*Glycine max* (L.) Merr.] are influenced by the timing of broadcast flaming. Field Crops Res 119:348–354
48. Knezevic, Ulloa (2007) Flaming: New potential tool for weed control in organically grown agronomic crops. J Agric Sci 52 2:95–104
49. Ascard J, Hatcher PE, Melander B, Upadhyaya, MK (2007) Chapter 10, Thermal weed control. In: Upadhyaya MK, Blackshaw RE (eds) Non-chemical weed management. CAB International, pp 155–175
50. Ulloa SM, Datta A, Bruening C, Gogos G, Arkebauer TJ, Knezevic SZ (2012) Weed control and crop tolerance to propane flaming as influenced by the time of day. Crop Prot 31:1–7
51. Knezevic S, Datta A, Bruening C, Gogos G (2012) Propane-fueled flame weeding in corn, soybean and sunflower training manual. Available from: http://www.agpropane.com/Content PageWithLeftNav.aspx?id=1916
52. Wszelaki AL, Doohan DJ, Alexandrou A (2007) Weed control and crop quality in cabbage [*Brassica oleracea* (capitata group)] and tomato (*Lycopersicon lycopersicum*) using a propane flamer. Crop Prot 26:134–144
53. Nemming A (1994) Costs of flame cultivation. Acta Hortic 372:205–212
54. Zimdahl, RL (1988) The concept and application of the critical weed-free period. In: Altieri MA, Liebman M (eds) Weed management in agroecosystems: ecological approaches. CRC Press, Boca Raton, pp 145–155
55. Hall MR, Swanton CJ, Anderson GW (1992) The critical period of weed control in grain corn (*Zea mays*). Weed Sci 40:441–447
56. Van Acker CR, Swanton CJ, Weise SF (1993) The critical period of weed control in soybean [*Glycine max* (L) Merr.]. Weed Sci 41:194–200
57. Zimdahl RL, Oregon State University, International Plant Protection Center (1980) Weed-crop competition—a review. Oregon State University Press, Cornvallis
58. Weaver SE (1984) Critical period of weed competition in three vegetable crops in relation to management practices. Weed Res 24:317–325
59. Weaver SE, Tan CS (1983) Critical period of weed interference in transplanted tomatoes (*Lycopersicum esculentum*): growth analysis. Weed Sci 31:476–481
60. Dalley DC, Kells JJ, Renner KA (1999) Weed interference in glyphosate resistant corn and soybean as influenced by time of weed removal and crop row spacing. In: Proceedings of the North Central weed science society, vol 54, p 65
61. Evans SP, Knezevic SZ. Critical period of weed control in corn as affected by nitrogen supply. In: Proceedings of the North Central weed science society, vol 55, p 151
62. Gower A, Loux MM, Cardina J (1999) Determining the critical period of weed management in glyphosate-tolerant corn. In: Proceedings of the North Central Weed Science Society, vol 54, p 66
63. Kalaher CJ, Stoller E, Young B, Roskamp G (2000) Proper timing of a single post-emergence glyphosate application in three soybean row spacings. In: Proceedings of the North Central weed science society, vol 55, p 113
64. Sellers BA, Smeda RJ (1999) Duration of weed competition and available nitrogen on corn development and yield. In: Proceedings of the North Central weed science society, vol 54, p 3
65. Higley LG (1994) Handbook of soybean insect pests. In: Boethel DJ (ed) Entomological Society of America, College Park
66. Zeiss MR, Pedigo LP (1996) Timing of food plant availability: effect on survival and oviposition of the bean leaf beetle (Coleoptera: Chrysomelidae). Environ. Entomology 25:295–302
67. Weber CR, Caldwell BE (1966) Effects of defoliation and stem bruising on soybeans. Crop Sci 6:25–27
68. Hunt TE, Higley LG, Witkowski JF (1994) Soybean growth and yield after simulated bean leaf beetle injury to seedlings. Agron J 86:140–146

69. Hunt TE, Higley LG, Witkowski JF (1995) Bean leaf beetle injury to seedling soybean: Consumption, effects of leaf expansion, and economic injury levels. Agron J 87:183–188
70. Gustafson T, Knezevic S, Hunt T, Lindquist J (2006a) Early-season insect defoliation influences the critical time for weed removal in soybean. Weed Sci 54:509–515
71. Gustafson T, Knezevic S, Hunt T, Lindquist J (2006b) Simulated insect defoliation and duration of weed interference affected soybean growth. Weed Sci 54:735–742
72. Headley JC (1972) Defining the economic threshold. In: Pest control strategy for the future. National Academy of Sciences, Washington, DC, pp 100–108
73. Coble HD, Mortensen DA (1992) The threshold concept and its application to weed science. Weed Technol 6:191–195
74. Cousens, R (1987) Theory and reality of weed controlled thresholds. Plant Prot Quart 2:13–20
75. Marra MC, Carlson GA (1983) An economic threshold model for weeds in soybeans (*G. max*). Weed Sci 31:604–609
76. Coble HD, Williams FM, Ritter RL (1981) Common ragweed interference in soybeans. Weed Sci 29:339–342
77. Bauer TA, Mortensen DA (1992) A comparison of economic and economic optimum threshold for two annual weed in soybeans. Weed Technol 6:228–235
78. Stern VM (1973) Economic thresholds. Annu Rev Entomol 18:259–80
79. Maxwell BD (1992) Weed thresholds: The space component and considerations for herbicide resistance. Weed Technol 6:205–212
80. Auld BA, Menz KM, Tisdell CA (1987) Weed control economics. Academic, London
81. Norris RF (1984) Weed thresholds in relation to long-term population dynamics. In: Proceedings of the West society of weed science, vol 37, pp 38–44
82. Weaver SE, Weersink A (1993) Weed economic thresholds. In: Integrated weed management. Department of Crop science, Univesity of Guelph, OMAF, Guelph
83. Cousens R, Moss SR, Cussans GW, Wilson BJ (1987b) Modelling weed populations in cereals. Rev Weed Sci 3:93–112
84. Neeser C, Dille JA, Krishnan G, Mortensen DA, Rawlinson JT, Martin AR, Bills LV (2004) WeedSOFT(R): a weed management decision support system. Weed Sci 52:115–122
85. Schmidt AA, Johnson WG (2004) Influence of early-season yield loss predictions from WeedSOFT and soybean row spacing on weed seed production from a mixed weed community. Weed Technol 18:412–418

Chapter 11
Integrated Weed Management in Horticultural Crops

Darren E. Robinson

Introduction

Integrated weed management (IWM) is defined here as the use of more than one broad category of weed control tactics (i.e., biological, chemical, cultural, and physical) to allow producers to: (1) minimize crop losses due to competition, allelopathy, and interference with crop management and harvest; (2) reduce production and survival of weed seeds and vegetative propagules; (3) prevent introduction of new problem species [1]; and (4) realize long-term gains in certain indicators of agroecosystem health (i.e., energy efficiency, water quality, and soil quality) [2]. Inherent in the design of IWM systems then should be the curtailing of weed outbreaks and the enhancement of the integrity of the agroecosystem. Though much of the research focus of weed science continues to be on the most effective use of chemical control measures in field crops [3], the literature contains many attempts to integrate two or three different weed control tactics that can be practicably applied in vegetable production systems.

Despite the efforts to develop IWM systems, there are only specific cases where growers are extensively implementing them [4], such as for the management of glyphosate-resistant annual ryegrass (*Lolium rigidum* Gaud.) in Australia. System models used to evaluate cropping systems based on IWM proposed that this is partially a result of grower perception that the cost to benefit ratio of applying multiple techniques is greater than reliance upon herbicides alone [5]. Closer analysis reveals that this perception is only true on a case-by-case basis. Two contrasting examples illustrate this: (1) In the northeastern USA, the conventional weed control method relative to IWM systems provided greater yield and net economic returns of corn and soybean [6] and (2) in the United Kingdom, spring cereals grown in a production system using IWM provided similar net return to that of spring cereals under conventional management [7]. Unfavorable grower perception of the economics

D. E. Robinson (✉)
Department of Plant Agriculture, University of Guelph, 120 Main Street East, Ridgetown Campus, Ridgetown N0P 2C0, ON, Canada
e-mail: drobinso@uoguelph.ca, darrenr@uoguelph.ca

B. S. Chauhan, G. Mahajan (eds.), *Recent Advances in Weed Management*,
DOI 10.1007/978-1-4939-1019-9_11, © Springer Science+Business Media New York 2014

of IWM systems is a key reason why they are not widely adopted. Weed scientists' success in facilitating adoption of IWM practices by producers can be improved by demonstrating the net economic effect of IWM measures.

In instances when economics favor the use of IWM over conventional weed control, the increased management complexity inherent in the application of IWM is still a challenge that must be met [5]. Producers have myriad concerns beyond managing weeds; when one examines the weed management techniques of growers with similar crops in different parts of the world, the use and effect of each technique must be considered in the context of the environmental conditions and production system [8]. One set of IWM practices cannot be prescribed in all situations and cropping systems, but rather there are several different techniques that must be considered for their potential to fit an existing production system. The selection and implementation of those practices in a manner that is both feasible economically and supports the primary objective of IWM is vital for producers to successfully adopt these systems [9].

Critical Period of Weed Control

The IWM system described for carrot, cole crops, and beets relies upon their ability to produce a canopy that will suppress weeds that would emerge from mid-season onward. The key cultural aspect of this system is the application of weed control techniques to prevent interference with these crops during their critical period of weed control. The critical weed control period is the interval in the lifecycle of the crop when it must be kept weed-free to prevent yield loss [10]. The critical period concept can be a useful tool for timing different weed control options; however, there are several factors that must be considered for their successful use. The beginning and duration of the critical period vary depending on rates of growth, morphology, height, and leaf area development of each crop [10]. Furthermore, cultural practices such as seeding date [11] and planting method (e.g., transplanting vs. direct seeding), which influence early growth rate and size relative to weeds, all significantly impact the critical period. Finally, crop response to variation in available moisture influence crop–weed interference is also an important determinant of the onset of the critical period of weed control [12].

The critical periods of weed control for carrot, cabbage, sugar beet, and red beet vary and are impacted by cultural management and environment. Since soil temperature, moisture, and planting date cause the actual dates of the critical weed-free period to vary substantially, the critical period in sugar beet is best delineated by crop growth stages and occurs from the four-leaf stage to canopy closure [13]. Similar to sugar beet, the maximum duration that carrots must be free of weeds, without compromising yield, is based upon leaf stage and is from emergence until the 12-leaf stage of crop growth [14]. Planting date influenced the critical period in carrot, such that delayed planting from late April to late May in southern Ontario reduced the length of the critical period to emergence until the four-leaf stage of crop [14]. Critical periods in cabbage and red beet have not been determined on the basis of leaf stages, but rather on time after crop emergence, and vary annually depending on the availability of moisture. For example, the critical period in direct-seeded cabbage began 2 or 4 weeks after emergence, depending on available moisture [15, 16]. The critical period of weed control in transplanted cabbage was consistently from emergence to 3 weeks after transplanting [17]. This consistency from year to year reflects the greater competitiveness and uniform growth of plants grown from transplants relative to seeded crops [18]. The critical period of red beet was also determined based on time from crop emergence. Kavaliauskaitė and Bobinas showed that red beet yield was not reduced as long as it was kept weed-free for the first 4 weeks after emergence [19].

Physical Control

Physical control during the critical period is a key component of IWM for cool season vegetable crops such as carrot, red beet, cabbage, and sugar beet. These crops produce a canopy that is slow to develop, and are planted at between row spacings of 30 and 70 cm, with much smaller spacings within the rows. As a result, control of weeds between the rows is much easier than controlling weeds inside rows. Whole-crop physical weed control is done by implements such as harrows and flame weeders; interrow weed removal with cultivators; and intrarow methods, such as is done with torsion or finger weeders, may be utilized to control weeds during the critical period. Each of these three types of physical weed control is appropriate to use during different times of the critical period; therefore, each will be separately discussed. Harrowing and flame weeding are most appropriate for use prior to crop emergence, while inter- and intrarow cultivation possess the selectivity to be safely used during early stages of crop development.

Harrowing and flame weeding prior to the emergence of cool season vegetable crops requires the crop to be sown deeper than from where weeds emerge, and are only effective on very small weeds. Harrowing is most effective on weeds from white thread to cotyledon stage [20], while flame weeding consistently caused a 90 % reduction in biomass of various weed species with between 5 and 11 leaves [21]. It is important to note that propane dosage is a significant determinant of the efficacy of flame weeding—a 76-kg ha^{-1} dose of propane was required to obtain this level of control [21]. Also, grass weed species are generally less sensitive to flaming [21–23]. A convenient and cost-effective method of nonchemical control is to use harrows to control very small weeds in these small-seeded vegetable crops before crop emergence [20]. Flame weeding can also be used to kill early emerging weeds prior to crop emergence, though the available time when this can be safely applied is limited in faster emerging crops such as cabbage, which may germinate before many weeds [24].

The use of flaming to destroy weeds without disturbing the soil may be done prior to or immediately following the planting of slower emerging crops like carrot to form a stale seedbed [25, 26]. Since this technique avoids soil disturbance, it minimizes weed emergence after the stale seedbed is formed [27]. Caldwell and Mohler compared weed density and biomass of a number of annual weed species in stale seedbeds formed with single or multiple applications of glyphosate, propane flaming, a spring tine weeder, a springtooth harrow, or a rotary tiller at approximately 4 weeks after an initial tillage operation and just prior to planting [28]. A single flaming or glyphosate application caused similar reductions in density and biomass of annual broadleaf weeds, and both were more effective than the spring tine weeder, springtooth harrow, and rotary tiller. None of the stale seedbed techniques provided season-long control of annual broadleaf weeds and none were effective against creeping perennial species such as yellow nutsedge (*Cyperus esculentus* L.) but could be integrated with in-crop weed control techniques to improve control and reduce the need for herbicides [28].

One of the drawbacks of forming a stale seedbed prior to planting is that the weed emergence in the planting rows is significantly greater than between them [28]. Modifications to planting techniques could further improve the effectiveness of stale seedbeds if they reduced the stimulation of weed germination in planting rows and subsequently reduced intrarow weeds [28]. A technique called punch planting shows utility in stale seedbed preparation, as it creates holes in the ground while minimizing soil disturbance outside the holes [29]. Punch planting is typically done using a dibber drill, which consists of wheels that run along the soil surface and several radially mounted plungers through which air is blown that pushes the plunger and seed into the soil [30]. Rasmussen examined the interaction between planting type (drilling vs. punch planting) and punch planting date after the formation of stale seedbeds by propane flaming on intrarow weed emergence in fodder beet [31]. Flaming reduced intrarow weed density by 30 % more in punch planting treatments than in drilled treatments across all planting times. Rasmussen also demonstrated that weed emergence in fodder beet was influenced by the length of time by which planting and flaming were delayed, such that waiting 2 and 4 weeks to plant and flame reduced emergence by 55 and 79 %, respectively [31]. Punch planting can reduce intrarow density, but its effectiveness is influenced by planting time in slower emerging crops such as fodder beet.

Similar experiments to those conducted by Rasmussen [31] were conducted on sugar beet and carrot [32]. In one experiment, intrarow weed emergence in stale seedbeds after flaming was compared between drilled and punch-planted treatments at five different planting dates. Punch planting alone did not reduce the average intrarow weed density in sugar beet. Across all planting dates, the effect of flaming and punch planting on intrarow weed density varied from 31 to 60 % in different years, depending on weed emergence patterns prior to flaming. Flaming reduced intrarow weed emergence by 89 % when planting was delayed to 4 weeks after seedbed preparation, compared to timely planting [32]. A second set of experiments in spring- and fall-seeded carrot compared drilling to punch planting, where flaming was used to kill weeds in the stale seedbed. Delayed planting did not affect weed emergence in any treatments in carrot. Punch planting and flame weeding did not reduce weed emergence in spring-seeded carrot, but they did reduce weed emergence by 52% in fall-seeded carrot. Crop tolerance also varied between sugar beet and carrot—though carrot density was similar in all treatments, sugar beet density was reduced in punch planting treatments compared to drilling [32]. This may have been influenced by flaming treatments delayed by rainfall at certain planting dates, though Rasmussen et al. attributed some of the reduction in sugar beet emergence to the punch planter itself [32].

Harrowing and flaming are unsuitable for the management of weeds *after* crop has emerged but before the critical period of weed control has passed. Roots of carrot, direct-seeded cabbage, red beet, and sugar beet plants up to the four-leaf stage possess less resistance to uprooting [33, 34] than is required to bury and/or remove certain early-germinating annual weed species enough to control them [35]. Even reducing moving speeds and setting harrow tines to shallow, less aggressive settings to minimize crop injury is not enough to adequately control weed seedlings [20]. Generally, broadleaf crops lack sufficient tolerance to avoid membrane disruption caused by the temperature levels experienced by plants during flaming [36]. This lack of selectivity during the early part of the critical period of these crops requires that measures other than harrowing or flaming be implemented to control these weeds soon after crop emergence.

Mechanical interrow cultivation is regularly conducted in carrot, cabbage, red beet, and sugar beet in organic and conventional production systems [37]. Various configurations of flexible tines (Fig. 11.1a) and rolling baskets (Fig. 11.1b) may be used with these crop species, with consideration of the between-row spacing needed, which will be fitted with different discs, plates, or hoods to protect the crop [38]. Steerage hoes must be used to ensure selective control of weeds using interrow cultivators, which is most commonly done by manual steering [37]. Electronic guidance systems were envisioned to be an important development to minimize crop damage when driving tillage implements along crop rows in crops such as carrot, cabbage, red beet, and sugar beet [39]. New machine vision guidance systems are being evaluated that utilize morphology, spectral characteristics, and visual texture to differentiate weeds from crops at positional errors of less than 30 mm for working speeds of 5–6 km hr^{-1} [40]. Though still under evaluation for different crops and moving speeds, the accuracy of computer vision technology has potential to increase the selectivity of interrow cultivation.

Computer vision technology is of even greater interest for use in intrarow cultivation, which, due to poor selectivity, is more difficult than interrow cultivation. Torsion, finger, and brush weeders were specifically developed for intrarow cultivation in high-value crops [37]. Torsion weeders have pairs of tines that are set on either side of crop rows; these provided the greatest level of cost-effectiveness for weed control compared with finger and brush weeders in sugar beet, carrot, and red beet [20], as well as in transplanted cabbage [37]. Steering control is crucial for the selectivity of these weeders, and research attempts to develop automated control is ongoing [41]. Row guidance, vision discrimination among plant species, within-row weed removal cutting implements, and global positioning systems to increase steering control are key components of this technology.

Fig. 11.1 (**a**) Torsion weeders with spiked discs, and (**b**) basket weeders are two types of mechanical weeders commonly used to control weeds in vegetable crops

The ability to visually discriminate a weed from a crop plant remains the greatest challenge to adoption of existing technology for intrarow weed control in cool season vegetable crops. In situations where weeds occupy a significant proportion of the crop rows, machine-guided sensors are unable to distinguish between crop and plants [40]. Mapping techniques that utilize real-time kinetic-global positioning systems (RTK-GPS) may eventually provide enough information about the position of individual crop plants to steer cultivator equipment with an acceptable level of selectivity for intrarow cultivation of cool season vegetable crops during the early part of the critical period of weed control [42]. Rasmussen et al. contrasted the accuracy of a rotor tine cultivator, or cycloid hoe, with and without crop plant RTK-GPS to move it around crop plants in the rows, and a flex tine harrow at the two- to four-leaf stage of sugar beet and carrot [41]. Rasmussen et al. also varied the crop establishment procedure using a normal planting drill and the punch planter described earlier on a stale seedbed prepared by flaming [41]. As was observed in a previous study [32], punch planting and flaming reduced crop emergence of sugar beet but not of carrot [41]. Interestingly, the cycloid hoe with RTK-GPS showed the same level of selectivity as the cycloid hoe without RTK-GPS and flex tine harrow. When averaged across the three treatments, intrarow weed control was 70 and 47 % in 2008 and 2009, respectively [41]. All three treatments caused a 20 % decrease in crop density compared to the untreated check, which was similar in both years. Further research on parameters that affect the accuracy of RTK-GPS, as well as the types of intrarow tillage implements, is needed to ensure crop safety in cool season vegetable crops, as producers would not accept this level of crop damage.

Early planting, tillage, and pesticide applications in cool season vegetable crops can increase soil compaction when these occur in cool, wet soils [43]. In addition, intensive tillage can reduce soil organic matter and damage soil structure such as is done for processing crops that are typically grown according to a strict production schedule. A potential avenue that the producers of these crops may use to offset these negative effects is to use modified reduced tillage systems to reduce the actual width of a field that is tilled in order to increase surface residues [44]. While cabbage showed yield reductions in no-till treatments due to poor establishment and cool soil temperatures, strip tilling a 20–30-cm-wide band of soil overcame these problems and resulted in yields similar to conventionally tilled cabbage [44]. Mochizuki et al. examined the interaction between the band tilled and tillage depth [45]. They showed that despite a 1 ℃ increase in soil temperature, increasing tillage width from 15 to 30 cm did not affect cabbage growth yield. Furthermore, growth and yield of cabbage were greatly improved—possibly due to decreased soil penetration resistance—by increasing tillage depth from 10 to 30 cm in the 15-cm-wide tilled zones, which still left between 60 and 80 % of the soil surface area undisturbed [45]. Weed control between the tilled zones of adjacent rows was done chemically, indicating that strip tillage must be integrated with other weed control techniques.

The presence of an oat (*Avena sativa* L.) cover crop in between tilled strips may augment herbicide use to suppress weeds between cabbage rows. The 25-cm-wide strips of preestablished oat are killed using a strip tiller, and cabbage is planted into the strips. The oats must then be killed with a herbicide at some time after transplanting; oat control timing is an important consideration in this type of strip-tillage system [46]. Oat reduced cabbage yield when oat control was delayed by 9 days after transplanting (DAT). An additional benefit of oat strips is an increase in certain beneficial arthropods that are natural enemies of key cabbage pests, such as imported cabbage worm (*Pieris rapae* L.), diamondback moth

(*Plutella xylostella* L.), and cabbage loopers (*Trichoplusia ni* Hübner) [46]. The authors concluded that the presence of the oat mulch enhanced habitat complexity enough to encourage the abundance of natural enemies to these pests.

Similar attempts to develop conservation tillage systems for carrot production have been made—unfortunately, reduced tillage across the entire bed interferes with stand establishment in carrot as it does for other vegetable crops [47]. Strip tillage has shown to have potential in carrot; however, the prevalence of perennial and winter annual weed species increased in strip tillage systems [48, 49]. When combined with compost applications, strip tillage may be used to provide acceptable weed control [50] and may also increase soil organic matter. A second option for control of weeds between rows in strip-tilled cool season vegetable crops includes living mulches grown in strips between crop rows [49].

Cultural Control

Previous review of diverse crop rotations generally indicates that more diverse rotations result in greater weed diversity but lower total weed density [51]. Dominance of a few different weed species decreases because different management practices associated with growing very different crops apply very different selection pressure on mixed weed populations [52]. For example, the selection pressures exhibited on weed populations by a cool season vegetable; a winter cereal; a competitive crop with many available and easily applied weed control options, such as corn and soybean; and a red clover (*Trifolium pretense* L.) cover crop will vary substantially. The cropping diversity effect will depend on physical and chemical control associated with each crop, as well as their respective times of emergence, and the overall competitiveness relative to the weeds. In certain instances, crop sequence may directly affect weed populations [53], and in others, the management practices used in a crop are the greatest determinant of future weed populations [54].

The inclusion of competitive cereal crops, such as barley (*Hordeum vulgare* L.), in rotations with vegetable crops such as carrot, have been shown to reduce weed populations and increase crop yields in some instances. Leroux et al. showed that seedling recruitment and biomass of nodding beggar-ticks (*Bidens cernun* L.) and Canada fleabane (*Erigeron canadensis* L.) were decreased the most when barley preceded a carrot crop [53]. In addition, the authors noted a significant reduction in root-knot nematode (*Meloidogyne hapla* Chitwood) injury in those rotations that included barley, which was attributed to a reduction in weeds that traditionally function as root-knot nematode hosts [55]. The combination of reduced competition and a reduction in root-knot nematode corresponded to a 35–50 % increase in carrot yield. The effect of competitive crops, such as spring cereals, may be used to reduce weed populations in subsequently grown cool season vegetable crops.

Research has also shown that incorporating a regularly mowed legume cover crop into rotation with cabbage can be a more effective means of managing future weed and other pest populations than a grass cover crop. Bellinder et al. showed that 2 years of frequently mowed alfalfa (*Medicago sativa* L.) and red clover grown prior to cabbage had the similar effect on weed seedbank density as tillage and a herbicide treatment in sweet corn grown before cabbage [56]. The allelopathic and competitive effect that grass cover crops, such as cereal rye, have on weed seedbanks [57] may not be as great as that of mowed alfalfa or red clover. Frequent mowing of legume cover crops may have reduced flowering and seed set, competed with weeds for available resources, and, in the absence of tillage, concentrated weed seeds on soil surface, which increases the risk of decay and predation than seed buried by tillage [58]. Bellinder et al. measured an increase in common ragweed (*Ambrosia artemisiifolia* L.) seed density after the cereal rye cover crop rotation only [56]. Average cabbage head size and yield were related negatively to density of common ragweed [59]. Not only is this increase in common ragweed seed density important from the standpoint of increased competition from weeds but also because common ragweed has been associated with increased incidence of white mold in cabbage [60].

Attempts have been made to overcome the effect of competition from interseeded cover crops in cool season vegetables such as cabbage. The presence of interseeded cover crops during the critical period of weed control reduced cabbage yield due to light interference [61], so delayed seeding of the cover crop was hypothesized to be a means of suppressing weeds after the critical period, without compromising crop yield. This requires that some other weed control measure, either chemical or physical, be employed during the critical period. Brainard et al. examined the effect of hairy vetch (*Vicia villosa* Roth.), lana vetch (*Vicia villosa* Roth ssp. *varia*), or oat overseeded at 10, 20, or 30 DAT [62]. A flex tine weeder and an S-tine cultivator were used to control weeds at 10, 10+20, or 10+20+30 DAT with or without cover crops sown at the time of the last cultivation. Cabbage yield was reduced when cover crops were interseeded at the earliest planting date, but either vetch species could be sown 20 or 30 DAT without reducing cabbage yield [62]. Weed control and cabbage yield were not different in the interseeded treatments than in those treatments that were cultivated in the absence of cover crops. As a result, the economics of overseeding cover crops from the perspective of weed control alone does not initially appear favorable because of costs associated with hairy vetch seed and management. Opportunities to utilize hairy vetch in cabbage grown in rotation with other high-value vegetable crops, such as tomato (*Solanum lycopersicum* L.), do exist. Sainju et al. estimated similar fruit yield, biomass, and N uptake in tomato grown after a fall cover crop of hairy vetch to tomato grown on bare ground and fertilized with 180 kg N ha^{-1} [63]. In addition to replacing fertilizer requirements, hairy vetch mulch suppresses many annual weeds, increases infiltration, and reduces runoff and sediment losses from fields [64]. The potential for improving soil fertility and structure may justify the extra cost and management associated with overseeding hairy vetch at 20 DAT in cabbage production systems that include high-value crops well suited to such a rotation.

In some cases, the effect of crop rotation on weed populations in cool season vegetable crops is due to crop management, rather than competition from rotational crops alone. For example, Ball and Miller reported that the density of kochia

(*Kochia scoparia* [L.] Schrad.) in sugar beet was more strongly influenced by tillage type (moldboard vs. chisel plowing) and herbicide rate than by rotational crop (corn vs. pinto beans) [65]. The importance of herbicide efficacy on future weed populations in rotations that include cool season crops like sugar beet was well illustrated by Bàrberi et al. [66]. The results showed that in low-herbicide-input systems, weed flora was more diversified 1 year after a less competitive crop like sugar beet. In high-herbicide-input systems, however, weed species diversity was a function of the selectivity of the herbicides used. Long-term research that evaluated the effects of several management practices over two rotational cycles, including rotational crop, herbicide program, tillage, inclusion of winter wheat, or inclusion of a red clover cover crop has also shown that rotational crop had much less effect than management practices on weed densities of summer annual weeds in transplanted cabbage. Brainard et al. found that rotations that included 3-year continuous field corn before cabbage did not produce greater summer annual weed densities than rotations that included winter wheat before cabbage [54]. Rather, the herbicide program used in cabbage significantly reduced summer annual seed production and spring tillage prevented winter annual survival and seed production. Including red clover had different effects depending on the rotation in which it was used [54]. When cabbage followed continuous field corn, red clover increased summer annual weed seed production or emergence; however, when red clover was used in a rotation that included sweet corn, peas, wheat, and cabbage, there was a 96 % reduction in seedbank density of winter annuals in the first, but not the second, cycle of the rotation [54]. Crop rotation, tillage, and herbicide use in each year of a rotation exert a strong effect on weed populations, and thus may be a useful means of seedbank manipulation in cool season vegetable crop rotations.

Row spacing may be used in cool season vegetable crops to increase sugar beet competitiveness with weeds and combined with herbicide application timing can impact weed control and crop yield. Armstrong and Sprague examined the effect of sugar beet planted in 38-, 51-, and 76-cm rows and herbicide applications to weeds of 5-, 10-, and 15-cm height on weed density and biomass, and sugar beet yield [67]. Weed density and biomass were less in 38- and 51-cm rows compared with 76-cm rows, as long as glyphosate was applied to weeds that were 10 cm in height or less. Furthermore, when averaged over row width, sugar beet root yields were reduced when glyphosate application was delayed until weeds were 15 cm in height. Sugar beet root and sugar yields were greater when planted in 38- and 51-cm rows than in 76-cm rows [67]. Sugar beet planted in narrow rows suppressed late-season weed growth more than when planted in wider rows. In this study, early season weeds were controlled with an initial application of glyphosate.

Chemical Control

Integration of chemical and physical weed control measures in carrot have been shown to provide effective control, while also reducing herbicide inputs [68]. Initial research in Switzerland showed that pre-sowing harrowing followed by interrow brushing with banded herbicide applications provided similar levels of weed control as broadcast chemical control treatment and reduced herbicide use by 50% [69]. Studies conducted in eastern Canada showed reductions in herbicide use by up to 66%, where herbicides were banded over crop rows, and side knives to control weeds on the sides of carrot beds were combined with S-tines to control weeds between crop rows [68]. This combination of banded herbicide with side knives and S-tines provided similar weed control to broadcast herbicide applications at a similar cost and without any reduction in yield. Treatments that included combinations of flaming, acetic acid, or stale seedbeds with side knives and S-tines were unable to provide similar levels of weed control and yield and cost significantly more [68].

Chemical control and tillage may interact to affect sugar beet emergence and subsequently canopy development and competitiveness. Water is transferred from soil aggregates to seed by direct contact; however, when aggregates are larger, seed-to-soil contact is reduced, which may delay crop germination and emergence [70]. Soil aggregates are larger in chisel-plowed than moldboard-plowed soil [71], which delayed sugar beet emergence in chisel-plowed treatments [72]. Furthermore, chisel and moldboard plowing affected weed and sugar beet response to preemergence (PRE) herbicides. Though PRE herbicides caused more visible injury to sugar beet in the chisel-plowed treatments, herbicide treatment had no effect on sugar beet plant stand, canopy closure, or final yield. Moldboard plowing did result in earlier sugar beet canopy closure and greater recoverable white sugar yield [72]. The inclusion of PRE herbicides with tillage increased control of common annual weed species, including common lamb's quarter (*Chenopodium album* L.), pigweed species (*Amaranthus* spp.), velvetleaf (*Abutilon theophrasti* Medik.), and giant foxtail (*Setaria faberi* Herrm.) compared to where tillage was used without a herbicide. The delay in crop emergence and canopy development caused by chisel plowing corresponds to a relative reduction in crop competitiveness compared with the moldboard-plowed system. Bollman and Sprague therefore recommended that sugar beet be grown in narrower rows when using chisel plowing [72].

Herbicide application can interact with tillage system (moldboard and chisel plowing and rotary tilling) and rotations that include continuous cotton or sugar beet grown after cotton. Vasileaiadis et al. showed that sugar beet yield was greatest in moldboard plowing treatments where PRE herbicides were used, and that the combination of herbicides and interrow cultivation used in the cotton cycle of the rotation reduced total weed density in the following sugar beet crop [73]. It should be noted that while density of important broadleaf weeds—redroot pigweed (*Amaranthus retroflexus* L.) and black nightshade (*Solanum nigrum* L.)—was lower in the herbicide-treated plots, barnyard grass (*Echinochloa crus-galli* [L.] Beauv.) density was higher, as the herbicides applied in sugar beet had little efficacy on this species. Black nightshade density was lower in the second year of the continuous cotton rotation than in the cotton–sugar beet rotation. Vasileaiadis et al. observed that most of the black nightshade emerged before cotton planting and was therefore killed during preparation of the cotton seedbed [73]. However, because sugar beet is planted much earlier than cotton, black nightshade emerged after sugar beet was planted but before the crop was able to develop a weed-suppressive canopy. PRE herbicide treatments in sugar beet integrated with moldboard tillage and interrow cultivation in previous cotton crop effectively controlled most weeds and improved crop yield; however, opportunities to incorporate other in-crop methods of weed control exist.

Conclusion

The findings made in the references cited here are of practical relevance to the development of IWM plans by producers, agronomists, and crop scouts to maintain weed populations at economically acceptable densities in cool season vegetable crops. The various measures that have been described detail several key decision points, including crop rotation, planting date, tillage selection, secondary and tertiary tillage, cover crops, planting density, row spacing, and herbicide use. While of practical use, this information can also be a resource to guide further research, whether it is to examine the possibility of adapting already currently developed systems to different regions where these crops are grown or to combine different aspects of different IWM systems. There are several areas of research on IWM in cool season vegetable crops that will improve our understanding of how weeds respond to novel management practices or new combinations of existing practices and how that response impacts the relationships between crops and weeds. Research on the effect of IWM practices on the soil environment and other biota is also needed to assess their potential impact on agroecosystem integrity [2].

One area of research that lacks focus is matching the very specific spectra of weeds different herbicides control to compliment specific weaknesses of the different tillage, crop rotation, cover crop, and other nonchemical methods detailed above. In a previous example, for instance, barnyard grass and black nightshade escapes could be managed with PRE applications of *S*-metolachlor in sugar beet [74], cabbage [75], or red beet and carrot [76]. It is equally important to understand that a PRE herbicide like *S*-metolachlor would be much less effective in a minimum-tillage system because surface residues would intercept much of the herbicide and thus reduce its efficacy [77]. This level of detail could help to improve the consistency of control of IWM strategies, particularly for weeds whose germination patterns, growth rate and habit, and adaptations to low levels of light, water, and nutrients allow them to escape nonchemical control treatments. This approach might seem to be in conflict with ecological weed management [9]. Alternatively, it could bridge the gap between chemically intensive weed control and cultural weed management by providing a starting point for encouraging growers who rely almost exclusively on herbicides for the control of weeds to begin integrating nonchemical control measures. More thoughtful inclusion of herbicides, based upon the weeds they control, into the development of IWM systems could therefore be one way of working with growers to increase the uptake of those systems.

In certain parts of the world (for example, in many parts of Europe), pesticide legislation and herbicide resistance have placed pressure on scientists to develop IWM systems that incorporate fewer pesticides [78] and increased tillage [79].

Though these limitations are not so thoroughly felt in other parts of the world, weed management in organic production systems faces a similar challenge regardless of political and social environments. Conventional tillage practices for cover crop management, seedbed establishment, and weed control are primarily practiced in organically grown cool season vegetables [48]; however, hybrid tillage systems, in which only the soil between the crop rows is tilled, are being adapted for vegetable production [49]. A key research issue to address in this type of system is the development of postemergence weed management tactics. This necessitates greater accuracy of differentiating weeds and crop plants for more effective physical intrarow weed control options [80], whether it be cutting implements [41], possibly shielded flame or steam weeders [81], and even air-propelled abrasives. Effective systems for IWM in cool season vegetable crops may be possible without pesticides or tillage over the entire field, but additional research to further optimize current systems will be required. While net economic returns and consistency of efficacy must be demonstrated to ensure that the potential of IWM systems can be successful, they must also be targeted for specific production systems.

References

1. Liebman M, Davis AS (2009) Managing weeds in organic farming systems: an ecological approach. Org Farm Ecol Syst 54:173–195
2. Swanton CJ, Murphy SD (1996) Weed science beyond the weeds: the role of integrated weed management (IWM) in agroecosystem health. Weed Sci 44:437–445
3. Harker KN, O'Donovan JT (2013) Recent weed control, weed management, and integrated weed management. Weed Technol 27:1–11
4. Llewellyn RS, Lindner RK, Pannell DJ, Powles SB (2004) Grain grower perceptions and use of integrated weed management. Aust J Exp Agric 44:993–100
5. Pardo G, Riravololona M, Munier-Jolain NM (2010) Using a farming system model to evaluate cropping system prototypes: are labour constraints and economic performances hampering the adoption of integrated weed management? Eur J Agron 33:24–32
6. Pimentel D, Hepperly P, Hanson J, Douds D, Seidel R (2005) Environmental, energetic, and economic comparisons of organic and conventional farming systems. Bioscience 7:573–582
7. Leake A (2000) The development of integrated crop management in agricultural crops: comparisons with conventional methods. Pest Manag Sci 56:950–953
8. Zoschke A, Quadranti M (2002) Integrated weed management. Quo vadis? Weed Biol Manag 2:1–10
9. Bastiaans L, Paolini R, Baumann DT (2008) Focus on ecological weed management: what is hindering adoption? Weed Res 48:481–491
10. Knezevic SZ, Evans SP, Blankenship EE, Van Acker RC, Lindquist JL (2002) Critical period for weed control: the concept and data analysis. Weed Sci 50:773–786
11. Martin SG, Van Acker RC, Friesen LF (2001) Critical period of weed control in spring canola. Weed Sci 49:326–333
12. Weaver SE, Kropff MJ, Groeneveld RW (1992) Use of ecophysiological models for crop-weed interference: the critical period of weed interference. Weed Sci 40:302–307
13. Scott RK, Wilcockson SJ, Moisey FR (1979) The effects of time of weed removal on growth and yield of sugar beet. J Agric Sci 93:693–709
14. Swanton CJ, O'Sullivan J, Robinson DE (2010) The critical weed-free period in carrot. Weed Sci 58:229–233

15. Roberts HA, Bond W, Hewson RT (1976) Weed competition in drilled summer cabbage. Ann Appl Biol 84:91–95
16. Miller AB, Hopen HJ (1991) Critical weed-control period in seeded cabbage (*Brassica oleracea* var. *capitata*). Weed Technol 5:852–857
17. Weaver SE (1984) Critical period of weed competition in three vegetable crops in relation to management practices. Weed Res 24:317–325
18. Perkins-Veazie PM, Cantliffe DJ, White JM (1987) Techniques to improve stands of direct-seeded cabbage. Acta Hortic (ISHS) 198:65–72
19. Kavaliauskaitė D, Bobinas C (2006) Determination of weed competition critical period in red beet. Agron Res 4:217–220
20. Van der Weide RY, Bleeker PO, Achten VTJM, Lotz, LAP, Fogelberg F, Melander B (2008) Innovation in mechanical weed control in crop rows. Weed Res 48:215–224
21. Ulloa SM, Datta A, Knezevic SZ (2010) Tolerance of selected weed species to broadcast flaming at different growth stages. Crop Prot 29:1381–1388
22. Cisneros JJ, Zandstra BH (2008) Flame weeding effects on several weed species. Weed Technol 22:290–295
23. Sivesind EC, Leblanc ML, Cloutier DC, Seguin P, Stewart KA (2009) Weed response to flame weeding at different developmental stages. Weed Technol 23:438–443
24. Melander B (1998) Interactions between soil cultivation in darkness, flaming and brush weeding when used for in-row weed control in vegetables. Biol Agric Hortic 16:1–14
25. Mohler CL (2001) Mechanical management of weeds. In: Liebman M, Mohler CL, Staver CP (eds) Ecological management of agricultural weeds. Cambridge University Press, Cambridge, pp 139–209
26. Boyd NS, Brennan EB, Fennimore SA (2006) Stale seedbed techniques for organic vegetable production. Weed Technol 20:1052–1057
27. Roberts HA, Potter ME (1980) Emergence patterns of weed seedlings in relation to cultivation and rainfall. Weed Res 20:377–386
28. Caldwell B, Mohler CL (2001) Stale seedbed practices for vegetable production. HortScience 36:703–705
29. Miles SJ, Reed JN (1999) Dibber drill precise placement of seed and granular pesticide. J Agric Eng Res 74:194–203
30. Gray D, Stechel JRA, Miles S, Reed J, Hiron RWP (1995) Improving seedling establishment by a dibber drill. J Hortic Sci 70:517–528
31. Rasmussen J (2003) Punch planting, flame weeding stale seedbed for weed control in row crops. Weed Res 43:393–403
32. Rasmussen J, Henriksen CB, Griepentrog HW, Nielsen J (2011) Punch planting, flame weeding and delayed sowing to reduce intra-row weeds in row crops. Weed Res 51:489–498
33. Ascard J, Bellinder RBB (1996) Mechanical in-row cultivation in row crops. Proceedings of the second international weed control congress, Copenhagen, Denmark, 1121–1126 pp
34. Fogelberg F, Dock Gustavsson AM (1997) Resistance against uprooting in carrots (*Daucus carota* L.) and annual weeds a basis for selective mechanical weed control. Weed Res 38:183–190
35. Kurstjens DAG, Kropff MJ (2001) The impact of uprooting and soil-covering on the effectiveness of weed harrowing. Weed Res 41:211–228
36. Knezevic SZ, Ulloa SM (2007) Flaming: potential new tool for weed control in organically grown agronomic crops. J Agric Sci 52:95–104
37. Melander B, Rasmussen IA, Bàrberi P (2005) Integrating physical and cultural methods of weed control: examples from European research. Weed Sci 53:369–381
38. Bond W, Grundy AC (2001) Non-chemical weed management in organic farming systems. Weed Res 41:383–405
39. Wiltshire JJJ, Tillett ND, Hague T (2003) Agronomic evaluations of precise mechanical hoeing and chemical weed control in sugar beet. Weed Res 43:236–244
40. Slaughter DC, Giles DK, Downey D (2008) Autonomous robotic weed control systems: a review. Comput Electron Agric 61:63–78

41. Rasmussen J, Griepentrog HW, Nielsen J, Henriksen CB (2012) Automated intelligent rotor tine cultivation and punch planting to improve the selectivity of mechanical intra-row weed control. Weed Res 52:327–337
42. Nørremark M, Griepentrog HW, Nielsen J, Søgaard HT (2008) The development and assessment of the accuracy of an autonomous GPS-based system for intra-row mechanical weed control in row crops. Biosyst Eng 101:396–410
43. Wolfe DW, Topoleski DT, Gundersheim NA, Ingall BA (1995) Growth and yield sensitivity of four vegetable crops to soil compaction. J Am Soc Hortic Sci 120:956–963
44. Hoyt GD (1999) Tillage and cover residue effects on vegetable yields. Horttechnology 9:351–358
45. Mochizuki MJ, Rangarajan A, Bellinder RR, Björkman T, van Es HM (2007) Overcoming compaction limitations on cabbage growth and yield in the transition to reduced tillage. HortScience 42:1690–1694
46. Bryant A, Brainard DC, Haramoto ER, Szendrei Z (2013) Cover crop mulch and weed management influence arthropod communities in strip-tilled cabbage. Environ Entomol 42:293–306
47. Holmstrom D, Sanderson K, Kimpinski J (2008) Effect of tillage regimens on soil erosion, nematodes, and carrot yield in Prince Edward Island. J Soil Water Conserv 63:322–328
48. Luna JM, Mitchell JP, Stresthra A (2012) Conservation tillage in organic agriculture: evolution toward hybrid systems. Renew Agric Food Syst 27:21–30
49. Brainard DC, Peachey RE, Haramoto ER, Luna JM, Rangarajan A (2013) Weed ecology and nonchemical management under strip-tillage: implications for Northern U.S. vegetable cropping systems. Weed Technol 27:218–230
50. Brainard DC, Noyes DC (2012) Strip tillage and compost influence carrot quality, yield and net returns. HortScience 47:1073–1079
51. Liebman M, Dyck E (1993) Crop rotation and intercropping strategies for weed management. Ecol Appl 3:92–122
52. Dorado J, Del Monte JP, Lopez-Fando C (1999) Weed seedbank response to crop rotation and tillage in semiarid agroecosystems. Weed Sci 47:67–73
53. Leroux GD, Beniot D-L, Banville S (1996) Effect of crop rotations on weed control, *Bidens cernua* and *Erigeron canadensis* populations, and carrot yields in organic soils. Crop Prot 15:171–178
54. Brainard DC, Bellinder RR, Hahn RR, Shah DA (2008) Crop rotation, cover crop, and weed management effects on weed seedbanks and yields in snap bean, sweet corn, and cabbage. Weed Sci 56:434–441
55. Belair G, Parent LE (1996) Using crop rotation to control *Meloidogyne hapla* Chitwood and improve marketable carrot yield. HortScience 31:106–108
56. Bellinder RR, Dillard HR, Shah DA (2003) Weed seedbank community responses to crop rotation schemes. Crop Prot 23:95–101
57. Weston LA (1996) Utilization of allelopathy for weed management in agroecosystems. Agron J 88:860–866
58. Liebman M, Davis AS (2000) Integration of soil, crop and weed management in low-external-input farming systems. Weed Res 40:27–47
59. Dillard HR, Bellinder RR, Shah DA (2004) Integrated management of weeds and diseases in a cabbage cropping system. Crop Prot 23:163–168
60. Dillard HR, Hunter JE (1986) Association of common ragweed with *Sclerotinia* rot of cabbage in New York State. Plant Dis 70:26–28
61. Masiunas JB, Eastburn DM, Mwaja VN, Eastman CE (1997) The impact of living and cover crop mulch systems on pests and yield of snap beans and cabbage. J Sustain Agric 9:61–88
62. Brainard DC, Bellinder RR, Miller AJ (2004) Cultivation and interseeding for weed control in transplanted cabbage. Weed Technol 18:704–710
63. Sainju UM, Singh BP, Whitehead WF (2001) Comparison of the effects of cover crops and nitrogen fertilization on tomato yield, root growth, and soil properties. Sci Hortic 91:201–214
64. Sainju UM, Singh BP (1997) Winter cover crops for sustainable agricultural systems: influence on soil properties, water quality, and crop yields. HortScience 32:21–28

65. Ball DA, Miller SD (1990) Weed seed population response to tillage and herbicide use in three irrigated cropping sequences. Weed Sci 38:511–517
66. Bàrberi B, Silvestri N, Bonari DF (1997) Weed communities of winter wheat as influenced by input level and rotation. Weed Res 37:301–313
67. Armstrong J-JQ, Sprague CL (2010) Weed management in wide- and narrow-row glyphosate-resistant sugarbeet. Weed Technol 24:523–528
68. Agriculture and Agri-Food Canada (2010) Carrot production on raised beds: reduced risk weed control strategies. AAFC Number 11169E. © Her Majesty the Queen in Right of Canada, 6 pp
69. Baumann DT, Slembrouck I (1994) Mechanical and integrated weed control systems in row crops. Acta Hortic 372:245–252
70. Brown AD, Dexter AR, Chamen WT, Spoor G (1996) Effect of soil macroporosity and aggregate size on seed-soil contact. Soil Tillage Res 38:203–216
71. Mikha MM, Rice CW (2004) Tillage and manure effects on soil and aggregate-associated carbon and nitrogen. Soil Sci Soc Am J 68:809–816
72. Bollman SL, Sprague CL (2009) Effect of tillage and soil-applied herbicides with micro-rate herbicide programs on weed control and sugarbeet growth. Weed Technol 23:264–269
73. Vasileiadis VP, Froud-Williams RJ, Eleftherohorinos IG (2012) Tillage and herbicide treatments with inter-row cultivation influence weed densities and yield of three industrial crops. Weed Biol Manag 12:84–90
74. Dexter AG, Luecke JL (2003) Dual and dual magnum on sugarbeet. Sugarbeet Res Ext Rep 34:79–83
75. Sikkema PH, Soltani N, Deen W, Robinson DE (2007) Effect of s-metolachlor application timing on cabbage tolerance. Crop Prot 26:1755–1758
76. Robinson DE, McNaughton KE (2012) Time of application of *s*-metolachlor affects growth, marketable yield and quality of carrot and red beet. Am J Plant Sci 3:546–550
77. Locke MA, Bryson CT (1997) Herbicide-soil interaction in reduced tillage and plant residue management systems. Weed Sci 45:307–320
78. Hillocks RJ (2012) Farming with fewer pesticides: EU pesticide review and resulting challenges for UK agriculture. Crop Prot 31:85–93
79. Morris NL, Miller PCH, Orson JH, Froud-Williams RJ (2010) The adoption of non-inversion tillage systems in the United Kingdom and the agronomic impact on soil, crops and the environment. Soil Tillage Res 108:1–15
80. Melander B, Munier-Jolain N, Charles R, Wirth J, Schwarz J, van der Weide R, Bonin L, Jensen PK, Kudsk P (2013) European perspectives on the adoption of nonchemical weed management in reduced-tillage systems for arable crops. Weed Technol 27:231–240
81. Melander B, Rasmussen G (2001) Effects of cultural methods and physical weed control on intrarow weed numbers, manual weeding and marketable yield in direct-sown leek and bulb onion. Weed Res 41:491–508

Chapter 12
Integrated Weed Management in Plantation Crops

Rakesh Deosharan Singh, Rakesh Kumar Sud and Probir Kumar Pal

Introduction

Plantation crops are long-term crops established for commercial interest. Major plantation crops are tea (*Camellia* spp.), coffee (*Coffea arabica* L.), oil palm (*Elaeis guineensis* Jacq.), areca nut (*Areca catechu* L.), cardamom (*Elettaria cardamomum* Maton and *Amomum subulatum* Roxb.), coconut (*Cocos nucifera* L.), cashew (*Anacardium occidentale* L.), cocoa (*Theobroma cacao* L.), and rubber (*Hevea brasiliensis* Mull. Arg.). Being long-term crops, and often grown as monocultures, plantation crops are severely infested with weeds. This chapter deals with the nature and effect of the weed menace in the above mentioned crops along with methods adopted for weed management. In the plantation crops, weeds are managed by physical, mechanical, and chemical methods similar to those generally adopted in arable/field crops. However, there are reports on the use of low-density polyethylene sheets for mulching interrow space and mowing between the rows to control weeds. Planting smother crops or leguminous cover crops and intercropping in the row space, and deploying grazing animals are the biological methods for weed management in some of these crops. Integrated approach involving a combination of cultural, me-

R. D. Singh (✉)
Department of Biodiversity, Council of Scientific and Industrial Research (CSIR)-Institute of Himalayan Bioresource Technology, Post Box # 6, Palampur, Himachal Pradesh 176061, India
e-mail: rdsingh@ihbt.res.in

R. K. Sud
Hill Area Tea Science Division, Council of Scientific and Industrial Research (CSIR)-Institute of Himalayan Bioresource Technology, PO Box #6, Palampur, Himachal Pradesh 176061, India

P. K. Pal
Natural Plant Products Division, Council of Scientific and Industrial Research (CSIR)-Institute of Himalayan Bioresource Technology, PO Box #6, Palampur, Himachal Pradesh 176061, India

B. S. Chauhan, G. Mahajan (eds.), *Recent Advances in Weed Management*,
DOI 10.1007/978-1-4939-1019-9_12, © Springer Science+Business Media New York 2014

chanical, and biological weed control methods is also adopted for combating weeds in an effective, economical, and eco-friendly manner.

Weed Menace in Plantation Crops

Nature of Weed Infestation

Tea

Weed management in tea is the second most expensive input after plucking [1]. In tea plantations, grasses generally predominate the weed flora followed by broadleaf weeds [2]. The major weeds in different tea-growing areas of the world are *Ageratum conyzoides* L., *A. houstonianum* Mill., *Artemisia vulgaris* L., *Arundinella benghalensis* (Spreng) Druce, *Axonopus compressus* (Sw.) P. Beauv., *Borreria alata* (Aubl.) DC., *B. hispida* (L.) K. Schum., *Commelina benghalensis* L., *Cynodon dactylon* (L.) Pers., *Eupatorium odoratum* L., *Imperata cylindrica* (L.) P. Beauv., *Mikania cordata* (Burm. f.) B. L. Robins, *M. micrantha* H.B.K., *Oxalis acetosella* L., *Panicum repens* L., *Paspalum conjugatum* Berg., *Paspalum scrobiculatum* L., *Pennisetum clandestinum* Hochst. ex Chiov., *Polygonum chinense* L., *Saccharum spontaneum* L., *Scoparia dulcis* L., and *Setaria palmifolia* (Koen.) Stapf. Sedges are not serious weeds in tea plantation [2]. Ferns, such as *Nephrodium* sp. and *Pteridium aquilinum* (L.) Kuhn., have also been reported to infest tea plantation. In tea gardens, owing to high humidity and limited sunny days throughout the year, mosses tend to cover soil surface under the canopy along with a large part of tea trunk and branches [3].

In India, studies in young tea plants have revealed that the critical period of weed competition is generally from spring to rainy season (April to September). Delay in weeding during this period adversely affects branching, growth, and yield of young tea plants [4]. In Sri Lanka, the critical period for weed competition in young tea was reported to be between 8 and 16 weeks after planting and the threshold period of competition was about 12 weeks after planting [5].

Coffee

In Cuba, total 266 weed species, belonging to 189 genera, were identified in coffee plantations [6, 7]. In central Cuba, *Elaterium carthaginense* Jacq. [8], a climbing weed, was reported to smother coffee plants and to be poisonous to cattle. Broadleaved species predominate the coffee plantations, and Asteraceae and Poaceae were the dominant families [9]. In Kenya, while *Bidens pilosa* L., *Chloris* sp., *Commelina benghalensis, Cynodon dactylon, Cyperus* sp., *Digitaria velutina* (Forssk.) P. Beauv., *Gnaphalium* sp., *Oxalis* sp., and *Parthenium hysterophorus* L. are major weeds in the coffee plantations [10], the tough-to-combat weeds are *Cynodon dactylon, Cyperus rotundus* L., *Digitaria scalarum* (Schweinf.) Chiov., *Oxalis* sp., and *Pennisetum clandestinum* [11].

Weeds commonly found in coffee in Costa Rica were *Bidens pilosa, Borreria latifolia* (Aubl.) Schum., *Drymaria cordata* Willd., *Emilia fosbergii* Nicolson, *Portulaca oleracea* L., and *Richardia scabra* L. [12]. In Brazil, *Amaranthus retroflexus* L., *Bidens pilosa, Brachiaria plantaginea* (Link) Hitchc., *Coronopus didymus* (L.) Smith., *Digitaria horizontalis* Willd., *Emilia sonchifolia* (L.) DC., *Galinsoga parviflora* Cav., *Ipomoea grandifolia* Lam., *Lepidium virginicum* L., and *Raphanus raphanistrum* L. have been reported to infest coffee plantations [13]. In Cuba, grass weeds, viz., *Brachiaria subquadriparia* (Trin.) Hitchc., *Digitaria sanguinalis* (L.) Scop., and *Eleusine indica* (L.) Gaertn. dominated the coffee crop. Among the broad-leaved species, *Amaranthus dubius* Mart. ex Thell. was the major species in the open fields while *Solanum nigrum* L. predominated the pockets under shade [14]. Other weed species reported in coffee in Cuba are *Alternanthera tenella* Colla syn. *A. ficoidea* (L.) R. Br., *Mikania cordifolia* (L. f.) Willd., *Paspalum conjugatum, Petiveria alliacea* L., *Phyla nodiflora* (L.) Greene, and *Pseudelephantopus spicatus* (Juss.) C.F. Baker. [15]. In Venezuela, *C. dactylon* had the highest frequency and abundance while broadleaf weeds were in majority in coffee [16].

A study on different cultivation regimes in coffee indicated that the type of cultivation practices adopted can be detected from the associated weed communities [17]. A study on floristic composition of weeds in coffee in Costa Rica revealed reduction in the relative frequency of climbing plants, Cyperaceae, and monocot species, and increase in broadleaf species and grasses [18].

In Monagas state, Venezuela, the critical period of weed interference in coffee was observed to be between May and September, coinciding with the fruiting stage [16]. Weed-free conditions increased the yield to 36 % compared to weedy plots during the same period. In Ethiopia, loss in coffee yield was recorded to be as high as 65 %, depending on the type and frequency of weeding operations [19]. In El Salvador, total weed-free conditions provided highest coffee yield followed by the plots remaining weed-free during the dry spell of November–April. Thus, the period from November to April was found to be critical from the weed management point of view [20].

Oil Palm

In oil palm plantation in Nigeria, Gill and Onyibe [21] revealed a total of 174 weeds comprising of five ferns, 52 monocotyledons, and 117 dicotyledons. A majority of the weeds, numbering 142 (81.6 %) were broad leaved, whereas 22 (12.6 %) were grasses, and remaining 10 (5.7 %) were sedges. *Chromolaena odorata* (L.) King and Robinson, *Panicum laxum* Swartz, and *Pueraria phaseoloides* (Roxb.) Benth. were the predominant weed species. In the plains of the eastern Himalayan region of West Bengal, India, a total of 20 angiosperm families were reported; of these, 17 belonged to dicots, and 3 to monocots while 5 were pteridophytes. Three species, viz., *Ageratum conyzoides, Oxalis corniculata* L., and *Vandelia* sp. were more widely distributed [22]. Two major problematic weed species in oil palm plantations reported from Selangor are *Calopogonium caeruleum* (Benth.) Sauv. and *Paspalum conjugatum* [23]. In Nigeria, *Asystasia coromandeliana* Wight ex Nees, *M. micrantha, Ottochloa nodosa* (Kunth) Dand, *P. conjugatum,* and some legumes constituted major weed flora [24]. In addition, *C. odorata* also poses problems [25]. Ikuenobe [26] and Ikuenobe and Utulu [27] also described *Aspilia africana* (Pers.) C. D. Adams, *C. odorata,* and *P. phaseoloides* as major weeds in the oil palm. In West Java, *Ischaemum timorense* Kunth, *M. micrantha, O. nodosa,* and *P. conjugatum* were the major species [28]. *Ottochloa nodosa* and *P. conjugatum* were also dominant weeds in young plantations in Malacca, Malaysia, whereas in mature plantations, *Ageratum conyzoides* and *Axonopus compressus* were the predominant weeds [29]. The weed flora in palm nursery comprised of *Acalypha ciliata* Forssk., *Ageratum conyzoides, Amaranthus spinosus* L., *Brachiaria miliiformis* (Presl) A. Chase, *Cyathula prostrata* (L.) Blume., *E. indica, Mariscus alternifolius* Vahl, and *P. oleracea* [30].

In the West Kalimantan region of Indonesia, *I. cylindrica* and *Melastoma malabathricum* L. were the widespread weed species in oil palm, whereas *A. coromandeliana, C. odorata, Mikania micrantha, Mimosa pigra* L., and *Pennisetum polystachion* (L.) Schult. were distributed in a limited area [31]. *Asystasia intrusa* Blume has been categorized as a noxious weed in oil palm [32]. *Rottboellia cochinchinensis* (Lour.) Clayton is another noxious weed in the young plantations, which interferes with agronomic operations like fertilization, spraying, pest and disease control, and harvesting, and causes reduction in oil palm yield [33]. Ismail et al. [34] reported *Asystasia gangetica* (L.) T. Anders., *B. alata, Cleome rutidosperma* DC., and *P. conjugatum* to account for more than 80 % of the total weed seeds in an oil palm field.

Coconut

In coconut, wide spacing favors a variety of weeds to grow, occupy the space, and compete with the crop [35]. In Brazil, a comprehensive weed survey revealed 201 weed species in coconut, mostly belonging to Poaceae, Amaranthaceae, Asteraceae, Euphorbiaceae, Leguminosae, and Malvaceae, with Poaceae being present in all areas of coconut cultivation. The weed species with the greatest frequency were *Amaranthus deflexus* L., *Cenchrus echinatus* L., *D. horizontalis,* and *Herissanthia crispa* (L.) Briz. [36]. In Kerala, India, Thomas and Abraham [37] identified 85 weed species in coconut gardens from 19 locations. In the eastern Himalayan region of West Bengal, India, *A. conyzoides, B. alata, Centella asiatica* (L.) Urban, *Gnaphalium* sp., *O. corniculata, S. nigrum,* and *Vandelia* sp. recorded 100 % frequency followed by other species, viz., *Clerodendron infortunatum* L., *Dryopteris* sp., *I. cylindrica, Melastoma* sp., *Mimosa pudica* L., *Selaginella* sp., and *Solanum nigrum* [22]. Major weed species reported in Sri Lanka are *Allmania nodiflora* (L.) R.Br. ex Wight., *Chloris barbata* Sw., *Chromolaena odorata, Croton hirtus* L'Herit, *Cynodon dactylon, Hyptis suaveolens* L. Poit., *I. cylindrica, Lantana camara* L., *Mimosa pudica, Mitracarpus villosus* (Sw.) DC., *Panicum maximum* Jacq., *Panicum repens, Pennisetum polystachion, Scoparia dulcis, Sida acuta* Burm. f., *Stachytarpheta jamaicensis* (L.) Vahl., *Tephrosia purpurea* (Linn.) Pers., *Tridax procumbens* L., *Urena lobata* L., and *Vernonia cinerea* (L.) Less. [35, 38].

Cashew

Similar to the case of coconut, the wide spacing between cashew trees provides ample scope to the weed menace in cashew farms. *Avena sativa* L., *Cynodon dactylon, Cyperus compressus* L., *I. cylindrica, Pennisetum polystachyon* (L.) Schult., and *Setaria glauca* (L.) P. Beauv. among grasses and sedges, and *Chromolaena odorata, Lantana indica* Roxb., *M. pudica, Smilax zeylanica* L., and *T. procumbens,* among broadleaf are the common weeds in cashew [39, 40]. Infestation of *Acanthus montanus* (Nees) T. Anderson, *Axonopus compressus, Chromolaena odorata, Commelina nudiflora* L., *Euphorbia heterophylla* L., *Fleurya aestuans* (L.) Gaudich., and *Tridax procumbens* has been reported in eastern Nigeria [41].

Cocoa

The predominant weed species in cocoa fields are *Chromolaena odorata, Cyperus* sp., *F. aestuans, I. cylindrica, L. camara, Monstera* sp., *P. repens, Setaria barbata* (Lam.) Kunth, and *Talinum triangulare* (Jacq.) Willd. Other common weeds in the cocoa fields are *Alternanthera sessilis* Br., *D. cordata, Panicum laxum,* and *Paspalum conjugatum* [42]. Mistletoes are plant parasites that live on other plants to obtain food, water, and support. At least six different species have been found on cocoa across the growing regions worldwide. Infestation of mistletoes may lead to the loss of vigor, reduced pod yield, and eventually death of the branch or the tree [43]. Two species of mistletoes, *Phragmanthera incana* (Schumach.) Balle and *Tapinanthus bangwensis* (Engl. & K. Krause) Danser are very common in West Africa [44, 45]. In Ghana, the cocoa farms are also extensively affected by *T. bangwensis* [45]. In cocoa, the young seedling and establishment stages are critical for weed control [46]. In Brazil, the competition of weeds with cocoa in agroforestry systems is severe during the winter months (June to November) [47].

Rubber

The weeds commonly found in the rubber plantations in Southeast Asian countries are *Asystasia intrusa, Axonopus* sp., *Borreria* sp., *Chromolaena odorata, Cynodon dactylon, Cyperus rotundus, I. cylindrica, Lantana aculeata* L., *Mikania micrantha, Mimosa pudica, Panicum repens, Paspalum* sp., and *Pennisetum* sp. Some ferns (e.g., *Adiantum, Nephrolepis,* and *Gleichenia linearis* (Burm. f.) C.B. Clarke) were also reported in rubber plantations in Sri Lanka [48].

Losses Caused

Tea

In tea plantations in India, the period of rainy season crop coincides with the period of active weed growth and infestation, necessitating more deployment of labor for plucking and weeding. In the nursery, the environmental conditions are congenial for plant growth, which facilitate rapid weed growth, competing with tea plants. It calls for extra labor for the nursery success. Weed infestation at the peak flowering stage causes maximum reduction in growth of tea plants [49].

In the newly planted tea in India, weed control during summer and rainy seasons (April–September) is essential for the establishment of the plants. The weed competition during this period has been reported to cause nearly 50% reduction in the number of primary branches and about 3.5 times decrease in the yield in the 2nd year [50]. Weeds also retard the efficiency of farm workers. Certain weeds, such as *B. pilosa* and *Rubus* sp., often reduce the plucking efficiency of workers. In the heavily infested tea sections, shoots of weeds are inadvertently harvested along with tea shoots, which adversely affects the quality of processed tea.

Coffee

In Brazil, Alcantara and Ferreira [51] reported reduction in the yield of processed beans to the extent of 178 kg/ha. At the young stage, coffee plants are sensitive to weed infestation as weeds cause reduction in the nutrient content in the coffee plant [52]. Dias et al. [53] reported that the foliar area and dry biomass of leaves were the most affected attributes in coffee by weed infestation in summer, while in winter, leaf number and dry stem biomass were significantly reduced. The critical periods of weed interference were 15–88 and 22–38 days after coffee seedling transplanting, during winter and summer, respectively. Weeds, including *Bidens pilosa, Brachiaria decumbens* Stapf, *Commelina diffusa* Burm. f., *Leonurus sibiricus* L., and *Richardia brasiliensis* Gomes, caused severe reduction in growth, mainly with increasing weed plant densities [54]. Weeds reduced root dry matter of coffee plants by 47–52% as compared to the weed-free treatment, regardless of the weed density. Crop and weed nutrient concentrations as well as competition degrees greatly varied, depending on both weed species and densities [55]. Weed competition during the dry period from November to April reduced parameters of coffee growth to a much greater extent than competition at any other time of the year. Of the parameters measured, stem diameter and the length of non woody branches were the best indicators of weed competition during the vegetative and generative stages of growth. Weed control between the rows using a rotary cultivator was more effective than mowing; however, the possible long-term damage to the soil due to the cultivator should also be taken into consideration [56].

Coconut

Based on the cumulative average yield of coconut for three consecutive years in Sri Lanka, Samarajeewa et al. [57] concluded that weed infestation may cause up to 54 % reduction in nut yield as compared to the weed-free conditions.

Cocoa

Adverse effects of weeds on the cocoa production are expected to be higher at the initial stages. Proper weed control is always beneficial for the growth and establishment of seedlings. Oppong et al. [58] reported favorable increment in seedling girth due to weed-free conditions over a long period. Weeds often climb up and twine around the plant and prevent it from unfolding.

Rubber

Weeds hinder the cultural operations, such as tapping, spraying, irrigation, and fertilizer application in rubber plantation. The cost of weed control operations in the young rubber plantations may be about 34 % of the total cost of cultivation [59].

Weed Management

Physical and Mechanical Methods

Tea

At the nursery stage, use of black polyethylene mulch provided satisfactory weed control [60]. Similarly, Singh et al. [61] concluded that in young China hybrid tea planted on slope, low-density polyethylene (LDPE) mulch totally suppressed weeds in the interrow spaces. LDPE mulch also enhanced plant growth and yield of tea compared to no mulch.

In northeast India, cultivation with a deep hoe on heavy soil on a flat terrain during the intermittent dormancy in June and winter dormancy in December provided higher average annual yield, while weed control throughout the year by cutting with a sickle provided the lowest yield [62].

Coffee

In Cuba, weed cover for 77 days in the nursery under shade improved the growth and development of coffee; however, in the open fields, the nursery could be left weedy only for 48 days without affecting its growth [63]. In Uganda, hoeing once a month reduced 75 % of weed seeds in the coffee field soils as compared to only 30 % reduction with slash weeding [64]. In Kerala, India, deep digging was done annually in winter season (October–November) for soil and moisture conservation measures, which was found injurious, particularly to the feeder roots, accelerated soil moisture loss, and was cost-ineffective [65]. On the other hand, scraping of soil with incorporation of weeds was the most beneficial for plant growth. It was therefore suggested that deep digging should be practiced only once in new clearings, and not annually. The combination of partial slashing and application of herbicides in patches was more effective in reducing the unwanted weed biomass and enhancing spread of the ground cover legumes, whereas the use of partial slashing enhanced spread of the grass weed *Oplismenus burmannii* (Retz.) P. Beauv. [66].

In the coffee plantations in Cuba, regular mowing between the rows encouraged low-growing grasses, particularly *Brachiaria subquadriparia,* while rotary cultivation favored the development of *C. rotundus* and broad-leaved species, such as *A. dubius, L. virginicum,* and *P. oleracea* [67].

Cardamom

Mulching around plants has been found to reduce weed infestation in large cardamom [68]. In small cardamom, sickle weeding during summer (May) and sickle weeding + forking + mulching during winter (October and January) provided the highest numbers of young tillers, mature tillers, and panicles, and the highest yield (270 kg/ha), as compared to the yield in non-treated control plots when averaged 161 kg/ha [69].

Coconut

A study in Sri Lanka reveals that manual weeding in coconut was not as effective as the chemical method [70]. Though plowing and harrowing operations in coconut reduces weed seed bank by burying them to deeper soil profiles, they lead to high seedling emergence from seeds brought from deeper to shallow depths [71]. Thus, slashing weeds with a tractor three times a year was less effective than establishing cover crop with *P. phaseoloides* and buffalo grazing once a month [38].

Cashew

In young cashew plantation, manual hoeing is done within 1.5–2.0-m diameter around trees followed by slashing the remaining weeds in the interspaces to ground level. Hand pulling of annual and perennial grasses and sickling of tall and perennial grasses particularly before application of herbicides are adopted wherever labor is available [39, 40].

In India, mechanical hoeing is recommended along the planted strip up to a width of about 4 m, with precautions to avoid root injury by leaving a strip of 2 m on both sides. Whereas in the new plantations where sufficient space exists, hoeing, plowing, and mowing may be adopted [39]. However, the mechanical hoeing, harrowing, or mowing cannot be as sustainable as those methods that promote soil covering through leguminous species or spontaneous vegetation in improving soil organic content [72]. Besides, these practices cause intense soil mobilization and fragmentation of biomass of cover plants and favor degradation of physical and chemical properties, resulting in the reduction of the soil quality [73].

Cocoa

In the cocoa plantation with wide row spacing and without shade trees, weeds are effectively controlled by tillage. However, deep tillage should be avoided to prevent damage to feeder roots. Hand weeding with hoe and sickle is recommended during the initial 6 months after planting when the plants are small. Weeding within 1-m radius around the trees by machete (a large cleaver-like knife) is commonly practiced in most of the cocoa plantations. Strip weeding 1 m each side of the trees in the row is also practiced. Small machines are used for mowing down weeds in the cocoa field where sufficient space is available between the rows for allowing the mower to pass. For mowing operations, the cocoa fields should be free from stone and boulders.

Rubber

Around 4–5 hand weedings are required during the initial 2 years of the crop. The hand weeding is generally done in strips 1.5–2.0 m wide along the rows or in a circle of about 1-m radius around the tree. This method is also recommended for selective weeding in the area where the cover crop has been established. However, much disturbance with heavy hoes is not recommended. Mowing in rubber plantation is recommended only for uniform land and the fields free from rocks. Burning is the traditional practice to control *I. cylindrica* in rubber plantations. However, this practice should not be continued because of concerns over environmental pollution.

Oil Palm

Rambe et al. [74] in Indonesia adopted circle weeding, initially with a radius of 1.6 m, which was increased to 2.0 m for the succeeding weeding operations done manually as well as mechanically. Compared with manual weeding, the motorized grass cutter resulted in a three times higher number of weeded circles.

Chemical Method

Tea

The common herbicides used in tea plantations are paraquat, glyphosate, simazine, 2,4-D sodium, 2,4-D amine, diuron, and dalapon. Linuron, methazole, metribuzin, dichlormate, dinoseb, oxadiazon, butachlor, and fluchloralin herbicides were also used in tea; however, they were trivial in performance. Though most of the herbicides approved are safe to tea, phytotoxicity on tea may occur due to reasons [4] including application of herbicides at rates higher than the recommended doses, improper or nontargeted spraying, spray drift, leaching of preemergence herbicides by heavy rains, and age of the tea bush. In tea nursery, simazine, atrazine, fluchloralin, oxadiazon, and methazole are recommended to be applied at the rate of 2 kg/ha in April, about 3 weeks before planting of clonal cuttings. Treatments may be repeated after hand weeding when weed cover exceeds 50 %. Mixtures of simazine or atrazine with oxadiazon or fluchloralin were found more effective.

Chemical weed control in young tea is distinct from that in mature tea, as young tea plants are relatively more susceptible to herbicide treatments and the weed flora is more diverse and intense. Ghosh and Ramakrishnan [75] observed that in young tea, oxyfluorfen at 0.125 kg/ha applied preemergence in May followed by oxyfluorfen (0.06 kg/ha) + either paraquat (0.24 kg/ha) or 2,4-D (0.8 kg/ha) as postemergence application using shield controlled most of the problem weeds throughout the season. The presence of moisture on the soil surface improves the bioefficacy of preemergence herbicides, in general. It is normally advised that young tea plants should be shielded from the herbicide spray since they are more likely to be adversely affected than older plants. Also, the application of dalapon and diuron in tea younger than 3 years is not recommended [4]. In Sri Lanka, oxyfluorfen at 0.29 kg/ha + paraquat at 0.17 kg/ha or glyphosate at 0.99 kg/ha + kaolin at 3.42 kg/ha provided better weed control than hand weeding in young tea [76]. Also, a combination of interrow mulching and the aforementioned herbicides followed by hand weeding at least every 6–8 weeks was found to be the most appropriate weed management system for young tea [76].

In mature tea, oxyfluorfen at 0.25 kg/ha provided effective control of broadleaf weeds without any phytotoxicity when it was applied to the clean soil or to growing weeds. Oxyfluorfen was comparable to simazine or diuron, each at 2 kg/ha [77]. Pendimethalin 0.75 kg/ha, oxyfluorfen 0.44 kg/ha, simazine 1.25 kg/ha, or atrazine 1.25 kg/ha were more effective preemergence treatments for suppressing seed-borne weeds [78, 79]. Subsequent weed growth in either of the cases could be controlled with spot treatments of 2,4-D and/or paraquat [80].

Kabir et al. [81] observed that glyphosate 0.92 or 1.23 kg/ha provided effective weed control in tea of the Darjeeling area in India. The United Planters' Association of Southern India (UPASI) [82] concluded that glyphosate is a promising herbicide against hardy perennial grasses and deep-rooted broadleaf weeds, and is not toxic to tea bushes even when sprayed directly on the bushes at a rate of 1.68 kg/ha [83].

However, studies at the Institute of Himalayan Bioresource Technology (IHBT), Palampur, India, had indicated that in case of seed-raised China hybrid tea plantations, use of glyphosate even at the rate of 1.03 kg/ha may cause phytotoxicity. The susceptible tea bushes showed loss of crop for one to two pluckings. Glyphosate was also found effective in controlling brush weeds like *L. camara*—a troublesome weed in tea plantations in Himachal Pradesh and Uttarakhand, India [84].

Glufosinate-ammonium at 0.38 kg/ha provided better weed control than paraquat dichloride at 2 weeks after spraying [85]. In an experiment conducted in Valparai, Tamil Nadu, India, napropamide at 2.0 kg/ha exhibited effective weed control only up to 60 days after the application of herbicide. Also, napropamide at 4.0 kg/ha was at par with oxyfluorfen at 0.25 kg/ha only up to 60 days [86]. Carfentrazone-ethyl at 4 and 8 g/ha combined with glyphosate at 620 or 720 g/ha exhibited 85 % weed control, while increments in the dose of carfentrazone-ethyl to 12, 15, and 20 g/ha combined with glyphosate at 620 or 720 g/ha provided total weed control and was comparable in its efficacy to the currently recommended tank mixture of glyphosate +2,4-D at 720+1120 g/ha [87]. Similar results were also reported by Rajkhowa et al. [88] with weed control up to 45 days after herbicide application.

Coffee

In Olancho, Honduras, terbuthylazine as preemergence followed by glyphosate or paraquat as postemergence treatments were more effective than terbuthylazine with weeding or weeding alone in coffee plantations [89]. Oxyfluorfen was the best treatment in maintaining the crop free of weeds for about 90 days compared to 50 days by the traditional controls [90].

In coffee plantations in Karnataka, India, glyphosate (1.25 kg/ha) provided lowest fresh weight of weeds (73 g/m^2) and significantly increased girth of the main stem by 252 %, and bush spread by 85 % [91]. Addition of urea (1 %) or ammonium sulfate (1 %) showed better weed control by glyphosate herbicides [92]. Also, addition of urea resulted in a considerable saving of glyphosate.

In young coffee from the recently transplanted stage to the 1-year-old, oryzalin and oxyfluorfen had better safety margins than simazine, atrazine, and diuron, which caused injury to young coffee at 4.5 kg/ha [93]. Oxyfluorfen, atrazine, simazine, and diuron provided better weed control than oryzalin. Oxyfluorfen failed to combat weeds of Asteraceae family, although these were totally controlled by atrazine and diuron.

In the coffee plantations in Kenya, application of paraquat alone or in combination with slashing or forking provided effective weed control [94]. In Tanzania, glyphosate at 0.729 kg/ha has been recommended for repeated applications for controlling weeds in this crop [95]. In Costa Rica, combined application of paraquat +2,4-D was not effective, which might be due to antagonism in this mixture [12]. In Brazil, ametryn plus simazine mixture applied at 2.5 and 3.0 kg/ha with glyphosate (960 g/ha) provided effective control of weeds followed by paraquat (300 g/ha). Performance of 2,4-D was not satisfactory in terms of yield of coffee beans [96].

Oil Palm

In oil palm at Selangor, Malaysia, effective control of major weeds, viz., *C. caeruleum* and *P. conjugatum* was reported with a tank-mixed application of metsulfuron and glyphosate or metsulfuron and paraquat [23]. Another study revealed that glyphosate at 720 g/ha and dicamba + glyphosate at 230 + 540 g/ha provided the most effective control of weeds [97]. Teng and Teh [24] reported that a translocative broad-spectrum herbicide Wallop, containing glyphosate 16.2 % and dicamba 8.1 %, was effective for weed control in both young and mature oil palm.

Paspalum fasciculatum Willd. ex Fluegge, the most prevalent weed in the oil palm, can be best managed with haloxyfop-methyl (Galant 240) at 100 g/ha [98]. Similarly, in Nigeria, the control of *C. odorata* in oil palm was achieved with glyphosate at 2.4 kg/ha and imazapyr at 0.5 kg/ha [25]. Bhanusri et al. [99] in India concluded that glyphosate had the highest percent of weed control efficiency on both dicots and monocots over paraquat and glufosinate ammonium. Efficiency of glyphosate can further be optimized through ammonium sulfate as surfactant [100]. Paraquat is considered to be the most effective herbicide with the fastest mode of action. In 2002, the Malaysian government banned the use of this hazardous herbicide. This led to a hunt for new effective herbicides for oil palm plantations in Malaysia. Wibawa et al. [101] evaluated the efficacy and ability of the less hazardous herbicides, viz., glufosinate ammonium and glyphosate, as an alternative to paraquat in controlling weeds in the immature oil palm (below 3 years old). The results showed that lower rates of glufosinate ammonium (200 g/ha) and glyphosate (400 g/ha) provided excellent weed control and the efficacy was much better than that of paraquat [102–104].

Nuertey et al. [105] reported that glyphosate at 0.47 kg/ha mixed with either 0.23 kg/ha of sodium chloride or 0.53 kg/ha of ammonium sulfate controlled the weeds up to 3 months after treatment. However, in the next year, these weed control measures were ineffective. So it may be inferred that the herbicidal effectiveness was controlled by the prevailing weather conditions. In Malaysia, a biotype of *Eleusine indica* showed resistance to glyphosate and invaded oil palm plantations in a vast area. Control of this resistant weed may need repeated applications of glyphosate, as many as eight times in a year [106]. Jalaludin et al. [107] recorded 82 % control of *E. indica* at a vegetable farm with glufosinate-ammonium at the recommended rate, but at another location in oil palm nursery, the same rate of herbicide failed to control the weed. Mortimer and Hill [108] studied weed species shifts in response to broad-spectrum herbicide use in oil palm and found that the use of a broad-spectrum herbicide changed the composition of the weed flora.

In 1988, an ultralow-volume applicator for herbicides with a spray volume of 10–50 l/ha showed a high level of consistency in efficacy. The equipment also reduced labor requirements by half and water requirements by 90 % [109]. A low-volume technique for weeding in a circle around oil palm using a spinning disc was also adopted [110]. Leng and Maclean [111] observed that controlled droplet applicator was effective in combating *Ischaemum muticum* L. in young oil palm when glyphosate was applied at 1.5 kg/ha. The battery-operated knapsack sprayer and controlled droplet applicator were suitable for the application of herbicides for the control of several problem weeds, including *Clidemia hirta* (L.) D. Don, *Dicranopteris linearis* (Burm. f.) Underw., *Melastoma malabathricum, Mikania micrantha,* and epiphytes [112]. However, Ikuenobe [26] recorded better weed control with greater spray volume than low-volume spray (70–93 % at 200 l compared to 10–75 % at 25 l). Eng et al. [113] in Malaysia used modified knapsack sprayers fitted with two different types of constant flow valves and found it more efficient in the area coverage and safe to the operator.

Haji Mustafa [114] in Malaysia advocated the use of a circle and rentice mechanical sprayer in mature oil palm plantations. The system comprised of a tractor-mounted collapsible boom mechanical sprayer unit. One operator setup usually covers 25–28 ha in a working day of 10 h. The sprayer resulted in five to eight times labor savings over manual knapsack sprayers with the advantage of a uniform area coverage. A Malaysian university has developed an automated sprayer system, using web camera in combinations with electromechanical system, sensor system, controllers, wireless data communication, and software [115].

Areca Nut

In West Bengal, India, glyphosate (1.7 kg/ha) applied alone or in combination with 2,4-D (6.86 kg/ha) provided broad-spectrum weed control in these plantations [116].

Coconut

In Sri Lanka, glyphosate at 1.44 kg/ha resulted in a significant reduction of weed biomass, specifically of *Imperata* sp., and a 25 % increase in nut yield over the uncontrolled weedy plots [35, 57]. In the dry zone in Sri Lanka, glyphosate at 1.08 kg/ha was found to be as effective as glyphosate at 1.44 kg/ha. The growth of coconut seedlings in terms of height and girth increased significantly at this rate, applied at the end of the nursery growth period [70].

Cashew

A study on the chemical weed control in India [117] showed that paraquat at 3.0 kg/ha was the most effective in terms of weed management, crop yield, and nutrient absorption by the cashew trees, followed by oxyfluorfen at 0.50 kg/ha. In south India, the standard recommendation for an effective control of all types of weeds is the application of paraquat at 0.4 kg/ha twice at bimonthly intervals, starting from July with alternate application of glyphosate at 0.8 kg/ha [39]. In Ghana, the use of herbicides and intercropping was compared [118]. In the young plantations, intercrops provided higher yields than the chemical weed control (glyphosate), but in the mature plants, chemical weed control slightly improved cashew nut yield. In

Brazil, Xavier et al. [72] observed substantial change in the floristic composition of spontaneous weed species with the application of herbicide.

Cocoa

Paraquat is commonly used for controlling weeds in young cocoa plantations. In mature cocoa, glyphosate is the most effective herbicide. Oppong et al. [119] reported that glyphosate at reduced rates (480–960 g/ha) can be used to control weeds without an adverse effect on the crop. Adeyemi [120] reported that a formulated mixture of glyphosate and terbuthylazine (2.10 to 3.15 kg/ha) was more effective as compared to hand slashing in a mature cocoa field. The herbicides should be applied 1 month before or 4 weeks after fertilizer application.

Rubber

In the rubber plantations, *I. cylindrica* and *P. repens* can be effectively controlled by glyphosate at 4.4 kg/ha [121]. In nursery, preemergence application of diuron at 3.0 kg/ha was found promising [48].

Biological Method

There is a conspicuous lack of efforts toward biological control of weeds deploying bio-herbicides or other biocontrol agents in plantation crops. However, there are some reports on the use of smother crops, the use of organic mulch materials, and grazing animals for weed management in these crops.

Use of Plant and Plant Materials

Tea

Slashing of intercrops or weeds (before flowering) and using them as mulch material has been reported to be effective in weed management. In young China hybrid tea planted on slopes, mulch of grassy weeds effectively controlled weeds in the interrow spaces of tea and it was statistically comparable to LDPE mulch in terms of yield [61].

Sandanam and Rajasingham [122] reported 89 and 51 t/ha soil loss with clean weeding during the 1st and 2nd years, respectively, compared with 7 and 1 t/ha with *Tripsacum laxum* Nash (Guatemala grass) mulching, and 5 and 2 t/ha with *Cymbopogon confertiflorus* (Steud.) Stapf (Mana grass) mulching, respectively. Soil loss with intercrop of *Crotalaria striata* DC. was 32 and 5 t/ha, and with intercrop of *Eragrostis curvula* (Schrad.) Nees, it was 11 and 2 t/ha as compared to 28 and 2 t/ha under selective manual weeding with minimum soil disturbance, in the 1st and 2nd year, respectively. Soil loss in the 3rd and 4th years was more than 2 t/ha with all the above treatments. It was also observed that tipping weights were highest with bare soil or selective weeding and lowest with *E. curvula*. Leaf yield in the 2nd year tended to be higher with Mana grass mulching. Mulching also increased soil moisture content. In China, intercropping with white clover and straw mulching were found to be effective ecological measures for weed control in tea plantations [123].

Coffee

In Nicaragua, Bradshaw and Lanini [124] found *Arachis pintoi* Krapov. & W.C. Gregory, *C. diffusa,* and *Desmodium ovalifolium* (Prain) Wall. ex Merr. having no potential role as long-term cover crop for smothering weeds in the coffee plantation. However, in Tanzania, in smallholder coffee plots, cut grass, sorghum, and corn residues are used as mulch material. The plant-based mulches are reported to increase growth and yield of coffee, as they control weeds, conserve soil moisture, improve soil fertility, and reduce runoff and soil losses [125]. In Puerto Rico, *Arachis kretschmeri* Krapov. & W.C. Gregory, *Axonopus compressus, Paspalum dilatatum* Poir., *P. notatum* Flugge, and *Urochloa subquadripara* (Trin.) R.D. Webster, as smother plants, showed significant reduction in weed infestation as compared to non-treated control in coffee [126]. In Ghana, raising *Canavalia ensiformis* (L.) DC., *Manihot utilissima* Pohl. (cassava), *Musa paradisiaca* L. (plantain), *Vigna unguiculata* (L.) Walp. (cowpea), and *Zea mays* L. (corn) as intercrop during the initial 2.5 years of establishment of young coffee was beneficial, though it also required 1–3 more manual weedings. In the 1st year, except the coffee + cowpea combination, intercropping provided higher net returns [127]. Cassava was not suitable as intercrop from 2nd year onward as it caused a reduction in coffee yields.

Oil Palm

In Malaysia, *C. caeruleum* and *P. phaseoloides* were used as cover crops for managing weed flora, viz., *B. latifolia, E. indica, O. nodosa,* and *P. conjugatum* in oil palm [128]. *Mucuna bracteata* DC. ex Kurz has also been evaluated as an effective cover plant for weed control [129] with potential to compete with the noxious weeds and persist until maturity stage in the oil palm plantation [130]. Besides, intercrops of soybean, corn, and cocoyam provided varying degree of success in the management of weeds in young oil palm [131]. In Monagas State, Venezuela, Barrios et al. [132] reported deployment of *Centrosema rotundifolium* Benth. and *D. ovalifolium* leguminous cover crops, in oil palm, exhibiting high covering index in the initial growth stages. Among these two species, *D. ovalifolium* showed higher competitive ability to spread and gradually displace the population of native weeds. Lee et al. [130] concluded that the leguminous cover crops can combat weeds, such as *Asystasia* and *Mikania.* Similarly, Samedani et al. [133] reported that *Axonopus compressus, Calopogonium caeruleum, Centrosema pubescens* Benth., *M. bracteata,* and *Pueraria javanica* (Benth.) Benth. were highly competitive cover crops against *Asystasia gangetica,* but none could compete against *Pennisetum polystachion.*

Coconut

Maintaining fast-growing cover crops is the other way to control competitive weeds in coconut [38, 134–137]. Cover crops are generally sown 1 year in advance to the coconut planting [134]. *Calopogonium caeruleum, Calopogonium mucunoides* Desv., *Centrosema pubescens, Moghania macrophylla* (Willd.) Kuntze, *Psophocarpus palustris* Desv., and *Pueraria phaseoloides* are suitable cover crops for different climates and soils in Indonesia [134]. The Coconut Research Institute of Sri Lanka [138] recommended *Calopogonium mucunoides, Centrosema pubescens,* and *Pueraria phaseoloides* as cover crops for wet areas; *Calopogonium mucunoides, Centrosema pubescens, Macroptilium atropurpureum* (DC.) Urb., and *Pueraria phaseoloides* for dry areas; and *Gliricidia maculata* (Kunth) Kunth or *Gliricidia sepium* (Jacq.) Kunth as bush cover crops. Broad guidelines on weed management through cultural and biological control and production of organic coconuts have also been described by Singh [139]. Growing *Crotalaria juncea* L. (sun hemp) thrice followed by hand weeding once is a better way to manage weeds in the coconut nursery. This also provided better nut germination, plant growth, and benefit-to-cost ratio [140].

Cashew

Mulching of the interrow space with straw, hay, farm wastes, weed residue, tree loppings, and sometimes coconut husk is done to smother weeds and also to conserve moisture during the dry spell [39, 40].

Cocoa

Cultural weed management includes all aspects of good crop husbandry used to minimize weed interference with crops [141]. At a young stage, a 1-m circle around each plant is cleaned and covered by 10-cm-thick mulch, viz., coconut husk, cocoa leaves, cut-bush, cut-grass, banana leaves, rice-straw, sawdust, and sugarcane bagasse. A small space of 20 cm from the trunk should be left so that the mulch does not touch the seedling. Different cover crops can be grown for weed management in the dense shade of well-established cocoa plantations. Leguminous cover plants, like *Desmodium asperum* (Poir.) Desv. and *Flemingia congesta* Roxb. ex W.T. Aiton, were extensively used in the cocoa fields for suppressing weeds and maintaining soil fertility. Fast-growing creeper legumes are potentially good for controlling the weeds that normally dominate in the thinly populated cocoa fields. However, care should be taken to ensure that they do not entangle the young cocoa plants. Also, intercropping is being adopted for weed management in cocoa plantations. Generally, sugarcane is planted in the gaps of cocoa fields to suppress weeds, but care should be taken to prevent its growth over young cocoa trees.

Rubber

For combating weeds in rubber plantations, banana, passion fruit, and pineapple can be successfully grown as intercrop without any adverse effect on the growth and yield of rubber. In addition, straw and crop residues are used as mulching material for suppressing weed growth. Planting tree legumes, such as *Flemingia, Gliricidia, Sesbania,* and *Tephrosia,* and formation of a mulch of leaf litter between planting rows by slashing and mulching is also a desirable method of weed control in the immature rubber plantations [121]. In rubber, controlled grazing by livestock such as sheep and goats has been adopted to control weeds. Wan Mohamed [142] reported that many weeds found in rubber plantations are highly nutritious and could be utilized to support sheep production. According to Tajuddin et al. [143], cost of weed management can be reduced by 18–36% by sheep grazing.

Use of Grazing Animals

Cattle were first deployed in some oil palm plantations way back in 1987 for combating weeds and augmenting farm income [144, 145]. In Malaysia, grazing by the cattle in oil palm plantations reduced overall weeding costs by about 30% and labor requirement by about 39% [146]. Yusoff [147] also reported grazing as a profitable way of controlling weeds in oil palm plantations. In the coconut plantations, growers use ruminants to prevent crop–weed competition and optimize the productivity of this system as a whole [38]. Senarathne and Gunathilake [38] compared the influence of buffalo grazing and *P. phaseoloides* cover crop with slashing of weeds through tractor on weed control and nut yield in coconut plantations in Sri Lanka. These methods were significantly effective for weed biomass reduction over slashing. Similarly, Seresinhe et al. [148] reported considerable reduction in the weeding cost and a nearly double coconut yield with grazing in coconut plots when compared with un-grazed plots. However, they cautioned that animal grazing could increase soil compaction affecting soil aeration, water infiltration, and other soil physical properties and thereby reduce the growth and productivity of the coconut. Harrowing the plots put under animal grazing appeared to be the best method to overcome soil compaction problem [38].

Integrated Approach

Integrated weed management (IWM) aims to minimize the weed population to a level at which weed infestation has no effect on yield and ecological functions. IWM is a knowledge-based decision-making process that coordinates the use of environmental information, weed biology and ecology, and all the other available technologies to control weeds by the most economical means, while posing the minimum possible risk to people and environment [149]. IWM is a combination of cultural, mechanical, and biological weed control methods to deploy in an effective, economical, and ecological manner.

Coconut

Senarathne and Perera [150] reported that cover cropping with *P. phaseoloides* and application of glyphosate was significantly effective over other treatments for reduction of weed biomass. Intercrops, viz., field crops, vegetable crops, fodder crops, fruit crops, and green manure crops could also be grown for a better utilization of the open ground space between the rows of the coconuts. The selection of intercrops should be based on the climatic requirement of the inter/mixed crop, irrigation facilities, soil type, market suitability, as well as the canopy size, age, and spacing of the coconut [151].

Cashew

Weed infestation during initial years after cashew planting is the main concern for the establishment of plants. According to the Food and Agriculture Organization (FAO) [40], the principal cover crops used in different cashew-growing areas include creeping cover crops, such as *C. pubescens* and *P. phaseoloides,* bush cover crops, viz., *G. maculata, Leucaena leucocephala* (Lam.) de Wit, and nitrogen-fixing trees, such as *Acacia mangium* Willd. Further, among intercrops, banana is popular in many cashew plantations. Pineapple, papaya, pomegranate, and coconut are also used as semi-perennial and perennial intercrops in some areas. The common annual crops grown in cashew plantations are legumes (cowpea, black gram, green gram), oil crops (sesame, groundnut), and condiments, such as hot pepper and onion. Intercropping with food crops reduces weed incidence, weed biomass, and frequency of weed control, resulting in increased financial returns [152].

In Nigeria, Adeyemi [41, 153] evaluated the effect of intercropping food crops, viz., corn, cassava, cowpeas, and banana plantain with cashews. The intercropped crop mixtures reduced biomass of the weeds by about 40 % and number of hand weedings by nearly half compared to sole cashews. Also, the degree of weed succession was influenced by some crop mixtures leading to alteration in the weed sequence and their population dynamics.

In Brazil, Ribeiro et al. [73] studied change in the physicochemical attributes of soil and their effect on growth and yield of cashew, as a consequence of distinct soil management practices. Higher and more stable cashew nut yields were recorded in the systems where the weed or vegetation around the trees was not removed. This was due to building up of higher organic matter and absence of soil disturbance. Studies of Xavier et al. [72] concluded that vegetation strips in the interrows between the cashew lines are ideal from the point of weed management, improvement in soil physical and chemical properties, and to check soil erosion. However, overgrowth of the strip should be cut regularly and the practice of localized weeding should be followed.

The Sri Lanka Cashew Corporation [154] suggested growing suitable intercrops and cover crop (e.g., *C. pubescence*) and resorting to the application of suitable herbicides alone or in combination with other manual and mechanical methods for an efficient weed management in cashew. Grazing of farm animals in off-season and growing vegetation in the interrow space in the plantation have also been identified as efficient methods, appropriate for long-term managements of weeds in cashew.

Rubber

Integrating spraying of herbicides and sheep grazing has been found effective for controlling some noxious weeds, such as *A. intrusa* and *M. micrantha* in rubber [155].

Conclusion: Future Aspects of Research

In order to improve efficiency and cost-effectiveness of the weed management in plantation crops with an eco-friendly approach, research on the following aspects becomes imperative.

Information on the biology of many serious weed species may have to be generated and collected for better and integrated weed management. Weed flora succession is another aspect that needs thorough understanding to regulate rotation, combination, and dosage of the herbicide(s). Harnessing allelopathic attributes of certain plants for weed management is another aspect that needs appropriate attention. For example, in Cuba, grounded pine (*Pinus caribaea* Morelet) needles resulted in the control of weeds in the coffee fields [156]. Research on bio-herbicides needs to be strengthened.

Concept of precision agriculture should be considered and appropriately developed to encompass weed management in the plantation crops. For example, Ishak and Rahman [115] highlighted the need of an intelligent sprayer that could regulate the usage of herbicides at the optimal level and identify in the real-time environment the prevalence of the existing weeds and apply herbicides automatically and precisely. The aim is to reduce chemical wastage, economize labor, reduce application cost, and prevent environment hazards. Similarly, Ghazali et al. [157] developed an intelligent real-time system for an automatic weeding strategy in oil palm plantation using image processing to identify and discriminate weed types, viz., narrow and broad. Similarly, Ishak et al. [158] discussed development of a variable rate automated sprayer for oil palm plantation based on a camera vision with color detection mechanism.

References

1. Sinha MP (1985) A perspective of weed control in tea. Two Bud 32(1–2):35–39
2. Singh RD, Sinha BK, Sud RK, Tamang MB, Chakrabarty DN (1994) Weed flora in tea plantations of Himachal Pradesh. J Econ Taxon Bot 18(2):399–418
3. Ronoprawiro S (1981) The possibility of using glyphosate to control mosses in tea. Ilmu Pertan 3(1):9–19
4. Sinha MP, Borthakur B (1992) Important aspects of weed control in tea. In: Field management in tea (compilation of lectures), Tocklai Experimental Station, TRA, pp 134–139
5. Prematilake KG, Froud Williams RJ, Ekanayake PB (1999) Investigation of period threshold and critical period of weed competition in young tea. Brighton crop protection conference: weeds. Proceedings of an international conference, Brighton, UK, 15–18 November 1999, 1:363–368
6. Alvarez Puente RJ (2002) Management of weeds in coffee plantations in Cuba. Cent Agric 29(3):16–20
7. Alvarez Puente RJ, Martinez Viciedo Y (2004) *Dendropemon claraensis* Leiva (Loranthaceae)—a new enemy of coffee trees. III Congreso 2004 Sociedad Cubana de Malezologia, Memorias, Jardin Botanico Nacional, Ciudad Habana, 28, 29 y 30 de Abril del, pp 108–110
8. Bozan JIR, Puente RA (1993) *Elaterium carthaginense* Jacq.- a weed of coffee plantations in the central region of Cuba. Cent Agric 20(1):95–96
9. Caro P, Muina M, Izquierdo JE (1987) Weeds in coffee plantations in eastern Cuba. Cienc Tec Agric Cafe Cacao 9(1):7–15
10. Njoroge JM, Kimemia JK (1993) Performance of sulfosate 480 g/L (Touchdown) and Basta 14 SL herbicides on weed control and yields of coffee in Kenya. Kenya Coffee 58(684):1625–1628
11. Nyabundi KW, Kimemia JK (1988) Difficult weeds in Kenya coffee—a review. Kenya Coffee 63(744):2747–2751
12. Echegoyen PE, Valverde B, Garita I (1996) Joint action of paraquat and 2,4-D on weeds associated with coffee in Costa Rica. Manejo Integr Plagas 41:8–15
13. Ronchi CP, Silva AA (2004) Weed control in young coffee plantations through post-emergence herbicide application onto total area. Planta Daninha 22(4):607–615
14. Relova RA, Pohlan YJ (1988) Different weed covering periods and its consequences in standing coffee. Cultiv Trop 10(4):30–37
15. Lopez N (2004) Weeds of coffee plantations in Guantanamo. Fitosanidad 8(2):13–16
16. Moraima GS de, Canizares A, Salcedo F, Guillen L (2000) A contribution to determine critical levels of weed interference in coffee crops of Monagas state, Venezuela. Bioagro 12(3):63–70
17. Weber G (1995) The weed flora of coffee plantations under different management regimes in Chiapas, Mexico. Feddes Repert 106(3–4):231–245
18. Ricci M dos SF, Virginio Filho E de M, Costa JR (2008) Diversity of weed community in agroforestry systems with coffee in Turrialba, Costa Rica. Pesqui Agropecu Brasil 43(7):825–834
19. Eshetu T (2001) Weed flora and weed control practices in coffee. 19eme Colloque Scientifique International sur le Cafe, Trieste, Italy. 14–18 Mai, pp 1–9

20. Merino Mejia CI, Ramirez Amador R (1996) Study on critical periods for interspecific competition, weeds—coffee. XVII simposio sobre caficultura latinoamericana, San Salvador, El Salvador 23–27 Octubre1995, 2:15
21. Gill LS, Onyibe HI (1988) Phytosociological studies of the weed flora of oil palm (*Elaeis guineensis* Jacq.) in Nigeria. J Plant Crops 16(2):88–99
22. Sit AK, Bhattacharya M, Sarkar B, Arunachalam V (2007) Weed floristic composition in palm gardens in plains of Eastern Himalayan region of West Bengal. Curr Sci 92(10):1434–1439
23. Chung GF, Chang SH (1990) Bioefficacy of herbicides in immature oil palm and rubber. Planter 66(768):143–150
24. Teng YT, Teh KH (1990) Wallop (glyphosate + dicamba): a translocative broad spectrum herbicide for effective general weed control in young and mature oil palm. BIOTROP Spec Publ 38:165–174
25. Ikuenobe CE, Ayeni AO (1998) Herbicidal control of *Chromolaena odorata* in oil palm. Weed Res Oxf 38(6):397–404
26. Ikuenobe CE (1991) Performance of post-emergence herbicides applied in different carrier volumes in oil palm (*Elaeis guineensis* Jacq.). Ann Appl Biol 118(Suppl):64–65
27. Ikuenobe CE, Utulu SN (1992) Evaluation of formulations of glyphosate and asulam for post-emerged weed control in oil palm. Tests Agrochem Cultiv 13:52–53
28. Wiroatmodjo J, Utomo IH (1990) Population dynamics of weeds as affected by herbicide applications in oil palm plantation. BIOTROP Spec Publ 38:95–105
29. Lam CH, Lim JK, Jantan B (1993) Comparative studies of a paraquat mixture and glyphosate and/or its mixtures on weed succession in plantation crops. Planter 69(812):525–535
30. Utulu SN, Ikuenobe CE (1990) Weed control in oil palm nursery with different herbicide mixtures. Niger J Weed Sci 3:35–41
31. Chiu SB, Chee KH (1998) Survey and control of some weeds in West Kalimantan. Planter 74(869):407–417
32. Nazeeb M, Loong SG (1992) *Asystasia* as ground cover in mature oil palm plantings. Planter 68(795):281–284, 287–290
33. Yan KC (2005) A very noxious weed in oil palm plantations—*Rottboellia cochinchinensis*. Planter 81(950):305–310
34. Ismail BS, Tasrif A, Sastroutomo SS, Latiff A (1995) Weed seed populations in rubber and oil palm plantations with legume cover crops. Plant Prot Q 10(1):20–23
35. Senarathne SHS, Samarajeewa AD, Perera KCP (2003) Comparison of different weed management systems and their effects on yield of coconut plantations in Sri Lanka. Weed Biol Manage 3(3):158–161
36. Kiill LHP, Lima PCF, Lima JLS de (2001) Invasive plants in orchards in Sao Francisco Valley. Documentos da Embrapa Semi Arido 170:29
37. Thomas CG, Abraham CT (1996) Weeds of coconut gardens in the central zone of Kerala. Indian Coconut J Cochin 26(11):8–10
38. Senarathne SHS, Gunathilake HAJ (2010) Weed management in mature coconut plantations in Sri Lanka. COCOS 19:93–100
39. Ikisan (2013) [Internet] Hyderabad, Andhra Pradesh, India: cashew, weed management. http://www.ikisan.com/Crop%20Specific/Eng/links/tn_cashewWeedManagement.shtml. Accessed 28 June 2013
40. FAO (2013) [Internet] Sri Lanka: Surendra GBB. Integrated production practices of cashew in Sri Lanka—[8]. http://www.fao.org/docrep/005/ac451e/ac451e08.htm. Accessed 18 June 2013
41. Adeyemi AA (1989) Cultural weed control in cashew plantations: use of intercrops to reduce weed incidence in cashew plots. Proceedings of integrated pest management in tropical and subtropical cropping systems 3; 8–15 Feburary 1989; Bad Durkheim, Germany, pp 827–842
42. Espino PB (1948) Effects of 2,4-D on some common. Plants Philipp Agric 32(1):60–64
43. Adenikinju SA, Abiola FA, Olaiya AO (2000) Observations on some epiphytes and mistletoes and mineral contents of infested cacao stems. Niger J Tree Crop Res 4:27–28
44. Parker C, Riches CR (1993) Parasitic weeds of the world. CABI, Oxford

45. Olalya AO, Atayese MO, Lawal IO (2012) Effects of spacing and intercropping on the rate of infestation of parasitic weed on cocoa plantation. Niger J Hortic Sci 17(2013):187–182
46. Zimdahl RL (2004) Weed–crop competition: a review. 2nd edn. Blackwell, Oxford, pp 109–129
47. Silva NPJ-da, Rocha NOG-da, Kato OR (2007) Cocoa tree growth and production in agroforestry systems according to weed management. Rev Cienc Agrar 48:99–112
48. Mathew M, Punnoose KI, Potty SN (1977) Report on the results of chemical weeds control experiments the rubber plantations in South India. J Rubber Res Inst Sri Lanka 54:478–478
49. Ilango RVJ, Satyanarayana N (1996) Influence of certain weed species on growth of young tea (*Camellia* spp.). Developments in plantation crops research, proceedings of the 12th symposium on plantation crops, PLACROSYM XII, Kottayam, India, 27–29 November 1996, pp 167–169
50. Rao VS, Singh HS (1977) Effect of weed competition in young tea. Proceedings 28th Tocklai Biennial conference, pp 15–19
51. Alcantara EN de, Ferreira MM (2000) Effects of different weed control methods on yield of coffee grown on a Dusky Red Latosol. Cienc Agrotecnol 24(1):54–61
52. Ronchi CP, Terra AA, Silva AA, Ferreira LR (2003) Nutrient contents of coffee plants under weed interference. Planta Daninha 21(2):219–227
53. Dias TCS, Alves PLCA, Lemes LN (2005) Interference periods of *Commelina benghalensis* after coffee establishment. Planta Daninha 23(3):397–404
54. Ronchi CP, Silva AA (2006) Effects of weed species competition on the growth of young coffee plants. Planta Daninha 24(3):415–423
55. Ronchi CP, Terra AA, Silva AA (2007) Growth and nutrient concentration in coffee root system under weed species competition. Planta Daninha 25(4):679–677
56. Friessleben U, Pohlan J, Franke G (1991) The response of *Coffea arabica* L. to weed competition. Café Cacao The 35(1):15–20
57. Samarajeewa AD, Senaratna RPBSHS, Perera KCP (2004) Effect of different control methods of *Imperata cylindrica* on coconut (*Cocos nucifera*) yield in low country dry zone of Sri Lanka. COCOS 16:37–42
58. Oppong FK, Osei Bonsu K, Amoah FM (2007) Appraisal of some methods of weed control during initial establishment of cocoa in a semi-deciduous forest zone of Ghana. Ghana J Agric Sci 40(1):67–62
59. Mani J, Pothen J (1987) Prospect of chemical weeding in rubber plantations. Rubber Board Bul 23(2):18–13
60. Smale PE (1991) New Zealand has its own green tea industry. Hortic N Z 2(2):6–9
61. Singh RD, Sohani SK, Singh B, Chakrabarty DN (1993) Influence of long term mulching on yield of young tea and weeds. Proceedings of an international symposium Indian society of weed science, 18–20 November 1993, Hisar (Haryana), India 3(S):34–36
62. Sarkar SK, Chakravartee J, Basu SD (1983) Effect of soil stirring on yield of tea in heavy soil. Two Bud 30(1–2):50–51
63. Relova R, Pohlan YJ (1988) Differences in weed population dynamics in coffee nurseries under controlled shade and sun. Cultiv Trop 10(1):84–90
64. Wetala MPE, Drennan DSH (1997) Weeding effects on coffee production and soil weed—seed numbers in establishing coffee in Uganda. Brighton crop conference: weeds proceedings of an international conference, Brighton, UK, 17–20 November 1997, 2:661–662
65. Reddy AGS, Keshavaiah KV, Siddaramappa SN, Naik TB (1995) Influence of deep digging (Kothu) on robusta coffee (*Coffea canephora* subvar. *robusta*) in Wayanad. Indian Coffee 59(9):9–12
66. Aguilar V, Staver C, Milberg P (2003) Weed vegetation response to chemical and manual selective ground cover management in a shaded coffee plantation. Weed Res Oxf 43(1):68–75
67. Friessleben V, Pohlan J, Mahn EG (1990) The dynamics of the weed community in a coffee plantation in Cuba in relation to mechanical weed control and varying periods of weediness. Z Pflanzenkrankh Pflanzenschutz 97(6):642–645

68. Rao YS, Anand K, Sujatha C, Naidu R, George CK, Kumar A, Chatterjee S (1993) Large cardamom (*Amomum subulatum* Roxb.)—a review. J Spices Aromat Crops 2(1–2):1–15
69. Siddagangaiah, Krishnakumar V, Vadiraj BA, Sudharshan MR (1998) Influence of cultural practices on growth and yield of cardamom (*Elettaria cardamomum* Maton). Developments in plantation crops research proceedings of the 12th symposium on plantation crops, PLACROSYM-XII, Kottayam, India, 27–29 November 1996, pp 195–198
70. Senarathne SHS, Perera KCP (2005) Effects of different concentrations of glyphosate on control of weeds in coconut nurseries and growth of coconut seedlings in the dry zone of Sri Lanka. Planter 81(953):515–519
71. Senarathne SHS, Sangakkara UR (2009) Effect of different weed management systems on the weed populations, and seed bank composition and distribution in tropical coconut plantations. Weed Biol Manage 9(3):209–216
72. Xavier FAS, Maia SMF, Ribeiro KA, Mendonça ES, Oliveira TS (2013) Effect of cover plants on soil C and N dynamics in different soil management systems in dwarf cashew culture. Agric Ecosyst Environ 165:173–183
73. Ribeiro KA, Oliveira TS de, Mendonça ES, Xavier FAS, Maia SMF, Sousa HHF (2007) Soil quality in soil management systems in dwarf cashew crops. Rev Brasil Cienc Do Solo 31(2):341–351
74. Rambe EF, Sinuraya Z, CheongKeng K, Sidhu M (2008) Circle weeding in young oil palm using a hand-held motorised grass cutter. Planter 84(983):79–84
75. Ghosh MS, Ramakrishnan L (1981) Study on economical weed management programme in young and pruned tea with oxyfluorfen. Proceedings eighth Asian-Pacific weed science society conference1981, pp 119–125
76. Prematilake KG, Froud Williams RJ, Ekanayake PB (2004) Weed infestation and tea growth under various weed management methods in a young tea (*Camellia sinensis* [L.] Kuntze) plantation. Weed Biol Manage 4(4):239–248
77. Rao VS, Kotoky B (1981) Oxyfluorfen new pre-emergence herbicide for tea. Two Bud 28(2):40–42
78. CSIR (1992) Annual report 1990–1991, CSIR complex Palampur (H.P.), India, p 9
79. CSIR (1993) Annual Report 1991–1992, CSIR complex Palampur (H.P.), India, pp 1,4–5
80. Singh RD, Shanker A, Gulati A (1992) Management strategies of weeds, insects and diseases in Kangra tea plantations. Kangra Tea Festival Souvenir, 5–7 June 1992, Palampur (H.P.), India, pp 13–14
81. Kabir SE, Chaudhury TC, Hajra NG (1991) Evaluation of herbicides for weed control in Darjeeling tea. Indian Agric 35(3):179–185
82. UPASI (1978) Annual report 1976 United Planters' Association of Southern India, pp 16–19
83. Sharma VS, Satyanarayana N (1976) Recent developments in chemical weed control in tea fields. UPASI Tea Sci Dep Bull 33:101–108
84. Singh RD, Singh B, Sud RK, Tamang MB, Chakrabarty DN (1997) Control of *Lantana* in non-cropped area and in tea plantation. Int J Pest Manage 43(2):145–147
85. Ilango RVJ, Sreedhar C (2001) Evaluation of glufosinate ammonium - a contact herbicide for weed control in tea (*Camellia* spp. L.). Indian J Weed Sci 33(1/2):79–80
86. Victor R, Ilango J, Pagalavan B, Ramasubramanian B (2001) Evaluation of napropamide for control of weeds in tea (*Camellia* spp. L.) Indian J Weed Sci 33(3/4):227–228
87. Ilango RVJ (2003) Evaluation of carfentrazone-ethyl for control of weeds in tea (*Camellia* spp. L.). Indian J Weed Sci 35(3/4):296–297
88. Rajkhowa DJ, Bhuyan RP, Barua IC (2005) Evaluation of carfentrazone-ethyl 40 DF and glyphosate as tank mixture for weed control in tea. Indian J Weed Sci 37(1/2):157–158
89. Sosa Lopez MH (1996) Study of a combination of chemical and manual methods for the control of weeds in young coffee. XVII simposio sobre caficultura latinoamericana San Salvador, El Salvador, 23–27 Octubre 1995, 2:15
90. Sanchez FL, Gamboa E (2004) Weed control by herbicides and mechanical methods on early stage of coffee plantations. Bioagro 16(2):133–136
91. Azizuddin M, Rao WK (1993) Evaluation of weedicides in coffee. Indian Coffee 57:11–15

92. Mir A, Rao WK, Manjunath AN, Hariyappa N (1994) Effects of additives on the bio-efficacy of glyphosate based weedicide. Indian Coffee 58(7):3–8
93. Nishimoto RK (1992) Evaluation of pre-emergence herbicides for establishing coffee. Trop Pest Manage 38(3):298–301
94. Njoroge JM, Kimemia JK (1990) A comparison of different weed control methods in Kenya. Kenya Coffee 55(644):863–870
95. Matowo PR, Msaky JWJ, Malinga S (1997) Evaluation of Agrocet 180, 360 EC and Round-up dry 420 solid granules, new roundup formulations in controlling weeds in coffee. Dix septieme colloque scientifique international sur le cafe Nairobi, Kenya, 20–25 Juillet 1997, pp 783–787
96. Ronchi CP, Silva AA, Terra AA, Miranda GV, Ferreira LR (2005) Effect of 2,4-dichloro-phenoxyacetic acid applied as a herbicide on fruit shedding and coffee yield. Weed Res Oxf 45(1):41–47
97. Kusnanto U (1991) Chemical and manual weed control in oil palm plantation; a study on the efficacy, application frequency and cost analysis. Bul Perkeb 22(3):147, 163–181
98. Fernandez O, Ortiz RA (1995) Gramineous herbicide evaluation for controlling (*Paspalum fasciculatum* Wild.) in oil palm (*Elaeis guineensis*). Agron Mesoam 6:15–22
99. Bhanusri A, Reddy VM, Rethinam P (2001) Weed management in oil palm plantations. Int J Oil Palm Res 2(1):45–46
100. Aladesanwa RD, Oladimeji MO (2005) Optimizing herbicidal efficacy of glyphosate iso-propylamine salt through ammonium sulphate as surfactant in oil palm (*Elaeis guineensis*) plantation in a rainforest area of Nigeria. Crop Prot 24(12):1068–1073
101. Wibawa W, Mohamad R, Omar D, Juraimi AS (2007) Less hazardous alternative herbicides to control weeds in immature oil palm. Weed Biol Manage 7(4):242–247
102. Wibawa W, Mohamad R, Juraimi AS, Omar D, Mohayidin MG, Begum M (2009) Weed control efficacy and short term weed dynamic impact of three non-selective herbicides in immature oil palm plantation. Int J Agric Biol 11(2):145–150
103. Wibawa W, Mohamad RB, Puteh AB, Omar D, Juraimi AS, Abdullah SA (2009) Residual phytotoxicity effects of paraquat, glyphosate and glufosinate-ammonium herbicides in soils from field-treated plots. Int J Agric Biol 11(2):214–216
104. Wibawa W, Mohamad RB, Omar D, Zain NM, Puteh AB, Awang Y (2010) Comparative im-pact of a single application of selected broad spectrum herbicides on ecological components of oil palm plantation. Afr J Agric Res 5(16):2097–2102
105. Nuertey BN, Tetteh FM, Opoku A, Afari AP, Asamoah TEO (2007) Effect of roundup-salt mixtures on weed control and soil microbial biomass under oil palm. J Ghana Sci Assoc 9(2):61–75
106. Doll J (2000) Glyphosate resistance in another plant. Res Pest Manage 11(1):5–6
107. Jalaludin A, Ngim J, Bakar BHJ, Alias Z (2010) Preliminary findings of potentially resistant goose grass (*Eleusine indica*) to glufosinate-ammonium in Malaysia. Weed Biol Manage 10(4):256–260
108. Mortimer AM, Hill JE (1999) Weed species shifts in response to broad spectrum herbicides in sub-tropical and tropical crops. The 1999 Brighton conference: weeds vol 2 proceedings of an international conference, 15–18 November 1999; Brighton, UK, pp 425–436
109. Chew ST, Teh KH, Teng YT (1992) An ultra low volume technique for general weed control with glyphosate herbicide. In: Ooi PAC, Lim GS (eds) Proceedings of the 3rd international conference on plant protection in the tropics; 20–23 March 1990; Genting Highlands, Ma-laysia; 2 pp 1–12
110. Hornus P (1990) Chemical weeding in adult oil palm circles. Low volume technique. Ol Paris 45(6):295–304
111. Leng T, Maclean RJ (1992) Control of *Imperata cylindrica* and *Ischaemum muticum* with glyphosate using CDA technique. In: Ooi PAC, Lim GS, Teng PS (eds) Proceedings of the 3rd international conference on plant protection in the tropics; 20–23 March 1990; Genting Highlands, Malaysia, pp 13–19
112. Chung GF, Balasubramaniam R, Cheah SS (2000) Recent development in spray equipment for effective control of pests and weeds. Planter 76(887):65–84

113. Eng OK, Omar D, McAuliffe D (1999) Improving the quality of herbicide applications to oil palm in Malaysia using the CF Valve—a constant flow valve. Crop Prot 18(9):605–607
114. Haji Mustafa M (2000) Mechanised weeding operations—a united plantation experience. Planter 76(887):87–96
115. Ishak WWI, Rahman KA (2008) Development of real-time color analysis for the on-line automated weeding operations. Proceedings of the 9th international conference on precision agriculture, Denver, Colorado, USA, 20–23 July, abstract 178
116. Sit AK, Sarkar B (2006) Effect of herbicides on control of weed flora in young areca garden under sub-Himalayan terai region of West Bengal. Environ Ecol 24S(Special 3A):864–867
117. Chattopadhyay N, Ghosh SN (1999) Evaluation of different herbicides for chemical control of weeds in cashew plantation. Environ Ecol 17(3):588–593
118. Opoku-Ameyaw K, Oppong FK, Akoto SO, Amoah FM, Swatson E (2012) Development of weed management strategies for cashew cultivation in Ghana. J Agric Sci Technol A 1:411–417
119. Oppong FK, Sarfo JE, Osei Bonsu K, Amoah FM (2006) Effects of sulfosate on weed suppression, cocoa pod set, and yield of cocoa and coffee. Ghana J Agric Sci 39(2):115–112
120. Adeyemi AA (2000) Evaluation of the formulated mixture of glyphosate and terbuthylazine in the control of weeds in mature cocoa plantation. Niger J Tree Crop Res 4(2):62–68
121. Samarappuli L (1994) Weed management in rubber cultivation. In: Labrada R, Caseley JC, Parker C (eds) Weed management in developing countries. Food & Agriculture Organization, pp 364–368
122. Sandanam S, Rajasingham CC (1982) Effects of mulching and cover crops on soil erosion and yield of young tea. Tea Q 51(1):21–26
123. Xiao RunLin, Xiang ZouXiang, Xu HuaQin, Shan WuXiong, Chen P, Wang GuiXue, Cheng X (2008) Ecological effects of the weed community in tea garden with intercropping white clover and straw mulching. Trans Chin Soc Agric Eng 4(11):183–187
124. Bradshaw L, Lanini WT (1995) Use of perennial cover crops to suppress weeds in Nicaraguan coffee orchards. Int J Pest Manage 41(4):185–194
125. Muralidhara HR, Raghuramulu Y (1997) A review on mulching practices in coffee plantations. Indian Coffee 61(9):10–13
126. Semidey N, Orengo Santiago E, Mas EG (2002) Weed suppression and soil erosion control by live mulches on upland coffee plantations. J Agric Univ Puerto Rico 86(3/4):155–157
127. Opoku Ameyaw K, Oppong FK, Amoah FM, Osei Bonsu K (1999) Preliminary investigations into the use of intercropping for weed management in young coffee in Ghana. Dix huitieme Colloque Scientifique International sur le Cafe, Helsinki, Finland, 2–8 Aout 1999, pp 441–444
128. Ooi GT, Leng T, Abdullah Z, Lee SC (1996) Field performance of pursuit 50A: a promising herbicide for legume establishment in oil palm planting. Planter 72(847):539–546
129. Mathews C (1998) The introduction and establishment of a new leguminous cover crop, *Mucuna bracteata* under oil palm in Malaysia. Planter 74(868):359–360, 363–368
130. Lee ChinTui, Chu KumChoon, Arifin I, Hashim I (2005) Early results on the establishment of *Mucuna bracteata* at various planting densities under two rainfall regimes. Planter 81(952):445–459
131. Erhabor JO, Aghimien AE, Filson GC (2002) The root distribution pattern of young oil palm (*Elaeis guineensis* Jacq) grown in association with seasoned crops in southwestern Nigeria. J Sustain Agric 19(3):97–110
132. Barrios R, Farinas J, Diaz A, Barreto F (2004) Evaluation of 11 leguminous accessions as cover crop in oil palm plantations in Monagas state, Venezuela. Bioagro 16(2):113–119
133. Samedani B, Juraimi AS, Anwar MP, Rafii MY, Awadz SAS, Anuar AR (2012) Competitive ability of some cover crop species against *Asystasia gangetica* and *Pennisetum polystachion*. Acta Agric Scand B Soil Plant Sci 62(7):571–582
134. Bourgoing R (1990) Choice of cover crop and planting method for hybrid coconut growing on smallholdings. Ol Paris 45(1):23–30
135. Teasdale JR (1996) Contribution of cover crops to weed management in sustainable agricultural systems. J Prod Agric 9:475–479

136. Griffin T, Liebman M, Jemison J (2000) Cover crops for sweet corn production in a short-season environment. Agron J 92:144–151
137. Sarrantonio M, Gallandt ER (2003) The role of cover crops in North American cropping systems. J Crop Prod 8:53–73
138. Coconut Research Institute of Sri Lanka (2012) [Internet]. Lunuwila, Sri Lanka: Coconut Research Institute—Sri Lanka [updated 20 March 2012]. http://www.cri.gov.lk/web/index.php?option=com_content&view=article&id=124&Itemid=90&lang=en. Accessed 18 June 2013
139. Singh HP (2001) Organic farming approach for sustainable coconut cultivation. Indian Coconut J 32(5):3–18
140. Marimuthu R, Giridharan S, Bhaskaran R (2002) Weed management in coconut nursery. Indian Coconut J 33(8):13–14
141. Akobundu IO (1987) Weed science in the tropic principles and practices. Wiley, Chichester, 522 pp
142. Wan Mohamed WE (1978) Utilisation of ground vegetation in rubber plantation for animal rearing. Proceedings of Rubber Research Institute of Malaysia planters conference 1977, Kuala Lumpur, pp 265–281
143. Tajuddin I, Najib LA, Chong DT, Abd Samat MS, Vanaja V (1990) Status report on RRIM Commercial Sheep Project 1985–1988, Rubber Research Institute of Malaysia
144. Nambiar G (1991) Cattle rearing under oil palm. Planter 67(789):598–605
145. Gopinathan N (1998) Cattle management in oil palm ESPEK's experience. Planter 74(870):503–514
146. Nor ZM, Ponijan K, Abdul-Kadir MY (2000) Implications of cattle integration on weeding in mature oil palm—YPJOPE's experience. Planter 76(890):279–286
147. Yusoff ARM (1992) Cattle rearing under oil palm. (Profitability of cattle fattening under oil palm). Planter 68(790):19–24
148. Seresinhe T, Marapana RAUJ, Kumanayaka L (2012) Role of local feed resources on the productivity and financial viability of a coconut-cattle integrated system in the southern Sri Lanka. Anim Nutr Feed Technol 12(2):145–156
149. Sanyal D (2008) Introduction to the integrated weed management revisited symposium. Weed Sci 56:140
150. Senarathne SHS, Perera KCP (2011) Effect of several weed control methods in tropical coconut plantation on weed abundance, coconut yield and economical value. Int Res J Plant Sci 2(2):25–31
151. Tamil Nadu Agriculture University [Internet]. (2013) Coimbatore, India: TNAU agritech portal; Horticulture, plantation crops, coconut. http://www.agritech.tnau.ac.in/horticulture/horti_plantation%20crops_coconut.html. Accessed 18 June 2013
152. Famaye AO, Adeyemi EA (2011) Effect of cashew/rice/plantain intercropped on weed incidence in Edo State, Nigeria. ARPN J Agric Biol Sci 6(6):62–65
153. Adeyemi AA (1998) Effects of intercropping on weed incidence in cashew (*Anacardium occidentale*) plantations. Niger J Tree Crop Res 2(1):83–94
154. Sri Lanka Cashew Corporation (2011) [Internet]. Sri Lanka: Sri Lanka Cashew Corporation, Ministry of Minor Export Crops Promotion [updated 16 September 2011]. http://www.cashew.lk/index.php?option=com_content&view=article&id=122&Itemid=127&lang=en. Accessed 18 June 2013
155. Chee YK, Faiz A (1991) Sheep grazing reduces chemical weed control in rubber. In: ACIAR proceedings on "forages for plantation crops", pp 120–123
156. Martinez Viciedo Y, Alvarez Puente RJ (2004) Allelopathic effects of *Pinus caribaea* Morelet var. Caribaea on coffee weeds under the sun. III Congreso 2004 Sociedad Cubana de Malezologia, Memorias, Jardin Botanico Nacional, Ciudad Habana, 28–29 y 30 de Abril del 2004, pp 119–122
157. Ghazali KH, Mustafa MM, Hussain A (2008) Machine vision system for automatic weeding strategy using image processing technique. American Eurasian. J Agric Environ Sci 3(3):451–458
158. Ishak WIW, Hudzari RM, Ridzuan MMN (2011) Development of variable rate sprayer for oil palm plantation. Bul Pol Acad Sci Tech Sci 59(3):299–302

Chapter 13
Management of Aquatic Weeds

Robert M. Durborow

Introduction

Ponds are ideal habitats for aquatic plants, which are a necessary component of pond ecosystems because they perform valuable functions. Photosynthesizing plants produce oxygen that is needed to sustain fish life. Also, plants assimilate ammonia that is excreted by fish, thereby helping to prevent the accumulation of potentially toxic concentrations of ammonia [1], and they are able to absorb toxins, such as chromium [2, 3] and arsenic [4, 5], from aquatic environments. Plants (e.g., water hyacinth) have even been investigated for their usefulness in nanoparticles derived from cellulose of the plant [6, 7] and for their nutrient value—for example, duckweed for use in fish feed [8]. Nevertheless, these plants can cause problems in ponds, and control measures often must be used to prevent them or reduce their abundance [1]. Many times the control of serious weed problems in large reservoirs requires the intervention of local or state governments. Ingwani et al. lamented the lack of an environmental ministry in Ethiopia to oversee the uncontrolled spread of water hyacinth (*Eichhornia crassipes* [Mart.] Solms) [9].

Types of Weeds

The plants that grow in ponds can be categorized into two groups. The algae are primitive plants that have no true roots, stems, or leaves and do not produce flowers or seeds [1]. Algae can be categorized as phytoplankton, filamentous algae, or muskgrass (which resembles a higher aquatic plant).

R. M. Durborow (✉)
Division of Aquaculture, College of Agriculture, Food Science and Sustainable Systems, Kentucky State University, Aquaculture Research Center, Frankfort, KY, USA
e-mail: robert.durborow@kysu.edu

B. S. Chauhan, G. Mahajan (eds.), *Recent Advances in Weed Management*,
DOI 10.1007/978-1-4939-1019-9_13,

Fig. 13.1 Blue-green algae (cyanobacteria) forming a surface scum

The higher aquatic plants are more advanced and usually have roots, stems, and leaves and produce flowers and seeds. Higher aquatic plants can either be submerged, emergent, or floating [1].

Algae

Phytoplankton

These algae are microscopic simple plants (sometimes referred to as microalgae) suspended in water or forming floating scums of near-microscopic colonies on pond surfaces. Communities of phytoplankton are called the "bloom." There are hundreds of species of phytoplankton and identification of the different species is difficult, requiring a microscope [1].

Phytoplankton are the most common type of plants found in ponds. Moderate densities of phytoplankton are desirable in ponds because they shade the pond bottom, preventing establishment of more troublesome types of weeds. This competitive relationship also works in the other direction; macrophytes such as alligator weed (*Alternanthera philoxeroides* [Mart.] Griseb) can have algicidal properties that prevent phytoplankton from establishing in a pond [10]. Phytoplankton develop into a weed problem when they become excessively abundant or when certain undesirable species become dominant in the community [1]. Excessive phytoplankton abundance causes serious water quality problems, such as frequent periods of dangerously low concentrations of dissolved oxygen; this low dissolved oxygen occurs when respiration of the heavy bloom exceeds its oxygen production, usually at night and during cloudy weather when photosynthesis is low. This problem is often associated with dense blooms of blue-green algae (Fig. 13.1). The blue-green bloom floats to the top of the pond forming a scum that can block sunlight and prevent proper photosynthesis. These blooms can also cause an off-flavor problem in fish raised in aquaculture ponds [1].

Fig. 13.2 *Pithophora* species filamentous algae held on a stick

Filamentous Algae

Most filamentous algae begin growing on the bottom of the pond and rise to the surface when gas bubbles become entrapped in the plant mass. They form mats of wooly/cottony or slimy plant material. These filamentous algae are also known as "pond scum," "string algae," or more commonly "moss." Positive identification of the different types of filamentous algae usually requires a microscope. Control methods are similar for all filamentous algae (typically copper-based algicides, such as copper sulfate) except *Pithophora* spp. (Fig. 13.2), which is resistant to most copper-based algicides and requires special treatment [1]. It is thus important to identify the species of filamentous algae present in the pond to ensure that the proper treatment is selected. The most common filamentous algae in ponds are:

- *Hydrodictyon* spp. (water net): Each cell is attached repeatedly to two others forming a repeating network of 5- or 6-sided mesh that looks like a "fish net" stocking (Fig. 13.3)
- *Spirogyra* spp.: It is usually a dark green slimy mass that can be pulled apart and drawn out into fine filaments (Fig. 13.4). This alga usually is easy to identify microscopically because the chloroplast is spiraled in a characteristic "corkscrew" along the inside of the cell wall [1].

Fig. 13.3 Water net (*Hydrodictyon* spp.)

Fig. 13.4 The filamentous algae *Spirogyra* sp. is slick and slimy and persists in ponds throughout the winter. It often goes away in hot summer temperatures

- *Pithophora* spp.: It is probably the most noxious and difficult filamentous algae to control. *Pithophora* spp. (Fig. 13.2) is irregularly branched, not slimy, and somewhat coarser than the masses of *Spirogyra* spp. [1]. A mass of *Pithophora* spp. feels like wet wool to the touch. The distinguishing microscopic characteristic is the presence of barrel-shaped spores along the filament.
- *Chara* spp.: It belongs to a more advanced group of algae that resembles submersed higher plants in growth habit. This plant is commonly called "muskgrass" because of the garlic or skunk-like odor released when it is crushed [1]. Masses of *Chara* spp. are serrated and feel rough or crusty when crushed in the hand. It is similar in appearance to coontail (*Ceratophyllum* spp.), a higher plant, but does not have the forked/branching leaves that coontail has.

Filamentous algae are aesthetically undesirable, giving the pond a "clogged-up" overgrown appearance, and they interfere with fishing by snagging the fisherman's hook or entangling the propeller on a motor boat's engine. In aquaculture, filamentous algae can prevent fish or shrimp from being harvested by seine nets. Seines may ride up over the mass of weeds, allowing fish to escape, and the weight of plant material caught in the seine may strain equipment or completely stop the harvesting process. Even if seining is possible, fish or shrimp may become entangled in the mass of weeds in the seine and will be stressed as workers pick through the weeds to recover them. This is particularly a problem with fingerlings and shrimp [1].

Higher Aquatic Plants

Submersed Plants

Submersed plants spend their entire lifetime beneath the surface of the water, although the flower and parts of the plant may extend above the surface. The plants are often rooted in the mud, but masses of plants may tear loose and float free in the water. These plants are objectionable because they interfere with fishing and fish harvest. The most common submersed higher aquatic plants in ponds are:

- *Najas guadalupensis* (Spreng.) Magnus (bushy pondweed): It is a rooted, submersed plant with slender branching stems and narrow ribbon-like leaves arranged oppositely or in whorls of three. Bushy pondweed is a common submersed weed problem in ponds.
- *Potamogeton pectinatus* L. (sago pondweed): It is a rooted, wholly submersed plant with long, narrow leaves tapering to a point. The stems are irregularly (and often highly) branched.
- *Potamogeton crispus* L. (curly leaf pondweed): It has curly/crinkled 1-cm-wide leaves, about 5–8 cm long arranged alternately on the stem.
- *Ceratophyllum demersum* (Cham.) Asch. (coontail): These plants have long thin stems that are not rooted. The leaves are in whorls and are forked.
- *Elodea* spp. (waterweed, pondweed, or *Anacharis*): These are often considered the generic waterweed or pondweed of North America and are often used in people's personal aquaria.

Emergent Plants

Emergent aquatic plants are rooted in the bottom mud and grow above the water. Many can also grow under strictly terrestrial conditions. The plants are rigid and not dependent on the water for support. Emergent plants usually infest only the pond margins and other shallow areas resulting from ponds being constructed inadequately, for example, with pond banks that are not steep enough creating a large shallow area along the pond's edge, water shortage conditions, or excessive bank erosion. If stands of emergent plants become too dense or widespread, they may interfere with fishing, seining, or feeding of fish. They can also create a habitat that harbors snakes. Fast-growing emergent plants, such as smartweed (*Polygonum pennsylvanicum* L.) inhabit shallow areas [1]. The most common emergent weeds in ponds are:

Fig. 13.5 Smartweed (*Polygonum* sp.) inhabits shallow areas of ponds

Fig. 13.6 Water primrose (*Ludwigia* spp.) on the pond's shoreline (note the yellow flowers)

- *Polygonum* spp. (smartweed): The leaves are alternate and elliptical on this plant. The stem is erect and jointed, with each swollen node covered by a thin sheath. Flowers are usually white or pink (Fig. 13.5).
- *Ludwigia* spp. (water primrose): These are herbs that inhabit wet environments and have alternate or opposite leaves that can be either entire or slightly toothed. The flower is most commonly yellow (Figs. 13.6 and 13.7), but herbs in this family can have white or rose-purple flowers.
- *Typha* spp. (cattails) have characteristic brown cigar-shaped flowers. These plants can overpopulate quickly in shallow areas along pond banks. Active transpiration from the long (up to 2.4 m) stout leaves can lower a pond's water level significantly when cattails are abundant, sometimes forming a swampy area that once was a pond (Fig. 13.8).
- *Salix* spp. (willows): These are shrubs or trees with simple, elliptical leaves in alternate arrangement. This plant can block access to ponds, and when allowed to grow on dams, their roots can eventually lead to water leakage from the pond.

Fig. 13.7 Water primrose (note the yellow flowers, leaf shape, and stem color)

Fig. 13.8 Cattails (*Typha* sp.) grow along a pond's edge in shallow water

Floating Plants

This category includes free-floating plants, such as duckweeds (*Lemna* spp.) and watermeal (*Wolffia* spp.), and floating-leaf plants, such as water lilies (*Nymphaea* spp.; Fig. 13.9) and lotus (*Nelumbo* spp.; Fig. 13.10). Small recreational ponds often have

Fig. 13.9 Water lilly (*Nymphaea* spp.) is a floating weed with a cleft in the leaf

Fig. 13.10 Lotus (*Nelumbo* sp.) with a seed pod in the foreground and a white flower. Some leaves float on the pond surface and others are elevated above the surface; they do not have a cleft

problems with duckweed and watermeal, especially when the pond is stagnant and sheltered from the wind. Larger ponds are often unsheltered from the wind, and duckweeds are continually washed ashore where they often dry up and die [1].

Occurrence of Weed Problems

Some plant life will always be present in ponds, but the type of aquatic plant community that establishes in a pond depends on the relative abilities of particular plants to compete for resources. The growth of phytoplankton is favored in waters with high concentrations of nitrogen, phosphorus, and other plant nutrients dissolved in the water. Phytoplankton are efficient at using dissolved nutrients and reproduce rapidly. Once established, the phytoplankton community competes effectively for nutrients and also restricts the penetration of light so that plants that germinate on the bottom do not receive enough light to continue growing [1].

Rooted submersed plants tend to establish in ponds with low supplies of nutrients in the water. These ponds often are clear with light penetrating to the bottom, and rooted plants can use the nutrients in the bottom mud for growth. Established stands of submersed weeds compete for nutrients and light and prevent phytoplankton from becoming established. Some higher plants also produce chemicals that inhibit the growth of phytoplankton [1, 10].

Emergent plants usually colonize only the margins of ponds where the water is less than 0.6–1 m deep. If levees or banks of the pond are eroded and have large areas of shallow water, expansive growths of emergent plants may be present. Emergent plants are rooted and can use nutrients in the mud. Thus, their establishment is also favored by low nutrient levels in the water [1].

Aquaculture ponds containing food-sized fish receive high levels of nutrients from daily feeding of formulated fish food. These ponds, therefore, rarely have submersed or emergent aquatic weed problems; the nutrients in the water promote an actively growing phytoplankton bloom, which, in turn, shades the pond bottom and prevents weed growth. Aquaculture fingerling production ponds, on the other hand, receive fairly low levels of nutrients because of the smaller biomass of fish; these ponds are much more likely to have a scant phytoplankton bloom and a problem with macrophytic plant growth, such as submersed and emergent weeds. Likewise, recreational ponds, especially ones that are not fertilized, also have a tendency to have an inadequate bloom and a macrophytic weed problem, such as *Najas* spp. (bushy pondweed) [1].

Environmentally sound and cost-effective aquatic weed management depends on the type of plant, the extent of plant coverage, the species and life stage of fish or crustaceans in the pond, water quality, time of year, and weather. Understanding these interactions, which differ for each weed problem, is largely a matter of experience. The spread of nuisance aquatic weeds from country to country has been documented. Hussner discusses significant problems caused by the spread of the submersed waterweed or pondweed, *Elodea* spp.; water hyacinth, *Eichhornia* spp.; water primrose, *Ludwigia* spp.; floating pennywort, *Hydrocotyle* spp.; and water milfoil, *Myriophyllum* spp. in Europe [11].

Prevention of Aquatic Weeds

Almost any plant can be tolerated as long as it does not become so abundant that it interferes with the intended use of the pond. It is, however, difficult to predict whether a small infestation of weeds will spread and become a problem, so most control measures are implemented when fairly large stands of weeds have already become established. At that time, using herbicides is usually the fastest way to eradicate weeds and reestablish a phytoplankton bloom, which is usually the most desirable plant form in ponds. Chemical weed control is, however, risky in fishponds because water quality deteriorates when dense stands of weeds are killed. Prevention of weeds is a preferred approach to aquatic plant management [1].

Certain management procedures can be used to minimize the chances of infestations of submersed and emergent plants and filamentous algae. Such procedures should become part of common pond management and may help avoid the use of chemical control measures [1].

Pond Construction

Most noxious weed growth starts in the shallow (less than 0.75 m deep) areas of ponds. If the area of the pond where light can penetrate to the bottom is reduced, rooted plants have less chance to become established. Pond levees and/or dams should have a fairly abrupt slope of about 3:1 (0.9 m out toward the center of the pond for every 0.3 m drop toward the pond bottom) or 4:1. A slope of greater than 4:1 would be too gradual and would create an excessive shallow area, while a slope of less than 3:1 would be too steep making the levee prone to erosion and sloughing-off. Plans should be made during construction for the shallowest part of the pond to be no shallower than 0.75 m when the pond is close to full (near the top of the drain pipe) [1]. In a study quantifying water milfoil biomass at various water depths, Wersal and Madsen reported 99 % greater water milfoil biomass when water depths were less than 0.77 m compared to growth at 1–1.37-m depth [12].

Refilling an Empty Pond

In many cases, a newly constructed pond will be completed by the end of summer (by the time the dry period of the year ends in the USA). The rainiest part of the year typically will occur during fall and winter, and the new pond will fill from rain runoff from the watershed during a time when weeds are less likely to grow. Ponds with a well water source are ideally filled during winter for this same reason. If they are filled during other times, it is best to fill the pond as quickly as possible from the well to attain an adequate water depth that will prevent aquatic weed growth (if one well serves four ponds, for example, one pond should be filled at a time to minimize the time needed to get the proper depth for weed control). Also, grass carp (*Ctenopharyngodon idella* Steindachner) may be stocked to prevent growth of nuisance weeds. About 18 triploid grass carp per hectare is a good preventive stocking rate (triploid grass carp are not able to reproduce and are, thus, usually required where grass carp stocking is legal in the USA). In addition, ponds can be left empty until the farmer plans to actually stock and begin feeding his/her fish, unless the pond is made of highly structured clay (stereotypically red clay with a high iron content); in this case, allowing the pond to dry out would promote deep cracking of the pond's clay lining, which could cause leaks when the farmer attempts to refill the pond at a later date. If water needs to be maintained in a pond, a fertilization program can help to prevent the water from being clear or aquatic dyes can be used for this same purpose. Fertilization promotes a healthy phytoplankton bloom that will shade out sunlight from reaching potential weeds attempting to germinate at the bottom of the pond [1].

Fertilization

The implementation and continued use of the proper fertilization program is perhaps the best method of preventing the growth of troublesome weeds in recreational ponds as well as fry nursery ponds. To avoid weed problems, establish a phytoplankton bloom as quickly as possible after filling the ponds. The best way to do this is to add inorganic fertilizers to the pond. The key ingredient in fishpond fertilizers is phosphorus. The most common phosphorus source in bagged, granular fertilizers is triple superphosphate (0-46-0). It should be noted, however, that when triple superphosphate is broadcast over ponds, it settles to the bottom because the granules are very insoluble. Most of the phosphorus reacts with the bottom mud and never reaches the water. Any phosphorus that dissolves while the granules settle through the water quickly reacts with calcium in the water and is changed into unavailable calcium phosphate. Granular fertilizers should be put on an underwater platform or in a porous container so they can dissolve slowly into the water before they have a chance to contact the mud bottom [1].

Liquid fertilizers are more effective than granular fertilizers at stimulating a phytoplankton bloom, especially in hard, alkaline waters. The phosphorus in liquid fertilizers is already in solution and immediately available for uptake by the phytoplankton. Although the phosphorus from liquid fertilizers also will eventually become unavailable due to reactions with calcium in the pond water, it remains in solution long enough to be taken up in adequate quantities by the phytoplankton [1].

The most common, and best, analysis for liquid fertilizers runs from about 10-34-0 to 13-38-0. This general analysis of about three times as much phosphorus (expressed as P_2O_5) as nitrogen (expressed as N) has been found to have an excellent balance. The rate used successfully by many commercial fish producers is about 2.4 L per ha applied every other day for about 4 days or until a noticeable phytoplankton bloom develops. Liquid fertilizer is heavier than water, so it should first be diluted in water before it is applied to the pond, preventing it from sinking into the bottom mud [1]. It can be sprayed from the bank or applied from a boat outfitted for chemical applications.

It should be noted that excessive water flow through ponds flushes plant nutrients from the water, favoring rooted weeds that can obtain nutrients from bottom soils. Ponds should not have watershed areas larger than necessary to maintain water level; excess runoff from large watersheds should be diverted away from ponds. Springs running into ponds can also dilute nutrients, and they too can be diverted away from the pond and can be allowed to enter the pond only when water is needed [1].

Manual Harvesting

Removing potentially noxious emergent weeds by hand is another management practice that may reduce the need for chemicals. As small areas of the pond margin become infested, plants are removed manually. Manual harvesting of weeds is only suited for controlling emergent vegetation in relatively small ponds, but often proves to be futile [1]. A study in Belgium noted that the invasive African elodea (*Lagarosiphon major* [Ridl.] Moss) could not be controlled by sediment dredging in large bodies of water populated with this weed [13]. However, Evans and Wilkie showed that mechanical harvest of hydrilla, *Hydrilla verticillata* (L.f.) Royle, may be successful in nutrient-poor waters [14]. Routine mowing of pond banks will help prevent the establishment of dense growths of shoreline plants, such as willows, and will also reduce habitat for snakes [1].

Water Drawdowns

Periodic water drawdowns are sometimes effective in killing or preventing aquatic weeds. The vegetation along the pond margin (the most common location for weed problems) is stranded and dies from drying up. However, one study done by Doyle and Smart showed that water drawdowns did not control hydrilla [15].

Biological Control of Aquatic Plants

Biological weed control in ponds involves the use of fish to consume unwanted aquatic vegetation. Grass carp are normally used in warm-water ponds. They are most often used to control submersed plants or filamentous algae. Koi (colorful common carp, *Cyprinus carpio koi*) are presently being evaluated at the Kentucky State University for their weed prevention potential [1]. Common carp and Nile tilapia, *Oreochromis niloticus,* successfully controlled aquatic weeds in Bangladeshi rice fields [16]. Morin et al. reviewed the effectiveness of biological weed control agents [17].

Grass Carp

The grass carp or "white amur" was introduced into the USA from Southeast Asia in 1963 and is now widespread, especially in the southeastern states of the USA The fish is banned in many US states, and some states allow only sterile, triploid grass carp. Where legal and available, this fish is a valuable tool to control nuisance aquatic weeds.

The controversy over the distribution and use of grass carp is based on the potential effect of this fish on native fish and wildlife. Considerable discretion should be used when planning to stock these fish into ponds and every effort should be made to prevent their escape into natural waters. To further diminish the likelihood that grass carp will reproduce and thrive in natural waters, it is recommended that only sterile, triploid carp are used [1].

The grass carp has several traits that make it a good species for recreational ponds and for polyculturing with channel catfish. Small grass carp (less than 0.5–1 kg) are almost completely herbivorous and will not compete to a significant degree with catfish for feed. Grass carp tolerate a wide range of environmental conditions: They can survive at water temperatures of 0–40 °C and are nearly as tolerant as catfish to low dissolved oxygen concentrations. The fish grows rapidly, as much as 2.3–4.5 kg a year. It must consume large quantities of plant material to grow and may consume two to three times its weight in plant material per day [1].

Grass carp prefer to eat succulent submersed plants such as *Najas* spp. and *Chara* spp. Fibrous plants such as grasses and smartweed are less preferred and grass carp will not eat these plants if more preferred plants are available. Food consumption by grass carp is greatest at water temperatures of 27–29 °C, and the fish stop eating when the temperature falls below about 13 °C.

In catfish nursery ponds, grass carp should be stocked prior to stocking the catfish fry in order to prevent weed growth. Likewise, in recreational ponds, grass carp should be stocked before weeds become a problem [1].

Grass carp also are used by some pond owners to control existing weeds. However, considerable time is required for grass carp to reduce weed infestations, particularly if coverage is extensive. Results may take a year to be realized. In aquaculture, food-fish ponds are usually not drained each year and grass carp become a permanent inhabitant of the pond [1]. Larger grass carp learn to feed on pelleted feeds, reducing their effectiveness in controlling weeds. Smaller grass carp can then be stocked if necessary, but sometimes when the weed problem is under control, a phytoplankton community develops, preventing further weed growth and reducing the need for carp.

The stocking rate for grass carp depends on the severity of the weed problem. When used to prevent the establishment of submersed weeds, 13–25 grass carp per ha should be stocked. They should be large enough to prevent largemouth bass or large catfish predators from eating them (25–28 cm, depending on the age of the pond and size of the predator fish). The same stocking rate is also adequate if the pond is lightly infested with weeds. For more severe weed problems, 25–38 fish per ha should be stocked. For heavily weed-infested ponds, stocking rates can be increased to 38–63 per ha or greater [1].

Koi

Koi have been shown to reduce the occurrence of submersed aquatic weeds and filamentous algae by keeping pond water turbid. The turbidity is caused by their activity in the pond bottom, which keeps the mud suspended in the water column and releases nutrients, supplying a food source for the phytoplankton bloom. Their "rooting-around" behavior on the pond bottom also prevents weeds from establishing there. Koi have been chosen for university demonstration projects over non-colorful common carp because of the side benefits of having an attractive addition

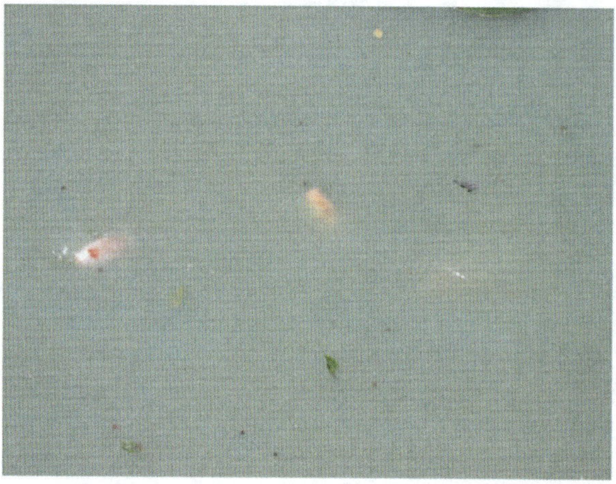

Fig. 13.11 Koi keep pond water turbid, which reduces the occurrence of nuisance aquatic weeds

Fig. 13.12 Koi added to an experimental pond at KSU "stirred up" the pond bottom suspending mud and nutrients in the water column. The nutrients supported a phytoplankton bloom, making the water column opaque, thus shading the pond bottom and preventing growth of macrophytes

to the pond and a fish that can be marketed for its ornamental value (Fig. 13.11). Figure 13.12 illustrates ponds with koi at the Kentucky State University Aquaculture Research Center [1].

Plant Pathogens and/or Insects

Dr. Raghavan Charudattan, a plant pathology researcher at the University of Florida, spent the majority of his career investigating the use of fungal plant pathogens as well as insects that prey on plants for the biological control of aquatic weeds [18, 19]. The fungi *Alternaria zonatum, A. eichhorniae,* and *Cercospora piaropi* along with weevils of *Neochetina* spp. showed promise for controlling water hyacinth even in the field [19]. More recently, Walsh et al. assessed the effectiveness of the weevil *Listronotus elongatus* and the fungus *Cercospora* spp. in controlling floating pennywort (*Hydrocotyle ranunculoides* L.f.) in Argentina [20].

Water milfoil was controlled with reduced levels of the herbicides fluridone, 2,4-dinitrophenol (2,4-D), and triclopyr when they were combined with the cellulolytic fungus *Mycoleptodiscus terrestris* [19]. And this same fungus, when combined with either fluridone, endothall, or diquat provided better control of hydrilla than the herbicides alone [19]. In Australia, Schooler et al. studied herbivory (consumption of plants) of alligator weed by a monophagous chrysomelid weed flea beetle, *Agasicles hygrophila* Selman and Vogt [21], and Telesnicki et al. examined the cytogenetic effects of this weed on the herbivorous beetle [22]. Also, in Australia, Stanley et al. examined the role of the insect *Xubida infusella* on controlling water hyacinth [23]. More recently, Gaskin et al. reviewed the use of molecular-based approaches to advance biological weed control and concluded that the collaboration of classic plant and insect taxonomists with molecular biologists is in jeopardy because of a shortage of taxonomists [24].

Control of Aquatic Plants with Herbicides

Control of aquatic weeds with herbicides is the most common means of eradicating weeds in ponds. Correct identification of weeds is critical because potential impacts and management differ for each plant. Strategies effective on one species may be ineffective even on similar species. In particular, herbicides are selective (some much more than others), and effective control depends on matching the weed with the most appropriate herbicide. Private consultants or experts at local universities can help identify the weed problem [1]. This section addresses synthetic herbicides. Natural herbicides are not within the purview of this chapter, but Flamini has done a thorough review of them [25].

In the USA, registration of chemicals for fishery use is granted by the Environmental Protection Agency (EPA) or the Food and Drug Administration (FDA) under the Federal Environment Pesticide Control Act (FEPCA) of 1972. The lack of registration does not necessarily mean that the chemical is harmful to the environment or that it is extremely toxic. Aquatic herbicide usage is considered minor by most chemical companies, and they are simply not willing to spend the large amount of money needed to compile the data necessary for registration review. However, some unregistered herbicides are toxic to fish or their use may result in chemical residues in the edible portion of the fish. For these reasons, only herbicides labeled for use in food-fish ponds should be used by pond owners, and label instructions should be followed carefully. Proper chemical usage can also minimize the effects on nontarget organisms inside and outside the pond. Skin and eye protection should be worn when working with all chemicals to prevent absorption into the body [1].

The following herbicides or herbicide groups are labeled for use in food-fish ponds. Tables 13.1 [26] and 13.2 [27–29] summarize herbicide use for common weeds in fishponds. Note that most aquatic herbicides specify on the label that application of the herbicide within 400–800 m of a potable water intake is not permitted unless the water intake can be turned off for a day or two immediately after application and substituted with an alternative (untreated) water source during this time. Imazamox is one exception, requiring no potable water intake valve to be shut off as long as the herbicide is at a concentration of less than or equal to 50 ppb at the potable water intake [30].

Bispyribac-Sodium (Tradewind®)

Bispyribac-sodium is an 80 % active ingredient powder that is mixed with water and applied to bodies of water with little or no outflow for controlling various floating (e.g., water hyacinth), submersed (e.g., hydrilla), and emergent (e.g., alligator weed, along with a nonionic surfactant) weeds (Table 13.1). It kills weeds systemically by inhibiting the action of a key plant amino-acid-synthesizing enzyme acetolactate synthase (ALS); this mode of action is very specific to plants and, therefore, there are no posttreatment restrictions against drinking the treated water or using it recreationally (e.g., fishing). Bispyribac-sodium is slow acting, controlling aquatic vegetation over a 30–60-day period. This herbicide should not be used in ponds where crustaceans, such as crayfish are being cultured [26, 31].

Carfentrazone-Ethyl (Stingray®)

Carfentrazone is a liquid contact herbicide used to treat floating weeds (e.g., duckweed) and some other weeds, including water milfoil (Table 13.1). When preparing the application (mixing with water), do not mix this herbicide with dirty/muddy water, which will reduce its effectiveness. Carfentrazone is activated by light; it inhibits the action of protoporphyrinogen oxidase (PPO), the enzyme that synthesizes chlorophyll. This process causes a buildup of phytotoxic intermediates that disrupt the weed's cell membranes, killing the weed within a few days [26, 32]. The herbicide works best on mature, actively growing weeds. Carfentrazone may not be applied to water within 400 m upstream of an active potable water intake, unless the intake can be turned off for 24 h immediately after the weed application and there is an alternative source for potable water [32].

Copper Sulfate (Various Trade Names)

Copper sulfate ($CuSO_4 \cdot 5H_2O$; copper sulfate pentahydrate) is available in various particle sizes from fine powder to large crystals. Chem One in Houston, TX, sells Fine 200 (powder); Fine 100 (size of table salt granules); Fine 20–30 (small rice grains); Small; Medium; and Large. The fine powder is more effective because it

Table 13.1 Treatment response of common aquatic plants to registered herbicides. (Reprinted from Masser et al. 2013 [26]. Aquatic weed management: herbicides. SRAC Publication No. 361. https://srac.tamu.edu/index.cfm/event/getFactSheet/whichfactsheet/66/)

Aquatic herbicide[a]	Bispyribac-sodium	Carfentrazone	Copper and copper ancomplexes, algicides	Copper complexes-herbicides	Diquat	Endothall	Flumioxazin	Fluridone	Glyphosate	Imazamox	Imazapyr	Penoxsulam	Sodium carbonate peroxyhydrate	Triclopyr	2,4-D
Aquatic group and vegetation															
Algae															
Chara/Nitella	P		E		P	G^{b}–p^{c}	P	P	P						P
Filamentous			E		G	G^{b}–p^{c}	G	P	P				G^{f}		P
Planktonic			E		P	G^{b}	F	P	P				G^{f}		P
Floating plants															
Azolla		G	P		G			E	F			E			F
Duckweeds		E	P		G	P	E	E	P		P	E			F
Salvinia	F	G	P		G		G	E	G	E		E			
Water hyacinth	E	G	P	G^{d}	E		P	E	G	E	E	E		E	E
Watermeal	F	G	P		F		E	G				G			F
Water lettuce	E	E	P	G^{d}	E		E	G	G		E	E		G	F
Submerged plants															
Coontail	P		P	G^{d}	E	E	G	E							G
Elodea			P	G^{d}	E	F	E	E				G			
Fanwort			P	P	G	F	G	E				G			F

Table 13.1 (continued)

Aquatic herbicide[a]	Bispyri-bac-sodium	Carfen-tra-zone	Copper and copper ancom-plexes, algicides	Copper com-plexes-herbi-cides	Diquat	Endo-thall	Flumiox-azin	Fluri-done	Glypho-sate	Imaza-mox	Imaza-pyr	Penox-sulam	Sodium carbon-ate peroxy-hydrate	Tricl-opyr	2,4-D
Hydrilla	E		P	G[d]	G	G	G	E		G		E			
Milfoils	G	E	P	G[d]	E	E	G	G		G		E		E	E
Naiads			P	G[d]	E	E	E	E				G			F
Parrot feather			P	P	E	E	G	E		G	G[e]	G		G	E
Pond-weeds	G		P	G[d]	G	E	G	E		E	G[e]	G			P
Emergent plants															
Alders			P		F	P		P	E		E			E	E
Alligator weed	E	F			P		G	F	G	G	E			E	F
Arrow-head	E		P		G	G	G	E	E	E	E				E
Button-brush			P		F	P		P	G		G				F
Cattails	P		P		G	P	P	F	E	E	E				F
Common reed			P		F		P	F	E	G	E				F
Frogbit	E			F[d]	E		G		F	E	E			E	E
Pickerel-weed	F			F[d]	G		P	P	F	E	E			G	G
Sedges and rushes	F		P		F		F	P	G		E[g]F[h]	G			F

Table 13.1 (continued)

Aquatic herbicide[a]	Bispyribac-sodium	Carfentrazone	Copper and copper ancomplexes, algicides	Copper complexes-herbicides	Diquat	Endothall	Flumioxazin	Fluridone	Glyphosate	Imazamox	Imazapyr	Penoxsulam	Sodium carbonate peroxyhydrate	Triclopyr	2,4-D
Slender spike rush			P		G		P	G	P		F				
Smartweed	G		P	F[d]	F		P	F	E	E	E	G		E	E
Southern water grass			P					G	E		E				P
Water lilies	F		P		P		F	E	G	G	G	G		G	E
Water pennywort	G		P		G		G	P	G		E	G		E	G
Water primrose		F	P		F	P	G	F	E	E	E			E	E
Water shield			P		P		G	G	G	G	E				E
Willows	P		P		F	P	P	P	E		E			E	E

[a] E = excellent control, G = good control, F = fair control, P = poor control, blank = unknown or no response
[b] Hydrothol formulations
[c] Aquathol formulations
[d] Specific copper complexes only, e.g., Nautique, Komeen (see label)
[e] Spray only emergent portion
[f] Best in blue-green algae (higher concentrations for green algae)
[g] E for sedge
[h] F for rush

dissolves faster. Copper sulfate is also available as an acidified solution for aquatic use and is sold under various trade names [26]. Copper sulfate should only be used to control algae, because rates necessary to kill other plants may also be toxic to fish. The filamentous algae *Pithophora* spp. are resistant to copper sulfate. Most algae are controlled more effectively if treatment with copper sulfate is made soon after plant growth has started [1] (Table 13.1).

In soft waters of low alkalinity, copper is extremely toxic to fish and it is recommended that copper sulfate not be used in waters with a total alkalinity of less than 50 parts per million (ppm) as $CaCO_3$. Conversely, copper sulfate is less effective as an algicide in hard, alkaline waters because the copper rapidly precipitates out of solution. The treatment rate increases with total alkalinity, and the formula used to calculate the treatment rate is:

$$\text{ppm copper sulfate} = (\text{ppm total alkalinity}) \div 100$$

In water with a total alkalinity greater than about 300 ppm as $CaCO_3$, copper from copper sulfate precipitates out of solution so rapidly that it is difficult to achieve an effective treatment [1].

Use of copper sulfate can lead to dangerously low oxygen concentrations, especially in the summer. Pond owners should have mechanical emergency aeration available to provide oxygen to fish in the pond when decomposing aquatic vegetation is actively removing dissolved oxygen from the water. This is especially advised during high summer temperatures when the warm water holds less oxygen.

To prepare the copper sulfate solution, mix 6 kg of the chemical for every 50 L of water (the numbers of pounds of copper sulfate to be used are dissolved in the same number of gallons of water before applying to the pond). It is best to apply copper sulfate in clear water above 16 °C and on a sunny day. It should also be noted that putting copper sulfate solution in galvanized containers causes the copper to chemically displace the galvanized lining. This removes copper from the treatment solution [1].

Chelated Copper (Cutrine®-Plus, Clearigate®, Cutrine®-Ultra, Mizzen®, K-Tea®, Algimycin®, Komeen®, Pondmaster®, Nautique®, Captain®)

These herbicides are available in both liquid and granular forms, but the liquid is most commonly used. The copper in these herbicides is bound in organic complexes so that the copper will not precipitate out of solution as rapidly as uncomplexed copper in hard, alkaline waters. Cutrine®-Plus, for example, has prolonged effectiveness because its chelating agent, ethanolamine, decomposes slowly in sunlight. Another chelated copper algicide, copper triethanolamine complex (Mizzen™), can be applied to ponds, fish hatcheries, and potable water reservoirs. No more than half of the pond should be treated at a time due to the threat of large-scale algae death resulting in oxygen loss and suffocation of fish and invertebrates. A 2-week interval between the partial treatments should be practiced. Treatment should begin along the shoreline and proceed toward the middle of the pond in increments to allow the fish to move away from the chemical. Lower doses of copper triethanolamine (0.2–0.5 mg/L) can control the blue-green algae (cyanobacteria) *Anabaena* spp., *Microcystis* spp., *Oscillatoria* spp., and *Aphanizomenon* spp.; the green algae *Spirogyra* spp., *Microspora* spp., and *Hydrodictyon* spp.; the diatom *Nitzschia* spp.; and the protists *Euglena* spp., *Glenodinium* spp., and *Cryptomonas* spp. Higher doses (0.5–1.0 mg/L) control the blue-green alga *Nostoc* spp.; the green algae *Pithophora* spp., *Chara* spp., *Chlorella* spp., *Oocystis* spp., and *Nitella* spp.; and certain hard-to-control diatoms and protists [29].

Although chelated copper herbicides usually are more effective than copper sulfate, they are considerably more expensive to use. Table 13.2 lists the aquatic weeds controlled by each of the chelated copper complexes. Copper herbicides have a reputation for effectively killing algae including phytoplankton, some filamentous algae, and *Chara* (muskgrass), which is also an alga. However, Cutrine®-Ultra [28] is specially designed to kill *Pithophora* spp. with a penetrating surfactant. Komeen®, Pondmaster®, Nautique®, and Captain®, unlike many other chelated coppers, are able to control higher aquatic plants such as *Najas,* coontail, *Elodea,* and sago pondweed (Tables 13.1 and 13.2). Additionally, chelated coppers are often combined with other herbicides such as Reward®, Aquathol®, or Sonar® to enhance their effectiveness and, in some cases, reduce the amount needed of both herbicides [1].

Diquat (Reward®, Harvester®, Tribune®, Weedplex Pro®, and Weedtrine D®)

Diquat is sold as a liquid and is a wide-spectrum contact herbicide that suppresses most filamentous algae, including *Pithophora* spp., and controls *Chara* spp.; submersed weeds, such as *Najas* spp., *Elodea,* milfoil, parrot feather, and coontail; and floating weeds, such as water hyacinth. It can be mixed with a surfactant and sprayed to control emergent weeds, such as cattail (Table 13.1). A copper-based algicide mixed with diquat may provide better weed control especially if algae are mixed with the submersed weed. Diquat should not be used in muddy water and mud should not be stirred up during application because diquat will bind tightly with clay particles suspended in the water rendering the herbicide ineffective at controlling plants growing beneath the surface. Diquat should be applied on a sunny day to actively growing weeds. Only one-third to one-half of the pond water area should be treated at one time with a 14-day interval between treatments. A 14-day withdrawal period is required by law after diquat use before treated water can be used for animal consumption, swimming, spraying, irrigation, or drinking [1, 33].

Table 13.2 Copper complex herbicides and the plants they control

Copper-based herbicides	Trade name[a]	Effective against
Copper sulfate pentahydrate	Various names	Planktonic algae Filamentous algae except for *Pithophora* sp. *Chara* sp. (musk grass)
Mixed copper–ethanolamine complexes	Cutrine®-Plus [27] Clearigate®	Planktonic algae Filamentous algae Chara/Nitella Hydrilla
Mixed copper–ethanolamine complexes in an emulsified formulation Contains an emulsified surfactant/penetrant for highly effective control of coarse (thick cell-walled) filamentous algae, *Pithophora* sp.	Cutrine®-Ultra [28]	Planktonic algae Filamentous algae including *Pithophora* sp. Chara/Nitella Hydrilla Egeria
Copper–triethanolamine complex and copper hydroxide	Mizzen® [29] K-Tea®	Green algae Blue-green algae Diatoms Flagellated protozoa
Copper citrate and copper gluconate	Algimycin®	Planktonic algae Filamentous algae except for *Pithophora* sp.
Copper–ethylenediamine complex and copper sulfate pentahydrate	Komeen® Pondmaster®	Hydrilla Water hyacinth Egeria Elodea Najas Coontail Water milfoil Sago pondweed American pondweed Water lettuce
Copper carbonate	Nautique® Captain®	Controls all of the above plants controlled by Komeen® and also controls: Curlyleaf pondweed Horned pondweed Thin Leaf pondweed Vallisneria Widgeon grass

[a] Does not imply endorsement of the product

Endothall, Dipotassium Salt (Aquathol K®, Aquathol Super K®)

The dipotassium salt of endothall is available in liquid (Aquathol K®) [34] or granular (Aquathol Super K®) [35] forms. It will not kill algae but will control a wide variety of submersed higher plants, including *Najas* spp., coontail, pondweeds (e.g., sago pondweed), and milfoil (Table 13.1). The granular formulation is relatively expensive but is particularly effective on *Najas* spp. Dipotassium salt of endothall is a contact killer. It is sprayed onto or injected below the water surface and can be sprayed at high concentrations directly on exposed weeds. For the best results, water temperatures should be 18 °C or warmer. When water temperatures are high and an increased danger of dissolved oxygen depletion exists, this herbicide should be applied to one-third to one-half of the pond per treatment with a 5–7-day interval between treatment applications.

Water treated with granular Aquathol Super K® must not be used for irrigation or for agricultural sprays on food crops or for domestic purposes within 7 days of treatment. More detailed restrictions exist for Aquathol® K. It may not be used for the aforementioned purposes as well as for watering livestock for 7 days after applying it up to 0.5 ppm; for 14 days after application up to 4.25 ppm; and for 25 days after application up to 5.0 ppm. In addition, water treated with Aquathol® K may not be used for swimming until 24 h after treatment [1, 34].

Endothall, Alkylamine Salt (Hydrothol® 191)

The alkylamine salt of endothall (Hydrothol® 191) is most commonly used in the liquid formulation. It is a more potent herbicide than the potassium salt (Aquathol® K) and will control most filamentous algae, including *Pithophora* spp. and *Chara* spp. (Table 13.1). Repeated treatments are recommended if algae growth reappears. Hydrothol® 191 is a relatively toxic herbicide to fish, and treatment rates required to treat submersed higher plants are generally too risky in commercial catfish ponds to justify its use. When the herbicide is used to treat filamentous algae, only a portion of the pond should be treated at one time. Fish avoid the treated area and are usually not killed. Hydrothol® 191 treatments as high as 0.3 ppm (often needed to kill *Pithophora* spp.) can be used, but higher rates will kill fish [1]. Water treated with Hydrothol®191 should not be used for watering livestock, preparing agricultural sprays for food crops, irrigation, or domestic purposes within 7 days after application, when up to 0.3-ppm Hydrothol® 191 is used [36].

Flumioxazin (Clipper® Herbicide)

Flumioxazin is an N-phenylphthalimide herbicide that disrupts cell membranes in the target plant and inhibits the activity of the PPO enzyme that is responsible for chlorophyll production in the plant's chloroplasts. As a result, the plant turns yellow and dies within a few days to 2 weeks after application. Flumioxazin is a granular contact herbicide that degrades quickly in water, especially at a pH greater than 8.5. It should be applied directly onto the targeted weed in the morning when pH tends to be lower. Flumioxazin's activity can persist in water for up to a full day at a neutral pH (7–8) but may last less than half an hour at pH greater than 9 [37].

Clipper has been available commercially only since 2011, and as of 2013, it is highly regarded among professionals who routinely treat aquatic weeds. It has a reputation of quickly controlling filamentous algae (*Pithophora* sp.), duckweed, and watermeal, as well as the submersed weeds coontail, hydrilla, southern naiad, sago pondweed, curlyleaf pondweed, and parrot feather. However, the Clipper label cautions that it is classified as a group 14 herbicide and therefore must not be used repeatedly or in successive years in the same body of water because weed species with acquired resistance to it could eventually dominate that body of water. The herbicide label recommends treating only half the pond at a time to avoid a whole-pond weed die-off, which could lead to low dissolved oxygen. Flumioxazin should not be used in crayfish farming [38].

Fluridone (Sonar®, Alligare Fluridone, Avast! SC)

Fluridone is available as an aqueous suspension or pellets. Fluridone will not kill phytoplankton or filamentous algae but controls a broad spectrum of submersed higher plants, including *Najas* sp., *Egeria, Elodea,* milfoil, hydrilla, and most pondweeds in the genus *Potamogeton*. It also controls the floating plant duckweed (*Lemna* spp.; Table 13.1). This herbicide is slow acting, and results may take 30–90 days to be noticeable. The slow-killing action (and thus slow weed decomposition) helps to prevent low dissolved oxygen that typically results from rapidly decomposing weeds. Fluridone should be applied to actively growing weeds, and the entire pond surface should be treated at once. Partial or spot treatments result in dilution of the herbicide with the untreated water. Various Sonar® formulations can be used depending on the outcome desired. The formulation used mostly by weed control consultants and contractors primarily for large bodies of water is Sonar® A.S., a concentrated fluridone solution (41.7 % a.i.; also the same a.i. for a competing brand Alligare Fluridone). Formulations used by individuals for private ponds include Sonar® RTU ("ready to use" in a 3.8 % a.i. solution) that can be applied to a pond three times (days 1, 21, and 42) to treat duckweed and several submersed weeds like naiads, pondweeds, milfoil, hydrilla, coontail, and water lily at a rate of 18.7 L/ha (as with all herbicides, check the label for specific directions in order to get maximum performance and to comply with the law). Another Sonar formulation SonarOne®, a granular herbicide (5 % a.i.) formulated for individual pond owners, may be a bit easier to use and combines the effects of Sonar® Q ("quick"-acting), Sonar® PR (precision release), and Sonar® SRP (slow-release pellets) into one herbicide. And yet another one Sonar® Genesis, an emulsified liquid (5 % a.i.), used by individual pond owners not only treats submersed vegetation but is designed also to work quickly on the floating plants, duckweed, and watermeal. Treated water should not be used for crop irrigation for 30 days after application [1, 39, 40].

2,4-D (Aquacide®, Aqua-Kleen®, Weed Rhap® A-4D, Weedtrine® D, Navigate®, Weedar® 64, etc.)

This herbicide is formulated for aquatic use as the dimethylamine salt or isooctyl ester. It is available in liquid or granular forms. The liquid formulations of 2,4-D are most effective on emergents (e.g., arrowhead, smartweed, water lilies, water primrose, and willows) and water hyacinth in spring when weeds start to grow [41]. The granular form is effective at controlling submersed higher plants, such as milfoils and parrot feather (Table 13.1). Acidic pH (6 and below) enhances its herbicidal activity, while a pH of 8 or above tends to make it less effective. Treating early in the morning when pH is usually lowest will increase the effectiveness of 2,4-D [1].

Glyphosate (Rodeo®, Aquamaster®, AquaPRO®, AquaNeat®, Refuge®, Eraser AQ®, etc.)

Glyphosate is sold as liquid and is for use mostly on emergent and shoreline plants. The herbicide is mixed with a nonionic surfactant and sprayed on the vegetation. Glyphosate is a broad-spectrum herbicide and is useful for the control of cattails, grasses, smartweed, and willows around pond margins (Table 13.1). Application when weeds are in the flowering or fruiting stage is more effective than earlier application. Visible results (wilting and yellowing) are usually not seen for 2–7 days after application. Rainfall occurring within 6 h of application reduces the effectiveness of Rodeo® [1, 42]. A study in South Africa examined the use of glyphosate to retard the vegetative growth of water hyacinth without killing the plant in order to sustain populations of the weevil beetles *Neochetina eichhorniae* and *N. eochetina bruchi* that keep the hyacinth population in check [43].

Imazamox (Clearcast)

Imazamox is a liquid (diluted or undiluted) herbicide used to treat submersed, emergent, or floating plants. It is effective against sago and curlyleaf pondweeds, water milfoil, hydrilla, water hyacinth; as well as the emergents cattails, smartweed, and water primrose; and the floating plant water lily. Surfactants are usually added to Clearcast when treating emergent and floating plants. Check the herbicide label for proper treatment concentrations for each weed [30]. Imazamox is a systemic herbicide that kills plants by inhibiting the function of an essential amino-acid-producing enzyme ALS. It can also be used as a preemergent or postemergent herbicide.

Imazapyr (Habitat®, AquaPier®)

Imazapyr is a liquid herbicide that is mixed with water and a surfactant or vegetable oil and sprayed on emergent or floating aquatic weeds (Table 13.1). When sprayed directly onto emergent leaves, the herbicide is translocated throughout the weed, concentrating in the roots where it causes the weed to die (which sometimes takes more than 2 weeks) and prevents future regrowth. Imazapyr is most effective if applied when the weed is actively growing. The effectiveness of imazapyr is reduced if it rains within an hour of application. The label notes that "Habitat® does not control plants which are completely submerged or have a majority of their foliage under water" [44].

Due to the risk of oxygen depletion from decomposing weeds, no more than half the pond's surface area should be treated at one time, and at least 10–14 days should separate the treatments. As with all herbicides, the most current herbicide directions (found in the leaflet label attached to the container) should override any other treatment advice, including that found in this book [1].

Imazapyr is relatively environmentally safe. Treated waters have no restrictions for recreation including swimming and fishing or for livestock consumption. However, imazapyr may not be applied to water within 0.8 km upstream of an active potable water intake (unless the intake can be turned off for 48 h). Also, water treated by imazapyr may not be used for irrigation for 120 days afterwards.

Triclopyr (Renovate 3®, Renovate OTF®, Garlon 3A®)

Renovate 3® is a liquid systemic herbicide used to control certain emergent, submersed, and floating aquatic plants (including alligator weed, milfoil, water hyacinth, water lily, and water primrose; Table 13.1) in bodies of water that have little or no continuous outflow. Mixing triclopyr with a nonionic surfactant is recommended to improve its effectiveness.

Renovate OTF® (ontTarget flakes) [45] settles directly on submersed weeds and, therefore, is effective at lower concentrations than liquid Renovate 3® [46] (60 % less a.i. can be used when applying Renovate OTF®, making it more environmentally friendly).

Triclopyr-treated water should not be used for irrigation for 120 days unless the triclopyr is not detectable by laboratory analysis. The most current herbicide directions (found in the leaflet label attached to the container) should override any other treatment advice, including that found in this book [1].

Penoxsulam (Galleon SC®)

Galleon SC® was approved for use as an aquatic herbicide in 2007. It controls duckweed, water hyacinth, *Cabomba, Egeria, Elodea,* Eurasian water milfoil, hydrilla, and sago pondweed [47] (Table 13.1). As a sidenote, Riis et al. performed some basic research in New Zealand on the effects of temperature, light availability, and water depth on the growth of *Egeria* and *Elodea* [48].

Galleon SC® partially controls weeds in the following list. Depending on dosage, time of year, growth stage, application method, and degree of water movement, the previously listed weeds can be controlled while leaving the following native species unharmed (or higher treatment rates can be used to control weeds in both lists). Partially controlled weeds include watermeal, alligator weed, arrowhead, parrot feather, smartweed, pondweed, *Najas,* and spike rush.

A surfactant must be added when emergent weeds are being controlled. As with Sonar®, Galleon SC® is slow acting and should be applied to the entire pond surface at once (partial or spot treatments result in dilution of the herbicide with the untreated water). Galleon SC® also can be applied as a preemergent in empty (dewatered) ponds. Check a current label to get the exact treatment rate for each type of weed.

Sodium Carbonate Peroxyhydrate (GreenCleanPRO®, Phycomycin®, Pak 27®)

Sodium carbonate peroxyhydrate (percarbonate) is a granular algicide/fungicide used to treat, control, and prevent a broad spectrum of algae and fungi (Table 13.1). For the most effective treatment, use percarbonate when algae growth first appears, and treat early in the day when sunny with little or no wind. Floating algae mats should be broken up before or during treatment. After treatment, dead algae can be removed from the water surface to prevent excessive nutrients from entering back into the water (during decomposition) and stimulating subsequent heavy phytoplankton blooms. The BioSafe Systems' technical bulletin points out that GreenCleanPRO® has no restrictions for use after it is used as a treatment and it is labeled for use in aquaculture [49]. Planktonic blue-green algae blooms are treated with 33–111 kg per ha-m (to convert kg per ha-m to the Imperial lb per ac-ft measurement, divide by 3.69) of water; the exact amount needed depends on the quantity of algae growth, light intensity, and water quality. The most current directions (found on the container) should override any other treatment advice, including that found in this book.

Carefully Follow Herbicide Labels

Herbicides sold in the USA must be registered with federal and state regulatory agencies. The printed information accompanying the herbicide container is called the "label" and constitutes a legal document. Failure to use herbicides according to label instructions can lead to severe penalties. From a practical standpoint, misuse of herbicides can result in poor weed control; risks to people, fish, or wildlife; or herbicide residue problems in fish.

The label provides information on the active ingredient (Table 13.3 [27, 28, 30–32, 34–36, 38, 40, 42, 44–47, 49, 50]), directions for correct use on target plant species, warnings and use restrictions, and safety and antidote information. Remember, state and local regulations may be more restrictive than federal regulations. Certain products are registered as "Restricted Use" herbicides and can be legally applied only by trained and certified applicators or by people under their direct supervision. Be sure to check federal, state, and local regulations prior to using herbicides.

Herbicide treatment rates are based on pond area or pond volume. Miscalculation will result in either overtreatment or undertreatment (which may require additional treatments to eradicate the weed). In either case, more chemical than needed will be applied to the pond. Carefully measure pond dimensions and keep up-to-date records of pond size and depth. Pond depth tends to decrease over time because of erosion of embankments and sedimentation of pond bottoms. The only way to be certain of average pond depth is to measure water depth before treatment at several dozen random locations.

Handle Herbicides Safely

Although aquatic herbicides are relatively safe to handle, it is nevertheless important for applicators to keep chemical exposure to an absolute minimum. Herbicide labels and material safety data sheets advise what protective clothing and equipment (e.g., a respirator) should be worn, any precautions the handler should follow, a statement of practical treatment in case of poisoning, statements concerning hazards to the environment, any physical or chemical hazards, and directions on proper storage and disposal. By law, copies of labels and any supplementary labels must be in the possession of the applicator at the application site for each herbicide used. Anyone who handles a pesticide must read and understand all label statements prior to using the product. Herbicide safety is reviewed in Southern Regional Aquaculture Center (SRAC) publication 3601 [51]. Advice in this SRAC fact sheet includes storing the volatile herbicide 2,4-D (particularly the ester form) separate from other chemicals.

Table 13.3 Aquatic herbicide generic and trade names

Product	Common trade names[a]
Bispyribac-sodium	Tradewind® [31]
Carfentrazone-ethyl	Stingray® [32]
Copper sulfate	Various trade names
Copper complexes (chelated copper)	Cutrine®-Plus [27], Aquatrine®, Clearigate®, Cutrine®-Ultra [28], K-Tea®, Algimycin®, Komeen®, Pondmaster®, Captain®, Nautique®
Diquat	Reward®, Harvester®, Tribune™, Tsunami DQ®, Diquat SPC 2 L, WeedPlex Pro, Weedtrine D®
Endothall, alkylamine salt	Hydrothol® 191 [36]
Endothall, dipotassium salt	Aquathol® K [34], Aquathol® Super K [35]
Flumioxazin	Clipper® [38]
Fluridone	Sonar® Genesis, Sonar® RTU, Sonar® A.S., Sonar® PR, Sonar® SRP, Sonar® Q, SonarOne®, Avast®, Avast SRP®, Alligare Fluridone [40]
Glyphosate	Rodeo® [42], Aquamaster®, AquaPRO®, AquaNeat®, Eraser AQ®, Refuge™, Eagre®, Glypro®, Aquastar®
Imazamox	Clearcast® [30]
Imazapyr	Habitat® [44], Arsenal®, Polaris® [50], AquaPier®
Penoxsulam	Galleon® [47]
Sodium carbonate peroxyhydrate	GreenClean®, GreenCleanPRO® [49], Pak 27®, Phycomycin®
Triclopyr	Renovate 3® [46], Renovate OTF® [45], Navitrol® DPF, Ecotriclopyr 3 SL, Garlon 3A®
2, 4-D	Navigate®, WeedRhap®, Weedar 64®, Aqua-Kleen®
Surfactant	Many brands including Cide Kick®, Magnify®, Kinetic®, Superb® HC, Cygnet Plus, Competitor®, Dyne-Amic®, R-11®, Kinetic® HV, and CLASS ACT® NG®
Dyes	Aquashade®, Aquashadow®, Admiral Liquid®, Admiral WSP®

[a] Does not imply endorsement of the product

Dispose of Herbicide Containers Properly

Improper disposal of herbicide containers can cause contamination of soil and water, and may result in fines or loss of an applicator's license. Empty herbicide containers must be triple rinsed, with each rinsing drained into the herbicide mix tank. If no mix tank is used, the rinse water from the container should be applied to the pond in the same manner as the herbicide in the container. Containers must then be punctured or crushed so that they cannot be reused. Empty bags must be rinsed or shaken clean and cut so that they cannot be used for other purposes. Laws regarding disposal of rinsed containers vary among states, so be sure to follow all state and local regulations regarding pesticide container disposal.

Two aquatic herbicides, 2,4-D and endothall, are regulated as hazardous materials under the federal law, and any waste generated during their use must be disposed of as hazardous waste. Triple-rinsed containers can be disposed of as with any other pesticide container. Any rinse water from cleaning of containers or application equipment must be applied as if it were the herbicide or disposed of at a hazardous waste disposal facility.

Consequences of Herbicide Use

When used according to the manufacturer's specifications, herbicides are seldom directly toxic to fish. However, the addition of any herbicide to a plant-infested body of water will alter water quality. Oxygen production by photosynthesis will be decreased and decomposition of the dead plant material will increase oxygen consumption. The result will be a noticeable decrease in dissolved oxygen concentrations compared to pretreatment levels. The extent to which dissolved oxygen levels are reduced depends on the amount of plant material killed, the amount of plant material unaffected by the herbicide, the rate at which death occurs, water temperature, and other factors. Decomposition of the dead plants will also raise carbon dioxide and total ammonia concentrations. The increase in total ammonia concentrations tends to decrease the pH, causing much of the ammonia to be in the nontoxic, ionized form. Phosphorus, potassium, and other minerals are also released upon plant decomposition, and concentrations of all essential plant nutrients will usually be higher after herbicide treatment. At some time after treatment, the concentration of herbicide will decrease to a nontoxic level and these nutrients will be available for new plant growth.

The deterioration in water quality following herbicide use can have serious consequences in fishponds. Obviously, if dissolved oxygen concentrations fall to very low levels, fish will be killed. Even if dissolved oxygen concentrations are maintained above lethal levels, the fish may be severely stressed and more susceptible to diseases. Stressed fish also feed poorly and decreased fish growth can be expected, particularly if water quality is affected for an extended length of time.

Control of Phytoplankton Abundance

Low concentrations of dissolved oxygen and development of off-flavor are very important water quality problems in aquaculture. Both problems are the result of uncontrolled phytoplankton growth in heavily fed ponds. Numerous efforts have been made to manage phytoplankton communities in fishponds, but most methods are ineffective and many actually further degrade water quality.

A variety of algicides have been used to reduce phytoplankton density, but the ultimate results are always undesirable. When sufficient algicide is added to a pond with a dense bloom, the sudden die-off usually causes severe oxygen depletion and high levels of carbon dioxide and ammonia. Phytoplankton repopulate the pond as soon as algicide levels decrease because nutrient levels remain high. Episodes of poor water quality resulting from this cycle of death and regrowth will stress fish and cause reduced growth or increased susceptibility to infectious diseases. Similar problems occur when algicides are used in attempts to eliminate specific noxious phytoplankton species. All of the effective algicides registered for use in food-fish ponds are broad spectrum in activity and cannot be used to selectively eliminate one species or one type of phytoplankton.

Biological control of phytoplankton growth is an alternative to the use of herbicides. Most efforts have involved the use of plankton-feeding fish, such as silver carp, bighead carp, or tilapia. In theory, the plankton-feeding fish continually harvests the bloom, improves water quality, and provides additional fish production. However, most attempts at biological control of phytoplankton growth have failed. Quite often, phytoplankton abundance increases when plankton-feeding fish are present because these fish effectively remove large phytoplankton and zooplankton, which compete with or consume small phytoplankton. The presence of plankton-feeding fish may thus change the structure of the plankton community but usually will not decrease overall phytoplankton density.

Decreasing nutrient levels by limiting daily feed allotments is the only reliable method available for reducing, on average, the incidence of phytoplankton-related water quality problems. Such problems are rare if maximum daily feeding rates are less than about 50 kg/ha, but this feeding rate is uneconomical in most commercial aquaculture enterprises. Keeping feeding rates below 100 kg/ha at least reduces the severity of phytoplankton overpopulation problems.

Nutrient-Reducing Pond Additives

Nutrient Reducers (Phoscontrol® and Sparklear®)

The company sponsoring Mizzen™ also sells PhosControl®, which it claims binds the nutrient phosphorus in ponds, thus preventing its use by plants and algae. The same company also sells SparKlear® as a product containing bacteria, enzymes, and trace minerals that reduce nutrients in ponds that would otherwise be used by plants and algae [52].

Note The mention of any product (trade) name does not imply endorsement of the product.

References

1. Durborow R, Tucker C, Gomelsky B, Onders R, Mims S (2008) Aquatic weed control in ponds. Kentucky State University Land Grant Program, 24 p
2. Elangovan R, Philip L, Chandraraj K (2008) Biosorption of chromium species by aquatic weeds: kinetics and mechanism studies. J Hazard Mater 152(1):100–112
3. Wang XS, Tang YP, Tao SR (2009) Kinetics, equilibrium and thermodynamic study on removal of Cr (VI) from aqueous solutions using low-cost adsorbent Alligator weed. Chem Eng J 148:217–225
4. Rahman MA, Hasegawa H (2011) Aquatic arsenic: phytoremediation using floating macrophytes. Chemosphere 83(5):633–646
5. Robinson B, Kim N, Marchetti M, Moni C, Schroeter L, van den Dijssel C, Milne G, Clothier B (2006) Arsenic hyperaccumulation by aquatic macrophytes in the Taupo volcanic zone, New Zealand. Environ Exp Bot 58:206–215
6. Mochochoko T, Oluwatobi S, Jumbam D, Songca S (2013) Green synthesis of silver nanoparticles using cellulose extracted from an aquatic weed; water hyacinth. Carbohydrate Polymers 98(1):290–294
7. Sundari M, Ramesh A (2012) Isolation and characterization of cellulose nanofibers from the aquatic weed water hyacinth—Eichhornia crassipes. Carbohydr Polym 87(2):1701–1705
8. Mukherjee AK, Kalita P, Unni BG, Wann SB, Saikia D, Mukhopadhyay PK (2010) Fatty acid composition of four potential aquatic weeds and their possible use as fish-feed neutraceuticals. Food Chem 123:1252–1254
9. Igwani E, Gumbo T, Gondo T (2010) The general information about the impact of water hyacinth on Aba Samuel Dam, Addis Ababa, Ethiopia: implications for ecohydrologists. Ecohydrol Hydrobiol 10(2–4):341–345
10. Zuo S, Mei H, Ye L, Wang J, Ma S (2012) Effects of water quality characteristics on the algicidal property of Alternanthera philoxeroides (Mart.) Griseb. in an aquatic ecosystem. Biochem Syst Ecol 43:93–100
11. Hussner, A (2012) Alien aquatic plant species in European countries. Weed Res 52:297–306
12. Wersal RM, Madsen JD (2010) Comparative effects of water level variations on growth characteristics of *Myriophyllum aquaticum*. Weed Res 51:386–393
13. Stiers I, Njambuya J, Triest L (2011) Competitive abilities of invasive *Lagarosiphon major* and native *Ceratophyllum demersum* in monocultures and mixed cultures in relation to experimental sediment dredging. Aquat Bot 95:161–166
14. Evans JM, Wilkie AC (2010) Life cycle assessment of nutrient remediation and bioenergy production potential from harvest of hydrilla (*Hydrilla verticillata*). J Environ Manage 91(12):2626–2631
15. Doyle RD, Smart RM (2001) Effects of drawdowns and desiccation on tubers of hydrilla, an exotic weed. Weed Sci 49(1):135–140
16. Frei M, Khan M, Razzak M, Hossain M, Dewan S, Becker K (2007) Effects of a mixed culture of common carp, *Cyprinus carpio* L., and Nile tilapia, *Oreochromis niloticus* (L.), on terrestrial arthropod population, benthic fauna, and weed biomass in rice fields in Bangladesh. Biol Control 41:207–213
17. Morin L, Reid A, Sims-Chilton N, Buckley Y, Dhileepan K, Hastwell G, Nordblom T, Raghu S (2009) Review of approaches to evaluate the effectiveness of weed biological control agents. Biol Control 51:1–15
18. Rosskopf EN, DeValerio JT, Elliott MS, Shabana YM, Ables CBY (2010). Influence and legacy of Raghavan Charudattan in biological control of weeds. Weed Technol 24:182–184
19. Shearer JF (2010) A historical perspective of pathogen biological control of aquatic plants. Weed Technol 24:202–207
20. Walsh G, Maestro M, Dalto YM, Shaw R, Seier M, Cortat G, Djeddour D (2013) Persistence of floating pennywort patches (*Hydrocotyle ranunculoides*, Araliaceae) in a canal in its native temperate range: effect of its natural enemies. Aquat Bot 110:78–83

21. Schooler S, Baron Z, Julien M (2006) Effect of simulated and actual herbivory on alligator weed, Alternanthera philoxeroides, growth and reproduction. Biol Control 36:74–79
22. Telesnicki MC, Sosa AJ, Greizerstein E, Julien MH (2011) Cytogenetic effect of *Alternanthera philoxeroides* (alligator weed) on *Agasicles hygrophila* (Coleoptera: Chrysomelidae) in its native range. Biol Control 57:138–142
23. Stanley JN, Julien MH, Center TD (2007) Performance and impact of the biological control agent *Xubida infusella* (Lepidoptera; Pyralidae) on the target weed *Eichhornia crassipes* (waterhyacinth) and on a non-target plant, *Pontederia cordata* (pickerelweed) in two nutrient regimes. Biol Control 40:298–305
24. Gaskin, J, Bon M, Cock M, Cristofaro M, De Biase A, De Clerck-Floate R, Ellison C, Hinz H, Hufbauer R, Julien M, Sforza R (2011) Applying molecular-based approaches to classical biological control of weeds. Biol Control 58:1–21
25. Flamini G (2012) Chapter 13—Natural herbicides as a safer and more environmentally friendly approach to weed control: a review of the literature since 2000. In: Studies in natural products chemistry, vol 38. pp 353–396
26. Masser MP, Murphy TR, Shelton JL (2013) Aquatic weed management: herbicides. SRAC Publication no. 361. https://srac.tamu.edu/index.cfm/event/getFactSheet/whichfactsheet/66/. Accessed 1 June 2013
27. Label for aquatic herbicide Cutrine®-Plus. http://kenspondandlake.com/Cutrine-Plus-Label.pdf. Accessed 16 Sept 2013
28. Label for aquatic herbicide Cutrine®-Ultra. http://www.lakesmanagement.com/files/PDF/products/CutrineUltra/CutrineUltraLabel.pdf. Accessed 16 Sept 2013
29. Label for aquatic herbicide Mizzen™. https://lakerestoration.com/Images/84868-1%20Mizzen%20Label%201-25-13.pdf. Accessed 16 Sept 2013
30. Label for aquatic herbicide Clearcast®. http://www.sepro.com/documents/Clearcast_Label.pdf. Accessed 16 Sept 2013
31. Label for aquatic herbicide Tradewind®. http://www.valent.com/Data/Labels/2011-TRA-0001%20Tradewind%20%20-%20form%201796-A.pdf. Accessed 16 Sept 2013
32. Label for aquatic herbicide Stingray®. http://www.helenachemical.com/specialty/Labels/Stingray_Aquatic_Herbicide12_04-C%20Label.pdf. Accessed 16 Sept 2013
33. Label for aquatic herbicide Reward®. http://www.syngentacropprotection-us.com/pdf/labels/SCP1091AL2C0605.pdf. Accessed 16 Sept 2013
34. Label for aquatic herbicide Aquathol® K. http://kenspondandlake.com/Aquathol-K-Label.pdf. Accessed 16 Sept 2013
35. Label for aquatic herbicide Aquathol® Super K. http://www.cygnetenterprises.com/Product.aspx?id=834&pid=0&mid=&cid=126. Accessed 16 Sept 2013
36. Label for aquatic herbicide Hydrothol® 191. http://kenspondandlake.com/Hydrothol-191-Label.pdf. Accessed 16 Sept 2013
37. University of Florida (2013) At the following Web site: http://plants.ifas.ufl.edu/manage/control-methods/details-about-the-aquatic-herbicides-used-in-florida. Accessed 16 Sept 2013
38. Label for aquatic herbicide Clipper® Herbicide. http://www.valent.com/Data/Labels/2012-CLP-0001%20Clipper%20-%20form%201791-B.pdf. Accessed 16 Sept 2013
39. Label for aquatic herbicide Avast! SC. http://www.sepro.com/documents/Avast!_Label.pdf. Accessed 16 Sept 2013
40. Label for aquatic herbicide Alligare Fluridone. http://www.midwestaquacare.com/product-labels/Alligare_Fluridone_Label.pdf. Accessed 16 Sept 2013
41. Label for aquatic herbicide Weed Rhap® A-4D. http://www.kellysolutions.com/erenewals/documentsubmit/KellyData%5CND%5Cpesticide%5CProduct%20Label%5C5905%5C5905-501%5C5905-501_WEED_RHAP_A_4D_3_10_2009_6_08_10_PM.pdf. Accessed 16 Sept 2013
42. Label for aquatic herbicide Rodeo®. http://www.aces.edu/dept/fisheries/rec_fishing/documents/rodeo_label.pdf. Accessed 16 Sept 2013

43. Jadhav A, Hill M, Byrne M (2008) Identification of a retardant dose of glyphosate with potential for integrated control of water hyacinth, *Eichhornia crassipes* (Mart.) Solms-Laubach. Biol Control 47:154–158.
44. Label for aquatic herbicide Habitat®. http://www.sepro.com/documents/Habitat_Label.pdf. Accessed 16 Sept 2013
45. Label for aquatic herbicide Renovate OTF®. http://www.sepro.com/documents/renovateotf_label.pdf. Accessed 16 Sept 2013
46. Label for aquatic herbicide Renovate 3®. http://www.sepro.com/documents/renovate_label.pdf. Accessed 16 Sept 2013
47. Label for aquatic herbicide Galleon SC®. http://www.sepro.com/documents/galleon_label.pdf. Accessed 16 Sept 2013
48. Riis T, Olesen B, Clayton J, Lambertini C, Brix H, Sorrell B (2012) Growth and morphology in relation to temperature and light availability during the establishment of three invasive aquatic plant species. Aquat Bot 102:56–64.
49. BioSafe Systems. http://www.biosafesystems.com/Product-PWT-GCPRO.asp. Accessed 16 Sept 2013
50. Label for aquatic herbicide Polaris®. http://www.cdms.net/LDat/ld8KR002.pdf. Accessed 16 Sept 2013
51. Avery JL. Aquatic Weed Management Herbicide Safety, Technology and Application Techniques. Southern Regional Aquaculture Center (SRAC) Publication 3601. https://srac.tamu.edu/index.cfm/event/getFactSheet/whichfactsheet/161/. Accessed 16 Sept 2013
52. http://www.lakerestoration.com/blog/. Accessed 16 Sept 2013

Chapter 14
Weed Management for Parasitic Weeds

Radi Aly and Neeraj Kumar Dubey

Introduction

Autotrophic plants are key factors in stabilizing the ecosystem of the Earth. During the course of evolution, these plants have evolved different interactions with various biotic and abiotic factors in their ecological niche. Among the biotic factors, different pathogens such as fungi, bacteria, and insects as well as some parasitic flowering plants influence the lifecycle of plants. Being a sessile organism, these plants cannot escape to incoming parasitic invaders. Heterotrophic or semi-autotrophic plants evolved themselves to use their host plants for their survival. Parasitism is a coexistence of two different organisms, of which one (the parasite) lives at the expense of the other (host). Parasitic plants are notorious pests for agricultural crops and cause serious yield loss. Recently, more than 20 families of parasitic plants have been recognized as serious pests, causing considerable economic damage. Parasitic weeds are among the most destructive weeds known [1]. These parasites adopt different forms to invade host plants. Some (dodders and mistletoes) invade aerial parts, while others (*Orobanche* and *Striga*) invade the underground roots [2]. Furthermore, they are widely varied in their degree of host dependence. Some parasitic plants are partially photosynthetic and have the ability to survive without a host, but are able to take advantage of an available host to augment their nutrition (facultative parasites, i.e., *Triphysaria* spp.). Other parasites have an absolute host requirement, but retain some photosynthetic capacity (obligate hemiparasites, i.e., *Striga* and *Alectra* spp., mistletoes, and some *Cuscuta* spp.). In the final category are parasites that lack any photosynthetic capacity (some have lost much of their chloroplast ge-

R. Aly (✉)
Department of Plant Pathology and Weed Research, Newe-Yaar Research Center,
P. O. Box 1021, Ramat Yishai, 30095 Haifa, Israel
e-mail: radi@volcani.agri.gov.il

N. K. Dubey
Unit of Weed Science, Newe-Yaar Research Center, P. O. Box 1021,
Ramat Yishai, 30095 Haifa, Israel
e-mail: neerajd19@yahoo.co.in

B. S. Chauhan, G. Mahajan (eds.), *Recent Advances in Weed Management*,
DOI 10.1007/978-1-4939-1019-9_14,

nomes) [3], and are completely reliant on the host for all nutritional needs. This last category (obligate holoparasites) represents the most extreme example of parasitism (*Orobanche, Phelipanche,* and some *Cuscuta* spp.).

Seven *Orobanche*/*Phelipanche* species were identified as threatening 16 million ha in Mediterranean and West Asia regions in a 1991 survey, and the problem has only gotten much worse [4]. The impact of *Orobanche* on food legumes is particularly significant because of the critical dietary role of these crops in human and animal nutrition and because of the soil-enhancing properties of symbiotic nitrogen-fixing rhizobia. *Vicia faba* (fava bean) is among the most important crops, occupying 21,000 ha with an estimated production of 55,000 tons (t) of seeds [5]. *Orobanche* infestation in fava crops in Morocco was first reported in 1994 in the Fez region; since then it has spread and reached the other regions in Morocco, threatening legume production in the Middle East. The biggest threat to fava is *O. crenata,* which can reduce yields from 20 to 100%, depending on the severity of infestation [6]. Heavy infestations of *O. crenata* in Egypt forced a 29% reduction in the cropping area of fava bean between 1968 and 1978, resulting in the upper Nile region becoming a net importer of fava bean [7]. *Orobanche* is also problematic in important vegetables, such as tomato, potato, carrot, and oilseed crops (e.g., sunflower and *Brassica*). Yield losses may be an imperfect measure because the typical farmer response to heavily infested fields is abandonment of the field. Another notoriously devastating parasitic weed in the same family is *Striga,* an obligate hemiparasite infesting more than 50 million ha of arable farmland in sub-Saharan Africa. *Striga* can completely destroy productivity of infested sorghum, maize, and cowpea fields, negatively affecting the food security of millions of Africans [8, 9]. Parasitic weeds, such as *Orobanche* and *Striga,* are difficult to control because they are closely associated with the host root, are concealed underground for most of their lifecycle, and they have the ability to produce a tremendous number of seeds that may remain viable in the soil for more than 15 years.

Biology and Development of Parasitic Weeds

The parasitic plant genera *Orobanche* and *Phelipanche* (Orobanchaceae) together consist of more than 100 species. The most harmful species, commonly referred to as broomrapes, are *Phelipanche aegyptiaca* (Pers.) Pomel, *Phelipanche ramose* (L.) Pomel, *Orobanche cumana* Wallr., *Orobanche minor* Sm., *Orobanche cernua* Loefl., and *Orobanche crenata* Forsk. They attack many dicotyledonous crops, including members of the Solanaceae, Fabaceae, Compositae, Brassicaceae, and Umbelliferae [1].

All parasitic plants (Orobanchaceae) directly invade and rob host plants via haustoria, multifunctional organs that attach the parasite to the host, physically penetrate host tissues, and provide a physiological bridge through which resources are translocated between the host and parasite [10]. The haustoria of parasitic plants directly connect them to the vascular system of the host plants [11, 12]. By developing a strong metabolic sink relative to the host, they channelize the flow of water and nutrients from the host to itself, thereby damaging crop development and reducing crop yield [13]. The haustorial connections of Orobanchaceae (*Orobanche* spp., *Phelipanche* spp., and *Striga*) are similar to graft junctions and thus the parasite has the ability to import and export molecules and macromolecules directly from and to the host plants. Increased knowledge concerning mobility and function of molecules trafficking between parasites and their hosts is expected to assist for an effective control strategy.

Broomrapes

Broomrapes (*Phelipanche/Orobanche* spp.) are a genus of more than 100 species, but only five (Table 14.1) [14–16] of them are economically significant pests [1, 17]. The plant of broomrape is small (10–60 cm tall, depending on species) and recognized by its yellow- to straw-colored stems, bearing yellow, white, or blue, snapdragon-like flowers. The leaves are merely triangular scales and both stem and leaf show absence of chlorophylls. The flower produces thousands of extremely small (0.15–0.5 mm long) tan-to-brown colored seeds, which blacken with age and can survive more than 15 years in a crop field. It shows a very complex type of lifecycle (Fig. 14.1). Since parasites are concealed underground most of their life cycle, it is very hard to detect these parasites on host plants until the flowering stage of parasites. The seed of the parasite remains dormant in the soil for many years and their germination is stimulated by certain compounds exuded by the host plant. Broomrapes have evolved sophisticated systems for detecting the presence of host plants and coordinating their development with the hosts [2, 18, 19]. The early stages of development are critical to parasite survival because an emerging seedling that fails to connect to a host will exhaust its energy reserves and die. Some broomrape seeds compensate for this by having strict protocols for germination and contact with the host. First, there is a period of preconditioning. Second, there is the requirement of specific root exudates (strigolactones) produced by host plants. Strigolactones are signaling molecules exuded by host plants and these molecules help in the recognition of host root by germination tubes of parasitic plants [20]. Cissoko et al. reported that the ratio of parasitic attack is proportional to strigolactones exudation [21]. The same class of molecules is also involved in the establishment of the symbiosis of plants with arbuscular mycorrhizal (AM) fungi. These exudates are secondary metabolites, and generally produced in low quantities by the hosts (and some nonhosts). Once these two steps are fulfilled, the germinating parasite produces a radicle that must contact a host root and establish a connection. The third step requires a haustorium initiation factor, which causes the radicle tip to redifferentiate into a haustorium that penetrates the host root [22]. The haustorium is the feature that separates parasitic from nonparasitic plants [10]. This organ forms the physical and physiological connection between parasite and host and its interaction with host tissues is important for translocation of molecules and macromolecules [23].

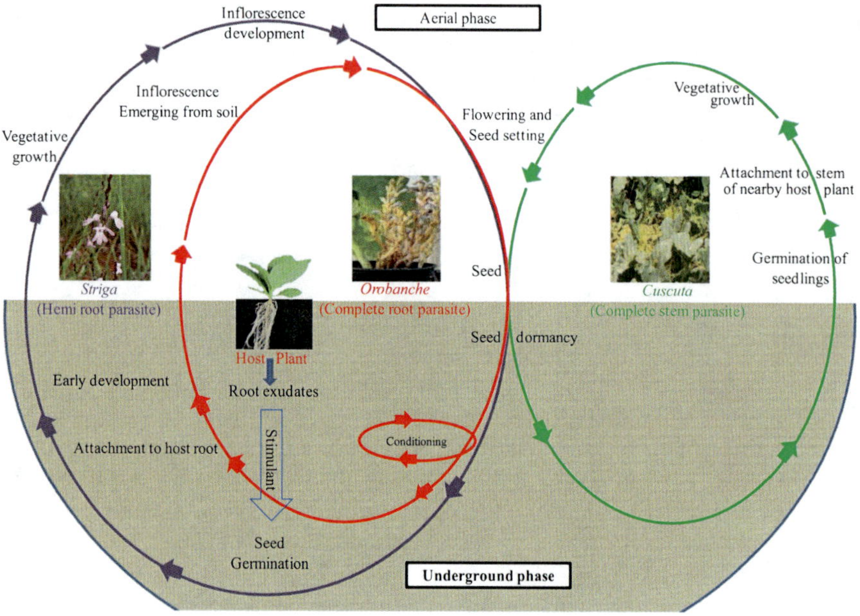

Fig. 14.1 Schematic representation of life cycle of *Orobanche*, *Striga*, and *Cuscuta*

The haustorium initially adheres to the host root by a secreted, mucilaginous substance and then penetrates by pushing between host cells [24]. Penetration is aided by digestive enzymes secreted by the parasite that include pectin methylesterase, polygalacturonase, and endocellulase [25–27]. Further, peroxidases produced by parasitic tubercles loosen the cell wall of the host in order to facilitate their penetration [28]. Following successful attachment, penetration, and vascular connection with the host, the broomrape tissue adjacent to the host root grows into a bulbous structure called a tubercle. Then a short, root-like organ, which in some cases is capable of forming secondary attachments to neighboring host roots, can emerge [29]. After approximately 4 weeks of growth, a floral meristem develops, which emerges aboveground to flower and disseminate seeds. A single plant can produce thousands of seeds, which can remain viable in the soil for more than 15 years [1].

Striga

Another devastating parasitic weed in the Orobanchaceae family is *Striga*, an obligate hemiparasite infesting more than 50 million ha of arable farmland in sub-Saharan Africa. *Striga*, also known as witchweed, grows naturally in parts of Africa, Asia, and Australia and it is considered to be the most devastating parasite in grain production in Africa (Table 14.1) [14–16]. Three species are considered serious weeds: *S. asiatica* and *S. hermonthica* in cereals and *S. gesnerioides* in legumes

Table 14.1 Geographical distribution of important parasitic weeds and their hosts. (Compiled from [14–16])

Parasite group	Important species	Host plants	Distribution	Yield loss (%)
Orobanche	*O. crenata* Forsk.	Papilionaceae, Umbelliferae	Mediterranean basin, S. Europe, the Middle East, Eastern Europe, and Australia	5–85
	P. aegyptiaca Pers.	Asteraceae, Fabaceae, and Solanaceae		
	P. ramosa L.	Solanaceae		
	O. cumana Wallr.	Solanaceae		
	O. cernua Loefl.	Compositae, Solanaceae, Asteraceae		
	O. minor SM.	*Trifolium pratense L*	USA	
	O. foetida Poir.	Legumes	N. Africa	
Striga	*S. hermonthica*	Cereals, maize, millet	Africa, Asia	30–90
	S. asiatica			
	S. gesnerioides	Cowpea, tobacco	Australia, USA	
Cuscuta	*C. campestris*	Vegetables, fruits, ornamentals, legumes	Worldwide	50–75
		Alfalfa, clover	Europe, Asia,	
	C. epithymum		N. America	
		Woody perennials, alfalfa	Asia	
	C. planiflora			
		Alfalfa	Asia	
	C. reflexa	Alfalfa	N. and S. America	
	C. indecora			

There are other parasitic higher plant species, but they cause relatively little economic damage

(Table 14.1) [14–16]. The genus *Striga* includes 11 species that parasitize crops. Economically important *Striga* species are reported from more than 50 countries, especially from East and West Africa and Asia. Crop losses due to *Striga* are currently estimated to be more than US$ 7 billion annually. *Striga* is considered as the greatest biotic constraint to food production in Africa, where the livelihood of 300 million people is significantly affected. In infested areas, yield losses associated with *Striga* damage are often significant, ranging from 30 to 90% [15] (Table 14.1). Due to infestation with *Striga* spp., corn yield dropped in sub-Saharan Africa from the world average of 4.2 to 1.3 t/ha [30]. Many control methods have been suggested for control of *Striga,* but the success has been limited.

Striga can completely destroy the productivity of infested sorghum, maize, and cowpea, negatively affecting the food security of millions of Africans [8, 9]. Witchweeds are characterized as obligate hemiparasites of roots and require a living host for germination and initial development, though they can then survive on their own.

The attacked host plant showed symptoms, like stunting, wilting, and chlorosis, similar to severe drought damage, nutrient deficiency, and vascular disease. Each *Striga* plant can produce up to 500,000 seeds, which may remain viable in the soil for more than 10 years. The exudates of host root contain strigolactones, signaling molecules that stimulate the *Striga* seed germination [31]. After successful parasitic attachment to host by haustoria, it grows 4–7 weeks underground, after which it emerges rapidly and produces flowers and seeds (Fig. 14.1). The dispersal of seed is mainly with human influence such as cloth and machinery tools used for agricultural practice.

Cuscuta

Cuscuta spp., commonly known as dodder, are important weeds in Europe, the Middle East, Africa, and North and South America [1]. *Cuscuta* are obligate parasitic plants with approximately 170 different species throughout the world [32]. All species of the genus *Cuscuta* are obligate parasites that attack stems and leaves of a wide variety of host species, including forage crops and vegetables (Table 14.1) [14–16], some tree crops (grapevine and coffee), and ornamentals [33]. Estimates of forage crop losses range from 20 to 57 %, and sugar-beet yields are reduced by 3.5–4 t/ha [34].

Cuscuta can be identified by its thin stem appearing leafless, with reduced leaves into minute scales. The flower color can be white, pink, yellow, and cream. The seeds are minute, produced in large numbers, have a hard coat, and can survive in the soil for 5–10 years. Dodder seeds sprout near the soil surface with or without host and then are attracted toward the nearby host plants by chemosensory mechanism. The exact mechanism of detection of a host plant is not known but some organic volatile compounds help in sensing the host plant for dodder germinated seedlings [35]. If it fails to find a suitable host within 5–10 days after germination, the seedling can die. Seedlings can survive 5–10 days on reserve food present in the seeds. After successful attachment to a host plant, it wraps itself to the host plant and produces haustoria to extract the nutrition (Fig. 14.1). The original root of the parasitic plant dies and now it becomes totally dependent on the host plant. In tropical areas, it can grow up to the canopy of a tree, but in temperate areas it dies with annual plants. It has the ability to transmit some viral diseases.

Management

Conventional Approaches to Control Parasitic Weeds

There are several control methods for parasitic weeds [1, 36–39]. A range of parasitic weed management practices have been developed (Table 14.2) [34] that can be broadly classified under the general themes of cultural (crop rotation, trap and catch

Table 14.2 Control options for the major parasitic weeds of global significance. (Compiled from [31, 34])

Technique	Parasite		
	Striga	*Orobanche*	*Cuscuta*
Preventive			
National quarantine	+	+	+
International quarantine	+	+	+
Cultural			
Crop rotation	+	+	+
Planting date	+	+	+
Mineral fertilizer	+	–	–
Flooding	+	+	–
Organic material	+	+	–
Managed fallow	+	+	+
Physical			
Cleaning of crop seed	–	–	+
Hand weeding	+	+	+
Burning	–	–	+
Deep plowing	+	+	–
Soil solarization	+	+	+
Chemical			
Fumigation	+	+	+
Germination compounds	+	+	–
Herbicides	+	+	+
Biological			
Insects	–	–	–
Fungi	–	–	–
Integrated control	+	+	+
Host resistance/tolerance	+	+	–

+ Effective, – Ineffective

crop, fallowing, hand pulling, fertilization, and time of planting), physical (solarization), biological (fungi and insects), chemical (herbicides, artificial stimulation of seed germination using ethylene, strigol, etc.), and host plant resistance (using resistant or tolerant crop varieties).

Cultural, Mechanical, and Physical Methods

Cultural, mechanical, and physical practices have been developed for parasitic weed control (Table 14.2) [34], crop rotation (trap and catch crops), fallowing, transplanting, hand pulling, nitrogen fertilization, time and method of planting, intercropping, and solarization [1, 34, 37, 38, 40]. The effectiveness of these methods is limited due to numerous factors in particular the complexity of the parasite life cycle, which reproduce by tiny seeds, and are difficult to diagnose until they irreversibly damage the crop. The main obstacle in the long term management of broomrape infested fields is the durable seed bank, which may remain viable for decades, and gives rise to only a very low annual germination percentage. The intimate connection between host and parasite also hinders efficient control by cultural methods [1]. It is also important to prevent the distribution of parasite seeds from infested to clean areas.

Crop Rotation and Trap and Catch Crops for Striga

Rotating susceptible cereal crops with crops that are nonhosts to *Striga,* particularly false hosts ("trap crops" that stimulate *Striga* to germinate but are not themselves parasitized), has been adapted to reduce *Striga* soil seedbank. A wide range of rotations can be effective in reducing *Striga* numbers and increasing yields in the subsequent cereal crop [41–46]. However, catch (parasite-susceptible) cropping is rarely used by small farmers to control *Striga* because the technique is not well known and should be adapted to a specific cropping system [47]. Trap and catch crops only have useful effects where the parasite soil infestation level is minimal. Crop rotations can be effective also in reducing *Cuscuta* infestation by growing cereals or other grass crops (false hosts for most dodder species) continuously for several years [48].

Transplanting

Following seedling establishment, crop plants are transferred to the field as a larger host plant might be able to resist the parasite better. The method is simple and requires a low skill level for implementation, and it can be performed by subsistence farmers and their families. In some areas of Africa and Asia, transplanting is a traditional practice [49]. Because of high labor requirement, transplanting maize under rainfed conditions is probably suitable for small areas (0.1 ha) highly infested with *Striga*; however, yield under transplanted crops can be more than doubled [50]. In a previous study, "underground" development of *Striga hermonthica* on established sorghum plants was low compared with directly sown sorghum [51].

Intercropping

Intercropping cereals with legumes and other crops is a common practice in most areas of Africa, and has been reportedly to reduce *Striga* infestation [37]. Intercropping maize with cowpea and sweet potato can significantly reduce the emergence of *Striga* in Kenya [52].

Seedbank Removal

The main obstacle in the long-term management of *Orobanche*- or *Striga*-infested fields is the seedbank, which may remain viable for decades [38]. Exploitation of summer sunlight to achieve high temperatures (55 °C) under clear polyethylene mulch by covering the soil for several weeks [53] is an approach to achieve destruction of the parasitic weed soil seedbank. Soil solarization was successfully applied in the Middle East in tomato, eggplant, faba beans, lentil, and carrot [54–56].

Hand Pulling

This method is effective in parasite removal, especially in fields with a relatively low infestation. However, in the case of *Striga* it is less effective because much of the damage to the host occurs while the parasite is still underground. In Kenya, some farmers have effectively controlled the *Striga* problem in their fields using the hand-pulling method. Removing mature *Striga* plants from an infested field will reduce the amount of seeds, but will not increase the host yield in the short term. For *Orobanche*, hand pulling is very effective since less damage is caused by the parasite underground [36]. Although removal of infested branches is useful, the best possible control measure for mistletoe (*Phoradendron macrophyllum*) is to replace severely infested trees with less susceptible species [57]. Removal of dodder by hand remains a viable approach when a small patch is infested, but is expensive if infestation is extensive [33]. In India, hand pulling has been recommended for controlling *O. cernua* in tobacco. In some crops, pulling may seriously disturb the crop root system as found for eggplant (*Solanum melongena*) [58].

Cleaning of Crop Seeds and Burning of Infected Crops

Removal of the minute seeds of *Striga* and *Orobanche* (*Alectra*) from crop seeds is normally not feasible, but where absolutely necessary, risks could be greatly reduced by thorough washing. It is somewhat easier in case of rough-surfaced dodder seeds and reduces the germination of parasitic seeds along with crop seeds. The burning method is only suitable with the *Cuscuta* control and the crop may also be damaged by this treatment. After the seed setting in *Cuscuta,* it has to be burned at the site where it grew to avoid any spread of viable seeds, and to kill seeds that have already dropped onto the soil surface.

Chemical Approaches

Fumigation

Fumigation with methyl bromide effectively controls *Orobanche* seeds in the soil [59, 60]. Unfortunately, methyl bromide use is being phased out by the international agreement to protect the global environment [61]. Other fumigants were tested as possible substitutes for methyl bromide, but are much less effective and more expensive [61–63]. All fumigants are expensive, labor intensive, and extremely environmentally hazardous.

Synthetic and Natural Germination Stimulants/Inhibitors

Striga and *Orobanche* species only germinate in the presence of a germination stimulant exuded by the roots of a potential host [1]. Host signals, such as strigolactones, that induce germination and haustorium formation of both *Orobanche* and *Striga* were found in the root exudates of various plant species [64]. Strigolactone was identified in the root exudates of sorghum, while strigol was identified in the root exudates of cotton (*Gossypium hirsutum* L.) and in the root exudates of a variety of other plants [65]. The application of germination stimulants to induce suicidal seed germination of parasitic weeds appears attractive for biosafety reasons, rapid soil decomposition, and high biological activity at very low application rates [50]. Compounds structurally related to strigolactone are potent synthetic germination stimulants for many *Striga* and *Orobanche* species [66, 67].

Ethylene induces germination of *Striga* seeds [68], and has been found a promising technology in the *S. asiatica* eradication program in the USA. In East Africa, however, reduced effectiveness of ethylene was observed for the suicidal germination of *S. hermonthica* [69]. It was suggested that ethylene can only stimulate suicidal germination of nondormant *Striga* seeds, as there may be significant dormancy in *S. hermonthica* [70].

Understanding the biology of the host–parasite interaction has been effectively applied to improve resistance to *Striga* and *Orobanche*. In sorghum, natural low *Striga* germination stimulant activity halts parasite development at an early stage. The low germination stimulant trait is independently inherited from the incompatible response [71]. Maize mutants that are tagged in several root-expressed terpene synthase genes have been currently studied for alterations in *Striga* germination stimulant production [19].

The herbicides fluridone [1-methyl-3-phenyl-5-(3-trifluoromethyl-phenyl)-4-(1-H)-pyridinone] and norflurazon [4-chloro-5-methylamino-2-(3-trifluoromethylphenyl) pyridazi n-3-one], inhibitors of carotenoid biosynthesis, were shown to induce gibberellic acid-like effects on the conditioning and germination of *O. minor*. Accordingly, soil application of carotenoid-biosynthesis inhibitors could potentially be used as a control method [72], which promotes seed conditioning of *O. minor* and other root parasites, and enhances the activity of germination stimulants to induce more effective suicidal germination. Further, strigolactones are the product of carotenoid pathway [31] and application of carotenoid inhibitor causes reduced attachment of *S. hermonthica* with rice plant [73].

Recently, it was reported that certain natural amino acids cause severe physiological disorders of germinating broomrape seeds [74]. In particular, methionine was able to inhibit almost total germination of *O. ramosa* seeds when applied at a concentration of 2 mM [74]. When methionine was applied to tomato roots, it strongly reduced the number of developing parasite tubercles. These findings suggest that appropriate amino acids applied exogenously to a root zone might result in the control of parasitic plants such as *Orobanche* spp.

Herbicides

In the recent decades, some chemicals have become available for parasitic weed control [75], although few herbicides are able to selectively control parasitic weeds [15, 63]. The chemical approach to control parasitic weeds poses some difficulties, such as lack of application technology, chemical damage to the host, continuous parasite seed germination throughout the season, marginal crop selectivity, environmental pollution, and low persistence. Crop damage by chemicals and availability of effective molecules are other major constraints that limit the successful usage of herbicides for parasitic weed control. Additionally, in developing countries, the income of subsistence farmers is usually too low to afford herbicides.

Two concepts are considered for chemical control of *Orobanche:* foliar herbicide application [76, 77] and soil herbicide application. Sulfonylurea herbicides effectively control imbibed and germinated seeds or young attachments of *Orobanche,* when applied directly to the soil in tomato, potato, and sunflower fields [78–81]. For broomrape control, glyphosate, imidazolinones, and sulfonylureas are being used currently. Some of these herbicides' chemistries showed a degree of selectivity, benefiting broomrape host plants [82–84]. Glyphosate is an inhibitor of enzyme 5-enolpyruvyl-shikimate-3-phosphate synthase (EPSP), which is a key enzyme in the biosynthesis of aromatic amino acids. Imidazolinones and sulfonylureas are inhibitors of acetolactate synthase (ALS or AHAS), a key enzyme in the biosynthesis of branched-chain amino acids. One ALS inhibitor, sulfosulfuron, has recently been recommended in Israel for the control of broomrape in tomato fields. However, this herbicide provides only partial control and its residues may damage subsequent crops. Transgenic carrots, resistant to the herbicide imazapyr, allow movement of unmetabolized herbicide through the crop to control the attached *P. aegyptiaca* [85]. Seed dressing with herbicides, using an ALS inhibitor, has also been used for *Striga* control in maize [86]. This technique involves the development of biodegradable formulations for seed dressing used together with small amounts of herbicide for parasite control. Slavov et al. transformed several tobacco cultivars with a mutant *AHAS3R* gene for resistance to the herbicide chlorsulfuron [87]. The herbicide was sprayed on plant leaves and translocated through the whole plant to the root system, killing the attached broomrape.

Dicamba and 2, 4-D are the most widely used herbicides against *Striga*. Dicamba is a systemic herbicide applied to the crop foliage about 35 days after crop emergence, whereas 2, 4-D is sprayed several times directly on the parasites during the growing season. Many other chemicals have been tested on *Striga* and some provided good parasite control. However, because of the cost and the technology needed, none of these chemicals are accessible to small-scale subsistence farmers in Africa [36].

Chemical control of *Cuscuta* is complicated as dodder species vary in susceptibility to herbicide treatment. Once *Cuscuta* attaches to crop plants, some yield loss will occur, regardless of the method of control [33]. Preemergence herbicides (soil- applied before crop emergence), such as the benzamide herbicide pronamide, have been used to prevent *Cuscuta* attachment to alfalfa, sugar beet, cranberry [88], and onion [89].

The plant growth regulator ethephon may be used to control mistletoe in dormant host trees. Spraying provides only temporary control, especially on well-established infestations, by causing some of the mistletoe plants to drop. The mistletoe will soon regrow at the same point, requiring re-treatment [57]. The growth regulator and herbicides like ethephon, 2, 4-D, and glyphosate did not provide resistance against leafy mistletoe, *Phoradendron tomentosum* (DC.) [90]. Watson et al. showed that naphthaleneacetic acid (NAA) and paint significantly inhibit the regrowth of removed mistletoe compared with control [90]. In Hungary, European mistletoe (*Viscum album* L.) causes infection of more than 3000 ha [91]. Varga et al. studied the effect of glyphosate isopropylamine salt, 2,4-dichlorophenoxyacetic acid, methsulfuron-methyl alone, and in combination with each other and reported that 2,4-dichlorophenoxyacetic acid is effective even at less concentration and is less damaging to the host tree while controlling the European mistletoe [91].

Biocontrol Agents for Parasitic Weed Control

This technique utilizes living organisms (insects, fungi, etc.) to suppress or reduce parasitic weeds. Pathogenicity towards nontarget plants is a major constraint; therefore, it is very important that host specificity and risk assessment should be made before the release of a control organism into the environment. Considerable attention and effort have been made in biological control, but until recently, the control of *Orobanche* in the field using insects or fungi as biocontrol agents failed. The results indicate that the biocontrol agents in most cases do not provide the level of control desired by farmers. Many insects (e.g., *Eulocastra argentisparsa* Hampson, *Smicronyx* spp., *Ophiomyia strigalis* Spencer, and *Phytomyza orobanchia* Kalt) have been collected on *Striga* and *Orobanche* in India and Africa, but most are not specific for these parasitic plant species [92–94]. The fly *Phytomyza orobanchia* Kalt is reported to be host-specific on *Orobanche,* but the distribution of its population is limited due to antagonists and deep plowing. Recently, fungal isolates were reported to be promising biocontrol agents for the control of *Orobanche* and *Striga*. Approximately 30 fungal genera are reported to occur on *Orobanche* spp. *Fusarium* isolates were most prominently associated with diseased *Orobanche* and *Striga* [38]. The Fusarium FOO (*F. oxysporum f. sp. orthoceras*) isolate exclusively attacks *O. cumana,* and susceptible biotypes of *O. aegyptiaca* [95, 96]. On the other hand, other *Fusarium* isolates (FOXY and FARTH) attack *O. aegyptiaca, O. cernua,* and *O. ramosa* [97]. The *Fusarium* isolate FOXY 2 significantly reduced the emergence of *S. hermonthica* and *S. asiatica,* whereas disease symptoms could only be observed on *S. hermonthica* [98]. *Fusarium oxysporum* Schlect (isolate PSM 197) could be also a potential mycoherbicide for controlling *Striga* spp. [99].

Novel approaches were recently developed to increase control by fungi, i.e., by a "multiple-pathogen strategy." In this strategy, two or more pathogens are combined and applied before or after parasite emergence. Some applied fungal mixtures caused a significant reduction of the number of emerging *O. cumana* [100]. Amsellem et al. [97] and Cohen et al. [101] observed reduction in *O. aegyptiaca* attached to tomato in greenhouse experiments using host-specific strains of *F. oxysporum* and *F. arthrosporioides*. Combined treatment of the herbicide Benzothiadiazole with the pathogen *F. oxysporum* f. sp. *orthoceras* successfully controlled *O. cumana,* and reduced parasite emergence up to 100% [102].

Other techniques such as formulation or encapsulation of fungal propagules in a solid matrix to prevent rapid desiccation or microbial competition have been developed [103, 104]. A successful example of granular formulation called "Pesta" showed high efficacy in controlling *S. hermonthica* and *O. cumana* in the greenhouse [105–107]. Another approach is the engineering of hypervirulence genes into weed-specific pathogens; e.g., genes that encode enzymes and the enzymes degrade metabolites involved in parasite defense mechanisms such as phytoalexines, or coding for enhanced virulence by the production of fungal toxins [15, 108].

Host Plant Resistance

The best long-term strategy for limiting damage by parasitic weeds is the development of resistant varieties [109, 110], but conventional breeding has yielded few varieties with stable resistance [111]. Significant progress has been made in developing screening methodologies for the identification of better sources of parasitic weed host resistance [40, 112]. Three *Striga*-resistant sorghum cultivars were officially released for wide cultivation in *Striga*-endemic regions of Ethiopia in 1999–2002 [50]. With *Orobanche,* the outstanding example has been the development of sunflower varieties resistant to *O. cernua* and *O. cumana.* Unfortunately, this resistance has often been overcome by new virulent "races" of *Orobanche* in many countries in the Mediterranean region, Eastern Europe, and the former Soviet Union [113]. Two cultivars of faba bean with a good level of resistance to *O. crenata* have been released in Middle and Upper Egypt [50]. Promising sources of resistance have been identified in wild *Pisum* species [114], which have hybridized with cultivated peas [115, 116]. Some resistance to *Cuscuta* spp. has been observed among sensitive crops. Tomato plants resistant or tolerant to *C. reflexa* have been reported [117]. Sensitivity to the highly virulent *C. pentagona* varied considerably in commercial tomato varieties [16, 118].

Do We Need Biotechnological Approaches for Parasitic Weed Control?

Parasitic weeds are not controlled effectively by traditional cultural or herbicidal weed control strategies and the best control method (fumigation) was phased out due to its expensiveness and hazard to the environment [61]. The control of *Striga* has been the aim of many research programs, but success has been limited. A few varieties of cereals have an inherent tolerance to the parasite [119]. However, the main concern with *Striga* is not just how many crop species it infects, but its potential to widen its host range. The development of herbicide-resistant crops has recently offered another *Orobanche* control approach, based on herbicide translocation through the host to the parasite [18, 120]. However, this approach depends on commercial availability of herbicide-resistant crops, requires correct chemical application, and may be countered by the development of herbicide-resistant populations of the parasite [121].

Effective means to control *Phelipanche* and *Orobanche* are scarce [34]. The best long-term strategy for controlling parasitic weeds would be through the identification and breeding of resistant genotypes, but despite many years of work by plant breeders, resistant cultivars of only a small handful of crops are available [109, 110]. A limited number of genes conferring resistance against *Phelipanche* have been identified and bred into sunflower and legumes [111, 122]. However, most genetic resistances have been overcome by new races of *Phelipanche* [111, 123]. Little work has been done on identifying *Cuscuta*-resistant varieties. Herbicides are of little use with parasitic mistletoes, and few host species show significant resistance useful in a breeding program. Despite many years of hard work by plant breeders, resistant cultivars of most crops are not available.

Genomic research on root parasites is likely to help in an overall understanding of some key aspects of parasitism. Model plants such as *Arabidopsis thaliana* and *Medicago truncatula* have been used for studies on host reactions to parasitic plant infection [11, 124] and, along with the model parasite *Triphysaria versicolor* [2], serve as valuable sources of genomic understanding of host–parasite interaction. Development of effective genetic engineering strategies for resistance to parasitic weeds requires identification of (1) genes whose products are selectively toxic and inhibit parasite growth and (2) promoter sequences that optimize expression of such toxins.

It is obvious that most recent advances in understanding of the host–parasite interaction have been best documented for *Orobanche* spp., and consequently will offer opportunities for using this approach to enhance resistance against other parasitic weeds.

New Biotechnological Approaches to Parasitic Weed Control

Transgenic Resistance

In the debate on the use of transgenic (genetically modified, GM) crops, it is clear that the direct application of these technologies to improve the efficiency of food production for small-scale farmers in developing countries [125] would be of the greatest moral value, and the least open to reproach. *Striga* is a devastating problem for sub-Saharan farmers. Most approaches adapted for weed control such as competitive plants expressing allelochemicals, plants with improved or modified mineral nutrition, and plants expressing herbicide-resistance genes [108] are not necessarily effective for parasitic weed control due to the parasite lifestyle discussed previously. Characterization of parasite-resistant crops suggests that the parasite life cycle can be interrupted at several critical stages. Unfortunately, the resistance mechanism(s) are still unclear [126, 127]. Investigation of the molecular regulation of the host-defense response to parasitic weed attack will enhance understanding of the interaction between host and parasite and provide tools necessary for engineering novel resistance against parasitic weeds [128].

Developing herbicide-resistant crops is not the only way for parasitic weed control. It is expected that there will be resistance genes to parasitic weeds as there are resistance genes to pathogens [108]. The gene *NRSA*-1 is homologous to a disease resistance gene expressed in roots of nonhost plants following parasitism by *S. asiatica*. It could possibly be a candidate gene for parasitic weed control [129].

In spite of a wide variety of approaches that have been aimed to control parasitic weeds during the last century, it is still difficult to eradicate parasitic weeds. Therefore, a thorough understanding of the host–parasite interaction is needed to develop novel control methods.

Regulation of the Trafficking Molecule Between the Host and Parasite at Haustorium Junction

The haustoria, formed at the junctions of parasite and host, open the way for translocation of a variety of molecules and macromolecules from the host to the parasite. At the same time, however, the haustoria also open opportunities for the development of methods to control parasitic plants. Improved understanding of the molecular exchange between host plants and their parasites is expected to lead to the development of state-of-the-art, effective approaches to parasitic weed management. Increased knowledge concerning mobility and function of molecules trafficking between parasites and their hosts can be expected to help plan an effective control strategy. Molecular translocation between host and parasite ranges from the movement of radiolabeled sugar [130], herbicides [18, 120, 131], plant viruses [132, 133], silencing signal (siRNA) [134, 135], and messenger MRNA (mRNA) transcripts [136] to the movement of proteins [23, 137, 138]. Translocation of the fluorescent dyes, Texas Red (TR) and 5, 6-carboxyfluorescein (CF), demonstrates the existence of a continuous connection between xylem and phloem of the host and parasite [139]. Some plant viruses may affect viability of the parasite seeds and thus could be used as a tool to control the parasite. By discovering new transportable parasitic weedicidal molecules from biodiversity as well as identification of unique genes involved in parasitic weed development and their silencing by using current knowledge of molecular exchange between host plants and their parasites can provide innovative methods to control these weeds.

Orobanche Control Based on Inducible Expression of Cecropin in Transgenic Plants

A novel strategy was designed to enhance host resistance to *Orobanche* based on parasite-induced expression of a selective sarcotoxin IA (one of three cecropin-type

Table 14.3 Transgenic plants generated against parasitic weeds. (Compiled and modified from [34])

Plant species	Target parasite	Mode of resistance	Reference
Maize, tobacco	*Orobanche spp.*	Acetolactate synthase (ALS) target site	Joel et al. [18]
Maize	*Striga hermonthica*	ALS target site	Berner et al. [140]
Tobacco, potato	*Orobanche spp.*	Asulam target site	Surov et al. [120]
Carrot	*P. aegyptiaca*	Imazapyr with mutated ALS	Aviv et al. [85]
Tobacco	*P. ramose*	Chlorsulfuron with mutated AHAS3R gene	Slavov et al. [87]
Tomato[a]	*P. aegyptiaca*	Toxic peptide	Aly et al. [138]
Tobacco[a]	*P. aegyptiaca* *P. ramosa*	Toxic peptide	Hamamouch et al. [137]
Tomato[a]	*P. aegyptiaca*	Silencing M6PR parasite gene	Aly et al. [135]
Maize[a]	*Striga hermonthica* (Delile) Benth	Silencing a parasite gene	de Framond et al. [141]
Lettuce[a]	*Triphysaria versicolor* Fisch	Silencing a reporter gene (GUS)	Tomilov et al. [134]
Tomato	*Cuscuta* spp.	Arabinogalactan protein	Albert M. et al. [142]
Tobacco[a]	*Cuscuta pentagona*	Silencing of *SHOOT MERISTEMLESS*-like gene	Alakonya et al. [143]

Examples here have been engineered with a gene giving resistance to a herbicide, to enable herbicide treatment of a parasitic weed

[a] The transgenic plants were engineered to be toxic to the parasite or to produce siRNA to silence a parasite target gene

proteins encoded by the sarcotoxin I gene) polypeptide (Table 14.3) [18, 34, 85, 87, 120, 134, 135, 137, 138, 140–143]. Sarcotoxin is an antimicrobial polypeptide of the cecropin family, produced by the flesh fly *Sarcophaga peregrine* [144]. The primary target of the cecropin family is the disruption of microbial membranes. Initial studies indicated that sarcotoxin IA peptide overproduced by yeast (*Saccharomyces cerevisiae*) inhibited *O. aegyptiaca* seed germination and radicle elongation [145]. Based on this study, Aly et al. showed for the first time enhanced resistance to *O. aegyptiaca* when the sarcotoxin IA gene was linked to the constitutive root-specific *Tob* promoter to generate sarcotoxin-expressing tomato plants [138]. However, this transgenic resistance to *Orobanche* was incomplete and did not provide adequate protection. Therefore, it was proposed to increase the efficacy of sarcotoxin-producing plants by regulating its expression with the *HMG2* promoter. The *HMG2* gene is involved in the isoprenoid biosynthesis pathway and is activated specifically during defense responses [146]. The *HMG2* promoter is specifically induced in host roots around the site of *Orobanche* penetration [128] and encodes a protein [12] associated with phytoalexin and sesquiterpene production.

Transgenic tobacco plants harboring the sarcotoxin IA gene, under the regulation of the *HMG2* promoter showed enhanced resistance to *Orobanche* resulting in higher numbers of aborted parasitization events, reduced *Orobanche* biomass, and greater host biomass following parasite inoculation compared to non-transgenic controls [137]. Protein stability may be the most critical limiting factor, because sarcotoxin IA is subject to rapid degradation in plants from extracellular proteases [147]. More research is needed to understand the mechanism of sarcotoxin IA selectivity toward *Orobanche,* and optimize this mechanism for engineering parasite-resistant crop species. Considering the importance of parasitic weeds to world agriculture and the difficulty in obtaining resistance by conventional methods, the developing strategy is superior to other methods in that it is effective, has low cost of implementation, and is environmentally safe. More than 1,000 lines of tobacco (and related species and mutants) have been screened for resistance to *O. aegyptiaca, O. ramosa,* and *O. cernua* over the past 30 years with little success [148]. In this context, the resistance to *Orobanche* of the sarcotoxin IA-expressing plants is remarkable because it was conferred by the addition of just a single gene [137]. However, the current technology will make this approach easier and more fruitful through identification of other *Orobanche*-responsive gene promoters and further toxic genes for *Orobanche* that could be useful in engineering-induced resistance.

Gene Silencing

Gene silencing is one of the most important biological discoveries of the last decade [149]. The introduction of double-stranded RNA (dsRNA) is a powerful tool for suppressing gene expression [150] through a process known as RNA interference (RNAi) in animals and posttranscriptional gene silencing (PTGS) in plants [151, 152]. The main element in the silencing process is a small RNA molecule, the short interfering RNA (siRNA) [153]. Gene silencing provides plants with a defense against various intercellular pathogens, and is a tool of immense importance for research on plant development [154]. Gene silencing by RNA is characterized by intercellular transfer of the silencing agent, and by long-distance systemic transport through the whole organism [155]. The gene silencing approach has already been demonstrated as an effective control method against nematodes [156] and viruses [157]. This approach also could be adapted for parasitic weed control. Mannitol content in the parasite *Phelipanche aegyptiaca* is regulated by the *M6PR* gene [158]. Recent research has shown that the key gene (M6PR) in *Phelipanche* spp. could be transmissibly silenced, thereby potentially providing the host plant with resistance [135]. The expression of M6PR-siRNA was detected in three independent transgenic tomato lines in the R1 generation, but was not detected in the parasite. qRT-PCR result showed that the expression of M6PR mRNA in the tubercles was suppressed up to 60–80% in comparison to tubercle grown on control non-transgenic plants. Further, the underground shoots of *O. aegyptiaca* were also reduced up to 80% grown on transgenic host plants compared to the controls. A significant decrease in the mannitol level and increase in the percentage of dead tubercle was also observed on the transgenic host plants (Fig. 14.2a, b). Although M6PR-siRNA was not detected in tubercles of parasitic plant grown on transgenic host plants,

Fig. 14.2 Comparison of *Orobanche* tubercles developed on (**a**) transgenic and (**b**) non-transgenic tomato roots following silencing of the parasite M6PR target gene. Dead tubercles become blackish in color. (**c**) Growth and appearance of the transgenic and non-transgenic tomato plants expressing M6PR-siRNA

the detection of mir390, which is involved with cytoplasmic dsRNA processing, proved the indication of the existence of gene-silencing mechanisms in *Orobanche* spp. [135]. Gene silencing mechanisms are probably involved with the production of decreased levels of M6PR mRNA in the parasites grown on the transformed tomato lines [135]. Under greenhouse conditions, no phenotypic differences were observed between the transgenic plants and the corresponding non-transgenic plants (Fig. 14.2c) during the vegetative or reproductive growth stages [135]. Similarly, RNAi strategy has also been tried to control *Striga hermonthica* (Delile) Benth in maize. Four key genes of different biosynthesis pathways like fatty acid, aromatic amino acid, adenosine monophosphate, and one gene for controlling vacuole morphogenesis were selected to knock down in *Striga asiatica* (L.) Kuntze. The transformed maize plants showed some differences in *Striga* growth but none of them were completely resistant to *Striga asiatica* [141]. Recently, the upregulated expression of *SHOOT MERISTEMLESS-like* (*STM*) homeobox transcription factors was demonstrated during haustoria formation in *Cuscuta* [143]. The transgenic tobacco expressing si-RNA of *STM* specific to *Cuscuta* showed reduced efficacy of dodder infection in transgenic tobacco plant and defects in haustorial connection, development, and establishment [143]. Potential host plants harboring such a silencing construct against a gene for an important metabolic activity in the parasite may develop a high-enough resistance level to allow for normal yield to be obtained. Recently, the Parasitic Plant Genome Project (PPGP) [159] has evaluated the transcriptome of different stages of some parasitic plants [160]. The potential target gene can be screened from the aforementioned database and can target by RNAi technology for development of parasitic weed resistance plant.

Engineered Herbicide Resistance in Crops

In the past five decades, crop yields increased due to chemical control of weeds, especially with selective herbicides. Recently, applications of herbicides have been gradually reduced due to toxicity, weed resistance to the herbicides, and environmental concerns [108]. Developing a new herbicide by chemical companies is difficult, time consuming, and very expensive. Accordingly, there is a pressing need for biotech-derived crops, not only crops with engineered herbicide resistance adapted over the past few years but also using newer technology based on genomic, proteomic, and metabolomic tools. Gressel described newer technologies that will assist in meeting the needs for herbicide-resistance crops [108]. Other approaches are likely to be unsuccessful, as even low doses of herbicides applied on tobacco plants may be phytotoxic [161]. Notably, herbicides that are metabolized by transgenic plants—i.e., glufosinate, which is metabolized by the *bar* gene in transformed plants before reaching the roots of transgenic-resistant crops—would be ineffective for parasitic weed control. Transgenic herbicide resistance may also pose food safety issues through the expression of the new gene in the crop plant. Concern may also arise regarding the possible gene transfer from transgenic crop plants to wild plants, although different ways to overcome these concerns have been proposed [108]. Therefore, these parameters should be taken in consideration while applying chemicals to herbicide-resistant crops.

It has been suggested that crops with target-site resistances would allow control of parasitic weeds by herbicides that inhibit metabolic pathways in the parasites [162]. The herbicide would pass through the plant and flow into the hidden parasite; it is essential for this mode of action that the host plants not metabolize (and consume) the herbicide. This concept has been successfully applied with several crops. Control of *P. aegyptiaca* without any significant effect on the crop or its yield was achieved using glyphosate on EPSPS-inhibitor-resistant oilseed rape. Oilseed rape (*Brassica napus*) infected with *Orobanche* and engineered with the *aroA* gene encoding a modified EPSPS completely prevented development of the parasite following glyphosate application to the transgenic plants. *Orobanche* was also effectively controlled by foliar application of chlorsulfuron on ALS-inhibitor-resistant tobacco [18]. Transgenic asulam-resistant potatoes infested with *Orobanche* and engineered with the herbicide resistance gene *sul,* which codes for a modified dihydropteroate synthase (DHPS)—the target site of the herbicide asulam, suppressed development of the parasite following application of the herbicide asulam [120] (Table 14.3). Aviv et al. engineered a mutant ALS gene into carrot, allowing control of broomrape by imazapyr (an imidazolinone ALS inhibitor) [85]. Several tobacco cultivars transformed with a mutant acetohydroxy acid synthase (AHAS) 3R gene (isolated from a sulfonylurea-resistant *Brassica napus* cell line) were resistant to the herbicide chlorsulfuron [87]. The effect of chlorsulfuron on broomrape was clearly demonstrated: A very low percentage (from 0.1 to 4 %) of its active ingredient that reached the plant roots was sufficient to kill the parasite at an early developmental stage after two treatments [87].

Parasitic weeds will rapidly evolve resistance to herbicides due to their prolific seed production. It is expected that resistance to glyphosate, asulam, chlorosulfuron, or imazapyr will eventually appear. Therefore, herbicide-resistant crops should be wisely used or combined with other control methods, and new resistant crops continually developed.

Chemical Control: New and Advanced Approaches

Refined Herbicidal Methods

Recently, other groups of herbicidal compounds have shown promise in broomrape control, i.e., sulfonylurea, imidazolinone, and other ALS inhibitors or acetohydroxy acid synthase (AHAS) inhibitors. Selective *Orobanche* control was achieved by applying some of these herbicides at low rates on non-engineered crops. Broomrape chemical control is possible [78–80, 163], depending mainly on the application method. Chemigation via sprinklers followed by excessive irrigation provided excellent control. Unfortunately, chemigation cannot be adapted for developing countries or countries suffering from lack of water.

Seed Dressing with Herbicides Using ALS Inhibitor

Slow release formulations of fertilizers, pesticides, and drugs are common. The principle of this technique is the development of biodegradable formulations for seed dressing with small amounts of herbicide for broomrape control. The slow-release herbicide formulations will achieve longer control of *Orobanche* with the ALS inhibitor imazapyr. The seed dressing allows imazapyr to spread throughout the crop root zone as the roots grow, prevents imazapyr from leaching away from the host rhizosphere and requires less herbicide [30]. A treatment of cowpea seeds with imazaquin was suggested for the control of *S. gesnerioides* and *Alectra vogelii* [164], and a similar approach is currently being tested for the control of *S. hermonthica* in sorghum and pearl millet (*Pennisetum glaucum* L.) [165]. This technology appears suitable for small-scale farmers in Africa.

Chemical Control Based on Growing Degree-Days

Following establishment of the parasitic weed on the host roots, degeneration and death of the parasite are the main factors that determine the host resistance. In sunflower, higher temperature was correlated with degeneration and death of more

Orobanche tubercles increasing resistance in some varieties [166]. A field study confirmed that growing degree-day (GDD) could be a predictive parameter for *O. minor* parasitism. Parasitism of *O. minor* in red clover could be predicted by GDD under controlled conditions [167]. This model was validated under field conditions [168]. Therefore, a predictive model may be a base for developing a decision-support system for chemical control (suitable timing for precise chemical control) of the parasite [166, 169]. Recently, technologies to improve chemical control were proposed [170]. Successful broomrape control could be achieved by both during the parasite's subsurface developmental stage and underground developmental stages. However, control efficacy and prevention of yield reduction are preferable during an early stage (underground developmental stages). This method will require a modeling approach to predicting the initial stages of parasitism in the proposed crops. The introduction of the minirhizotron video camera and its adaptation for nondestructive in situ monitoring of broomrape development in the soil subsurface allowed the development of a robust thermal time model and its validation under field conditions. This new technology considerably enhanced and optimized the efficacy of chemical control of parasitic weeds, enabling defining the stages at which broomrape is most sensitive to herbicides.

Chemical Mutagenesis

In parallel to the plant breeding and transgenic strategies, a mutagenesis approach has also been widely used to produce new resistant varieties against parasitic weeds [171]. Ethyl methane sulfonate (EMS) is a commonly used chemical mutagen because of its high effectiveness in generating new and desired traits. This method has been widely used in the generation of male sterility and herbicide resistance in plants [171]. EMS creates single nucleotide substitution in the plant genomic DNA at a rate of 5×10^{-4} to 5×10^{-2} per gene without substantial killing of the plant cell. Optimal concentration of EMS treatment to develop a new trait is about 50% seed mortality in some plants like chickpea. The beauty of a herbicide resistance trait is that it can be used to selectively kill the harmful weeds in agricultural fields. It has already been demonstrated that broomrape is sensitive to ALS-inhibiting herbicides [163]. Being a strong sink, parasitic weeds suck the herbicides together with nutrients. The translocated herbicide selectively kills the underground root of the parasitic plant and the host plant is safe because of the herbicide-resistant property of the mutagenized crops [18]. In this concern, imidazolinone is a good herbicide because of its highly efficient absorbance by the host plant and it is easily transported to the root-attached parasitic weeds.

Conclusion

The nature of parasitic weeds makes their control extremely difficult, costly, or environmentally hazardous. Different potential approaches for parasitic weed control have been discussed. Unfortunately, none of the conventional methods currently used proves to be very successful in controlling parasitic weeds in the field. It was claimed that integrated approaches combining several techniques could be more effective. However, these integrated programs are practiced only on a small scale in a few countries, because of cost and technical problems. While avoidance of dispersal of parasitic weeds, crop resistance, and prevention control methods could be effective and the most economical methods to reduce parasitic weed infestations in agricultural fields, the potentially simplest and most effective approach to parasitic weed control—host resistance—remains an unrealized goal. Optimal parasitic weed control could be achieved by the use of either parasite-resistant crops (from conventional breeding) or crops genetically engineered for resistance. Advantages of these approaches are: no chemical applications, no need for additional labor or complicated management, and no expensive equipment or instrumentation. Additionally, crop resistance approaches are superior to other methods in effectiveness, low cost, and environmental safety, and may also deplete the parasite soil seedbank. So far, only a few crop varieties with stable resistance have been developed after decades of conventional plant breeding, and genetic resources for resistance genes are limited.

Recently, progress has been made in the genomic and genetic molecular research of host–parasite interaction, and the first *Orobanche*-resistant crop was engineered [137, 138]. It would be highly desirable to have crop plants naturally or artificially resistant to parasitic plants. It is also recommended to develop active links between farmers and researchers for the transfer of the available innovative technologies, as technology transfer is currently a limiting constraint on their use. The availability of resistant plant varieties would lessen or eliminate the need for alternative parasitic plant eradication measures, while increasing crop yields. Farmers who have adopted genetically modified organisms (GMO)—i.e., Round-Up Ready and BT crops—have experienced lower costs of production, workers' safety, simplicity, and flexibility in farm management, and obtained higher yields because of more cost-effective weed control. Therefore, it is reasonable to hypothesize that GMO approaches will be adopted for parasitic weed control in the near future.

Acknowledgments The authors gratefully acknowledge Mrs. Jacklin Abu-Nassar for her help in preparation of this book chapter. Contribution from the Agricultural Research Organization, the Volcani Center, Bet Dagan, Israel, no. 532/13.

References

1. Parker C, Riches CR (1993) Parasitic weeds of the world: biology and control. CAB Int., Wallingford, 332 pp
2. Yoder JI (1997) A species-specific recognition system directs haustorium development in the parasitic plant *Triphysaria* (Scrophulariaceae). Planta 202:407–413
3. dePamphilis CW, Palmer JD (1990) Loss of photosynthetic and chlororespiratory genes from the plastid genome of a parasitic flowering plant. Nature 348:337–339
4. Parker C (2009) Observations on the current status of *Orobanche* and *Striga* problems worldwide. Pest Manag Sci 65:453–459

5. Dahan R, Mourid ME (2003) Integrate management of Orobanche in food legumes in the Near East and North Africa. Proceeding of the expert consultation on IPM for *Orobanche* in food legume systems in the Near East and North Africa, Rabat, Morocco, p 128
6. Linke KH, Sauerborn J, Saxena MC (1991) Host-parasite relationships-effect of *Orobanche crenata* seed banks on development of the parasite and yield of faba bean. Angew Botanik 65:229–238
7. Grenz JH, Sauerborn J (2006) Predicting the potential geographical distribution of parasitic weeds, a contribution to risk management: case study of *Orobanche crenata* (Eugen Ulmer Gmbh Co), pp 281–288
8. Ejeta G (2007) Breeding for *Striga* resistance in sorghum: exploitation of an intricate host-parasite biology. Crop Sci 47:216–227
9. Scholes JD, Press MC (2008) *Striga* infestation of cereal crops-an unsolved problem in resource limited agriculture. Curr Opin Plant Biol 11:180–186
10. Kuijt J (1969) The biology of parasitic flowering plants. University of California Press, Berkeley, 246 pp
11. Westwood JH (2000) Characterization of the *Orobanche-Arabidopsis* system for studying parasite-host interactions. Weed Sci 48:742–748
12. Joel DM, Portnoy VH (1998) The angiospermous root parasite *Orobanche* L. (Orobanchaceae) induces expression of a pathogenesis related (PR) gene in susceptible tobacco roots. Ann Bot 81:779–781
13. Joel DM (2000) The long-term approach to parasitic weeds control: manipulation of specific developmental mechanisms of the parasite. Crop Prot 19:753–758
14. Aly R, Westwood J, Cramer C (2003) Crop protection against parasites/pathogens through expression of sarcotoxin-like peptide. Patent no. WO02094008
15. Gressel J, Hanafi A, Head G, Marasas W, Obilana AB, Ochanda J, Souissi T, Tzotzos G (2004) Major heretofore intractable biotic constraints to African food security that may be amenable to novel biotechnological solutions. Crop Prot 23:661–689
16. Lanini T (2004) Economical methods of controlling dodder in tomatoes. Proc Calif Weed Sci Soc 56:57–59
17. Raynal-Roques A. (1996) A hypothetical history of *Striga*- A perelimnary draft. In: Moreno MT, Cubero JI, Berner D, Joel DM, Musselman LJ, Parker C (eds) Advances in parasitic plant research, Proceedings of the 6th international parasitic weeds symposium, Cordoba, Spain, pp 106–111
18. Joel DM, Kleifeld Y, Losner-Goshen D, Herzlinger G, Gressel J (1995) Transgenic crops against parasites. Nature 374:220–221
19. Bouwmeester HJ, Matusova R, Zhongkui S, Beale MH (2003) Secondary metabolite signalling in host-parasitic plant interactions. Curr Opin Plant Biol 6:358–364
20. Cardoso C, Ruyter-Spira C, Bouwmeester HJ (2011) Strigolactones and root infestation by plant-parasitic *Striga, Orobanche* and *Phelipanche* spp. Plant Sci 180:414–420
21. Cissoko M, Boisnard A, Rodenburg J, Press MC, Scholes JD (2011) New Rice for Africa (NERICA) cultivars exhibit different levels of post-attachment resistance against the parasitic weeds *Striga hermonthica* and *Striga asiatica*. New Phytol 192:952–963
22. Keyes WJ, Taylor JV, Apkarian RP, Lynn DG (2001) Dancing together. Social controls in parasitic plant development. Plant Physiol 127:1508–1512
23. Aly R, Hamamouch N, Abu-Nassar J, Wolf S, Joel DM, Eizenberg H, Kaisler E, Cramer C, Gal-On A, Westwood JH (2011) Movement of protein and macromolecules between host plants and the parasitic weed *Phelipanche aegyptiaca* Pers. Plant Cell Rep 30:2233–2241
24. Joel DM, Losner-Goshen D (1994) Early host-parasite interaction: models and observations of host root penetration by the haustorium of *Orobanche*. In: Pieterse AH, Verkleij JAC, ter Borg SJ (eds), Proceedings of the third international workshop on *Orobanche* and related *Striga* research, Royal Tropical Institute, Amsterdam, The Netherlands, pp 237–247
25. Ben hod G, Losner D, Joel DM, Mayer AM (1993) Pectin methylesterase in calli and germinating seeds of *Orobanche aegyptiaca*. Phytochemistry 32:1399–1402
26. Shomer-Ilan A (1993) Germinating seeds of the root parasite *Orobanche aegyptiaca* Pers. excrete enzymes with carbohydrase activity. Symbiosis 15:61–70

27. Losner-Goshen D, Portnoy VH, Mayer AM, Joel DM (1998) Pectolytic activity by the haustorium of the parasitic plant *Orobanche* L. (Orobanchaceae) in host roots. Ann Bot 81:319–326
28. Mor A, Mayer AM, Levine A (2008) Possible peroxidase functions in the interaction between the parasitic plant, *Orobanche aegyptiaca*, and its host, *Arabidopsis thaliana*. Weed Biol Manage 8:1–10
29. Kuijt J (1977) Haustoria of phanerogamic parasites. Ann Rev Phytopathol 17:91–118
30. Kanampiu F, Ransom J, Gressel J, Jewell D, Friesen D, Grimanelli D, Hoisington D (2002) Appropriateness of biotechnology to African agriculture: *Striga* and maize as paradigms. Plant Cell Tiss Organ Cult 69:105–110
31. Matusova R, Rani K, Verstappen FW, Franssen MC, Beale MH, Bouwmeester HJ (2005) The strigolactone germination stimulants of the plant-parasitic *Striga* and *Orobanche* spp. are derived from the carotenoid pathway. Plant Physiol 139:920–934
32. Holm L, Doll J, Holm E, Panch JV, Herberger JP (1997) World weeds: natural histories and distribution. Wiley, New York, 1129 pp
33. Lanini WT, Kogan M (2005) Biology and management of *Cuscuta* in Crops. Cien Inv Agri 32:127–141
34. Aly R (2007) Conventional and biotechnological approaches for control of parasitic weeds. Vitro Cell Dev Biol Plant 43:304–317
35. Runyon JB, Mescher MC, De Moraes CM (2006) Volatile chemical cues guide host location and host selection by parasitic plants. Science 313:1964–1967
36. Verkleij JAC, Kuiper E (2000) Various approaches to controlling root parasitic weeds. Biotechnol Dev Monit 41:16–19
37. Kroschel JA (2001) Technical manual for parasitic weed research and extension. Kluwer Academic, Dordrecht, 256 pp
38. Joel DM, Hershenhorn Y, Eizenberg H, Aly R, Ejeta G, Rich PJ, Ransom JK, Sauerborn J, Rubiales D (2006) Biology and management of weedy root parasites. In: Janick J (ed) Horticultural reviews, vol 33. Wiley, Hokobken, pp 267–349
39. Aly R (2012) Advanced technologies for parasitic weed control. Weed Sci 60:290–294
40. Omanya GO (2001) Variation for indirect and direct measures of resistance to *Striga* (*Striga hermonthica* (Del.) Benth.) in two recombinant inbred populations of Sorghum (*Sorghum bicolor* (L.) Moench). Grauer, Beuren, 142 pp
41. Odhiambo GD, Ransom JK (1994) Preliminary evaluation of long-term effects of trap cropping on *Striga*. In: Pieterse AH, Verkleij JAC, ter Borg SJ (eds) Biology and management of *Orobanche*, Proceedings of the third international workshop on *Orobanche* and related *Striga* Research, Royal Tropical Institute, Amsterdam, The Netherlands, pp 505–512
42. Sauerborn J, Sprich H, Mercer-Quarshie H (2000) Crop rotation to improve agricultural production in Sub Saharan Africa. J Agr Crop Sci 184:67–72
43. Oswald A, Ransom JK (2001) *Striga* control and improved farm productivity using crop rotation. Crop Prot 20:113–120
44. Schulz S, Hussaini MA, Kling JG, Berner DK, Ikie FO (2003) Evaluation of integrated *Striga hermonthica* control technologies under farmer management. Expt Agr 39:99–108
45. Ahonsi MO, Berner DK, Emechebe AM, Lagoke ST (2004) Effects of ALS-inhibitor herbicides, crop sequence, and fertilization on natural soil supressiveness to *Striga hermonthica*. Agric Ecosyst Environ 104:453–463
46. Hess DE, Dodo H (2004) Potential for sesame to contribute to integrated control of *Striga hermonthica* in the West African Sahel. Crop Prot 23:515–522
47. Oswald A, Ransom JK, Kroschel J, Sauerborn J (1999) Developing a catch-cropping technique for small-scale subsistence farmers. In: Kroschel J, Mercer-Quarshie H, Sauerborn J (eds) Advances in parasitic weed control at on-farm level. Joint action to control *Striga* in Africa, vol 1. Margraf, Weikersheim, pp 181–187
48. Dawson JH (1987) *Cuscuta* (Convolvulaceae) and its control. In: Proceedings of the 4th international symposium on parasitic flowering plants, Marburg, Germany, pp 137–149
49. Rehm S (1989) Hirsen. In: Rehm S (ed) Spezieller Pflanzenbau in den Tropen und Subtropen. Eugen Ulmer Verlag, Stuttgart, pp 79–86

50. Elzein AEM, Kroschel J (2003) Progress on management of parasitic weeds. In: Labrada R (ed) Series title: FAO plant production and protection. Weed management for developing countries, Papers 120 Add, Food and Agriculture Organization of the United Nations, Rome, pp 109–143
51. Dawoud DA, Sauerborn J, Kroschel J (1996) Transplanting of *Sorghum*: a method to reduce yield losses caused by the parasitic weed *Striga*. In: Moreno MT, Cubero JI, Berner D, Joel DM, Musselman LJ, Parker C (eds) Advances in parasitic plant research. Proceedings of the sixth international parasitic weed symposium, Cordoba, Spain, pp 777–785
52. Oswald A, Ransom JK, Kroschel J, Sauerborn J (2002) Intercropping controls *Striga* in maize based farming systems. Crop Prot 21:367–374
53. Katan J, deVay JE (1991) Soil solarization: historical perspectives, principles, and uses. In: Katan J, deVay JE (eds) Soil solarization. CRC, Boca Raton, pp 23–37
54. Abu-Irmaileh BH (1991) Soil Solarization controls broomrapes (*Orobanche* spp.) in host vegetable crops in the Jordan valley. Weed Technol 5:575–581
55. Jacobsohn R, Greenberger A, Katan J, Levi M, Alon H (1980) Control of Egyptian broomrape (*Orobanche aegyptiaca*) and other weeds by means of solar heating of the soil by polyethylene mulching. Weed Sci 28:312–316
56. Sauerborn J, Linke KH, Saxena MC, Kock W (1989) Solarization: a physical control method for weeds and parasitic plants (*Orobanche* spp.) in Mediterranean agriculture. Weed Res 29:391–397
57. Lichter JM, Reid MS, Berry AM (1991) New methods for control of leafy mistletoe (*Phoradendron* spp.) on landscape trees. J Arboric 17:127–130
58. Misra A, Tosh BN (1982) Preliminary studies on broomrape (*Orobanche* spp.)—a parasitc weed on brinjal. Abstract of papers, annual conference of Indian society of weed science, p 45
59. Goldwasser Y, Kleifeld Y, Golan S, Bargutti A, Rubin B (1995) Dissipation of metham-sodium from soil and its effect on the control of *Orobanche aegyptiaca*. Weed Res 35:445–452
60. Jacobsohn R (1994) The broomrape problem in Israel and an integrated approach to its control. In: Pieterse AH, Verkleij JAC, ter Borg SJ (eds) Biology and management of *Orobanche*. Proceedings of the third international workshop on *Orobanche* and related *Striga* Research, Royal Tropical Institute, Amsterdam, The Netherlands, pp 652–658
61. McDonald D (2002) Fumigants and soil sterilants: alternatives to methyl bromide. Int Pest Contr 44:118–122
62. Foy CL, Jain R, Jacobsohn R (1989) Recent approaches for chemical control of broomrape (*Orobanche* spp.): a review. Weed Sci 4:123–152
63. Goldwasser Y, Kleifeld Y (2004) Recent approaches to *Orobanche* management: a review. In: Inderjit (ed) Weed biology and management. Kluwer, The Netherlands, pp 439–466
64. Yasuda N, Sugimoto Y, Kato M, Inanaga S, Yoneyama K (2003) (+)—*Strigol*, a witchweed seed germination stimulant, from *Menispermum dauricum* root culture. Phytochemistry 62:1115–1119
65. Yoneyama K, Takeuchi Y, Sato D, Sekimoto H, Yokoka T (2004) Determination and quantification of strigolactones. Proceedings of the 8th international parasitic weeds symposium, Durban, South Africa, p 9
66. Zwanenburg B, Wigchert SCM (1998) The molecular inception of *Striga* and *Orobanche* seed germination. In: Wegmann K, Musselman LJ, Joel DM (eds) Current problems of *Orobanche* researches. Proceedings of the 4th international *Orobanche* workshop, Albena, Bulgaria, pp 25–31
67. Wigchert SC, Kuiper E, Boelhouwer GJ, Nefkens GH, Verkleij JA, Zwanenburg B (1999) Dose-response of seeds of the parasitic weed *Striga* and *Orobanche* towards the synthetic germination stimulants GR 24 and Nijmegen 1. J Agric Food Chem 47:1705–1710
68. Egley GH, Eplee RE, Norris RS (1990) Discovery and development of ethylene as a witchweed seed germination stimulant. In: Sand PF, Eplee RE, Westbrooks RC (eds) Witchweed research and control in the United States. WSSA, Champaign, IL, USA, pp 56–67
69. Ransom JK, Njoroge J (1991) Seasonal variation in ethylene induced germination of *Striga hermonthica* in western Kenya. In: Pansom JK, Musselman LJ, Worshman AP, Parker C

(eds) Proceedings of the 5th international symposium on parasitic weeds. CIMMYT, Nairobi, Kenya, pp 391–396

70. Gbehounou G (1998) Seed ecology of *Striga hermonthica* in the Republic of Benin: host specificity and control potentials. PhD thesis, Vrije Universiteit, Amsterdam, The Netherlands, 126 pp
71. Grenier C, Rich PJ, Mohamed A, Ellicott A, Shaner C, Ejeta G (2001) Independent inheritance of lgs and IR genes in *Sorghum*. In: Fer A et al (eds) Proceedings of the 7th international parasitic weed symposium, Nantes, France, pp 220–223
72. Chae SH, Yoneyama K, Takeuchi Y, Joel DM (2004) Fluridone and norflurazon, carotenoid-biosynthesis inhibitors, promote seed conditioning and germination of the holoparasite *Orobanche minor*. Physiol Plant 120:328–337
73. Jamil M, Charnikhova T, Verstappen F, Bouwmeester H (2010) Carotenoid inhibitors reduce strigolactone production and *Striga hermonthica* infection in rice. Arch Biochem Biophys 504:123–131
74. Vurro M, Boari A, Pilgeram AL, Sands DC (2006) Exogenous amino acids inhibit seed germination and tubercle formation by *Orobanche ramosa* (Broomrape): potential application for management of parasitic weeds. Biol Control 36:258–265
75. Garcia-Torres L (1998) Reflection on parasitic weed control: available or needed technology. In: Wegmann K, Musselman LJ, Joel DM (eds) Current problems of *Orobanche* researches. Proceedings of the 4th international *Orobanche* workshop, Albena, Bulgaria, pp 323–326
76. Mesa-García J (1985) García-Torres L. *Orobanche crenata* (Forsk) control in *Vicia faba* (L.) with glyphosate as affected by herbicide rates and parasite growth stages. Weed Res 25:129–134
77. Sauerborn J, Saxena MC, Meyer A (1989) Broomrape control in faba bean (*Vicia faba* L.) with glyphosate and imazaquin. Weed Res 29:97–102
78. Hershenhorn J, Goldwasser Y, Plakhine D, Lavan Y, Herzlinger G, Golan S, Chilf T, Kleifeld Y (1998) Effect of sulfonylurea herbicides on Egyptian broomrape (*Orobanche aegyptiaca*) in tomato (*Lycopersicon esculentum*) under greenhouse conditions. Weed Technol 12:115–120
79. Hershenhorn J, Plakhine D, Goldwasser Y, Westwood JH, Foy CL, Kleifeld Y (1998) Effect of sulfonylurea herbicides on early development of Egyptian broomrape (*Orobanche aegyptiaca*) in tomato (*Lycopersicon esculentum*). Weed Technol 12:108–114
80. Kleifeld Y, Goldwasser Y, Plakhine D, Eizenberg H, Herzlinger G, Golan S (1998) Selective control of *Orobanche* spp. in various crops with sulfonylurea and imidazolinones herbicides. In: Joint action to control *Orobanche* in the WANA-Region: experiences from Morocco. Proceedings, regional workshop, Rabat, Morocco, 26 pp
81. Aly R, Goldwasser Y, Eizenberg H, Hershenhorn J, Golan S, Kleifeld Y (2001) Broomrape (*Orobanche cumana*) control in sunflower (*Helianthus annuus*) with Imazapic. Weed Technol 15:306–309
82. Kleifeld Y, Goldwasser Y, Herzlinger G, Plakhine D, Golan S, Chilf T (1996) Selective control of *Orobanche aegyptiaca* in tomato with sulfonylurea herbicides. In: Cubero JI, Berner D, Joel DM, Musselman LJ, Parker C (eds) Advances in parasitic plant research. Proceedings of the sixth international parasitic weed symposium, Cordoba, Spain, pp 707–715
83. Plakhine D, Goldwasser Y, Hershenhorn J, Kleifeld Y (1996) Effect of sulfonylurea herbicides on early development stages of *Orobanche aegyptiaca*. In: Moreno MT, Cubero JI, Berner D, Joel DM, Musselman LJ, Parker C (eds) Advances in parasitic plant research. Proceedings of the sixth international parasitic weed symposium, Cordoba, Spain, pp 716–725
84. Jurado-Exposito M, Castejon-Munoz M, Gracia-torres L (1996) Broomrape (*Orobanche crenata* Forsk.) control with Imazetapyr applied to Pea (*Pisum sativum* L.) seeds. Weed Technol 10:774–780
85. Aviv D, Amsellem Z, Gressel J (2002) Transformation of carrots with mutant acetolactate synthase for *Orobanche* (broomrape) control. Pest Manage Sci 58:1187–1193

86. Kabambe VH, Kanampiu F, Ngwira A (2008) Imazapyr (herbicide) seed dressing increases yield, suppresses Striga asiatica and has seed depletion role in maize (Zea mays L.) in Malawi. Afr J Biotechnol 7:3293–3298
87. Slavov S, Valkov V, Batchvarova R, Atanassova S, Alexandrova M, Atanassov A (2005) Chlorsulfuron resistant transgenic tobacco as a tool for broomrape control. Transgenic Res 14:273–278
88. Bewick TA, Binning LK, Dana MN (1989) Control of swamp dodder in cranberry. HortScience 24:850
89. Rubin B (1990) Weed competition and weed control in *Allium* crops. In: Rabinowitch HD, Brewster JL (eds) Onions and allied crops, vol 2. CRC Press Inc., Boca Raton, pp 63–84
90. Watson WT, Martinez-Trinidad T (2006) Strategies and treatments for Leafy Mistletoe (*Phoradendron tomentosum* (DC.) Engelm ex. Gray) suppression on Cedar Elm (*Ulmus crassifolia* Nutt.). Arboric Urb For 32:265–270
91. Varga I, Nagy V, Baltazár T, Mátyás KK, Poczai P, Molnár I (2012) Study of the efficiency of different systemic herbicides against European mistletoe (*Viscum album*) and their antifungal activity against hyperparasitic mistletoe fungus. Növényvédelem 48:507–517
92. Kroschel J, Jost A, Sauerborn J (1999) Insects for *Striga* control: possibilities and constraints. In: Kroschel J, Mercer-Quarshie H, Sauerborn J (eds) Advances in parasitic weed control at on-farm level. Vol I. Joint Action to Control *Striga* in Africa. Margraf, Weikersheim, pp 117–132
93. Traoré D, Vincent C, Stewart RK (1999) *Smicronyx guineanus* Voss and *S. umbrinus* Hustache (Coleoptera: Curculionidae): potential biocontrol agents of *Striga hermonthica* (Del.) Benth. (Scrophulariaceae). In: Hess DE, Lenné JM (eds) Report on the ICRISAT sector review for *Striga* control in *Sorghum* and Millet. International Crops Research Institute for the Semi-Arid Tropics, Bamako, Mali, pp 105–115
94. Klein O, Kroschel J (2002) Biological control of *Orobanche* spp. with *Phytomyza orobanchia*, a review. BioControl 47:245–277
95. Bedi JS (1994) Further studies on control of sunflower broomrape with *Fusarium oxysporum* f.sp. orthoceras—a potential mycoherbicide. In Pieterse AH, Verkleij JAC, ter Borg SJ (eds) Proceedings of the third international workshop on *Orobanche* and related *Striga* research. Royal Tropical Institute, Amsterdam, The Netherlands, pp 539–544
96. Thomas H (1998) Das Potential von Pilzen zur Kontrolle von *Orobanche* spp. unter Berücksichtigung von Anbausystemen im: Tera, Nepal. PLITS, vol 16. Margraf, Weikersheim, 110 pp
97. Amsellem Z, Kleifeld Y, Kerenyi Z, Hornok L, Goldwasser Y, Gressel J (2001) Isolation, identification and activity of mycoherbicidal pathogens from juvenile broomrape plants. Biol Control 21:274–284
98. Elzein AEM, Kroschel J (2004) *Fusarium oxysporum* "Foxy 2" shows potential to control both *Striga hermonthica* and *S. asiatica*. Weed Res 44:433–438
99. Marley PS, Kroschel J, Elzein A (2005) Host specificity of *Fusarium oxysporum* Schlect (isolate PSM 197) a potential mycoherbicide for controlling *Striga* spp. in West Africa. Weed Res 45:407–412
100. Charudattan R (2001) Biological control of weeds by means of plant pathogens: significance for integrated weed management in modern agro-ecology. Biocontrol 46:229–260
101. Cohen BA, Amsellem Z, Lev-Yadun S, Gressel J (2002) Infection of tubercles of the parasitic weed *Orobanche aegyptiaca* by mycoherbicidal *Fusarium* species. Ann Bot 90:567–578
102. Müller-Stöver D, Buschmann H, Sauerborn J (2005) Increasing control reliability of *Orobanche cumana* through integration of a biocontrol agent with a resistance-inducing chemical. Eur J Plant Pathol 111:193–202
103. Amsellem Z, Zidack NK, Quimby PC Jr, Gressel J (1999) Long-term dry preservation of active mycelia of two mycoherbicidal organisms. Crop Prot 18:643–649
104. Quimby PC Jr, Zidack NK, Boyette CD, Grey WE (1999) A simple method for stabilizing and granulating fungi. Biocontrol Sci Technol 9:5–8

105. Kroschel J, Mueller-Stoever D, Elzein A, Sauerborn J (2000) The development of mycoherbicides for the management of parasitic weeds of the genus *Striga* and *Orobanche*—a review and recent results. In: Spencer NR (ed) Proceedings of the X international symposium on biological control of weeds, Bozeman, Montana, USA, p 139
106. Müller-Stöver D (2001) Possibilities of biological control of *Orobanche crenata* and *O. cumana* with *Ulocladium botrytis* and *Fusarium oxysporum* f. sp. orthoceras. Agroecology 3. Apia, Laubach, 174 pp
107. Elzein AEM (2003) Development of a granular mycoherbicidal formulation of *Fusarium oxysporum* "Foxy2" for the biological control of *Striga hermonthica* (Del.) Benth. Dissertation, Margraf Publishers, Weikersheim, Germany, 172 pp
108. Gressel J (2002) Molecular biology of weed control. Taylor & Francis, London, pp 362–390
109. Cubero JI, Hernández L (1991) Breeding faba bean (*Vicia faba* L.) for resistance to *Orobanche crenata* Forsk. In: Cubero JI, Saxena MC (eds) Present status and future prospects of faba bean production and improvement in the Mediterranean countries. CIHEAM, Zaragoza, pp 51–57
110. Ejeta G, Butler LG, Hess DE, Vogler RK (1991) In: Ransom JK, Musselman LJ, Worsham AD, Parker C (eds) Proceedings of the 5th international symposium of parasitic weeds, CIMMYT, Nairobi, Kenya, pp 539–544
111. Rubiales D (2003) Parasitic plants, wild relatives and the nature of resistance. New Phytol 160:459–461
112. Haussmann BIG, Hess DE, Reddy BVS, Mukuru SZ, Seetharama N, Kayentao M, Omanya GO, Welz HG, Geiger HH (2000) QTL for *Striga* resistance in *Sorghum* populations derived from IS 9830 and N 13. In: Haussmann BIG, Koyama ML, Grivet L, Rattunde HF, Hess DE (eds) Breeding for *Striga* resistance in cereals. Proceedings of a workshop, IITA, Ibadan, Nigeria, 18–20 Aug 1999. (Margraf, Weikersheim, Germany, pp 159–171)
113. Fernández-Martínez JM, Velasco L, Pérez-Vich B (2005) Resistance to new virulent *O. cumana* races. In: Murdoch A (ed) Abstr COST849 meeting on broomrape biology, Control and Management, Department of Agriculture, University of Reading, UK 15–17 Sept 2005, pp 17–18, 27–28
114. Rubiales D, Pérez-de-Luque A, Aparico MF, Sillero JC, Román B, Kharrat M, Khalil S, Joel DM, Riches C (2006) Screening techniques and sources of resistance against parasitic weeds in grain legumes. Euphytica 147:187–199
115. Rubiales D, Sillero JC, Cubero JI (1999) Broomrape (*Orobanche crenata*) as a major constraint for pea cultivation in southern Spain. In: Cubero JI, Moreno MT, Rubiales D, Sillero JC (eds) Resistance to *Orobanche*: the state of the art. Junta de Andalucía, Sevilla, pp 83–89
116. Pérez-de-Luque A, Rubiales D, Cubero JI, Press MC, Scholes J, Yoneyama K, Takeuchi Y, Plakhine D, Joel DM (2005) Interaction between *Orobanche crenata* and its host legumes: unsuccessful haustorial penetration and necrosis of the developing parasite. Ann Bot 95:935–942
117. Loffler C, Sahm A, Wray V, Czygan FC, Proksch P (1995) Soluble phenolic constituents from *Cuscuta reflexa* and *Cuscuta platyloba*. Biochem Syst Ecol 23:121–128
118. Goldwasser Y, Lanini WT, Wrobel RL (2001) Tolerance of tomato varieties to lespedeza dodder. Weed Sci 49:520–523
119. Yoder JI, Scholes JD (2010) Host plant resistance to parasitic weeds; recent progress and bottlenecks. Curr Opin Plant Biol 13:478–484
120. Surov T, Aviv D, Aly R, Joel DM, Goldman-Guez T, Gressel J (1998) Generation of transgenic asulam-resistant potatoes to facilitate eradication of parasitic broomrapes (*Orobanche* spp.), with the sul gene as the selectable marker. Theor Appl Genet 96:132–137
121. Gressel J, Segel L, Ransom JK (1996) Managing the delay of evaluation of herbicide resistance in parasitic weeds. Int J Pest Manag 42:113–129
122. Labrousse P, Arnaud MC, Serieys H, Berville A, Thalouarn P (2001) Several mechanisms are involved in resistance of *Helianthus* to *Orobanche cumana* Wallr. Ann Bot 88:859–868

123. Antonova TS (1998) The interdependence of broomrape virulence and sunflower resistant mechanisms. Proceedings of the fourth international workshop on *Orobanche* Research, Albena, Bulgaria, pp 147–153
124. Rodríguez-Conde MF, Moreno MT, Cubero JI, Rubiales D (2004) Characterization of the *Orobanche*—*Medicago truncatula* association for studying early stages of the parasite-host interaction. Weed Res 44:218–223
125. Daniell H (1999) GM crops: public perception and scientific solutions. Trends Plant Sci 4:467–469
126. Lane JA, Bailey JA, Butler RC, Terry PJ (1993) Resistance of cowpea (*Vigna unguiculata* (L.) Walp.) to *Striga gesnerioides* (Wild.) Vatke, a parasitic angiosperm. New Phytol 125:405–412
127. Vogler RK, Ejeta G, Butter LG (1996) Inheritance of low production of *Striga* germination stimulant in sorghum. Crop Sci 36:1185–1191
128. Westwood JH, Yu X, Foy CL, Cramer CL (1998) Expression of a defense-related 3-hydroxy-3-methylglutaryl CoA reductase gene in response to parasitization by *Orobanche* spp. Mol Plant-Microbe Interact 11:530–536
129. Gowda BS, Riopel JL, Timko MP (1999) NRSA-1: a resistance gene homolog expressed in roots of non-host plants following parasitism by *Striga asiatica* (witchweed). Plant J 20:217–230
130. Aber M, Fer A, Salle G (1983) Etude du transfert des substances organiques de l'hôte (*Vicia faba*) vers le parasite (*Phelipanche crenata* Forsk.). Transfer of organic substances from the host plant *Vicia faba* to the parasite *Orobanche crenata* Forsk. Z Pflanzenphysiol 112:297–308
131. Nandula VK, Foy C, Orcutt DM (1999) Glyphosate for *Orobanche aegyptiaca* control in *Vicia sativa* and *Brassica napus*. Weed Sci 47:486–491
132. Hosford RM (1967) Transmission of plant viruses by dodder. Bot Rev 33:387–406
133. Gal-On A, Naglis A, Leibman D, Ziadna H, Kathiravan K, Papayiannis L, Holdengreber V, Guenoune-Gelbert D, Lapidot M, Aly R (2009) Broomrape can acquire viruses from its hosts. Phytopathology 99:1321–1329
134. Tomilov AA, Tomilova NB, Wroblewski T, Michelmore R, Yoder JI (2008) Trans-specific gene silencing between host and parasitic plants. Plant J 56:389–397
135. Aly R, Cholakh H, Joel DM, Leibman D, Steinitz B, Zelcer A, Naglis A, Yarden O, Gal-On A (2009) Gene silencing of mannose 6-phosphate reductase in the parasitic weed *O. aegyptiaca* through the production of homologous dsRNA sequences in the host plant. Plant Biotechnol J 7:487–498
136. Roney JK, Khatibi PA, Westwood JH (2007) Cross-species translocation of mRNA from host plants into the parasitic plant dodder. Plant Physiol 143:1037–1043
137. Hamamouch N, Westwood JH, Banner I, Cramer CL, Gepstein S, Aly R (2005) A peptide from insects protects transgenic tobacco from a parasitic weed. Transgen Res 14:227–236
138. Aly R, Plakhin D, Achdari G (2006) Expression of sarcotoxin IA gene via a root-specific tob promoter enhanced host resistance against parasitic weeds in tomato plants. Plant Cell Rep 25:297–303
139. Birschwilks M, Sauer N, Scheel D (2007) Neumann S. *Arabidopsis thaliana* is a susceptible host plant for the holoparasite *Cuscuta campestris*. Planta 226:1231–1241
140. Berner DK, Ikie FO, Green JM (1997) ALS-inhibiting herbicide seed treatments control *Striga hermonthica* in ALS-modified corn (*Zea mays*). Weed Technol 11:704–707
141. de Framond A, Rich PJ, Mcmillan J, Ejeta G (2007) Effects on *Striga* parasitism of transgenic maize armed with RNAi constructs targeting essential *S. asiatica* genes. In: Ejeta G, Gressel J (eds) Integrating new technologies for *Striga* control. World Scientific Publishing Co., Singapore, pp 185–196
142. Albert M, Belastegui-Macadam X, Kaldenhoff R (2006) An attack of the plant parasite *Cuscuta reflexa* induces the expression of attAGP, an attachment protein of the host tomato. Plant J 48:548–556

143. Alakonya A, Kumar R, Koenig D, Kimura S, Townsley B, Runo S, Garces HM, Kang J, Ynez A, Schwartz RD, Machuka J, Sinha N (2012) Interspecific RNA Interference of SHOOT MERISTEMLESS-like disrupts *Cuscuta pentagona* plant parasitism. The Plant Cell 24:3153–3166
144. Kanai A, Natori S (1989) Cloning of gene cluster for sarcotoxin I, antibacterial proteins of *Sarcophaga peregrina*. FEBS Lett 258:199–202
145. Aly R, Granot D, Mahler-Slasky Y, Halpern N, Nir D, Galun E (1999) *Saccharomyces cerevisiae* cells, harboring the gene encoding Sarcotoxin IA secrete a peptide that is toxic to plant pathogenic bacteria. Protein Expr Purif 16:120–124
146. Cramer CL, Weissenborn D, Cottingham CK, Denbow CJ, Eisenback JD, Radin DN, Yu X (1993) Regulation of defense-related gene expression during plant-pathogen interactions. J Nematol 25:507–518
147. Okamoto M, Mitsuhara I, Ohshima M, Natori S, Ohashi Y (1998) Enhanced expression of an antibimicrobial peptide sarcotoxin IA by GUS fusion in transgenic tobacco plants. Plant Cell Physiol 39:57–63
148. Alonso LC (1998) Resistance to *Orobanche* and resistance breeding: a review. In Wegmann K, Musselman LJ, Joel DM (eds) Proceedings of the Fourth International Workshop on *Orobanche*. Institute for Wheat and Sunflower "Dobroudja", Albena, Bulgaria, pp 233–257
149. Galun E (2005) RNA silencing in plants. In Vitro Cell Dev Biol Plant 41:113–123
150. Fire A, Xu S, Montgomery MK, Kostas SA, Driver SE, Mello CC (1998) Potent and specific genetic interference by double-stranded RNA in *Caenorhabditis elegans*. Nature 391:806–811
151. Brummelkamp TR, Bernards R, Agami R (2002) A system for stable expression of short interfering RNAs in mammalian cells. Science 296:550–553
152. Yoo BC, Kragler F, Varkonyi-Gasic E, Haywood V, Archer-Evans S, Moo Lee Y, Lough TJ, Lucas WJ (2004) A systemic small RNA signaling system in plants. Plant Cell 16:1979–2000
153. Denli AM, Hannon GJ (2003) RNAi: an ever-growing puzzle. Trends Biochem Sci 28:196–201
154. Hunter C, Poethig RS (2003) Missing links: miRNA and plant development. Curr Opin Genet Dev 13:372–378
155. Mlotshwa S, Voinnet O, Mette MF, Matzke M, Vaucheret H, Ding SW, Pruss G, Vance VB (2002) RNA silencing and the mobile silencing signal. Plant Cell 14:S289–S301
156. Atkinson HJ, Urwin PE, McPherson MJ (2003) Engineering plants for nematode resistance. Annu Rev Phytopathol 41:615–639
157. Abhary MK, Anfoka GH, Nakhla MK, Maxwell DP (2006) Post-transcriptional gene silencing in controlling viruses of the tomato yellow leaf curl virus complex. Arch Virol 151:2349–2363
158. Delavault P, Simier P, Thoiron S, Veronesi C, Fer A, Thalouarn P (2002) Isolation of mannose 6-phosphate reductase cDNA, changes in enzyme activity and mannitol content in broomrape (*Orobanche ramosa*) parasitic on tomato roots. Physiol Plant 115:48–55
159. Westwood JH, dePamphilis CW, Das M, Fernández-Aparicio M, Honaas LA, Timko MP, Wafula EK, Wickett NJ, Yoder JI (2012) The Parasitic Plant Genome Project: new tools for understanding the biology of Orobanche and Striga. Weed Sci 60:295–306
160. Westwood JH, Yoder JI, Timko MP, dePamphilis CW (2010) The evolution of parasitism in plants. Trends Plant Sci 15:227–235
161. Raju CA (1996) Studies on chemical control of *Orobanche cernua* in tobacco fields. In: Cubero JI, Berner D, Joel D, Musselman LJ, Parker C (eds) Advances in parasitic plant research. Proceedings of the sixth international parasitic weed symposium, Cordoba, Spain, pp 739–745
162. Gressel J (1992) The needs for new herbicide-resistant crops. In: Denholm I, Devonshire AL, Hollomon DW (eds) Resistance '91: achievements and developments in combating pesticide resistance. Elsevier, London, pp 283–294

163. Hershenhorn J, Goldwasser Y, Plakhine D, Aly R, Blumenfeld T, Bucsbaum H, Herzlinger G, Golan S, Chilf T, Eizenberg H, Dor E, Kleifeld Y (1998) *Orobanche aegyptiaca* control in tomato fields with sulfonylurea herbicides. Weed Res 38:343–349
164. Berner DK, Cardwell KF, Faturoti BO, Ikie FO, Williams OA (1994) Relative roles of wind, crop seeds and cattle in dispersal of *Striga* spp. Plant Dis 78:402–406
165. Dembélé B, Dembélé D, Westwood JH (2005) Herbicide seed treatments for control of purple witchweed (*Striga hermonthica*) in *Sorghum* and millet. Weed Technol 19:629–635
166. Eizenberg H, Hershenhorn J, Plakhine D, Shtienberg D, Kleifeld Y, Rubin B (2003) Effect of temperature on susceptibility of sunflower varieties to *Orobanche cumana* and *O. aegyptiaca*. Weed Sci 51:279–286
167. Eizenberg H, Colquhoun J, Mallory-Smith CA (2004) The relationship between temperature and small broomrape (*Orobanche minor*) parasitism in red clover (*Trifolium pratense*). Weed Sci 52:735–741
168. Eizenberg H, Colquhoun J, Mallory-Smith CA (2005) A predictive degree-days model for small broomrape (*Orobanche minor*) parasitism in red clover (*Trifolium pratense*) in Oregon. Weed Sci 53:37–40
169. Eizenberg H, Plakhine D, Hershenhorn J, Kleifeld Y, Rubin B (2003) Resistance to broomrape (*Orobanche* spp.) in sunflower (*Helianthus annuus* L.) is temperature dependent. J Exp Bot 54:1305–1311
170. Cochavi A, Rubin B, Eizenberg H (2011) Developing a predictive model based on temperatures for *P. aegyptiaca* parasitism in carrots. In: The 11th world congress of parasitic plants, Martina Franca, Italy, 7–12 June 2011, p 118
171. Kostov K, Batchvarova R, Slavov S (2007) Application of chemical mutagenesis to increase the resistance of tomato to *Orobanche ramosa* L. Bulg J Agric Sci 13:505–513

Chapter 15
Herbicide Resistance in Weeds and Crops: Challenges and Opportunities

Hugh J. Beckie

Herbicide Resistance in Weeds: Extent of the Problem

Globally, to date, there are 400 herbicide-resistant (HR) weed biotypes—defined as species by herbicide site(s) of action (SOA)—representing 217 species: 129 dicots and 88 monocots (Fig. 15.1) [1]. An average of nine HR weed biotypes are reported each year [1]. Of these 400 biotypes, one third are resistant to acetolactate synthase (ALS) inhibitors (group B/2; Table 15.1) [1], commercially introduced in 1982. There are 71 biotypes (18 % of total) resistant to photosystem-II inhibitors (group C1/5), and 42 biotypes (11 %) resistant to acetyl-CoA carboxylase (ACC) inhibitors (group A/1). Therefore, these three SOAs constitute 61 % of all HR biotypes. Currently, there are 24 weed species resistant to glyphosate (group G/9), the most popular herbicide worldwide following its commercial introduction in 1974.

Cross-resistance in a weed biotype is characterized as being resistant to two or more herbicides of the same or different chemical families or SOA due to a single resistance mechanism [2]. In contrast, multiple resistance in an individual HR plant or population is commonly defined by the occurrence of two or more resistance mechanisms. The mechanism can be target site-based resistance (TSR), i.e., mutation at the site of herbicidal action, or nontarget site resistance (NTSR), such as altered metabolism or translocation. Both enhanced metabolism and reduced translocation in HR biotypes prevent phytotoxic levels of herbicide from reaching the SOA. Enhanced metabolism is generally responsible for cross-resistance across herbicide SOA, whereas cross-resistance attributed to altered target site or translocation is usually restricted to herbicides with the same SOA.

Multiple resistance in an HR population is usually the result of sequential herbicide SOA selection, or accumulation of resistance alleles in progeny as a result of pollen flow in allogamous or outcrossing species, such as *Lolium rigidum* Gaud., *Alopecurus myosuroides* Huds., and a number of *Amaranthus* spp. Herbicides can

---

H. J. Beckie (✉)
Saskatoon Research Centre, Agriculture and Agri-Food Canada,
107 Science Place, Saskatoon, SK S7N 0X2, Canada
e-mail: hugh.beckie@agr.gc.ca

B. S. Chauhan, G. Mahajan (eds.), *Recent Advances in Weed Management*, 347
DOI 10.1007/978-1-4939-1019-9_15, © Springer Science+Business Media New York 2014

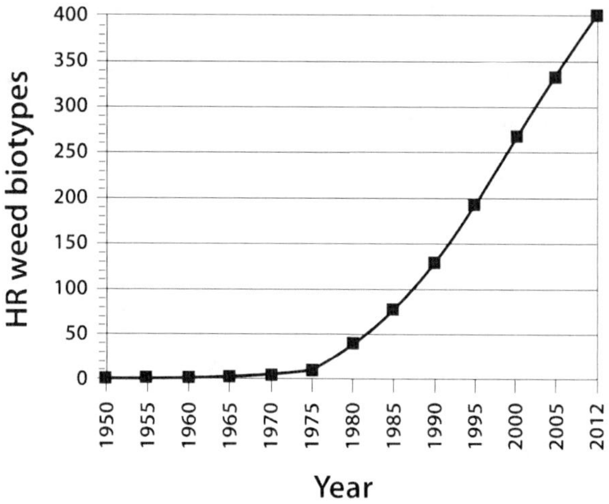

Fig. 15.1 Chronological increase in the number of herbicide-resistant (HR) weed biotypes globally. (Adapted from [1])

select for any preexisting mechanism conferring resistance in weed populations, which has been repeatedly demonstrated in the aforementioned weed species. As the frequency of different resistance mechanisms in a weed population under selection increases, so does the probability of outcrossing between plants with different mechanisms. The incidence of intergroup (i.e., across-group or SOA) HR, due to cross-resistance (i.e., metabolism-based mechanism) or multiple resistance, is continually increasing. Globally, there are more than 50 weed species with populations exhibiting intergroup HR (Fig. 15.2). Since 1990, about two species with intergroup HR have been reported annually.

Metabolic resistance has been reported much more frequently in grass (monocots) than in broadleaf (dicot) weeds [2]. Two major enzyme systems have been implicated in HR due to increased detoxification: cytochrome P450 monooxygenases and glutathione *S*-transferases (GSTs). Cases of NTSR (often attributed to metabolic detoxification) are more frequent than those conferred by TSR (gene mutation) in populations of *A. myosuroides, Avena* spp., and *Lolium perenne* L. ssp. *multiflorum* (Lam.) Husnot resistant to ACC inhibitors or other herbicides in the UK; in European populations of *A. myosuroides* resistant to ACC inhibitors, ALS inhibitors, or chlorotoluron (group C2/7); and in European populations of grass species resistant to ALS inhibitors [3–7]. In Canadian populations of ALS inhibitor-HR *Avena fatua* L., metabolism-based resistance was also much more prevalent than TSR [8].

Most cases of metabolic resistance in weeds were selected by photosystem-I (group D/22) or -II, ACC inhibitors, or ALS inhibitors, conferring cross-resistance to other herbicides with these SOA as well as dinitroanilines (group K1/3, i.e., pendimethalin). Cross-resistance can frequently occur between ACC and ALS inhibitors, or between photosystem-II inhibitors and ACC inhibitors [9]. However, different patterns of cross-resistance can occur in different species [10].

There are additional metabolism-based mechanisms of resistance. Weed resistance to propanil, an amide photosystem-II inhibitor (group C2/7), is attributed to

Table 15.1 Herbicide-resistant weed biotypes globally, by site of action (group). (Adapted from [1])

	Herbicide group	Site of action	Dicots	Monocots	Total
B/2	ALS inhibitors	Inhibition of acetolactate synthase	82	49	131
C1/5	PS-II inhibitors	Inhibition of photosystem II (site A)	49	22	71
A/1	ACC inhibitors	Inhibition of acetyl-CoA carboxylase	0	42	42
O/4	Synthetic auxins	Synthetic auxins	23	7	30
D/22	Bipyridiliums	Photosystem-I electron diversion	18	10	28
G/9	Glycines	Inhibition of EPSP synthase	11	13	24
C2/7	Ureas, amides	Inhibition of photosystem II	8	14	22
K1/3	Dinitroanilines	Inhibition of microtubule assembly	2	9	11
N/8	Thiocarbamates	Inhibition of lipid synthesis	0	8	8
E/14	PPO inhibitors	Inhibition of protoporphyrinogen oxidase	6	0	6
F3/11	Triazoles, ureas, isoxazolidiones	Inhibition of carotenoid synthesis	1	4	5
C3/6	Nitriles	Inhibition of photosystem II (site B)	3	1	4
K3/15	Chloracetamides	Inhibition of very long chain fatty acids	0	4	4
F1/12	Carotenoid synthesis inhibitors	Inhibition of phytoene desaturase	2	1	3
F2/27	HPPD inhibitors	Inhibition of hydroxyphenylpyruvate dioxygenase	2	0	2
H/10	Glutamine synthase inhibitors	Inhibition of glutamine synthetase	0	2	2
Z/25	Arylaminopropionic acids	Unknown	0	2	2
Z/26	Unknown	Unknown (e.g., chloroflurenol)	0	2	2
K2/23	Mitosis inhibitors	Inhibition of mitosis	0	1	1
L/20	Cellulose inhibitors	Inhibition of cellulose synthesis	0	1	1
Z/17	Organoarsenicals	Unknown	1	0	1
Total			*208*	*192*	*400*

enhanced metabolism by aryl acylamidase, the same enzyme responsible for catalyzing propanil metabolism in rice (*Oryza sativa* L.) [11]. Additionally, glycosyl transferases have been implicated in metabolism-based chlorotoluron resistance in *A. myosuroides* [12].

About half of the intergroup-HR weed species are grasses (reviewed in Beckie 2014 [13]). This proportion is relatively high, given that grass species account for about 25% of prominent weeds [14]. The proclivity for metabolic resistance in grass species undoubtedly contributes to this disproportionate representation. Most of the reported cases are in crop versus non-crop situations. Based on the number of countries where documented, the dominant species are *L. rigidum* and *L. perenne* subsp. *multiflorum,* followed by *A. fatua* and *A. myosuroides*. *Lolium* and *Alopecurus* spp. are obligate outcrossers, whereas *Avena* spp. are highly self-pollinated [15].

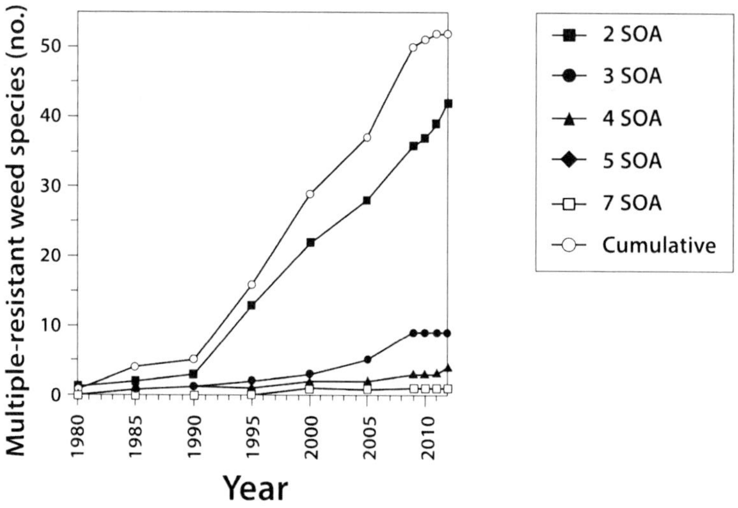

Fig. 15.2 Global weed species exhibiting intergroup herbicide resistance across sites of action (SOA). (Updated from Beckie and Tardif 2012 [2])

Lolium rigidum has been documented with biotypes resistant to a maximum of seven SOA herbicides, *A. myosuroides* to five, *A. fatua* to four, and *L. perenne* subsp. *multiflorum* to three SOA herbicides (among other species). The three most common SOA to which biotypes exhibit resistance are ACC inhibitors and ALS inhibitors, followed by photosystem-II inhibitors (ureas, amides; group C2/7). Of these grass species, four have biotypes (total of 18 cases) resistant to glyphosate: *L. rigidum, L. perenne* subsp. *multiflorum,* and *Eleusine indica* (L.) Gaertn., and *Sorghum halepense* (L.) Pers. Biotypes of *L. perenne* subsp. *multiflorum* and *E. indica* are also resistant to another nonselective herbicide glufosinate (group H/10).

Similar to grass species, most of the reported biotypes of broadleaf weed species occur in crop situations (reviewed in Beckie 2014 [13]). *Amaranthus* spp. comprise a large share of broadleaf species with intergroup HR. Within this genus, *A. tuberculatus* (Moq.) Sauer (syn. *rudis*) in the midwestern USA has the greatest number of reported biotypes. The outcrossing (e.g., dioecious) mating system of species in this genus facilitates intergroup HR. Biotypes of *A. tuberculatus* and *A. palmeri* (S.) Wats. are resistant to up to four and three SOA herbicides, respectively. Similar to *A. tuberculatus,* a biotype of *Raphanus raphanistrum* L. in Australia is resistant to four SOA herbicides [16].

Other significantly widespread species include *Conyza canadensis* (L.) Cronq. across large areas of the midwestern or eastern USA and Canada, *Ambrosia artemisiifolia* L. in the same regions, and *Kochia scoparia* (L.) Schrad. in western USA and Canada. The two most common SOA herbicides to which biotypes exhibit resistance are ALS inhibitors and photosystem-II inhibitors (group C1/5).

Glyphosate resistance (GR) has been reported in biotypes of *Conyza* spp., *A. tuberculatus, Ambrosia* spp., and *A. palmeri.* A common biotype of *C. canadensis* in the USA and Canada is resistant to ALS inhibitors and glyphosate [1]. A midwestern US survey of multiple resistance in *A. tuberculatus* found that all populations resistant to glyphosate exhibited resistance to ALS inhibitors, with 40% of populations also resistant to protoporphyrinogen oxidase (PPO) inhibitors (group E/14) [17]. Of more than 500 *A. tuberculatus* plants in another survey in the midwestern USA, 5% were ALS inhibitor-, PPO inhibitor-, and glyphosate-HR; of 120 fields, 10% had populations with this triple-SOA resistance [18]. Clearly, increasing incidence of cross- (metabolism-based) or multiple resistance in weed populations can greatly restrict alternative herbicide options for growers. When one of the compromised herbicides is glyphosate, the consequences can be serious.

HR in Field Crops: Commercialized Cultivars and Traits

In 2012, 18 countries grew at least 50,000 ha of genetically modified (GM) or transgenic crops [19]. The top five countries with transgenic crops were USA, Brazil, Argentina, Canada, and India, constituting 90% of global area (Table 15.2). The USA accounted for 41% of the area cultivated to transgenic crops, Brazil 22%, Argentina 14%, Canada 7%, and India 6%. Maize (*Zea mays* L.), soybean (*Glycine max* (L.) Merr., and cotton (*Gossypium hirsutum* L.) were the main transgenic field crops grown in the USA, whereas soybean dominated the cultivated area in Brazil and Argentina. In Canada, oilseed rape (canola) (*Brassica napus* L.), soybean, and maize are the main transgenic crops, while cotton is the only transgenic crop grown in India (Table 15.2). These four crops account for most of the transgenic crop area (Table 15.2).

Two traits still dominate commercialized transgenic field crops: herbicide and insecticide resistance (*Bacillus thuringiensis, Bt*). Single-trait HR accounted for 59% of transgenic crop area, single-trait *Bt* 15%, and stacked traits (HR + *Bt*) 26% (Table 15.2). Therefore, 17 years after the introduction of transgenic crops, the HR trait still dominates the cultivated area (85% of total).

Because the imidazolinone-HR trait in crop cultivars is non-transgenic, the area grown to these crops is not well documented. Indeed, these cultivars, which were developed by chemical mutagenesis, are only regulated in Canada [20]. This trait has been incorporated into oilseed rape, maize, lentil (*Lens culinaris* L.), rice, sunflower (*Helianthus annuus* L.), and wheat (*Triticum aestivum* L.; Table 15.3). The adoption of crop cultivars with this trait has generally been hampered by the high incidence of ALS inhibitor-HR weeds (Table 15.1).

Regardless of the HR trait, volunteers present in the next crop grown in rotation often need to be controlled by alternative herbicides to protect crop yield and quality [21]. The relative weediness of HR crop volunteers in cropland, disturbed areas, and natural areas is summarized in Table 15.3. HR volunteers are common weeds, and weediness depends on species, genotype, seed shatter prior to harvest and disbursement of seed at harvest, management practices, and environment. Chemical control options may be more limited if the crop volunteers are HR. In agroecosys-

Table 15.2 Transgenic crops grown in 2012, listed by country (>50,000 ha), trait, and crop. (Adapted from [19])[a]

	Area (million ha)	Area (%)	Crops grown (order of decreasing area)
By country (18):			
USA	69.5	40.8	Maize, soybean, cotton, canola, sugar beet, alfalfa, papaya, squash
Brazil	36.6	21.5	Soybean, maize, cotton
Argentina	23.9	14.0	Soybean, maize, cotton
Canada	11.6	6.8	Canola, maize, soybean, sugar beet
India	10.8	6.3	Cotton
China	4.0	2.4	Cotton, papaya, poplar, tomato, sweet pepper
Paraguay	3.4	2.0	Soybean, maize, cotton
South Africa	2.9	1.7	Maize, soybean, cotton
Pakistan	2.8	1.6	Cotton
Uruguay	1.4	0.8	Soybean, maize
Bolivia	1.0	0.6	Soybean
Philippines	0.8	0.5	Maize
Australia	0.7	0.4	Cotton, canola
Burkina Faso	0.3	0.2	Cotton
Myanmar	0.3	0.2	Cotton
Mexico	0.2	0.1	Cotton, soybean
Spain	0.1	<0.1	Maize
Chile	<0.1	<0.1	Maize, soybean, canola
Total area	*170.3*		
By trait:			
Herbicide resistance (HR)	100	59	
Bt (Bacillus thuringiensis)	26	15	
HR + *Bt*	44	26	
By crop:			
Soybean	80	47	
Maize	55	32	
Cotton	24	14	
Canola	10	6	
Other	<1	<1	

[a] Imidazolinone-HR crops are non-transgenic and therefore excluded; alfalfa, *Medicago sativa* L.; canola, *Brassica napus* L.; cotton, *Gossypium hirsutum* L.; maize, *Zea mays* L.; papaya, *Carica papaya* L.; soybean, *Glycine max* (L.) Merr.; squash, *Cucurbita moschata* L.; sugar beet, *Beta vulgaris* L

Table 15.3 Herbicide-resistant field crops registered in Canada and the USA, and weediness in cropped land, non-cropped disturbed areas (including roadsides and waste ground), and natural areas. (Updated from Beckie and Owen 2007 [21])

					Weediness		
Species	Herbicide resistance	Variety registration	Regulatory approval	Breeding system	Cropland	Disturbed areas	Natural areas
Alfalfa (*Medicago sativa*)	Glyphosate	NA[a]	Yes	Highly outcrossing	Yes	Yes	Yes
Canola (*Brassica napus*)	Glyphosate Glufosinate Imidazolinone	Yes Yes Yes	Yes Yes Yes	ca. 30 % outcrossing	Yes	Yes	No
Maize (*Zea mays*)	Glyphosate Glufosinate Imidazolinone Sethoxydim	NA NA NA NA	Yes Yes Yes Yes	Highly outcrossing	Yes	No	No
Cotton (*Gossypium hirsutum*)	Glyphosate Glufosinate	NA NA	Yes Yes	Selfing	Rarely	No	No
Creeping bentgrass (*Agrostis stolonifera*)	Glyphosate	NA	Yes[b]	Highly outcrossing	Yes	Yes	Yes
Lentil (*Lens culinaris*)	Imidazolinone	Yes	Yes	Highly selfing	Yes	No	No
Rice (*Oryza sativa*)	Imidazolinone Glufosinate	Yes No	Yes No	Highly selfing	Yes	No	No
Soybean (*Glycine max*)	Glyphosate Glufosinate	Yes Yes	Yes Yes	Highly selfing	Rarely	No	No
Sugar beet (*Beta vulgaris*)	Glyphosate Glufosinate	NA NA	Yes Yes	Selfing	Rarely	No	No
Sunflower (*Helianthus annuus*)	Imidazolinone	Yes	Yes	Selfing	Yes	No	No
Wheat (*Triticum aestivum*)	Imidazolinone	Yes	Yes	Highly selfing	Yes	No	No

[a] NA not required
[b] Currently restricted due to US District Court order

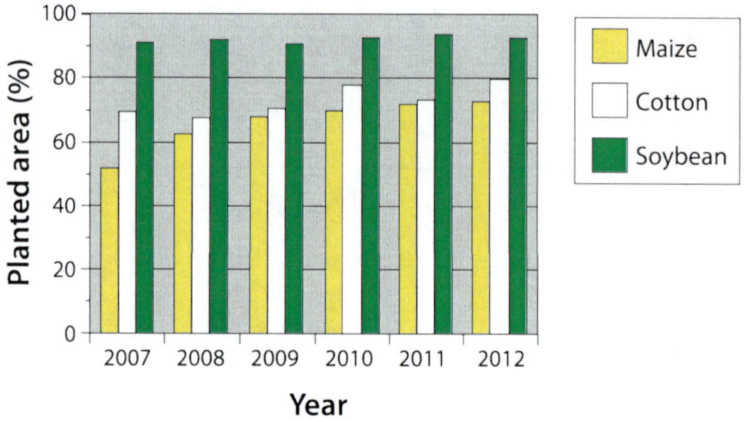

Fig. 15.3 Adoption of transgenic herbicide-resistant maize, cotton, and soybean in the USA. (Updated from Beckie and Owen 2007 [21]; USDA 2013 [72])

tems, there generally are no marked changes in volunteer weed problems associated with HR crops, except in no-tillage systems when glyphosate is used alone to control volunteers. However, in cropping systems where GR crops, such as oilseed rape or maize, are frequently grown or are expected to expand in cultivated area, increasing occurrence of unintended HR in volunteers due to gene flow has occurred and will continue to occur [21]. Control of such volunteers may require additional use of herbicides with alternative SOA and/or different tactics. A diverse rotation consisting of both non-HR and HR crops will mitigate HR crop volunteer problems and potential increased herbicide use.

GR crops account for about 85 % of transgenic crops grown worldwide [22]. In the USA, the adoption rate of GR soybean, cotton, and maize in 2012 was 93, 80, and 73 %, respectively (Fig. 15.3). Many of the farmers who adopted this technology used it year after year, with glyphosate as the only herbicide for weed management [23]. Consequently, the intense selection pressure resulted in the evolution of GR weeds only 4 years after the introduction of GR crops [24]. By 2010, 11 of 21 GR weed species worldwide had evolved in GR crop systems in the USA, Brazil, and Argentina [25]. The lesson that glyphosate is not immune to herbicide resistance and thus cannot be a standalone herbicide for sustainable weed management should have been learned following the discovery of the first GR weed, *L. rigidum,* in Australia in 1995 [26]. In general, glyphosate-based weed control programs worldwide are in need of additional partners for weed control and resistance management to reduce selection pressure on a single herbicide SOA [27].

However, cultivation of a GR crop over millions of hectares annually does not necessarily lead to the evolution of GR weeds. GR oilseed rape has been grown in Canada since 1995 and more recently in 2008 in Australia [28]. Most oilseed rape in Canada is grown in the western Prairies on more than 6 million ha annually. GR oilseed rape has comprised 40 % or more of the crop area since 2000 (Fig. 15.4). In 2012, transgenic cultivars (glufosinate- and glyphosate-resistant) constituted 97 %

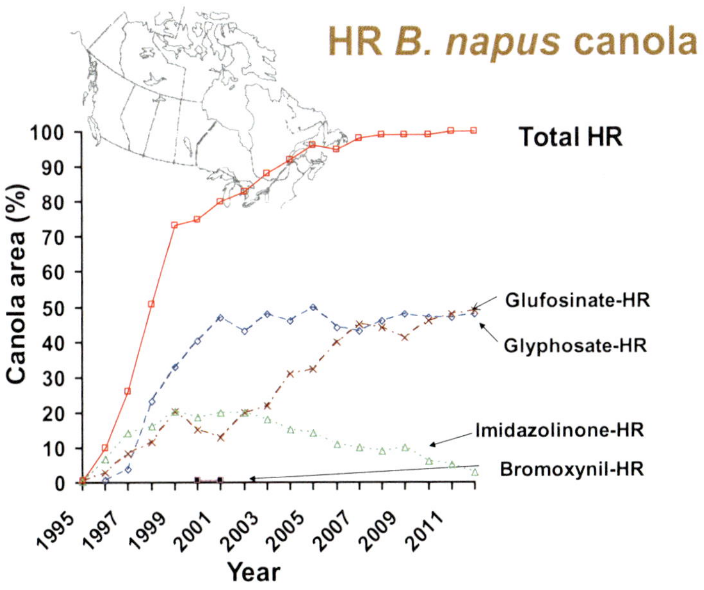

Fig. 15.4 Adoption of herbicide-resistant (HR) oilseed rape (canola) in Canada. (Updated from Beckie 2011 [29])

of crop area. To date, evolved GR weeds are absent in oilseed rape, attributed to moderate level of adoption of the trait (48% in 2012; Fig. 15.4) and frequency of growing the crop in rotation, typically once every 3–4 years [29]. In contrast to all other crops with GR cultivars, oilseed rape in Canada is the only case where another HR trait (glufosinate-resistant) is more popular. Moreover, many oilseed rape growers do not plant only cultivars with the same HR trait in a field over time. The majority of GR oilseed rape growers apply a glyphosate burndown treatment each spring, which may be tank-mixed with another herbicide for some residual weed control, and two in-crop glyphosate applications in GR oilseed rape. With favorable oilseed rape prices in the past few years, and as a means of controlling HR weeds, some growers are planting GR oilseed rape once every 2 years in their rotation. This high frequency of GR oilseed rape cultivation can increase the risk of evolution of GR weeds [30].

To facilitate proactive management of GR in weeds, a Web-based decision-support tool[1] has been developed for farmers in the Northern Great Plains of North America to assess the relative risk of resistance on a field-by-field basis [30]. Practices with the greatest risk weighting are lack of crop rotation diversity (growing mainly oilseeds) and a high frequency of GR crops in the rotation when only glyphosate is used for weed management. A similar tool at the same Web site was developed by the university research and extension personnel in eastern Canada and the USA midwestern region for maize and soybean growers. A risk calculator for

[1] http://www.weedtool.com.

Table 15.4 Some multiple or stacked herbicide-resistant crops commercialized or under development. (Adapted from [51])

Herbicide types[a]	Crops
Glyphosate, glufosinate	Soybean, maize, cotton
Glyphosate, ALS inhibitors	Soybean, maize
Glyphosate, glufosinate, 2,4-D	Soybean, cotton
Glyphosate, glufosinate, dicamba	Soybean, maize, cotton
Glyphosate, glufosinate, HPPD inhibitors	Soybean, cotton
Glyphosate, glufosinate, 2,4-D, ACC inhibitors	Maize
Glufosinate, dicamba	Wheat

[a] See Table 15.1 for abbreviations and description of herbicide sites of action

GR in *L. rigidum,* based on the frequency of use of glyphosate and tillage, was also developed for use by Australian growers [31]. Risk assessment decision-support software helps educate growers and dealers on the impact of management practices on selection for GR weeds.

In contrast to Canada, glyphosate usage is not recommended the year following GR oilseed rape in Australia to reduce glyphosate selection pressure [28, 32]. With the increasing incidence of GR weeds, the role and utility of herbicide-use regulations versus incentives in proactively or reactively managing GR weeds is the topic of much debate [29]. Duke has suggested that the paucity of *Bt* resistance worldwide may be related to greater regulatory oversight, that is, minimum non-*Bt* refuge area in a field, than that for herbicide resistance [33]. Regardless of the approach to herbicide and HR crop stewardship, the lesson to be learned from the experience of transgenic oilseed rape in Canada is that weed management diversity by means of a "real" rotation of (1) crops, (2) cultivars with and without HR traits or those with different HR traits, and (3) herbicide SOA can mitigate GR weed evolution.

Proactive or reactive HR weed management can be aided by crop cultivars with alternative single- or stacked-HR traits, which will become increasingly available to growers in the future. Because no major new SOA herbicide has been introduced to the marketplace for over 20 years [22], more efficient use of our existing arsenal is required to combat weed resistance. Combinations of HR traits, including glyphosate, glufosinate, ALS inhibitors, hydroxyphenylpyruvate dioxygenase (HPPD) inhibitors (group F2/27), and synthetic auxins (2,4-D, dicamba; group O/4), can be stacked in crop cultivars [34, 35]. Cultivars with dual (glyphosate plus glufosinate)-stacked HR traits are already available for soybean, maize, and cotton [36] (Table 15.4). These traits may be further stacked, for example, with a synthetic auxin or an HPPD inhibitor.

In the absence of new SOA herbicides, this is the only strategy available to industry in the short to medium term, that is, make better use of existing, albeit relatively old, herbicides. This strategy is generally viewed as giving enhanced flexibility to growers to cost-effectively manage weed resistance through mixtures, sequences, or rotations [37], provided that sufficient herbicide SOA diversity is maintained in rotations involving crops with stacked traits [38–41]. The latter caveat is critical to the sustainability of crops with stacked-HR traits. Weed populations resistant to glufosinate or HPPD inhibitors were first reported in 2010 or 2011 [42–46]. If crops with stacked-HR traits are managed similarly as many of the current GR crops, the same problems of weed shifts and evolved herbicide resistance will occur [41].

Challenges

The capture of a large portion of the herbicide market by glyphosate with the widespread adoption of GR crops led to significantly diminished herbicide discovery efforts since the mid-1990s [22]. For example, the number of herbicide active ingredients (ai) used on at least 10 % of USA soybean area declined from 11 in 1995 to just 1 in 2002: glyphosate [36]. The extensive (i.e., millions of hectares) and intensive (multiple applications within a crop) selection pressure exacerbated the evolution of GR. The incidence of multiple-HR weed biotypes in the past 20 years reflects growers' reliance on a few SOA herbicides (including glyphosate) to manage their weed populations. This reliance is the result of limited alternative herbicide options for many growers because of the crops grown in rotation (or monoculture) or the presence of existing HR weeds in their fields. Multiple resistance in weeds, particularly those resistant to glyphosate plus other herbicide SOA, can greatly reduce a grower's options for effective and economical weed management.

The introduction of HR crops in the mid-1990s was supposed to reduce the herbicide load (kg ai ha $^{-1}$) in the environment. Though fewer types of herbicides have been applied since the adoption of GR crops in the USA, the overall amount of herbicide ai has not necessarily decreased. The actual amount of ai applied per hectare increased from 1996 to 2007 in soybean and cotton, but decreased over the same period in maize [41]. Now with the increasing incidence of GR and intergroup-HR weeds worldwide, it is inevitable that herbicide use and overall cost of weed management will increase to effectively manage these populations. This consequence of poor HR crop stewardship is a reality for an increasing number of growers worldwide.

Growers with severe infestations of GR and/or intergroup-HR weeds have returned to more time-consuming and costly weed management practices used before the advent of GR crops. These strategies involve the use of several herbicides in a crop, as well as cultural and mechanical methods. For example, herbicide options to control multiple-HR *Amaranthus palmeri* and *A. tuberculatus* are limited, so growers are intensifying the use of tillage and even using expensive hand weeding in some situations [47]. To manage GR *A. palmeri,* growers rely heavily on herbicides, tillage, and hand weeding [48]. Herbicide use has increased sharply, with 2.5 times more herbicide ai applied in cotton now relative to before resistance. Although growers spend US$ 170 ha^{-1} on herbicides, control is not adequate. Generally six or seven SOA herbicides are used in a crop [48, 49]. Moreover, 92 % of Georgia cotton growers are hand weeding, with an average cost of US$ 28 per hand-weeded hectare. Tillage is used on 20–30 % of the cotton area, and increasing over time. Thus, the gains in conservation tillage, facilitated by HR crops, are being eroded.

On the other hand, this situation has improved the market for glufosinate-HR crops [33]. However, there is evidence that history is repeating itself, with some farmers in southern USA "abusing" glufosinate-HR crops as a means of controlling GR *A. palmeri* [49].

Multiple- or stacked-HR crop research is focused on six SOAs: ACC and ALS inhibitors, synthetic auxins, glyphosate, glufosinate, and HPPD inhibitors (Table 15.4). These SOAs have been used extensively in the past, and currently represent 77 % of the herbicide market [36]. All of the HR crops being actively developed and likely to be introduced in the next 5 years are resistant to herbicides with SOAs that have been used for decades [22]. Furthermore, there are already weed populations that are resistant to these herbicides. Where intergroup resistance to these SOA has already occurred in weed populations, the utility of some of these new transgenic-HR crops is questionable [22]. Crops with stacked-HR traits will incrementally help growers by enabling more herbicide options and expand the utility of existing herbicides, but will not be total solutions [50]. In fact, the consensus from the Global Herbicide Resistance Challenge conference in Australia, 2013, was that multiple-HR crops were an interim or short-term solution, not a long-term fix for HR weed management (e.g., [51, 52]). An important component of future crop stewardship is a rapid field assay for monitoring GR weeds, which is under development [53].

At the Global Herbicide Resistance Challenge conference in Australia in 2013, there was a strong, unified message of the urgent need for growers to adopt more diversified cropping systems and proactive weed management practices. For some HR weeds with efficient and extensive HR gene dispersal via seed or pollen, such as *Kochia scoparia,* a collective regional response is really required because of HR gene movement via seed and, to a lesser extent, via pollen [54]. However, the requirement for a regional response may hinder individual growers to be proactive. Prevention and mitigation strategies for HR weed management are well understood, but there has been little will to implement them; this may change if the severity of GR and intergroup-HR weed problems intensifies [33]. There is little doubt about that scenario unfolding. There are a number of excellent online resources for proactive and reactive HR weed management strategies, tactics, and practices. In many cases, such information has been extensively transferred to extension specialists, consultants, pesticide dealers, crop commodity organizations, and growers. However, growers' management of herbicides typically does not align with what weed scientists recommend as optimal HR management [55]. What then will it take for growers to finally heed the message? There is an urgent call for action.

Opportunities

The increasing evolution and spread of GR and intergroup-HR weeds have caused the agrochemical industry to reinvest in herbicide discovery [22]. Herbicides with new SOA are needed to help manage key weeds in our major field crops. A number of new HR crops are due to be launched in the next 5 years [22], and will be an important component of future weed management systems [56] despite the limitations outlined previously. If used properly, they will provide farmers new tools to delay and mitigate HR weeds. Dual-stack HR crops with glyphosate and glufosinate resistance are already available in maize, soybean, and cotton [36]. Broadleaf crops that are resistant to auxin herbicides will likely be the next HR crop technology to have a significant impact (e.g., soybean and cotton); resistance to HPPD-inhibiting herbicides also could have a big impact in the next decade [36].

To avoid history repeating itself, there is a need for best management practices (e.g., [29, 57, 58]) to be implemented when new multiple-HR crops are introduced into the market place. This implementation will require industry and government incentives, as well as a strong education and awareness campaign directed at all stakeholders [59]. In a 2005 survey of nearly 1200 growers in six USA states, the top three sources of information on weed resistance issues were farm publications (41 %), dealers/retailers (17 %), and university/extension (14 %) [60]. In a follow-up survey in 2010, involving about 1650 growers in 22 USA states, these three sources were 41, 22, and 20 %, respectively [61]—similar results as the previous survey. Therefore, targeting weed resistance messaging at these top sources of information for growers may be the most cost-effective strategy.

There are some promising technologies under development that may help manage HR weeds. Advances in weed-detecting technology for use in fallow land and potentially field row crops may result in better HR weed management, reduced herbicide use, and potentially reduced selection pressure [62]. The potential benefit of using RNAi gene silencing technology to manage GR weeds, such as *A. palmeri* in GR crops, was recently described (Monsanto BioDirect™ [63]). It would certainly rank as a "game changer" if a postemergence spray application of a xenobiotic developed with this technology could make GR weeds susceptible to glyphosate. We must always be cognizant, however, that nature always has a way of circumventing new technology. There are no silver bullets in HR weed management.

In the past couple of years, there has been much more emphasis by weed scientists in the midwestern and southern USA in better managing the soil seed bank. A tool under development for reducing weed seed return to the soil seed bank is the Harrington Seed Destructor [64]. Targeting weed seeds at harvest—now primarily via chaff carts or narrow windrow burning—is now a major focus of growers in Western Australia, and the majority of these growers are optimistic about the future of grain cropping despite high incidence of HR weeds [65]. The Harrington Seed Destructor, narrow windrow burning, and chaff cart treatments each reduced *L. rigidum* emergence by 55 % compared with nontreated controls [64]. Substituting windrow burning with alternative methods of weed seed destruction or capture would certainly help the environment.

Attempting to manage herbicide resistance solely with herbicides is doomed to fail [66]. Sustainability will only be achieved if there is diversity in both the agroecosystem and the herbicide and non-herbicide tools employed for weed control [65, 67, 68]. True integrated weed management systems will, of necessity, become more popular as weed resistance to popular herbicides continues to increase [69].

There are encouraging signs that more integrated weed management is starting to happen in traditional monoculture, monoherbicide GR cropping systems. For managing GR *A. palmeri* in southern US cotton and soybean, preemergence residual herbicides, competition through narrow-row seeding, introduction of cover crops, hand weeding of weed escapes, and moldboard plowing to bury weed seeds are techniques now used to lessen selection pressure on existing postemergence herbicides [49]. Sustainable weed management in GR crop-based systems was demonstrated in a multistate, large-scale field study in the midwestern and southern USA. The study showed that alternative weed management tactics and practices in the glyphosate-based systems maintained or reduced weed communities, mitigated or reduced GR in weed populations, and resulted in net economic returns either positive or neutral relative to local grower management practices [70].

Herbicide resistance in weeds, particularly those exhibiting intergroup resistance, is the greatest threat to sustained agriculture production in the industrialized countries. Cost-effective solutions to many of our HR weed problems already exist, but largely lay untouched on the shelf. Moreover, research and development has increased since 1995 on non-herbicidal weed management strategies as well as strategies that integrate other weed management systems with herbicide use [71]. However, uptake of these technologies by most growers has been poor. In the future, weed management by growers will require more knowledge, planning, time, cost, and risk than in the past, despite ever-increasing farm size. Those growers who are able to adapt to this reality will have farm enterprises that survive and are profitable. How many growers will accept the challenge?

References

1. Heap IM (2013) International survey of herbicide resistant weeds (Internet). http://www.weedscience.org. Accessed 10 July 2013
2. Beckie HJ, Tardif FJ (2012) Herbicide cross resistance in weeds. Crop Prot 35:15–28
3. Moss SR, Cocker KM, Brown AC, Hall L, Field LM (2003) Characterisation of target-site resistance to ACCase-inhibiting herbicides in the weed *Alopecurus myosuroides* (black-grass). Pest Manage Sci 59:190–201
4. Claude J-P, Didier A, Favier P, Thalinger PP (2004) Development of a European database for the evolution follow-up of resistant black-grass (*Alopecurus myosuroides* huds.) populations in cereal crops. Proceedings of the fourth International Weed Science Congress; Durban, South Africa. International Weed Science Society, Davis, CA, p 48
5. Marshall R, Moss S (2004) Resistance to acetolactate inhibiting herbicides in UK black-grass (*Alopecurus myosuroides*) populations. Weed Sci Soc Am Abstr 44:15
6. Délye C, Menchari Y, Guillemin J-P, Matejicek A, Michel S, Camilleri C, Chauvel B (2007) Status of black grass (*Alopecurus myosuroides*) resistance to acetyl-coenzyme A carboxylase inhibitors in France. Weed Res 47:95–105
7. Délye C, Michel S, Berard A, Chauvel B, Brunel D, Gullemin J-P, Dessaint F, le Corre V (2010) Geographical variation in resistance to acetyl-coenzyme A carboxylase-inhibiting herbicides across the range of the arable weed *Alopecurus myosuroides* (black-grass). New Phytol 186:1005–1017

8. Beckie HJ, Warwick SI, Sauder CA (2012) Basis for herbicide resistance in Canadian populations of wild oat (*Avena fatua*). Weed Sci 60:10–18
9. Preston C (2004) Herbicide resistance in weeds endowed by enhanced detoxification: complications for management. Weed Sci 52:448–453
10. Preston C, Mallory-Smith CA (2001) Biochemical mechanisms, inheritance, and molecular genetics of herbicide resistance in weeds. In Powles SB, Shaner DL (eds) Herbicide resistance and world grains. CRC, New York, pp 23–60
11. Valverde BE, Chaves L, Garita I, Ramirez F, Vargas E, Carmiol J, Riches CR, Casely JC (2001) Modified herbicide regimes for propanil-resistant junglerice control in rain-fed rice. Weed Sci 49:395–405
12. Brazier M, Cole DJ, Edwards R (2002) *O*-Glucosyltransferase activities toward phenolic natural products and xenobiotics in wheat and herbicide-resistant and herbicide-susceptible black-grass (*Alopecurus myosuroides*). Phytochemistry 59:149–156
13. Beckie HJ (2014 In press) Weed resistance to herbicides with different sites of action. In: Ward S (ed) Herbicide resistance in agroecosystems. Wiley Blackwell, New York
14. Holm LG, Plucknett DL, Pancho JV, Herberger JP (1991) The world's worst weeds. Distribution and biology. The University Press of Hawaii, Honolulu, p 609
15. Beckie HJ, Francis A, Hall LM (2012) The biology of Canadian weeds. 27. *Avena fatua* L. (updated). Can J Plant Sci 92:1329–1357
16. Walsh MJ, Owen MJ, Powles SB (2007) Frequency and distribution of herbicide resistance in *Raphanus raphanistrum* populations randomly collected across the Western Australian wheatbelt. Weed Res 47:542–550
17. Tranel PJ, Riggins CW, Bell MS, Hager AG (2011) Herbicide resistances in *Amaranthus tuberculatus*: a call for new options. J Agric Food Chem 59:5808–5812
18. Tranel PJ (2013) Update on herbicide resistance in waterhemp (*Amaranthus tuberculatus*). In: Proceedings of the global herbicide resistance challenge; Feb 18–21; Perth, Australia. Australia Herbicide Resistance Initiative, p 31. http://www.herbicideresistanceconference.com.au
19. James C (2013) ISAAA Brief 44-2012: Executive summary. Global status of commercialized biotech/GM crops: 2012 (Internet). http://www.isaaa.org/resources/publications/briefs/44/executivesummary/default.asp. Accessed 10 July 2013
20. Beckie HJ, Harker KN, Hall LM, Warwick SI, Legere A, Sikkema PH, Clayton GW, Thomas AG, Leeson JY, Seguin-Swartz G, Simard M-J (2006) A decade of herbicide-resistant crops in Canada. Can J Plant Sci 86:1243–1264
21. Beckie HJ, Owen MDK (2007) Herbicide-resistant crops as weeds in North America. CAB Rev: Perspect Agric Vet Sci Nutr Nat Resour 2(044):22
22. Duke SO (2012) Why have no new herbicide modes of action appeared in recent years? Pest Manage Sci 68:505–512
23. Wilson RG, Young BG, Matthews JL, Weller SC, Johnson WG, Jordan DL, Owen MDK, Dixon PM, Shaw DR (2011) Benchmark study on glyphosate-resistant cropping systems in the United States. Part 4: weed management practices and effect on weed populations and soil seedbanks. Pest Manage Sci 67:771–780
24. VanGessel MJ (2001) Glyphosate-resistant horseweed from Delaware. Weed Sci 49:703–705
25. Heap IM (2011) Global distribution of glyphosate resistant weeds (Internet). Weed Sci Soc Am Abstr No 20. http://wssaabstracts.com/public/index.php?conf=4. Accessed 10 July 2013
26. Pratley J, Baines P, Eberbach P, Incerti M, Broster J (1996) Glyphosate resistance in annual ryegrass. In: Virgona J, Michalk D (ed) Proceedings of the 11th annual conference of the Grasslands Society of New South Wales; Grasslands Society of NWS, Wagga Wagga, Australia, p 122
27. Wright T (2013) Auxinic herbicide resistance. In: Proceedings of the global herbicide resistance challenge; 2013 Feb 18–21; Perth, Australia. Australia Herbicide Resistance Initiative, p 46. http://www.herbicideresistanceconference.com.au
28. Preston C (2010) Glyphosate-resistant rigid ryegrass in Australia. In Nandula VK (ed) Glyphosate resistance in crops and weeds. New York: Wiley, pp 233–247.

29. Beckie HJ (2011) Herbicide-resistant weed management: focus on glyphosate. Pest Manage Sci 67:1037–1048
30. Beckie HJ, Harker KN, Hall LM, Holm FA, Gulden RH (2011) Risk assessment of glyphosate resistance in western Canada. Weed Technol 25:159–164
31. Stanton RA, Pratley JE, Hudson D, Dill GM (2008) A risk calculator for glyphosate resistance in *Lolium rigidum* (Gaud.). Pest Manage Sci 64:402–408
32. Anonymous (2010) Stewardship for roundup ready canola—a focus on the components that reduce potential selection pressure for resistance to glyphosate (Internet). Canberra, Australia: Grains Research and Development Corporation. http://www.grdc.com.au/director/events/researchupdates?item_id=C495D546A6D33BB694E6923B47113293&pageNumber=8. Accessed 10 July 2013
33. Duke SO (2011) Comparing conventional and biotechnology-based pest management. J Agric Food Chem 59:5793–5798
34. Feng PCC, CaJacob CA, Martino-Catt SJ, Cerny RE, Elmore GA, Heck GR, Huang J, Kruger WM, Malven M, Miklos JA, Padgette SR (2010) Glyphosate-resistant crops: developing the next generation products. In: Nandula VK (ed) Glyphosate resistance in crops and weeds. Wiley, New York, pp 45–65
35. Green JM, Castle LA (2010) Transitioning from single to multiple herbicide-resistant crops. In: Nandula VK (ed) Glyphosate resistance in crops and weeds. Wiley, New York, pp 67–91
36. Green JM (2011) Outlook on weed management in herbicide-resistant crops: need for diversification. Outlooks Pest Manage 22:100–104
37. Beckie HJ, Reboud X (2009) Selecting for weed resistance: herbicide rotation and mixture. Weed Technol 23:363–370
38. Green JM, Hazel CB, Forney DR, Pugh LM (2008) New multiple-herbicide crop resistance and formulation technology to augment the utility of glyphosate. Pest Manage Sci 64: 332–339
39. Carpenter JE, Gianessi LP (2010) Economic impact of glyphosate-resistant weeds. In: Nandula VK (ed) Glyphosate resistance in crops and weeds. Wiley, New York, pp 297–312
40. Culpepper AS, Webster TM, Sosnoskie LM, York AC (2010) Glyphosate-resistant Palmer amaranth in the United States. In: Nandula VK (ed) Glyphosate resistance in crops and weeds. Wiley, New York, pp 195–212
41. Owen MDK (2010) Herbicide-resistant weeds in genetically-engineered crops (Internet). Summary report to the Subcommittee on Domestic Policy, Committee on Oversight and Government Reform, U.S. House of Representatives. http://www7.nationalacademies.org/ocga/testimony/t_Herbicide-Resistant_Weeds_in_GE_Crops.asp. Accessed 10 July 2013
42. Jalaludin A, Ngim J, Bakar BHJ, Alias Z (2010) Preliminary findings of potentially resistant goosegrass (*Eleusine indica*) to glufosinate-ammonium in Malaysia. Weed Biol Manage 10:256–260
43. Seng CT, Lun LV, San CT, Sahid IB (2010) Initial report of glufosinate and paraquat multiple resistance that evolved in a biotype of goosegrass (*Eleusine indica*) in Malaysia. Weed Biol Manage 10:229–233
44. Avila-Garcia WV, Mallory-Smith C (2011) Glyphosate-resistant Italian ryegrass (*Lolium multiflorum*) populations also exhibit resistance to glufosinate. Weed Sci 59:305–309
45. Hausman NE, Singh S, Tranel PJ, Riechers DE, Kaundun SS, Polge ND, Thomas DA, Hager AG (2011) Resistance to HPPD-inhibiting herbicides in a population of waterhemp (*Amaranthus tuberculatus*) from Illinois, United States. Pest Manage Sci 67:258–261
46. McMullan PM, Green JM (2011) Identification of a tall waterhemp (*Amaranthus tuberculatus*) biotype resistant to HPPD-inhibiting herbicides, atrazine, and thifensulfuron in Iowa. Weed Technol 25:514–518
47. (CAST) Council for Agricultural Science and Technology (2012). Herbicide-resistant weeds threaten soil conservation gains: finding a balance for soil and farm sustainability. Paper No 49, Ames IA: CAST

48. Culpepper AS, Sosnoskie L (2013) Glyphosate-resistant Palmer amaranth increases herbicide use, tillage, and hand weeding in Georgia cotton (Internet). Weed Sci Soc Am Abstr No 270. http://wssaabstracts.com/public/17/abstract-270.html. Accessed 10 July 2013
49. Norsworthy J (2013) Glyphosate-resistant Palmer amaranth (*Amaranthus palmeri*) in southern USA row-crop production: impact and current management strategies. In: Proceedings of the global herbicide resistance challenge; 2013 Feb 18–21; Perth, Australia. Australia Herbicide Resistance Initiative, p 92. http://www.herbicideresistanceconference.com.au
50. Green JM, Owen MDK (2011) Herbicide-resistant crops: utilities and limiations for herbicide-resistant weed management. J Agric Food Chem 59:5819–5829
51. Green JM (2013) State of herbicides and herbicide traits at the start of 2013. In: Proceedings of the global herbicide resistance challenge; 2013 Feb 18–21; Perth, Australia. Australia Herbicide Resistance Initiative, p 27. http://www.herbicideresistanceconference.com.au
52. Heap IM (2013) Overview of global herbicide resistance cases and lessons learnt. In: Proceedings of the global herbicide resistance challenge; 2013 Feb 18–21; Perth, Australia. Australia Herbicide Resistance Initiative, p 28. http://www.herbicideresistanceconference.com.au
53. Sammons D, Shaner D, Rumecal D, Kretzmer K, DeJarnette R (2013) A rapid enyzme assay for shikimate to detect glyphosate resistance in the field. In: Proceedings of the global herbicide resistance challenge; 2013 Feb 18–21; Perth, Australia. Australia Herbicide Resistance Initiative, p 42. http://www.herbicideresistanceconference.com.au
54. Beckie HJ, Blackshaw RE, Low R, Hall LM, Sauder CA, Martin S, Brandt RN, Shirriff SW (2013) Glyphosate—and acetolactate synthase inhibitor-resistant kochia (*Kochia scoparia*) in western Canada (Internet). Weed Sci Soc Amer Abstr No 231. http://wssaabstracts.com/public/17/abstract-270.html. Accessed 10 July 2013
55. Llewellyn R (2013) Considering the socio-economics of herbicide resistance management decisions. In: Proceedings of the global herbicide resistance challenge; 2013 Feb 18–21; Perth, Australia. Australia Herbicide Resistance Initiative, p 28. http://www.herbicideresistanceconference.com.au
56. Green JM (2012) The benefits of herbicide-resistant crops. Pest Manage Sci 68:1323–1331
57. Beckie HJ (2006) Herbicide-resistant weeds: management tactics and practices. Weed Technol 20:793–814
58. Norsworthy JK, Ward SM, Shaw DR, Llewellyn RS, Nichols RL, Webster TM, Bradley KW, Friswold G, Powles SB, Burgos NR, Witt WW, Barrett M (2012) Reducing the risks of herbicide resistance: best management practices and recommendations. Weed Sci (Spec Issue): 31–62
59. Soteres JK (2013) Monsanto global strategies for the management of herbicide-resistant weeds. In: Proceedings of the global herbicide resistance challenge; 2013 Feb 18–21; Perth, Australia. Australia Herbicide Resistance Initiative, p 32. http://www.herbicideresistanceconference.com.au
60. Givens WA, Shaw DR, Newman ME, Weller SC, Young BG, Wilson RG, Owen MDK, Jordan DL (2011) Benchmark study on glyphosate-resistant cropping systems in the United States. Part 3: Grower awareness, information sources, experiences and management practices regarding glyphosate-resistant weeds. Pest Manag Sci 67:758–770
61. Prince JM, Shaw DR, Givens WA, Newman ME, Owen MDK, Weller SC, Young BG, Wilson RG, Jordan DL (2012) Benchmark study: II. A 2010 survey to assess grower awareness of and attitudes toward glyphosate resistance. Weed Technol 26:531–535
62. Cook T (2013) Weed-detecting technology: an excellent opportunity for advanced glyphosate resistance management. In: Proceedings of the global herbicide resistance challenge; 2013 Feb 18–21; Perth, Australia. Australia Herbicide Resistance Initiative, p 94. http://www.herbicideresistanceconference.com.au
63. Sammons D, Wang D, Morris P, Duncan B, Griffith G, Findley D (2013) Strategies for countering herbicide resistance. In: Proceedings of the global herbicide resistance challenge; 2013 Feb 18–21; Perth, Australia. Australia Herbicide Resistance Initiative, p 106. http://www.herbicideresistanceconference.com.au

64. Aves C, Walsh M (2013) The Harrington seed destructor and harvest weed seed control in South Eastern Australia. In: Proceedings of the global herbicide resistance challenge; 2013 Feb 18–21; Perth, Australia. Australia Herbicide Resistance Initiative, p 101. http://www.herbicideresistanceconference.com.au
65. Newman P (2013) Successful control of multiple herbicide-resistant *Lolium* with integrated strategies in Western Australia. In: Proceedings of the global herbicide resistance challenge; 2013 Feb 18–21; Perth, Australia. Australia Herbicide Resistance Initiative, p 100. http://www.herbicideresistanceconference.com.au
66. Preston C (2013) What has herbicide resistance taught us? In: Proceedings of the global herbicide resistance challenge; 2013 Feb 18–21; Perth, Australia. Australia Herbicide Resistance Initiative, p 29. http://www.herbicideresistanceconference.com.au
67. Délye C (2013) Non-target-site-based resistance to herbicides: what do we know, and how can we know more? In: Proceedings of the global herbicide resistance challenge; 2013 Feb 18–21; Perth, Australia. Australia Herbicide Resistance Initiative, p 66. http://www.herbicideresistanceconference.com.au
68. Walsh MJ (2013) Herbicide resistance management in cereal cropping systems. In: Proceedings of the global herbicide resistance challenge; 2013 Feb 18–21; Perth, Australia. Australia Herbicide Resistance Initiative, p 93. http://www.herbicideresistanceconference.com.au
69. Harker KN, O'Donovan JT, Blackshaw RE (2013) Weed control systems in HR canola—a resistance reprieve? In: Proceedings of the global herbicide resistance challenge; 2013 Feb 18–21; Perth, Australia. Australia Herbicide Resistance Initiative, p 94. http://www.herbicideresistanceconference.com.au
70. Owen MDK, Shaw DR, Weller SC, Dixon PM, Young BG, Jordan DL, Wilson RG (2013) The Benchmark study-a field scale project demonstrating the sustainability of glyphosate-based crop production. In: Proceedings of the global herbicide resistance challenge; 2013 Feb 18–21; Perth, Australia. Australia Herbicide Resistance Initiative, p 33. http://www.herbicideresistanceconference.com.au
71. Harker KN, O'Donovan JT (2013) Recent weed control, weed management, and integrated weed management. Weed Technol 27:1–11
72. (USDA) United States Department of Agriculture (2013) Acreage report (Internet). United States Department of Agriculture, National Agricultural Statistics Service. http://www.nass.usda.gov. Accessed 10 July 2013

Chapter 16
Challenges and Opportunities in Weed Management Under a Changing Agricultural Scenario

K. K. Barman, V. P. Singh, R. P. Dubey, P. K. Singh, Anil Dixit and A. R. Sharma

Introduction

The world's population is expected to reach 9 billion by 2050 [1]. Meeting the food requirements of this huge population will not be easy. The farmers around the world will have to produce higher yields, and simultaneously will have to give attention to a fragile environment and conserve the valuable resources of land and water. Further, population growth and economic development will result in more demand for meat and other animal products as well as fruits and vegetables [2]. Presently, about one-third of global cereal production is used as animal feed to obtain eggs, dairy products, and meat [3], and due to this increased demand for animal products, the world will face an increased pressure on cropland, fossil fuel energy, and water [4]. It is estimated that food production will need to increase by 50–100 % to support the growing and changing population [5].

Agriculture is characterized by unique combinations of soil, climate, topography, hydrology, and biological diversity, as well as a diversity of crops and production systems. A single farming system or approach will not be able to best feed the

A. R. Sharma (✉) · K. K. Barman · V. P. Singh · R. P. Dubey · P. K. Singh · A. Dixit
Directorate of Weed Science Research, Indian Council of Agricultural Research, Jabalpur, Madhya Pradesh 482004, India
e-mail: sharma.ar@rediffmail.com

K. K. Barman
e-mail: barmankk@gmail.com

V. P. Singh
e-mail: vpsinghnrcws@gmail.com

R. P. Dubey
e-mail: dubeyrp@gmail.com

P. K. Singh
e-mail: drsinghpk@gmail.com

A. Dixit
e-mail: dranildixit@in.com

B. S. Chauhan, G. Mahajan (eds.), *Recent Advances in Weed Management*,
DOI 10.1007/978-1-4939-1019-9_16,

planet, while also protecting the environment, because of the enormous variation in agroecological circumstances across the planet as well as unpredictable weather and market conditions. A wide diversity of crops, livestock, and farming systems will help promote resilience, and will likely play a key role in future food and ecosystem security. Hence, like any other farming activities, weed management under diversified farming systems will require flexible, adaptive, and localized management systems that cannot be covered by one-size-fits-all policies. This chapter deals with the probable future agricultural scenario and consequent challenges in weed science research.

What do Weeds Cause?

Weeds have been known to humans since the very beginning of civilization. The term "weed" is used to describe a plant considered undesirable within a certain context, and usually applied to unwanted plants in human-controlled settings, viz. farm fields, gardens, lawns, and parks. The word weed does not carry any significance in relation to botanical classification, since a plant that is a weed in one context is not a weed in another context where it is wanted. For example, Bermuda grass (*Cyanodon dactylon*), unlike in crop fields, is not a weed in lawns where it is grown and nurtured. In an agricultural field, all other plants except those grown with an aim to harvest are termed as weeds. Thus, a weed may be defined as "any plant that is objectionable or interferes with the activities or welfare of man" [6]. Despite several modern weed control technologies developed with an aim to keep weeds under control, they are still a threat to agricultural productivity [7]. Weed management is more than control of existing weed problems and places greater emphasis on preventing weed reproduction, reducing weed emergence after crop planting, and minimizing weed competition with the crop [8, 9].

Weed science is an integrative, applied scientific discipline typical of most other pest management and production-oriented disciplines of modern agriculture [10]. It combines fundamental and applied sciences to study weeds, and focuses on mitigating the negative impacts of weeds in human-controlled settings, especially in agricultural production systems. Purdue University described weed science as "the study of vegetation management in agriculture, aquatics, horticulture, and right-of-way, essentially anywhere plants need to be managed. It involves the study of all the tools available for this purpose, such as cropping systems, herbicides, management techniques, and seed genetics. It is not just the controlling of plants, but the study of these plants. This includes plant ecology, physiology, and the genetics of plants species that have been identified to have impact on the economy and our ecology" [11]. Weed scientists focus their research on basic biological and ecological characteristics of weeds, and develop tools and tactics to reduce weeds and their effects in crops, rangelands, forest plantations, roadsides, and aquatic environments [12].

Weeds can effectively compete for nutrient, water, space, and light and thereby can irreversibly harm the desired plants in agricultural and horticultural farming systems. Besides that, weeds may directly or indirectly affect the management of all the terrestrial and aquatic resources and interfere with the values and activities of

Table 16.1 The negative impacts of weeds in managed ecosystems

Decrease in crop yield
Interference in harvesting operation
Increase in production cost of crops
Reduced quality of crop yield
Steal shelter and food from animals by invading the grazing areas
Inflict injury or death of animals
Act as a potential source of fire hazard in forests
Impart odors to milk and meat
Act as alternative hosts for insects and pathogens
Interfere in fisheries/aquaculture and navigation
Reduce the aesthetic and recreational value of water bodies, public parks, etc.
Interfere in irrigation water management by hindering free flow of water through canals
Cause health hazards like skin allergy, fever, asthma, nasal diseases, etc.
Restrict visibility of signs, intersections, and traffic signals along road and railways
Create hindrances in electricity installations and security operations

people belonging to various segments of society, viz. foresters, ranchers, etc. Some routinely encountered negative impacts of weeds in human-controlled settings and managed ecosystems are listed in Table 16.1.

The competitive ability of weeds is determined by several plant characteristics. One of the most common traits of a weed species is its tendency to be an annual or biennial, rather than a perennial; this allows the species a faster reproduction rate leading to a higher fecundity [13]. Another characteristic that determines the "weediness" of a species is the ability to colonize under high sunlight and low soil moisture conditions. Plants that have capabilities of dealing with herbivores as well as plants that have allelopathic traits tend to be better at outcompeting surrounding plant species. Some non-native species of plants are considered to be very weedy in nature, as they can grow faster and bigger, increase reproduction rates, and can have increased survival rates when outside of their native habitat. This may be due, in part, to the loss of environmental checks needed to keep these plants in balance within their natural habitat. Genetic makeup also determines the ability of a plant to become weedy in nature; however, a genetic pattern has yet to be described [14]. In India, out of the total 826 reported weed species, 80 species are considered as very serious and 198 as serious weeds.

History of Weed Science

Man has been plagued by unwanted plants among cultivated fields since the Biblical times. Importance of controlling weeds for better yield and use of tools for removing unwanted plants were depicted in ancient writings and archeological artifacts [10]. However, weed control received little attention or research efforts until the late 1800s and early 1900s, and for centuries, weed control has been accomplished as a by-product of seedbed preparation. Agricultural mechanization efforts largely ignored weed control implements until 1914 when the rodweeder was introduced primarily for weed control [15]. Even the modern hoe, which is synonymous with weed control, was specifically designed by Jethro Tull to break up the soil to make nutrients more readily available to the crop's roots [15]. Early methods of weed control include labor-intensive hand hoeing and hand pulling of weeds as well as cultural practices, such as crop rotation. Although *hoe-hands* are rare in developed countries, hand removal of weeds remains the dominant form of weed control in many undeveloped nations. Rotation practices were largely replaced by monoculture systems and chemical weed control by the 1940s [16]. However, in recent times, crop rotation has again become an integral part of weed management in organic farming as well as integrated weed management (IWM) practices in conventional farming systems.

Herbicidal action of some compounds for weed control was first highlighted in 1885 [17]. In fact the study of weeds as a science began with the introduction of phytotoxic chemicals for the control of weeds in the early 1900s [12]. The first chemical used to control weeds was inorganic copper salt, which was then followed by sulfuric acid. Thus, the history of weed science parallels the history of modern agriculture and is hardly 100 years old [10]. Planned weed-controlled opportunities, and thus the birth of weed science as a discipline, took place with the synthesis of 2,4-dichlorophenoxyacetic acid (2,4-D) in 1941 by Pokorny, followed by the discovery of its plant growth-regulating and herbicidal properties by Hammer and Tukey in 1944 [12]. This is the first account of a synthesized organic chemical used to control weeds [18]. Weed science received a major boost as a valid scientific discipline with the commercial acceptance of 2,4-D as an effective herbicide. Until this point, research was limited in funding as well as in interest by the scientific community; those who did dare tackle questions about weed control did so neither with the chance of recognition nor with insight from previous research. When 2,4-D appeared in the market, it offered users a cheaper option of weed control that could be applied at relatively low rates and in many agricultural settings [19]. The characteristics of 2,4-D offered hope that chemical weed control could revolutionize global food production, in turn, drawing a great deal of attention to weed control research. The success story of 2,4-D led to an explosion of synthesized herbicides during the 1940s and 1950s. By 1950, there were roughly 25 herbicides available for use [15]. By the 1960s, more than 120 effective herbicides were available for weed management, which were enough to ensure that chemical weed control was a viable replacement of labor-intensive mechanical weed removal. Thus, weed science was guaranteed a spot among respected subsets of agricultural sciences.

Introduction of glyphosate to the herbicide market in the year 1974, and subsequent development of glyphosate-resistant soybean and its commercialization in 1996, initiated a new era in modern weed science, similar to that of 2,4-D discovery [20]. This technology allowed the use of a non-selective herbicide within a row crop setting without injury to the resistant crop. This gave farmers the freedom of using a hassle-free means to control weeds in their fields as and when it was desired. Presently, attempts are being made to design an herbicide-resistant crop that contains resistance to multiple non-selective herbicides. If it becomes a reality, this feat would allow farmers greater flexibility in herbicide choice, reduce dependency on a single herbicide, and also reduce the apprehension with respect to probable evolution of glyphosate-resistant "super weed" species.

Modern Weed Management Strategies

Much advancement has been achieved in weed control since the beginning of modern weed science research. These achievements came through several complications and defeats; however, advancements have still been made and improved weed control methods have allowed farmers to witness dramatic increases in crop yield. In view of the continuous increase in world population and diminishing availability of agricultural land, it is imperative that the research in weed management progress further with the changing agricultural needs to guarantee adequate food for ourselves and posterity.

The main reason behind widespread adoption of herbicides in the industrially developed countries was socialistic, through a reduction in the need for labor and the concomitant release of people from farming [17]. Chemical weed control offered several benefits to farmers by reducing weeds, enabling early planting, reducing need for soil tillage, and providing economic advantages through reduced cost of production. However, it was not the only tool to manage weeds. The disadvantage associated with herbicide techniques is the development of herbicide resistance in weeds. The wheat growers in the Indian states of Punjab and Haryana suffered a lot during the late 1990s and early years of the last decade due to the development of isoproturon resistance in *Phalaris minor*, a major weed of this region. The problem persisted until alternative herbicides to control this weed became available in the market. The researchers developed and fine-tuned several other strategies to manage weeds to deal with various social, cultural, environmental, and economic issues. All those weed management strategies are typically grouped into five categories: preventive, cultural, mechanical (physical), biological, and chemical.

Preventive Strategies

Among all the weed control strategies, prevention is an important component, which needs greater attention. It comprises methods used for avoiding the introduction and spread of weeds, i.e., avoiding weed seed introduction into new areas including contaminated crop seeds; movement of seeds and plant parts, tillage, harvest, and processing equipment; livestock; manure and compost; irrigation and drainage water; and forage and food grains [21]. Prevention of weeds can be successful, depending upon the weed species, means of dissemination, and farm size [7]. Preventive weed management programs are successful when undertaken at a community level. Use of certified seeds by the farmers and enforcement of weed laws can make weed prevention programs successful.

Cultural Strategies

Cultural weed control comprises the principles of using plant competition or cropping practices to suppress weeds, through the use of either smother or competitive crops and crop rotation. Cultural methods may include crop sowing time and spatial arrangement, crop genotype, cover crops, intercropping, and crop fertilization.

Crop Sowing Time and Spatial Arrangement

Making modifications in crop sowing dates and sowing patterns can either reduce weed emergence or increase the competitive ability of the crop [22]. Increasing the seed rate may not only increase the competitive ability of a crop against weeds but also cause reduction in crop yield and quality of produce [23]. However, an optimum spacing may provide the benefit of both competitive ability of crop and better yield, showing the importance of closer spacing as a weed management strategy. A lower uptake of nutrients by weeds and higher weed control efficiency in closer spacing have also been reported [24].

Crop Genotype Choice

Crop genotypes may have higher or lower competitive ability against weeds. Genotypes having faster seedling emergence and quick canopy establishment [25] can reduce the need for direct weed control measures; however, the expression of competitive advantage of a genotype may vary depending upon the prevalent environmental conditions [26]. Some traits (for example, plant height) are known to provide competitive advantage against weeds [27]; however, they may not be exploited due to some other associated disadvantages (e.g., lodging). Allelopathy in some cultivars may be exploited as a part of cultural weed control [28].

Fertilization

Soil nutrition influences the crop-weed competition; hence, specific methods to use fertility management as part of IWM are needed. Management strategies that maximize nutrient uptake by crops may reduce the harmful effects of weeds to some extent and minimize nutrient availability to weeds [29]. Fertilizers applied in close proximity to the crop row can improve weed management as the probability of the crop to capture nutrients (especially nitrogen) increases [25]. Band placement of fertilizer lowered weed density, biomass, and N uptake and resulted in increased wheat yield [30]. Other methods to alter the relative nutrient availability to crops and weeds can also be manipulated by change in timing of fertilizer applications [31], altering nutrient sources [32], and by using materials, such as nitrification inhibitors [33].

Nutrient availability can also be altered by applying organic amendments, especially for nitrogen and phosphorous. Soil nutrient concentrations strongly influence the germination and early growth of many weed species [29, 34].

Crop Rotation

Crop rotation is considered as an important component of weed management. Growing similar crops in rotation over the years favors weed species that are similar to the crop. However, a diversified crop rotation disrupts the growing cycle of weeds and prevents selection of the flora toward increased abundance of problem species [35]. Environmental conditions specifically created by crop rotations affect weed survival, propagule production, and germination in the soil, and thereby subsequent weed population dynamics [36].

Intercropping

Intercropping compared to crop monocultures can influence the competitive suppression of weeds. Intercrops of differing growth forms, phenologies, and physiologies can create different patterns of resource availability, especially light, to weeds [37]. As resource availability influences weed occurrence the most [10, 38], increased resource utilization under intercropping can provide better opportunities for IWM. Intercrop sown in a row-by-row layout, besides increasing the ecological diversity in a field, decreases relative soil cover of weeds, and may result in increased total crop yield [39].

Cover Crops

Cover crops may be grown for weed control, thereby replacing an unmanageable weed population with a manageable cover crop [40]. There are at least two major types of cover crops that can be used for weed control [7]: Off-season cover crops may be taken to produce sufficient plant residue or allelochemicals to create an unfavorable environment for weed seed germination and establishment, while a smother crop displaces weeds from the harvested crop through resource competition. Basic understanding of the mechanisms by which cover crops change weed population dynamics is required for improving the effectiveness. The effect of cover crop on weeds depends upon cover crop species and composition of weed community [41]. It has been reported that small-seeded weed species are more sensitive to physical as well as to allelochemical effects of cover crops compared to large-seeded weed species [42].

Mechanical Strategies

Mechanical measures may include physical removal of growing weeds from the field by hand weeding, hoeing, mowing, burning, tilling, etc. Annual and biennial weeds and non-creeping perennials can be removed by pulling them out. This is best done when the soil is moist and before seed is produced. However, it may not be suitable for large acreages.

Mulching is done to exclude light from the top of the weeds until the reserve food supply in the roots is exhausted and the weeds wither away. Mulches may include crop straw, hay or manure, sawdust, and transparent or black plastic.

Soil solarization technique is employed to kill weed seeds through solar heating. To make the solarization effective, the soil surface must be evenly prepared and contain enough moisture to favor heat transfer throughout the profile to damage reproductive structure of weeds, resulting in reduced weed seed germination [43].

Soil tillage influences the weed flora through changes in seed distribution in the soil, effects on seed predators, and effects on weed control practices [44, 45]. It is important to change the tillage practices in component crops year after year so that weed density is reduced greatly [41]. For example, conventional tillage (CT)—zero tillage (ZT) rotation was found better than CT–CT or ZT–ZT rotation in terms of weed management in a rice–wheat system [46].

Chemical Strategies

Several factors that must be looked into, while formulating chemical options of weed control, are the effectiveness of the chemical methods, such as application methods, stage of application, and selection of suitable herbicides on the basis of the nature of weeds. Faulty herbicide application methods may cause injury to the crop. Environmental factors, herbicide residues in the farm produce, residual effects of persistent herbicide in soil, compatibility problems with other pesticides, and occupational hazard to the applicator should be studied in detail. Development of herbicide-resistant weeds in recent years and its possible consequences on weed management suggest that over-reliance on chemical methods alone may not be the best strategy.

Biological Strategies

Biological control may be defined as the actions of parasites, predators, and pathogens in maintaining another organism's density at a lower average than would occur in their absence [47, 48]. It uses natural agents such as insects, nematodes, pathogens, herbivorous fish, and even grazing animals for the control of weeds. The objective in biocontrol is to reduce a weed's density to non-economic levels, not its eradication. Biocontrol is mostly followed for non-cultivated lands with troublesome biennial or perennial weeds. It is usually not practiced in cultivated lands as the weed (food source) for the biotic agent is removed periodically. An exception is the discovery of a specific fungus that controls round-leaved mallow in wheat fields [49].

Integrated Weed Management

The goal of a weed management program should be to keep the competition offered by weeds under check and not the complete removal or eradication from the ecosystem. To achieve this, a comprehensive action plan utilizing preventive methods, scientific knowledge, management skills, monitoring procedures, and efficient use of control practices should be devised, making conditions unfavorable to the weeds and their survival [7].

A successful IWM program must include prevention of weeds from invading, knowing the identity and details of the weed species, mapping its distribution and damage, formulating control strategy based on knowledge of potential damage, cost of control method, and environmental impact of the weed, using a combination of control strategies to reduce the weed population to an acceptable level, and, finally, evaluating its effectiveness. In a study, for example, integrating cultural and mechanical weed management practices was superior to the use of individual practices because they additively control weeds in an organic cropping system [50].

Challenges to Weed Management

Human population is still increasing at a faster rate, necessitating increased production of food grains in successive years. The food consumption patterns are also likely to change drastically. Economic development of a society also increases its consumption of fruits and vegetables. Thus, the future demand for increased production of fruits, vegetables, oilseeds, and fodders will be much higher than that of cereals over their existing production level. So far, major emphasis has been placed on the development of weed management technologies for cereals. It is time for weed scientists to change their focus and place increased emphasis toward the development of improved weed management technologies for oilseeds, vegetables, fruits, and fodder crops.

Weed problems are dynamic in nature, and these are likely to be more serious in the coming years due to high-input agriculture, climate change, globalization, and a host of other factors. Future weed science is likely to encounter the following challenges:

Economic Thresholds and Weed Management

From an economical perspective, there is no reason to apply control measures unless the weed population inflicts crop damage greater than the cost of the control measure. The economic threshold is the weed density at which the cost of control equals the value of the crop that would be lost if weeds are not controlled. According to this principle, weeds are not to be controlled if their densities are below the economic threshold. But, in some instances, the decision to control a weed will have to be made even when the cost of control may be more than the immediate damage inflicted by the weed. However, the concept of economic thresholds does not take into account the future effect of weed seed production. No use of control measures at below economic threshold densities of velvetleaf lead to rapid increase in its soil seed bank and subsequent densities [51–53]. Further, the yield loss caused by a specific weed infestation may vary, depending upon the environment and crop production practices. The distribution of weeds within agricultural fields is generally not uniform; usually, they occur in patches having a high relative density surrounded by areas with low density [54]. Hence, predicting yield losses assuming a regular distribution of weeds is of little value and often results in an over-estimation of weed-related yield losses [55]. Developing a mathematical model, taking into account the irregular distribution of weeds in a field, for using the economic threshold concept in precision agriculture is an issue that requires attention.

Weed Dynamics in High-Input-Intensive Production Systems

The scope for increasing area under crops is limited, and therefore enhanced food production will necessarily have to come from vertical growth, i.e., by increasing productivity per unit area per unit time. This will require a more intensive cultivation of crops with high doses of fertilizers, irrigation, and other inputs. While these interventions will put a greater constraint on the available natural resources, the weed problems are likely to shift in unpredictable ways. It is evident that with the discontinuation of some of the traditional practices such as crop rotations, intercropping, mulching, organic manuring, etc., the soil health as well as weed scenario has undergone a sea change in many parts of the India. The ability of weed communities to shift in response to control practices suggests the need for more integrated and diverse approaches to weed management [56]. It is therefore expected that future weed problems due to adoption of modern cultivation systems will be far more complex and challenging.

Interactions of Weeds with Other Pests

The interaction of weeds with insects and diseases plays an important role in formulating integrated pest management (IPM) program. For example, weeds serve as alternative hosts for plant-parasitic nematodes, thereby reducing the success of certain nematode management strategies [57]. Herbicides used for weed control may exert an effect on plant diseases, as weeds may serve as alternate hosts to pathogenic fungi and nematodes in fields [58, 59] that damage crops. Further, herbicides may also alter the ability of crop plants toward their response to pathogens. For example, sub-lethal rates of acetolactate synthase

(ALS)-inhibiting herbicides, imazamox and propoxycarbazone-sodium, could alter severity of injury symptoms caused by *Rhizoctonia solani* in barley [60]. According to Norris and Kogan [61], there are three types of interaction mechanisms: (1) weeds act as a food source for insect-pests or predators; (2) weeds may alter habitat, which may thus increase or suppress insect infestations; and (3) changes in non-target pest populations owing to control strategies. Most major weeds and plant-parasitic nematodes are place-bound organisms and passively dispersed. Weed–nematode interactions in agricultural production systems may be more intricate and complex than the simple function of weeds as alternative hosts [62]. It is a challenge to identify effective, compatible IPM strategies that address weed and nematode management collectively.

Crop–Weed Interaction under Changing Climate

Climate change is expected to influence weed communities, and management approaches must be adapted to take this into account. Global climate change is likely to cause a widespread shift in patterns of photosynthetic limitation in higher plants [63]. In a recent review, Yamori et al. [64] found that the inherent ability for temperature acclimation of photosynthesis was different: (1) among C3, C4, and crassulacean acid metabolism (CAM) species and (2) among functional types within C3 plants. These authors have concluded that C3 plants generally had a greater ability for temperature acclimation of photosynthesis across a broad temperature range; CAM plants acclimated day and night photosynthetic process differentially to temperature, and C4 plants adapted to warm environments. Hence, the long-term threat of increasing temperature and CO_2 concentration on crop–weed interaction should be viewed seriously, since a majority of crops belong to C3, whereas large numbers of weeds belong to the C4 category. C4 plants will have an advantageous position over C3 plants (e.g., rice) under higher temperatures and limited water availability. On the contrary, elevated CO_2 levels will improve the competitiveness of C3 crops relative to C4 weeds. Increased atmospheric CO_2 levels may also improve tolerance of rice against parasitic weeds, while prevalence of parasitic species may be amplified by soil degradation and more frequent droughts or floods [65]. Climate change is expected to promote a proliferation of new weed species and cause shifts in the composition of weed flora, especially in the tropics and subtropics. As weeds are highly dynamic and adapt quickly to new conditions, the management solutions have to address an ever-changing scenario. Some reports are available on the individual effects of CO_2 and temperature on crop–weed interaction. However, the combined effect of these two factors is yet to be studied in depth. Therefore, it is essential to undertake basic and strategic research, including physiological, biochemical, and molecular aspects, to evolve weed management technologies in the context of climate change. There is a need to generate information with respect to herbicide bio-efficacy, herbicide resistance development, behavior of bio-agents, and herbicide persistence vis-à-vis climate change.

Weeds in Conservation Agriculture Systems

It is widely believed that adoption of modern agricultural practices, such as intensive tillage, clean cultivation, fixed crop rotations, and other faulty management practices, including imbalanced fertilizer application and indiscriminate use of irrigation water, has led to serious resource degradation problems. In view of these, conservation agriculture (CA) technologies involving minimum soil disturbance, permanent soil cover through crop residues or cover crops, and dynamic crop rotations are being advocated for achieving higher and sustainable productivity. Globally, the concepts and technologies for CA are being practiced in about 128 million ha area, with the major countries being the USA, Brazil, Argentina, Canada, and Australia [66]. The area is further expanding rapidly due to their potential benefits on crop productivity and farm profitability. Farmers have been benefited due to the adoption of this technology in many ways, viz.: (1) reduction in cost of production [67, 68]; (2) enhancement of soil quality, i.e., soil physical, chemical, and biological conditions [69, 70]; (3) enhancement in C sequestration and buildup in soil organic matter in the long-term [71], which is important for mitigation of climate change effects; (4) reduction in incidence of *P. minor*, a major weed in wheat [67]; (5) enhancement in water- and nutrient-use efficiency [71, 72]; (6) enhancement in production and productivity [73]; (7) advances in sowing date [67]; (8) greater environmental sustainability [74]; (9) no loss of nutrients and no environmental pollution as crop residues are not burnt [75]; (10) opportunities for crop diversification and intensification [76]; (11) enhanced resource-use efficiency through residue decomposition, soil structural improvement, increased recycling and availability of plant nutrients [69]; and (12) moderate soil temperature, reduced evaporation, and improved biological activity through residue mulch [70, 77].

Changes from conventional to conservation farming practices often lead to weed flora shift in the crop field, which in turn also dictate the requirement of a new weed management technology. As the density of certain annual and perennial weeds increases under CA, effective weed control techniques are required to manage weeds successfully. The development of post-emergence broad-spectrum herbicides immensely ushered the way of controlling weeds in CA-based systems. However, weeds are still a big constraint toward the adoption of CA, and there is a need for developing more effective and economic IWM practices in diversified cropping systems by including various approaches, viz. preventive measures, cultural practices, and herbicides. There is a need to carry out an analysis of factors affecting adoption and acceptance of no-tillage agriculture among farmers. A lack of information on the effects and interactions of minimal soil disturbance, permanent residue cover, planned crop rotations, and IWM, which are key CA components, can hinder CA adoption [78]. This is because these interactions can have positive and negative effects, depending on regional conditions. The positive impacts should be exploited through system research to enhance CA crop yields. Information has mostly been generated on the basis of research trials, and more on-farm-level research and development is needed. Farmers' involvement in participatory research and demonstration trials can accelerate adoption of CA, especially in the areas where CA is a new technology.

Management of Herbicide Resistance in Weeds

Herbicide resistance in weeds is a major limiting factor to food security in global agriculture. Herbicide-resistant biotypes emerged in many regions of the world as a consequence of the intensive use of herbicides. Isoproturon resistance in *P. minor* in some parts of India was a costly lesson learnt, as the weed devastated the wheat crop and threatened the sustainability of the rice–wheat system for nearly a decade until some new alternate herbicides were introduced. This kind of phenomenon may continue to be a problem in the foreseeable future as well. The adoption of zero tillage is expected to further increase the use of non-selective herbicides, viz. glyphosate, glufosinate, and paraquat as a pre-plant application. There are currently 400 unique cases (species × site of action) of herbicide-resistant weeds globally, with 217 species (129 dicots and 88 monocots) [79]. Weeds have evolved resistance to 21 of the 25 known herbicide sites of action and to 148 different herbicides. Herbicide-resistant weeds have been reported in 65 crops in 61 countries. Therefore, it is important to monitor the impact of the evolution of resistance against nonselective herbicides under zero-till conditions and develop management strategies. Instead of depending on one particular technique, weed management methods are to be rotated and suitably integrated. Formation of broad-based special resistance management groups, involving both herbicide industries and core scientists, to monitor the resistance development and solutions is becoming imperative.

Minimizing herbicide resistance represents a big challenge that will require great research efforts to develop alternative control strategies. As pointed out by Busi et al. [80], weed scientists, plant ecologists, and evolutionary biologists should join forces and work toward an improved and more integrated understanding of resistance across all scales to facilitate the design of innovative solutions to the global herbicide resistance challenge. These authors have also noted that future research should integrate questions about standing genetic variation versus *de novo* resistance mutations, fitness benefits, and costs under herbicide selection and links between metabolic resistance and general detoxification pathways involved in stress-response dynamics.

Herbicide-Tolerant Crops and Evolution of Super Weeds

There has been a boom in the adoption of genetically modified (GM) crops over the past 15 years as the total area covered with GM crops has increased from 1.7 m ha in 1996 to more than 175 m ha in 2013. However, concerns are being raised about the possible environmental impact of this technology. Yet, few studies have conducted a critical needs analysis to assess the potential of specific GM traits in light of issues, such as climate change, increased environmental legislation (e.g., EU Water Framework, Nitrates Directive, proposed reform to the Pesticide Directive and Common Agricultural Policy reform), mitigating biodiversity loss, and sustainable biofuel production [81].

The potential for weed resistance to specific herbicide is always a concern with herbicide programs. It is more of a concern when talking about herbicide-tolerant crops (HTCs), as weed management in these crops depends on a specific herbicide only. On the other, some HTCs are becoming volunteer weeds and causing segregation and introgression of herbicide-resistant traits in weed populations [82]. Beckie and Warwick [83] reported that oilseed rape transgenes can survive in the environment for several years even if all cultivars with the conferred trait are removed from the area. There are also other apprehensions about HTCs as follows:

- Increase in use of a specific herbicide that may promote the development of herbicide-resistant weeds because of over-reliance on a single herbicide or a group of closely-related herbicides. *Conyza canadensis* has been reported to develop resistance against glyphosate in zero-till roundup ready corn–soybean rotations in the USA [84].
- Adverse effect on the biodiversity of the farm
- Gene-drift from HTCs to similar species may confer the resistance to their wild relatives, which can become serious weed in the crop
- Possibility of the development of "super weeds" due to introduction of these crops

Therefore, the HTCs should not be considered as a stand-alone component of weed management. Further, adoption of HTC has risen dramatically since their commercial introduction, but there is still no evidence of associated production cost reductions or enhanced yields [85], but the anticipated concerns about their actual benefits and effects on the environment are yet to be fully addressed.

Growing Infestation of Parasitic Weeds

Parasitic plants are problems mostly in the Mediterranean and tropical agriculture in major crops. The most economically damaging parasitic weeds are members of the genera *Striga* (witchweeds), *Orobanche* (broomrapes), and *Cuscuta* (dodder). For example, serious infestations of *Orobanche* in many tomato, mustard, tobacco, and potato-growing areas of India are causing huge losses in productivity. The weed emerges from soil in the middle and later stages of growth, by which time, it has already caused enough damage to the host plant. Biology of these weeds is not well understood, and there is no simple solution for their management worldwide. In spite of several efforts, the major problems of parasitic weeds have not been reduced to any significant degree [86] and in the case of *Striga,* there may even continue to be some spread and intensification of the problems [87].

The main focus of research on parasitic weeds has been around agronomic practices and the use of herbicides, although success has been marginal. In addition, global environment change together with changing land-use patterns means that some geographical areas and farming systems that do not currently suffer from parasitic weeds could become affected within coming decades [88]. It is, therefore, necessary to develop management technologies for these weeds, which are spreading to newer areas and parasitizing many other host plants. Biocontrol approach is expected to make valuable contributions to manage parasitic weeds, especially *Striga*. Increasing soil fertility is perhaps the only way to manage *Striga* as of now [87].

Environmental Impact of Herbicides

Herbicides have the capacity to move in the environment away from the target area and to cause damage to non-target plants and animals. More than 95 % of herbicides reach a destination other than their target species, including non-target species, air, water, bottom sediments, and food [89]. Hence, it is a big challenge to use herbicides in the safest way for ensuring food and biological security.

The impact of herbicides on soil, however, differs depending upon the soil type, experimental conditions, herbicide in question and its dose, and the sensitivity of the non-target species or strains. No severe ill effect on soil flora, soil biochemical indices, and soil fauna has been observed so far at recommended doses of herbicides under field conditions [90], but the adverse effects of their overdose or long-term use cannot be discounted. Systematic research on long-term herbicide usage on soil health and water bodies is needed. Widespread and increasing use of herbicides is likely to cause greater concern about potential ecological effects. Hence, how herbicide use offsets the delicate ecological balance should also be an area of priority. To avoid the potential ill effects, strict registration and stringent regulatory mechanisms are to be developed.

Monitoring herbicide residues in the environment and food chain should continue to be an important activity as new chemicals are expected to be introduced into the market. Permanent herbicide trials have to be planned in major cropping systems under different agroecological regions, which would yield a wealth of information on the long-term implications of herbicide use, including effect on crop productivity, weed flora shifts, resistance of weeds, etc. In addition, degradation pathways and mitigation strategies of herbicide residue hazards need to be developed to lessen their effect on the environment.

Weeds in Organic Farming Systems

Growing concern for human health and sustainability of agricultural production are giving way to organic farming in some parts of the world. However, weed management is a major concern for organic farmers and is seen as a major obstacle for the conversion toward organic farming [91]. Effective weed management strategies are limited in organic cropping systems owing to the prohibition of herbicide use. Organically cultivated fields show higher levels of weed infestation compared to conventional agriculture [92], and it is a big challenge to make the non-chemical methods of weed control effective and economical. Mechanical approaches, generally used to manage weeds in this system, provide lower weed control efficiency than herbicides [93]. But at some instances, weed harrowing may provide yields similar to weed-free situations [94]. Soil solarization may be a useful tool in nurseries and in high-value crops under organic agriculture; it is not yet a practicable option for field crops due to high cost. Although *P. minor* was controlled in wheat to some extent by using ZT technology in the Indo-Gangetic Plains, such success has not been achieved in other crops and weed species. Plant allelochemical or essential oil-based organic herbicides are available commercially, but these are very expensive and are utilized mainly for spot applications in a field to deal with a localized infestation of noxious weeds [95]. Currently, no bioherbicides based on specific plant pathogens are available commercially. In maize, growing cowpea as an intercrop for fodder or green manure has been found to suppress the weeds significantly. In mustard, better weed control and higher total productivity can be obtained by intercropping with berseem. Incorporation of *Sesbania* grown as an intercrop (brown manuring) in upland direct-seeded rice can be adopted for managing weeds and obtaining higher productivity. Enhancing a crop's competitive ability by integrating both cultural and mechanical weed control methods is a key strategy in organic systems, but the relative efficacy of different cultural and mechanical strategies and their interactions and additive effects when combined is not well known [50]. There is ample scope of developing system-based approaches and mechanical tools as part of IWM strategies in organic farming systems.

Obnoxious Weeds

Invasive weeds are an important problem for natural and agronomic systems and a major threat to global biodiversity [96]. According to the evolution of increased competitive ability (EICA) hypothesis, plants in invasive range allocate more to growth than to defense [97], and consequently the invasive plants perform better than plants of the same species from the native range. Abela-Hofbauerová and Münzbergová [98] observed that the plants from the invasive range have higher ability to use resources and are thus able to perform well even in nutrient-poor conditions. Further, the invasive potential of some alien invasive weed species may be enhanced due to absence of natural enemies [99, 100].

Obnoxious weeds, such as *Lantana, Parthenium, Ageratum, Chromolaena, Mikania,* and *Mimosa,* have invaded vast areas of forest, grasslands, wastelands, orchards, and plantation crops across the world. *Parthenium,* one of the seven most difficult weeds of the world, was previously a problem on roadsides and non-cultivated areas in India and is now entering into the field crops. *Chromolaena odorata was* earlier restricted to the north-eastern region and Western Ghats but it is now fast spreading to other areas. Similarly, *Mikania micrantha,* which is popularly called mile-a-minute weed on account of its rapid growth, is a big nuisance in forestry and plantation crops in northeast and south India [101]. *Lantana camara* has invaded large areas of non-crop lands in the north-western Himalayan region. *Ag-*

Table 16.2 World's major weeds that have not yet been recorded in India. (Source: Holm et al. [103])

Country	No. of weed species
Australia, New Zealand	195
African countries	181
SE Asia, Far East	150
Middle East	118
South America	102
Europe	90
Central America	86
North America	33
Former Soviet Union	20
Total	*975*

eratum has become a big nuisance in both crops and non-cropped areas. Widespread infestation of these weeds has threatened not only agricultural production systems but also biodiversity and human and animal health.

There are several barriers to the effective control of obnoxious weeds. For instance, a lack of public awareness about the invasiveness and ill effects of these weeds lead to limited public and legislative support; this consequently leads to insufficient human and fiscal resources to contain the weed problem. Due to insufficient resources, weed control efforts often lack planning and monitoring for effectiveness. Preventing the spread of these weeds before the situation gets more serious requires a great deal of money and people's participation.

Globalization and New Weed Problems

Weeds are spread internationally as contaminants through trade, travel, and illegal activities. For example, *Chromolaena odorata,* introduced from the West Indies in the ballasts of cargo boats [102], and *M. micrantha,* from Central and South America after the Second World War to camouflage airfields, have become great problems for plantations and forests in eastern and southern parts of India. Similarly, *Parthenium,* a menace in civic amenities, and *P. minor,* a major weed in wheat along the Indo-Gangetic Plains, were introduced in India through imported wheat grain from the USA.

Although the risk of entry is minimized by quarantine arrangements, an increased exchange of grains and seeds following globalization of agricultural trade is expected to further enhance the probability of entry of weeds in a new territory. For example, there are several weeds of invasive nature existing in different parts of the world, but they are not in India (Table 16.2) [103]. Increasing trade and globalization coupled with liberalization policies will, however, increase the risk of invasion by these weeds in India. The sanitary and phytosanitary agreement of the World Trade Organization (WTO) suggests that the countries should not only update their quarantine laws but also incorporate the elements of pest-risk analysis for making regulatory decisions for both import and export. Therefore, there is an urgent need to analyze the risk factor associated with different exotic weeds to design safeguards and to lower the risk of their entry. Many countries like Australia, New Zealand, and the USA have developed strong protocols for weed risk analysis and for identification of quarantine weeds. Similarly, other countries of the world should strengthen their capacity on weed risk analysis and develop more stringent guidelines and standards for prevention of introduction of alien, invasive weeds into the respective countries.

Dissemination of Weed Management Technologies

Improved weed management technologies have not reached the Indian farmers and elsewhere at the same pace as it happened in case of high-yielding varieties, fertilizers, and insecticides. Compared to the other improved agricultural practices in cereals, adoption of chemical weed management technologies by farmers is very dismal in India [104]. Similarly, adoption gap of sugarcane technologies was more in weed control followed by plant protection measures, time of sowing, irrigation, sowing methods, high-yielding varieties, and seed rate [105]. Lack of awareness and technical know-how among the farming community are the reasons for poor adoption of weed management technologies. Sometimes, extension agents and other traditional information dissemination mechanisms, such as using community decision leaders, neighbors, and seminars, are largely ineffective in the dissemination of weed management technologies [106]. The use of weeds as livestock feeds, fuel wood, construction material, and as medicines is also one of the deterrents toward non-adoption of new weed management technologies. In some places, the herbicides are not available locally to those farmers who are interested in using use them. Intensive training programs [107] and TV programs [108, 109] could be the effective extension techniques to enhance adoption of chemical weed control practices. Information about safe use of herbicides, herbicide application technology for higher efficacy, and integrating chemicals with other methods of weed management are also to be disseminated. However, overcoming the challenge of lesser attention of the growers toward adoption of new weed management practices than other production technologies, viz. seeds and fertilizers, is a matter of concern.

Site-Specific Weed Management

The concept of site-specific agriculture [110] is applicable to weed management, owing to the spatial and temporal heterogeneity of weed populations across agricultural fields [54, 111]. The uniform application of herbicides over heterogeneously distributed weed populations may lead to inefficiency in weed management [51]. Site-specific weed management may result in savings of herbicides and ecological and economical benefits.

Site-specific management of weeds involves locating specific areas of infestation and identification of weeds in a field for necessary herbicide treatments depending on the weed species present [112]. This will require a more precise application of weed management principles and biology to determine where, when, and what control practices are to be applied. Patchy weed distributions are the result of efforts made to manage weeds uniformly; it will also be important to notice whether site-specific management will change the nature of weed populations in fields.

Opportunities

With the advent of chemical weed control in the early 1940s, the contribution of weed science has been immense in increasing and sustaining the global food production. Herbicides became the mainstay for management of weeds more particularly in developed countries. In view of the changing climate, new cropping systems, weed shifts, changing land use, and environmental concerns, new opportunities in weed science exist that need to be exploited at a faster pace. The emerging problems could only be addressed when weed science works hand-in-hand with other disciplines on complex issues in vegetation management, viz. ecological weed management, molecular biology and physiology of weedy traits, invasion biology, and ecosystem restoration.

The following opportunities in weed science need to be exploited for efficient and safer weed management in future:

- Safer low-dose synthetic molecules of various modes of action will be introduced to replace more conventional herbicides. New formulations and spraying technologies of herbicides will be developed.
- Alternate weed control strategies involving mechanical, cultural, and biocontrol will also be given importance. A search for bioactive botanicals and microbial metabolites, which may act as lead molecules for herbicide development, is important.
- Breakthroughs made in biotechnology could be taken to advantage, leading to development of new HTCs and strains of bio-agents for specific weed control.
- The changing global climate may create new opportunities for the introduction of alien, invasive weed species. Immediate action to thwart their introduction to newer areas will help in protecting the biodiversity of native species.
- Climate change research would provide further insights into crop–weed association, herbicide, and bio-agent efficacy for developing effective weed management technologies.
- Research on nano-composite-based controlled release formulation is essential for precision weed management. The controlled release of herbicide molecules in application zones provides long-term control of weeds, avoiding repeated application of herbicides. These formulations minimize herbicide residues in the

environment, increase the efficacy and longevity of the herbicide by protecting it from environmental degradation, and decrease the application cost.
- Using remote sensing technologies for site-specific weed monitoring and their management under precision agriculture will greatly help in avoiding wastage of herbicides and minimizing residue hazards.
- A growing demand for cheap and effective non-chemical weed control measures, i.e., mechanical, cultural, bioherbicides, and biocontrol agents in the era of environmental awareness is observed.
- Innovative production systems such as CA are being developed for enhancing resource-use efficiency, crop productivity, and environmental sustainability. Weed management in such systems would require greatly enhanced knowledge and application. New-generation machines for tillage, sowing, interculture, spraying, harvesting, and residue management are being developed, which will provide cost-effective means of weed management.
- New tools aimed at more effective transfer of technology for weed management are available in the era of ICT. Management Information Systems (MIS) are required for researchers and farmers to obtain quick access of weed management technology.
- Efficient diagnostic techniques for monitoring herbicide residues would lead to safer chemical weed control and a cleaner environment. Effective decontamination techniques for active and transformation products will provide opportunities for mitigation of residue hazards.
- Solar energy-aided microwave-generating device may be helpful for the control of target weeds. The success of it may reduce herbicide consumption manifold. This device coupled with sensor technology may become the part of precision and automated weed control technology.
- Robotic science may also come in aid of weed science for environmentally safe weed management.
- Weed utilization techniques are available for effective conversion of weed biomass into enriched compost, medicinal use, bioremediation, and industrial application.

Conclusion

The dynamic nature of weed populations makes them a never-ending problem in crop fields. The cropping environment and the production practices—viz. crop rotation, tillage, fertilization, crop spacing, herbicides, irrigation, etc.—together dictate the nature and intensity of weed infestation. Accordingly, various approaches of weed management have evolved in the history of weed science. Because of the complexity and diversity of weed communities, application of a given control tactic leads to a weed population shift, thereby compelling the grower to use another tactic, and the cycles go on. This situation demands the use of integrated approaches involving more than one control tactic to favor the crop in its competition over weeds for natural resources.

The growing demand for food grains and other agricultural products on one hand and the shrinking availability of agricultural land on the other hand are already the burning problems being faced by agriculture. Further, climate change is predicted to affect precipitation rates and patterns, which will consequently affect temperature, growing season, soil moisture levels, and other critical agricultural production factors. All these developments are expected to force the growers to shift toward highly intensive production systems using newer production technologies. Weeds, being highly complex and competitive, and due to their wild and dynamic nature, are expected to adapt and remain a problem in the future production systems, and will necessarily create demand for newer integrated control tactics. Moreover, the way in which the interaction of weeds with crops and other pests will move under the changing climate is yet a domain of unknown probabilities. Developing innovative and economical weed management tactics to make more diverse and integrated approach of weed management for the future cropping systems is a great challenge and a continuous process for weed scientists.

Availability of herbicides simplified the weed management and benefitted the agricultural community in many ways, viz. timely weeding, overcoming the problem of labor shortage, reducing production cost, etc. However, the over-reliance on herbicides has already shown its consequences in the form of weed resistance to herbicides, and adoption of HTCs may further exuberate such a situation. Hence, the challenge is to manage herbicides in a manner that prevents adapted weed species from reaching troublesome proportions. Development of site-specific weed management systems is another challenge to be sorted out to reduce herbicide consumption and also to reduce the environmental impact of herbicides by preventing herbicide load where it is not required.

Present-day agriculture is also facing the problem of transborder movement of weed seeds being accompanied by the growing international trade of agricultural produce. It is of greater concern if the alien species are obnoxious and invasive in nature. Risk assessment and developing management tactics for such weeds in a newer environment are always a challenge.

Technology dissemination is as important as technology development. Minimum attention has been paid by the growers toward adoption of new weed management practices as compared to the adoption of other production technologies (viz. seeds and fertilizers), which is a matter of concern for the weed scientists.

References

1. UNPD (2011) United nations population division, department of economic and social affairs. World population prospects: the 2010 revision. New York: United Nations. http://esa.un.org/unpd/wpp/Excel-Data/population.htm. Accessed 24 May 2014
2. Von Braun J (2007) The world food situtation: new driving forces and required actions. International Food Policy Research Institute, Washington, DC

3. Reijinders L, Soret S (2003) Quantification of the environmental impact of different dietary protein sources. Am J Clin Nutr 78(suppl):664–668
4. Pimentel D, Pimentel M (2003) Sustainability of meat-based and plant-based diets and the environment. Am J Clin Nutr 78(suppl):660–663
5. World Bank (2007) World development report 2008: agriculture for development. World Bank, Washington, DC
6. WSSA (2002) Herbicide handbook. Weed Science Society of America, Lawrence, 462 pp
7. Buhler DD (2002) Challenges and opportunities for integrated weed management. Weed Sci 50:273–280
8. Buhler DD (1996) Development of alternative weed management strategies. J Prod Agric 9:501–505
9. Zimdahl RL (1991) Weed science—a plea for thought. US Department of Agriculture, Cooperative State Research Service, Washington, DC, 34 pp
10. Radosevich SR, Holt JS, Ghersa CM (1997) Weed ecology, implications for management. Wiley, New York, pp 3–4, 37–40
11. Purdue University Weed Science (2013) https://ag.purdue.edu/btny/weedscience/Pages/default.aspx. Accessed 24 Oct 2013
12. Hamill AS, Holt JS, Mallory-Smith CA (2004) Contributions of weed science to weed control and management. Weed Technol 18:1563–1565
13. Sutherland S (2004) What makes a weed a weed: life history traits of native and exotic plants in the USA. Oecologia 141:24–39
14. Ward SM, Gaskin JF, Wilson LM (2008) Ecological genetics of plant invasion: what do we know? Invasive Plant Sci Manage 1:98–109
15. Timmons FL (2005) A history of weed control in the United States and Canada. Weed Sci 53:748–761
16. Appleby AP (2005) A history of weed control in the United States and Canada—a sequel. Weed Sci 53:762–768
17. Das TK (2008) Weed science—basics and applications. Jain Brothers, New Delhi
18. Stephenson GR, Solomon KR, Frank R, Hsiung T, Thompson DG (2001) Pesticides in environment, section 3. University of Guelph, Guelph, pp 1–12
19. Ross MA, Lembi CA (1999) Applied weed science, 2nd edn. Prentice Hall, Upper Saddle River
20. Green JM, Hazel CB, Forney DR, Pugh LM (2008) New multiple herbicide resistance and formulation technology to augment the utility of glyphosate. Pest Manage Sci 64:332–339
21. Walker RH (1995) Preventative weed management. In: Smith AE (ed) Handbook of weed management systems. New York: Marcel Dekker, pp 35–50
22. Mohler CL (1996) Ecological bases for the cultural control of annual weeds. J Prod Agric 9:468–474
23. Litterick AM, Redpath J, Seel W, Leifert C (1999) An evaluation of weed control strategies for large-scale organic potato production in the UK. In: Proceeding of the 1999 Brighton Conference—Weeds, Brighton, UK, pp 951–956
24. Sindhu PV, George CT, Abraham CT (2013) Plant population and SRI management on crop-weed competition in rice in the humid tropics of Kerala, India. Indian J Agric Res 47:288–295
25. Rasmussen K, Rasmussen J (2000) Barley seed vigour and mechanical weed control. Weed Res 40:219–230
26. Lemerle D, Verbeek B, Orchard B (2001) Ranking the ability of wheat varieties to compete with *Lolium rigidum*. Weed Res 41:197–209
27. Benvenuti S, Macchia M (2000) Role of durum wheat (*Triticum durum* Desf.) canopy height on *Sinapis arvensis* L. growth and seed production. In: Proceeding of XIème Colloque International sur la Biologie des Mauvaises Herbes, Dijon, France, pp 305–312
28. Olofsdotter M (2001) Rice—a step toward use of allelopathy. Agron J 93:3–8
29. DiTomaso JM (1995) Approaches for improving crop competitiveness through the manipulation of fertilization strategies. Weed Sci 43:491–497
30. Kirkland KJ, Beckie HJ (1998) Contribution of nitrogen fertilizer placement to weed management in spring wheat (*Triticum aestivum*). Weed Technol 12:507–514

31. Anderson RL (1991) Timing of nitrogen application affects downy brome (*Bromus tectorum*) growth in winter wheat. Weed Technol 5(3):582–585
32. DeLuca TH, DeLuca DK (1997) Composting for feedlot manure management and soil quality. J Prod Agric 10:235–241
33. Teyker RH, Hoelzer HD, Liebl RA (1991) Maize and pigweed response to nitrogen supply and form. Plant Soil 135:287–292
34. Karssen CM, Hillhorst HWM (1992) Effect of chemical environment on seed germination. In: Fenner M (ed) Seeds: the ecology of regeneration in plant communities. CAB International, Wallingford, pp 327–348
35. Karlen DL, Varvel GE, Bullock DG, Cruse RM (1994) Crop rotations for the 21st century. Adv Agron 53:1–45
36. Liebman M, Gallandt ER (1997) Many little hammers: ecological management of crop-weed interactions. In: Jackson LE (ed) Ecology in agriculture. Academic, San Diego, pp 291–343
37. Ballare CL, Casal JJ (2000) Light signals perceived by crop and weed plants. Field Crops Res 67:149–160
38. Harper JL (1977) The limiting resources in the environment. In: The population biology of plants. Academic, London, pp 305–346
39. Baumann DT, Kropff MJ, Bastiaans L (2000) Intercropping leeks to suppress weeds. Weed Res 40:359–374
40. Teasdale JR (1998) Cover crops, smother plants, and weed management. In: Hatfield JL, Buhler DD, Stewart BA (eds) Integrated weed and soil management. Ann Arbor, Chelsea, pp 247–270
41. Bàrberi P, Mazzoncini M (2001) Changes in weed community composition as influenced by cover crop and management system in continuous corn. Weed Sci 49:491–499
42. Liebman M, Davis AS (2000) Integration of soil, crop, and weed management in low-external-input farming systems. Weed Res 40:27–47
43. Temperini O, Bàrberi P, Paolini R, Campiglia E, Marucci A, Saccardo F (1998) Solarizzazione del terreno in serra-tunnel: effetto sulle infestanti in coltivazione sequenziale di lattuga, ravanello, rucola e pomodoro. In: Proceeding of XI SIRFI Biennial Congress, Bari. 12–13 November. (Italian), Italy, pp 213–228
44. Brust GE, House GJ (1988) Weed seed destruction by arthropods and rodents in low-input soybean agroecosystems. Am J Altern Agric 3:19–25
45. Buhler DD (1995) Influence of tillage systems on weed population dynamics and management in corn and soybean in the central USA. Crop Sci 35:1247–1258
46. Mishra JS, Singh VP (2012) Effect of tillage sequence and weed management on weed dynamics and productivity of dry-seeded rice (*Oryza sativa*)-wheat (*Triticum aestivum*) system. Indian J Agron 57:14–19
47. DeBach P (1964) The scope of biological control. In: De Bach P (ed) Biological control of insect pests and weeds. Chapman and Hall, London, pp 3–20
48. McFayden RE (1998) Biological control of weeds. Ann Rev Entomol 43:369–393
49. Mortensen K (1988) The potential of an endemic fungus, *Colletotrichum gloeosporioides*, for biological control of round-leaved mallow (*Malva pusilla*) and velvetleaf (*Abutilon theophrasti*). Weed Sci 36:473–478
50. Benaragama D, Shirtliffe SJ (2013) Additive weed control by integrating cultural and mechanical weed control strategies in an organic cropping system. Agron J 105:1728–1734
51. Cardina J, Johnson GA, Sparrow DH (1997) The nature and consequences of weed spatial distribution. Weed Sci 45:364–373
52. Hartzler RG (1996) Velvetleaf (*Abutilon theophrasti*) population dynamics following a single year's seed rain. Weed Technol 10:581–586
53. Zanin G, Sattin M (1988) Threshold level and seed production of velvetleaf (*Abutilon theophrasti*) in maize. Weed Res 28:347–352
54. Cardina J, Sparrow DH, McCoy EL (1996) Spatial relationships between seedbank and seedling populations of common lambsquarters (*Chenopodium album*) and annual grasses. Weed Sci 44:298–308

55. Wiles LJ, Wilkerson GG, Gold HJ, Coble HD (1992) Modeling weed distribution for improved post-emergence control decisions. Weed Sci 40:546–553
56. Buhler DD, Liebman M, Obrycki JJ (2000) Theoretical and practical challenges to an IPM approach to weed management. Weed Sci 48:274–280
57. Duncan LW, Noling JW (1998) Agricultural sustainability and nematode integrated pest management. In: Barker KR, Pederson GA, Windham GL (eds) Plant and nematode interactions. Agronomy monograph 36. ASA, Crop Science Society of America, Soil Science Society of America, Madison, pp 251–287
58. Grönberg L, Andersson B, Yuen J (2012) Can weed hosts increase aggressiveness of phytophthora infestans on potato? Phytopathology 102:429–433
59. Boydston RA, Mojtahedi H, Bates C, Zemetra R, Brown CR (2010) Weed hosts of *Globodera pallida* from Idaho. Plant Dis 94:918
60. Lee H, Ullrich SE, Burke IC, Yenish J, Paulitz TC (2012) Interactions between the root pathogen *Rhizoctonia solani* AG-8 and acetolactate-synthase-inhibiting herbicides in barley. Pest Manage Sci 68:845–852
61. Norris RF, Kogan M (2000) Interactions between weeds, arthropod pests, and their natural enemies in managed ecosystems. Weed Sci 48:94–158
62. Thomas SH, Schroeder J, Murray LW (2005) The role of weeds in nematode management. Weed Sci 53:923–928
63. Sage RF, Kubien DS (2007) The temperature response of C3 and C4 photosynthesis. Plant Cell Environ 30:1086–1106
64. Yamori W, Hikosaka K, Way DA (2013) Temperature response of photosynthesis in C3, C4, and CAM plants: temperature acclimation and temperature adoption. Photosynth Res. doi:10.1007/s11120-013-9874-6
65. Rodenburg J, Meinke H, Johnson DE (2011) Challenges for weed management in African rice systems in a changing climate. J Agric Sci 149:427–435
66. FAO (2013) Food and agriculture organization of the United Nations, 2012. http://www.fao.org/ag/ca/6c.html. Accessed 24 May 2014
67. Malik RK, Gupta RK, Singh CM, Yadav A, Brar SS, Thakur TC, Singh SS, Singh AK, Singh R, Sinha RK (2005) Accelerating the adoption of resource conservation technologies in rice wheat system of the Indo-Gangetic plains. In: Proceedings of project workshop, directorate of extension education, Chaudhary Charan Singh Haryana Agricultural University, June 1–2, 2005, Hisar, India
68. RWC-CIMMYT (2005) Agenda notes. 13th regional technical coordination committee meeting. RWC-CIMMYT, Dhaka, Bangladesh
69. Jat ML, Gathala MK, Ladha JK, Saharawat YS, Jat AS, Kumar V, Sharma SK, Kumar V, Gupta R (2009a) Evaluation of precision land leveling and double zero-till systems in rice-wheat rotation: water use, productivity, profitability and soil physical properties. Soil Tillage Res 105:112–121
70. Gathala MK, Ladha JK, Saharawat YS, Kumar V, Kumar V, Sharma PK (2011b) Effect of tillage and crop establishment methods on physical properties of a medium-textured soil under a seven-year rice−wheat rotation. Soil Sci Soc Am J 75:1851–1862
71. Saharawat YS, Ladha JK, Pathak H, Gathala M, Chaudhary N, Jat ML (2012) Simulation of resource-conserving technologies on productivity, income and greenhouse gas emission in rice-wheat system. J Soil Sci Environ Manage 3:9–22
72. Jat ML, Malik RK, Saharawat YS, Gupta R, Mal B, Paroda R (2012) Proceedings of regional dialogue on conservation agricultural in South Asia, New Delhi, India, 1–2 November, 2011. APAARI, CIMMYT, ICAR, p 32
73. Gathala MK, Ladha JK, Kumar V, Saharawat YS, Kumar V, Sharma PK, Sharma S, Pathak H (2011a) Tillage and crop establishment affects sustainability of south asian rice-wheat system. Agron J 103:961–971
74. Pathak H, Saharawat YS, Gathala M, Mohanty S, Chandrasekharan S, Ladha JK (2011) Simulating the impact of resource conserving technologies in rice-wheat system on productivity, income and environment—part I. Green house gases. Sci Technol 1:1–17

75. Sidhu HS, Singh M, Humphreys E, Singh Y, Singh B, Dhillon SS, Blackwell J, Bector VM, Singh S (2007) The happy seeder enables direct drilling of wheat into rice straw. Aust J Exp Agric 47:844–854
76. Jat ML, Singh S, Rai HK, Chhokar RS, Sharma SK, Gupta RK (2005) Furrow irrigated raised bed planting technique for diversification of rice-wheat system of Indo-Gangetic plains. J Jpn Assoc Int Coop Agric For 28:25–42
77. Jat ML, Singh RG, Saharawat YS, Gathala MK, Kumar V, Sidhu HS, Gupta R (2009b) Innovations through conservation agriculture: progress and prospects of participatory approach in the Indo-Gangetic plains. In: Pub lead papers, 4th world congress on conservation agriculture, 4–7 February, 2009. India, New Delhi, pp 60–64
78. Farooq M, Flower KC, Jabran K, Wahid A, Siddique KHM (2011) Crop yield and weed management in rainfed conservation agriculture. Soil Tillage Res 117:172–183
79. Heap I (2013) The international survey of herbicide resistant weeds. http://www.weedscience.org. Accessed 21 Dec 2013
80. Busi R, Vila-Aiub MM, Beckie HJ, Gaines TA, Goggin DE, Kaundun SS et al (2013) Herbicide-resistant weeds: from research and knowledge to future needs. Evol Appl 6:1218–1221
81. O'Brien M, Mullins E (2009) Relevance of genetically modified crops in light of future environmental and legislative challenges to the agri-environment. Ann Appl Biol 154:323–340
82. Owen MDK, Zelaya IA (2004) Herbicide-resistant crops and weed resistance to herbicides. Pest Manage Sci 61:301–311
83. Beckie HJ, Warwick SI (2010) Persistence of an oilseed rape transgene in the environment. Crop Prot 29:509–512
84. Mueller TC, Massey JH, Hayes RM, Main CL, Stewart CN Jr (2003) Shikimate accumulates in both glyphosate-sensitive and glyphosate-resistant horseweed (*Conyza canadensis* L. Cronq.). J Agric Food Chem 51:680–684
85. Martinez-Ghersa MA, Worster CA, Radosevich SR (2003) Concerns a weed scientist might have about herbicide-tolerant crops: a revisitation. Weed Technol 17:202–210
86. Parker C (2009) Observations on the current status of *Orobanche* and *Striga* problems worldwide. Pest Manage Sci 65:453–459
87. Parker C (2012) Parasitic weeds: a world challenge. Weed Sci 60:269–276
88. EWRS (2014) Parasitic weeds working group. http://www.ewrs.org/parasitic_weeds_details.asp. Accessed 4 Jan 2014
89. Miller GT (2004) Sustaining the earth, 6th edn. Thompson Learning, Pacific Grove, pp 211–216
90. Barman KK, Varshney JG (2008) Impact of herbicides on soil environment. Indian J Weed Sci 40:10–17
91. Sumption P, Firth C, Davies G (2004) Observation on agronomic challenges during conversion to organic field vegetable production. In: Hopkins A (ed) Organic farming: science and practice of profitable livestock and cropping. Proceedings of the BGS/AAB/COR conference. British Grassland Society, pp 176–179
92. Romero A, Chamorro L, Sans FX (2008) Weed diversity in crop edges and inner fields of organic and conventional dry-land winter cereal crops in NE Spain. Agric Ecosyst Environ 124:97–104
93. Lundkvist A (2009) Effect of pre- and post- emergence weed harrowing on annual weeds in peas and spring cereals. Weed Res 49:409–416
94. Armengot L, Jos-Maria L, Chamorro L, Sans FX (2013) Weed harrowing in organically grown cereal crops avoids yield losses without reducing weed diversity. Agron Sustain Dev 33:405–411
95. Schonbeck M (2011) Principles of sustainable weed management in organic cropping systems, 3rd edn, September 2011. Mark Schonbeck, independent sustainable agriculture consultant. http://www.carolinafarmstewards.org/wp-content/uploads/2012/12/1-Schonbeck-Principles-of-Sustainable-Weed-Management-in-Organic-Cropping-Systems.pdf dt 08.01.2014
96. Wilcove DS, Rothstein D, Dubow J, Phillips A, Losos E (1998) Quantifying threats to imperiled species in the United States. BioScience 48:607–615

97. Blossey B, Notzold R (1995) Evolution of increased competitive ability in invasive non-indigenous plants: a hypothesis. J Ecol 83:887–889
98. Abela-Hofbauerová I, Münzbergová Z (2011) Increased performance of *Cirsium arvense* from the invasive range. Flora morphology, distribution. Funct Ecol Plants 206:1012–1019
99. Cappuccino N, Carpenter D (2005) Invasive exotic plants suffer less herbivory than non-invasive exotic plants. Biol Lett 1:435–438
100. Colautti RI, Ricciardi A, Grigorovich IA, MacIsaac H (2004) Is invasion success explained by the enemy release hypothesis? Ecol Lett 7:721–733
101. Yaduraju NT, Prasadbabu MBB, Gogoi AK (2003) Green invaders—a growing threat to agriculture and environment. In: Abstracts, national seminar alien invasive weeds in India, April 27–29, 2003, AAU, Jorhat, pp 1–9
102. Biswas K (1934) Some foreign weeds and their distribution in India and Burma. Indian For 60:861–865
103. Holm L, Pancho J, Herberger J, Plucknett D (1979) A global atlas of world weeds. Wiley, New York, 391 pp
104. Bala B, Sharma SD, Sharma RK (2006) Knowledge and adoption level of improved technology among rural women owing to extension programmes. Agric Econ Res Rev 19:301–310
105. Dashora P, Verma AK, Rokadia P, Punia SS (2011) Transfer of technology to bridge the yield gap in sugarcane of south east Rajasthan. J Progress Agric 2:59–62
106. Wanjala MJ, Lloyd DJ, Nichols JD (2003) Information sources and dispersal channels in the extension of pasture weed management technologies in southern eastern Kenyan rangelands. In: Proceedings from APEN National Forum 2003, 26–28 November, Hobart, Tasmania
107. Wijeratne M, Abeydeeral RN (1994) Knowledge gap on agrochemical use in rice farming. Int Rice Res Notes 19(4):26–27
108. Chundi J, Srivastva A (1999) Role of media in dissemination of information in rural areas. Interaction 17(1, 2):82
109. Goswami G, Dhawan D, Bareth LS (2003) Impact of T.V. programme on rural women. Allahabad Farmer 57:21–27
110. Robert PC, Rust RH, Larson WE (1994) Preface. In: Robert PC, Rust RH, Larson WE (eds) Site-specific management for agricultural systems. American Society of Agronomy, Madison, p 3
111. Clay SA, Lems GJ, Clay DE, Forcella F, Ellsbury MM, Carlson CG (1999) Sampling weed spatial variability on a field-wise scale. Weed Sci 47:674–681
112. Swinton SM (2005) Economics of site-specific weed management. Weed Sci 53:259–263

Chapter 17
Strengthening Farmers' Knowledge for Better Weed Management in Developing Countries

Narayana Rao Adusumilli, R. K. Malik, Ashok Yadav and J. K. Ladha

Introduction

Of more than 3 billion people (nearly half of the world's population) who live in rural areas, around 2.5 billion derive their livelihoods from agriculture [1], which remains crucial to developing countries and their economies for meeting the demands of affordable food, feed, energy, and the security of their populations. Approximately, three quarters of the world's agricultural value is generated in developing countries and, in many of these, the agriculture sector contributes as much as 30% to gross domestic product (GDP). It has been observed that GDP growth from agriculture benefits the incomes of poor people two to four times more than the

---

Developing regions, which are referred to throughout the chapter, consist of Africa; the Americas excluding Northern America, Latin America, and the Caribbean; Asia excluding Japan; and Oceania excluding Australia and New Zealand. Developed regions are Northern America, Europe, Japan, Australia, and New Zealand

---

N. R. Adusumilli (✉)
Resilient Dryland Systems and International Rice Research Institute (IRRI), International Crop Research Institute for the Semi-Arid Tropics (ICRISAT), Room 123 A, Building 303, Patancheru, Hyderabad 502324, India
e-mail: a.narayanarao@cgiar.org; anraojaya1@gmail.com

Plot: 1294 A, Road: 63 A, Jubilee Hills, Hyderabad 500033, India

R. K. Malik
Cereal Systems Initiative for South Asia (CSISA) Hub for Eastern U.P. and Bihar, K.P. Towers, block A, Digha Road, Patna 800025, India
e-mail: RK.Malik@cgiar.org

A. Yadav
Weed Science, Department of Agronomy, Chaudhary Charan Singh Haryana Agricultural University, National Highway 65, Hisar 125004, India

J. K. Ladha
International Rice Research Institute (IRRI)/India Office, 1st Floor, CG Block, NASC Complex, DPS Marg, New Delhi 110012, India
e-mail: j.k.ladha@irri.org

B. S. Chauhan, G. Mahajan (eds.), *Recent Advances in Weed Management*,
DOI 10.1007/978-1-4939-1019-9_17,

GDP growth in other sectors of the economy. The agricultural sector has the greatest potential for improving rural livelihood and eradicating the poverty of developing countries, as a significant number of the rural population in developing countries depends primarily upon small-scale, subsistence-oriented, agriculture-based family labor. However, they have limited access to technologies, in addition to essential resources, alternative livelihood, and production options.

By 2050, the world population is projected to reach 9.1 billion, up by 32 % from 2010. In absolute terms, the world's population is expected to grow by 2.2 billion. Over 85 % of population growth is expected in large urban centers and megacities in developing countries [2]. Of those additional people, almost 1 billion will live in Africa. Asia's population will increase by more than 1 billion, including 400 million more people in India. In comparison, China's slowing and ensuing negative growth will add only 63 million people [3]. Demand for food is predicted to rise 60 % globally by 2050, relative to 2009 levels. The majority of extra food demand is anticipated to reflect rising population and incomes in Asia. Rising incomes in China are predicted to be a major driver of this demand, accounting for 43 % of the global increase, while India accounts for 13 % [4].

To meet the demands of increasing population, it is essential to double the yields of smallholder farmers in developing countries of the world by improving the input efficiency and reliability of agricultural production. This is possible largely by scaling up best practices of currently available technologies and farming systems. The rural farming communities in developing countries are home to the most hard-working and self-reliant farmers looking for newer technologies for improving crop productivity, their income, and livelihood. A substantial increase in agricultural yield and output is expected to be realized by implementing interventions aimed at speeding up the assimilation and adoption of improved agricultural technologies and management practices of the research stations to the farmers' fields.

In this chapter, we made an attempt to: (1) put forward the importance of weeds and their management in enhancing the needed crop productivity to meet the demands of increasing population, (2) identify the weed management technologies that need special attention in upscaling them to larger numbers of farmers, and (3) list possible means and approaches for enhancing the farmers' knowledge for better weed management in agro-ecosystems of developing countries.

Importance of Weed Management to Attain Optimal Crop Productivity

Global estimated loss potential of weeds in rice, wheat, and maize indicates that weeds account for 46.2–61.5 % of potential losses and 27.3–33.7 % of actual losses caused by all pests together [5]. In most of the farming systems and for most of the crops of smallholder farms in developing countries, large yield gaps were identified [5–12]. Hence, significant scope exists for the improvement of crop yields by identifying and alleviating the constraints. Several studies were conducted to identify constraints causing the yield gaps, and among biotic constraints, the most important constraint in Africa is weeds (Table 17.1). Competition from weeds and shortcomings in weed management were severe in several of the developing nations of Asia and Africa. The shortage of labor is affecting the timely weed management in all cropping systems.

Continuous research efforts are being made to manage weeds in different crops, and cropping systems and technologies are available for managing weeds effectively and economically. A study in India revealed that the overall average gap in weed management practices in rice and wheat crops was 25 % and 25.8 %, respectively [13]. The maximum average technological gap of 31.4 % in wheat crop was found in case of chemical weeding followed by integrated weed management (20.3 %). Waddington et al. [12] observed that among the ten farming systems in South Asia, East Asia, and sub-Saharan (SS) Africa, inadequate farmer knowledge/training of different crops was reported as the major constraint in attaining optimum yields in the following farming systems:

1. SS Africa—cereal-root crop mixed; South Asia—rainfed mixed; South Asia—dry rainfed; East Asia P—lowland rice; and East Asia P—upland intensive mixed farming systems for rice
2. SS Africa—highland temperate mixed; SS Africa—maize mixed; and South Asia—rainfed mixed farming systems for sorghum
3. SS Africa—root crop and SS Africa—agro-pastoral millet/sorghum farming systems for cowpea
4. SS Africa—highland temperate mixed and South Asia—rice for chickpea
5. SS Africa—maize mixed for cassava

The closing of yield gaps signals effective knowledge transfer to farmers for successfully fostering the adoption of effective weed management. The exchange of information between scientists and farmers will be essential to reduce the time lag between development and implementation of more sustainable weed management practices. Hence, there is an urgent need to create awareness on the available and appropriate weed management practices among farmers in developing countries of Asia and Africa to tackle the weed menace and boost the crop production.

Weed Management Technologies that Need Special Attention

Continuous awareness creation and knowledge enhancement of the farming community are needed to benefit from weed management technological innovations. Recently, several technological advances occurred in the field of weed management. Some weed management technologies, about which the farmers' knowledge should be strengthened, include the following.

Table 17.1 Estimated smallholder farm yield gaps and an initial breakdown of yield losses associated with four categories of constraint for five food crops in Asian and African farming systems. The contribution of weed competition to yield gap is given in the parenthesis under the column biotic stresses. (Source data: [12])

Crop	Region	Farming system	Highest small-holder farmyield[a] (t/ha) (SD)	Average small-holder farmyield (t/ha) (SD)	Small-holder yield gap (t/ha) (SD)	Yield losses by constraint category (percent of total yield gap)[b]			
						Socio-econozmic	Abiotic	Biotic (weeds)	Manage-ment-related
Wheat	Sub-Saharan Africa	Highland temperate mixed	4.14(1.51)	2.02(1.11)	2.12(0.69)	28	20	19 (6)	32
	South Asia	Highland mixed	3.80(1.62)	2.05(0.99)	1.76(0.89)	23	30	21 (5)	27
		Rice–Wheat	4.81(1.16)	2.46(0.74)	2.38(0.89)	20	28	20 (6) (6)	31
		Rainfedmixed	4.96(1.11)	2.39(0.57)	2.54(0.80)	20	28	22 (7)	28
		Dry Rainfed	5.32(0.99)	2.16(0.42)	3.10(0.72)	23	30	18 (5)	30
	East Asia Pacific	Lowland rice	8.18(2.32)	5.12(1.76)	3.06(1.00)	20	29	20	30
		Upland intensive mixed	7.81(3.43)	3.99(1.51)	3.82(3.02)	24	30	17	29
		Temperate Mixed	8.74(1.71)	5.84(2.67)	3.13(0.98)	20	34	20	27
	Crop mean		5.97	3.25	2.74	22	29	20	29
Rice	Sub-Sahara Africa	Root crop	4.54(1.85)	2.88(1.36)	1.72(1.01)	38	18	25 (5)	18
		Cereal–root cropmixed	4.83(2.31)	2.74(1.31)	2.09(1.34)	51	14	16 (3)	19
	South Asia	Highland mixed	4.70(1.60)	2.52(0.57)	2.18(0.59)	27	29	21 (9)	23
		Rice	6.98(1.69)	3.74(1.17)	2.85(1.16)	27	25	22 (6)	26
		Rice–Wheat	6.23(1.29)	3.12(1.44)	2.76(0.82)	26	26	20 (6)	27
		Rainfedmixed	5.04(1.12)	2.95(0.65)	2.09(0.99)	27	22	19 (7)	31
		Dryrainfed	6.58(1.77)	3.88(1.53)	2.57(0.76)	27	24	23	26
	East Asia Pacific	Lowland rice	8.95(2.72)	5.93(2.02)	2.94(1.20)	28	21	17	35
		Upland intense mixed	9.94(1.20)	6.88(1.27)	3.05(1.23)	25	29	19	27
		Temperate mixed	10.35(1.11)	6.85(0.74)	3.54(0.99)	29	24	23	24
	Crop mean		6.81	4.15	2.58	31	23	21	26

Table 17.1 (continued)

Crop	Region	Farming system	Highest small-holder farmyield[a] (t/ha) (SD)	Average small-holder farmyield (t/ha) (SD)	Small-holder yield gap (t/ha) (SD)	Yield losses by constraint category (percent of total yield gap)[b]			
						Socio-econozmic	Abiotic	Biotic (weeds)	Manage-ment-related
Sorghum	Sub-Saharan Africa	Highland temperate	3.44(1.28)	1.80(0.58)	1.64(1.44)	15	25	28 (7 + Striga: 5)	29
		Cereal–root crop mixed	2.93(1.44)	1.13(0.55)	1.80(1.18)	23	23	26 (6 + Striga: 8)	24
		Maize-mixed	2.30(1.67)	1.09(0.78)	1.23(1.34)	30	29	19 (7)	22
		Agropastoralmillet/sorghum	2.15(1.33)	0.66(0.37)	1.50(1.12)	24	32	18 (Striga: 5)	27
	South Asia	Rainfedmixed	3.11(1.50)	1.62(0.31)	1.69(1.36)	23	25	22 (4)	30
		Dryrainfed	4.47(1.27)	1.85(0.65)	2.86(1.27)	27	30	20 (6)	23
	Crop mean		3.07	1.36	1.78	24	27	22	26
Cowpea	Sub-Saharan Africa	Root-crop	0.80(0.45)	0.38(0.19)	0.41(0.48)	31	16	31 (4)	21
		Cereal–root crop	0.97(0.62)	0.54(0.43)	0.50(0.42)	22	21	37	21
		Cropmixed Maize mixed	1.36(0.90)	0.42(0.41)	0.92(0.77)	27	25	28 (4)	20
		Agro pastoral-millet/	1.63(1.14)	0.62(0.52)	1.01(0.81)	21	21	30 (7)	29
	Crop mean		1.19	0.49	0.71	25	21	32	23
Cassava	Sub-Saharan Africa	Rootcrop	22.67(11.21)	13.77(6.26)	8.24(5.81)	32	23	22 (5)	24
		Cereal–rootcropmixed	21.35(12.66)	12.88(2.27)	9.00(5.91)	31	20	23 (5)	26
		Maizemixed	19.87(7.32)	8.74(3.78)	12.22(5.14)	25	20	25 (4)	29
		Lowlandrice	40.00(7.07)	21.00(1.41)	19.00(5.66)	15	25	25	35
		Uplandintensive	27.13(17.26)	15.80(3.89)	19.20(5.98)	28	27	17 (5)	28
	Crop mean		26.20	14.44	13.53	26	23	22	28

SD standard deviation

[a] The highest farm yield, average farm yield, and yield gap estimated by panelists in the round 1 questionnaire were adjusted independently by respondents during round 2. Thus, in some cases the reported average yield plus the yield gap does not sum to the highest farm yield

[b] Initial (round 1) percent estimate for four categories of constraint

Best Management Practices in Integrated Weed Management to Reduce Weed Menace

Farmers' knowledge on the ecology of weeds, weed seed production, prevention, or minimization, and ecological integrated weed management during the critical period of crop–weed competition must be enhanced [14, 15]. Farmers should be made aware of emerging problems, such as weedy rice, and the proper ways and means of managing them in an integrated manner. Weedy rice, and its development as an important problem, is associated with lowland rice ecology of eastern Uttar Pradesh and the adjoining parts of Bihar of India as in many parts of the developing countries. The stale seedbed-technique should be an effective strategy to exhaust the existing seed bank and the use of hybrid seeds to solve seed contamination problem. This will help facilitate the adoption of zero-tillage (ZT) in direct-seeded rice in the region. Like any other technology, the practicalities may get in the way forward. If we look at the whole system, the use of pre-seeding herbicides can be an efficient tool, which can lead to flexibility in respect of using or not using post emergence herbicides. This also makes it possible to boost the early crop canopycover.

Proper Application and Use of Herbicides and Other Weed Management Tools

Herbicides are becoming increasingly popular in developing countries because of the increasing cost and non-availability of manual labor used traditionally for hand weeding by the farming community. Herbicide use provides a pro-poor technology for both rainfed and irrigated crop production in developing countries, where farmers are striving to cut production costs and increase crop output as well as income. There are five key recommendations that will improve spray efficiency: (1) selecting the correct nozzle, (2) using appropriate pressure, (3) using multiple boom nozzles, (4) avoiding adverse weather conditions, and (5) keeping up with technologies [16]. Another key to keeping up with herbicide application education is understanding new spraying techniques. Farmers must be well trained in the proper use of herbicides and other best weed management practices to effectively control weeds and avoid the development of resistance in weeds.

Innovative channels are being used to deliver improved weed management knowledge to farmers, including primary schools in Tanzania and the herbicide supply chain in Bangladesh [17]. A series of training workshops on herbicide application techniques were organized in India and Nepal in 2000 [18]. The workshops focused on teaching the participants how to use and fabricate multiple-nozzle booms, the importance of flat-fan nozzles, calibration, drift avoidance, and applicator safety. These workshops helped in improving the efficiency of herbicides and also facilitated a major shift from application of herbicides by mixing in sand and broadcasting to the adoption of recommended spraying method.

Similar effort was made recently in Africa [19] on:

1. The production of farmer-to-farmer instruction videos on efficient and safe use of herbicides and on the use of an affordable, hand-operated, rotary weeder
2. Testing two rotary weeder types against best weed management practice and hand weeding
3. Training local blacksmiths in manufacturing locally adjustable rotary weeders
4. Enhancing weed science capacities in Tanzania by training R&D professionals and agronomy/weed science students in accessing and using relevant information and tools for developing optimal weed management strategies

Herbicide Resistance Management

Over the past several years, there was a steady increase in herbicide resistance, that is, the evolved capacity of a previously herbicide-susceptible weed population to withstand a herbicide and complete its life cycle when the herbicide is used at its normal dose in an agricultural situation [20]. Several important weeds have evolved resistance in developing countries, having an important economic impact on specific crops, which were reviewed by Valverde [21]. The most recent information on the occurrence of herbicide-resistant (HR) weeds may be found on the website www.weedscience.org, maintained by Herbicide Resistance Action Committee of Weed Science Society of America (WSSA).

Herbicide resistance was the most serious problem in wheat in the rice–wheat cropping system during the early 1990s. Efforts on herbicide resistance management before 1996–1997 were concentrated around alternate crops [22]. The problem of resistance was so serious that farmers in the state of Haryana (India) started sowing sunflower to exhaust the seed bank of *Phalaris minor* Retz. (wild canary grass). Crop rotation was possible only in a small area and farmers needed a viable technology for herbicide-resistance management.

Emergence of very heavy population during the early phases of crop cycles can be prevented with the use of ZT technology. ZT in wheat reduces the emergence rate of *P. minor* compared to conventional tillage [23]. In a study conducted by Franke et al. [23] at farmers' fields in Haryana, correlating the number of germinable *P. minor* seeds in soil with the number of *P. minor* seedling emerged, it was found that ZT reduced the emergence rate of first flush of *P. minor* by 50 % (Fig. 17.1a). The rate of emergence of second and third flushes was also lower in ZT plots compared to conventionally tilled plots (Fig. 17.1b, c). The first flush of *P. minor* is more damaging to the crops compared to later flushes, and ZT was found relatively more effective in reducing the first flush than other flushes.

ZT made it possible to achieve three major objectives, leading to create competition in the favor of crop. The objectives are optimum plant population, seeding at a time that is not conducive to *P. minor* emergence, and accurate fertilizer placement. Reduced population of this weed does not mean that the *Phalaris* problem will be solved by ZT alone. It also does not mean that farmers will stop using herbicides. Long-term trials at five sites in different villages indicate that farmers can skip her-

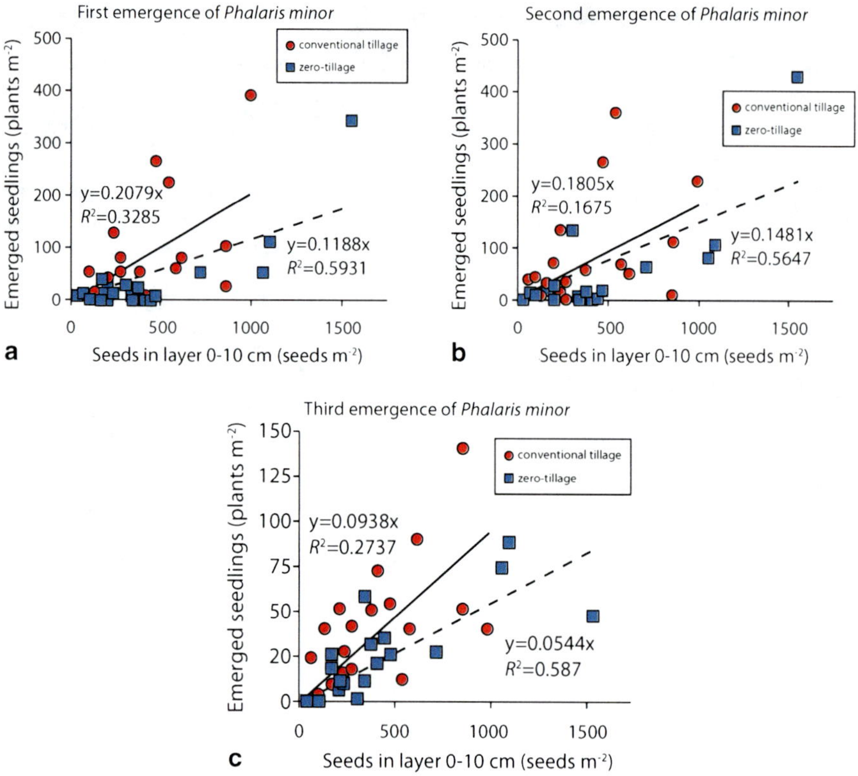

Fig. 17.1 Emergence rate of the first (**a**), second (**b**), and third (**c**) flush of *Phalaris minor* under conventional (•, *solid line*) and zerotillage (□, *dashed line*) in wheat. (Source: [23])

bicide once in 3–4 years. There is a constant danger that this weed will constantly evolve resistance to new herbicides. Using herbicides alone is not a long-term solution for managing resistance. Details of resistance development and its management using integrated approach with focused attention on ZT have been published [22, 23].

It is possible to continuously use effective weed management tools by the adoption of weed management strategies aimed at the prevention of herbicide resistance of weeds. Best management practices suggested by Norsworthy et al. [24] are applicable to developing countries also.

Management of Herbicide-Resistant Crops

Genetically modified crops have become extremely popular since their introduction in 1996. Currently, they are grown on more than170-mha area in 29 countries involving more than 17 million farmers of whom about 15 million are smallholder

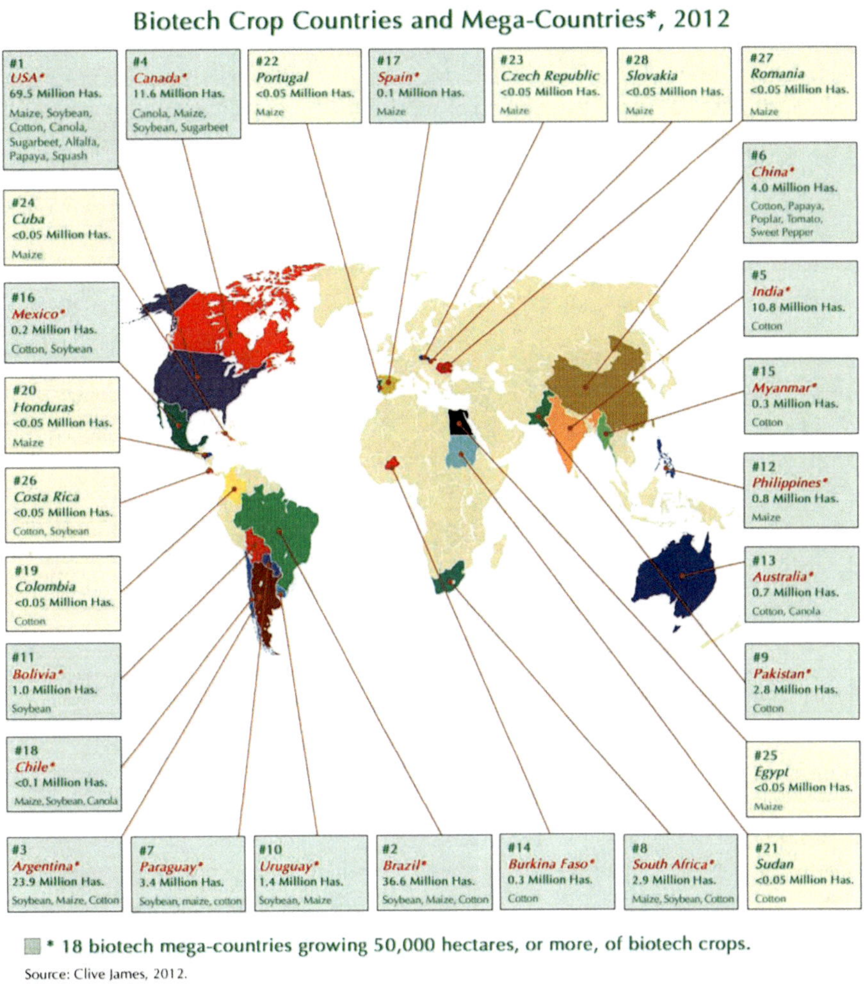

Fig. 17.2 Biotech-crop-growing countries. (Reprinted with permission from James, Clive. 2012. Global Status of Commercialized Biotech/GM Crops: 2012. ISAAA Brief No. 44.ISAAA: Ithaca, NY. http://www.isaaa.org/resources/publications/briefs/44/executivesummary/)

and resource-poor farmers (Fig. 17.2). Tolerance to herbicides is the most predominant trait, contributing nearly 70% of the total area. India with 10.8 mha and China with 4.0 mha are ranked fifth and sixth, respectively, in terms of total area under genetically modified crops. Pakistan, Philippines, Australia, and Myanmar are a few other countries that are growing genetically modified crops. Glufosinate-resistant soybean, corn, cotton, and canola are now commercialized in certain countries, and, in the near future, crops resistant to the herbicides 2,4-D, dicamba, hydroxyphenyl pyruvate dioxygenase (HPPD) inhibitors, and possibly to the PPO-inhibiting herbicides are expected to reach the marketplace [25]. Further, transgenic crops with resistance to more than one herbicide mode of action (i.e., stacked traits) have also been commercialized in recent times. As new HR crops become available, management of novel HR weeds will be a major challenge. However, the introduction of HR crops also prompted concerns about potential transfer of herbicide resistance to weed populations via crop-to-weed gene flow [26–30].

Clearfield rice, an imidazolinone (IMI)-resistant rice derived from conventional breeding technique, has been in cultivation in Malaysia mainly for managing weedy rice [31]. The possible evolution of resistance to ALS-inhibiting herbicides in weedy rice and the risk of weedy rice acquiring resistance to herbicide following introgression of resistant gene from the HR rice are the major concerns that need to be addressed adequately. In the near future, transgenic crop technology would be brought to the farming community in most of the developing countries as well [32]. Farmers need to be adequately trained on proper use of the HR cultivars before their introduction in developing countries.

Ways to Strengthen Farmers' Knowledge and Ability to Manage Weeds Ecologically, Economically, and Effectively

Effective Extension

In order to ensure that farmers are equipped with the knowledge of the best weed management technologies to optimize long-term agricultural productivity, effective extension should be available. Effective extension would enable increased rates of adoption of improved weed management technology by the farming community. The essential ingredients for an effective extension were summarized in another context [33] that are applicable to weed management also. They are:

- Building the credibility and trust in extension officers by avoiding short-term funding, rapid staff turnovers, and staff who are inexperienced or lack technical farming expertise.
- Recruiting high-caliber personnel on the ground as extension agents who should ideally have authority and technical expertise, be perceived by farmers as similar to them, have a local profile; possess good communication skills; have personal relationships with landholders, and be able to acknowledge and empathize with the problems and circumstances of landholders.
- The use of multiple methods—for example, print articles, verbal presentations, group extension, and advertisements—enhances effectiveness.
- Although group extension work is useful, one-on-one on-farm advice is critical.
- Counseling assistance may aid extension in some circumstances, as those in the most difficult circumstance are also often reluctant to seek help. Integration of counseling with extension programs may help identify those in need of assistance.
- Extension efforts should be based on farmers' needs.

Farmers' Participatory Evaluation of Weed Management Technologies

As farming is risky, farmers' willingness to adopt improved weed management technology depends on demonstrated benefits of the technology. If the benefits are demonstrated with farmers' participation, the chances of farmers' adoption and thus receiving the benefits would be greater.

Implementation of improved weed management technologies will be knowledge-intensive, and, as a result, there is a need for better linkage between farmers and agricultural researchers in order to couple the farmers' location-specific experience with scientists' subject expertise. This linkage should involve information flowing in both directions during research and in extension. As the research is being designed and conducted, interaction between farmers and researchers will help ensure that the location-specific land, soil, and climate conditions are taken into account. To increase the adoption rate of existing and new technologies, farmers should be fully involved in the development of the technologies. Thus, the farmers' participatory process of evolving technologies is one of the ways to strengthen the knowledge of farmers.

Partnership with International Institutes

Weed management is a complex process and it needs combined efforts from several organizations (national and international) for enhancing the farmers' knowledge. The partnership between the state, non-state organizations (private sector), and global scientific research organizations is essential to achieve dissemination of new technologies to the end users and to achieve faster progress [34]. Partnership between these organizations is critical to let cost-efficient weed management technologies disseminate to the end users. Each of the organizations have their own strengths and could complement each other's efforts in taking research from laboratory to field with new institutional mechanism as well as enabling policies. Partnership with global scientific organizations could lead to faster progress as well as behavioral/attitude (work ethics as well as commitment) changes among state actors (bureaucrats and policy makers) as was observed in the Bhoochetana project implementation by ICRISAT and the Government of Karnataka in India [35].

Involving Women in Technology Development Transfer

Women are very actively involved in rice farming and rice processing in both Asia and Africa. Thus, the technology development and extension should have a gender focus in order to ensure that research is effective and efficient. This will greatly enhance the efficiency and impact of research as well as reduce gender inequalities in access to technologies. One recent example that proves it is the rapid adoption of

NERICA varieties across the African continent through the participatory variety selection work involving female and male farmers [36]. Participatory learning and action research methods have facilitated wide adoption of improved technologies for inland valley swamp development in Côte d'Ivoire, Mali, Ghana, and Madagascar.

There is a need to present research knowledge in formats that are easily digested by farmers and other prospective users. Africa Rice has acquired some experience in the use of videos in conveying certain messages to farmers and provoking village-level discussions on issues related to rice cultivation and rice processing. These videos have been translated into 33 local African languages [37–39].

Involving the Private Sector

Technological popularization among the farmers should be a convergent process involving farmers, private sector, department of agriculture officials, university staff, and the scientists from national institutions. Involvement of the private sector would enable the sector to make sure the availability of different components of integrated weed management, such as improved competitive cultivars-adopted to specific locations, herbicides, location-specific mechanical weeders, and other implements and inputs. Involvement of different private sectors with farmers would not only ensure higher production by the farmers through effective weed control but also ensure better marketing of the produce by the farming community.

Farmer Field Schools

The farmer field schools (FFS) training approach was based on active participation of farmers sharing knowledge with each other. Farmers learn new concepts through the experiential learning cycle in a process of learning by doing. The FFS facilitators help farmers to learn from practical experience. Since the initiation of the first FFS in 1989/1990 in Indonesia for educating farmers on the principles of "integrated pest management" for managing major outbreaks of the brown plant hopper, the concept has spread to other Asian countries [40]. This concept may be used for improving the farmers' knowledge of weed management.

Utilizing Information Technology

Most developing countries have started using Internet-based information technology (IT). India, in particular, now has Internet connectivity down to the district level throughout the country. Organizations, such as ITC and Mahindra ShubhLabh Services, have "e-centers" to assist farmers. Many of the agricultural universities have Web sites incorporated with weed management technologies. The information can be accessed by extension staff and passed on to farmers. In addition to existing communication and knowledge dissemination systems, IT may be used simultaneously for transferring knowledge and enriching farmers technologically. Several organizations are incorporating technological information within the new information systems. CAB International manages a wide range of information resources of existing agricultural information through publications, CD-ROMs, and research studies. The National Innovations Foundation (NIF) in India has been established to build linkages between excellence in formal scientific systems and informal knowledge systems [41]. The rapid extension of the Internet, mobile phones, and other communication networks will provide new opportunities. But, in certain developing countries, such progress is not there. However, in the future they will need to use IT to effectively pass on the weed management technology to the farming community.

Conclusion

In these current days of enormous challenges—including climate change, soil degradation, and resource scarcity—there is an urgent need for capacity building of the farming community to combat the menace caused by ever-adapting dynamic weeds. Enhancing farmers' knowledge with timely, relevant, and accurate technological information from time to time is crucial. For strengthening the knowledge of farmers on effective weed management, it is essential to have a networking of weed scientists and other people interested in weeds for rapid knowledge and information sharing among each other.

Acknowledgments A. N. Rao extends his thanks to the Government of Karnataka for the financial support.

References

1. FAO (2013) FAO statistical yearbook 2013. World Food and Agriculture. Food and Agriculture Organization of the United Nations, FAO, Rome, p 289
2. OECD/FAO (2011) OECD–FAO agricultural outlook 2011–2020, OECD Publishing and FAO. http://dx.doi.org/10.1787/agr_outlook-2011-en. Accessed 26 Nov 2013
3. Moir B, Morris P (2011) Global food security: facts, issues and implications. Issue 1–2011. Science and economic insights. Australian Bureau of Agricultural and Resource Economics and Sciences (GPO Box 1563 Canberra, ACT 2601), Australia, p 19
4. FAO (2009) How to feed the world in 2050: high-level expert forum. Food and Agriculture Organization of the United Nations, Rome, p 35
5. Oerke EC (2006) Crop losses to pests. J AgricSci 144:31–43
6. Widawsky DA, O'Toole J (1996) Prioritizing the rice research agenda for eastern India. In: Evenson RE, Herdt RW, Hossain M (eds) Rice research in Asia: progress and priorities. CAB International in association with the International Rice Research Institute, Wallingford, pp 109–129
7. Cassman KG (1999) Ecological intensification of cereal production systems: yield potential, soil quality, and precision agriculture. Proc Natl Acad Sci U S A 96(11):5952–5959

8. Peng S, Cassman KG, Virmani SS, Sheehy J, Khush GS (1999) Yield potential trends of tropical rice since the release of IR8 and the challenge of increasing rice yield potential. Crop Sci 39:1552–1559
9. Hobbs PR, Sayre KD, Ortiz-Monasterio JI (1998) Increasing wheat yield sustainably through agronomic means. NRG Paper 98-01. CIMMYT, México D.F., México, p 22
10. Evans LT, Fischer RA (1999) Yield potential: its definition, measurement, and significance. Crop Sci 39:1544–1551
11. Rao AN, Johnson DE, Sivaprasad B, Ladha JK, Mortimer AM (2007) Weed management in direct-seeded rice. Adv Agron 93:153–255
12. Waddington SR, Li X, Dixon J, Hyman G, de Vicente MC (2010) Getting the focus right: production constraints for six major food crops in Asian and African farming systems. Food Secur 2:27–48
13. Singh S, Lall AC (2001) Studies on technological gaps and constraints in adoption of weed management practices for rice-wheat cropping system. Indian J Weed Sci 33:116–119
14. Rao AN, Nagamani A (2010) Integrated weed management in India-revisited. Indian J Weed Sci 42:1–10
15. Rao AN, Ladha JK (2011) Possible approaches for ecological weed management in direct-seeded rice in a changing world. In: Proceedings of the 23rd Asian-Pacific weed science society conference, The Sebel, Cairns, Australia, 26–29 Sept 2011, pp 444–453
16. http://agro.basf.us/stewardship/herbicide-best-practices.html. Accessed 26 Nov 2013
17. Riches CR, Mbwaga AM, Mbapila J, Ahmed GJU 2005. Improved weed management delivers increased productivity and farm incomes from rice in Bangladesh and Tanzania (Pathways out of Poverty). Asp Appl Biol 75:127–138
18. Bellinder RR, Miller AJ, Malik RK, Ranjit JD, Hobbs PR, Brar LS, Singh G, Singh S, Yadav A (2002) Improving herbicide application accuracy in South Asia. Weed Technol 16:845–850
19. Rodenburg J (2012) Building local capacities in weed management for rice-based systems—Narrative Technical Report document. African Rice Center, International Institute of Tropical Agriculture, Nigeria. (www.africa-rising.net)
20. Heap I, LeBaron H (2001) Introduction and overview of resistance. In: Powles SB, DL Shaner (eds) Herbicide resistance in world grains. CRC Press, Boca Raton, pp 1–22
21. Valverde BE (2003) Herbicide-resistance management in developing countries. In: Labrada R (ed) Weed management for developing countries. Addendum 1. FAO plant production and protection paper 120 Add. 1. Food and Agriculture Organisation, Rome, Italy
22. Malik RK, Yadav A, Singh S, Malik RS, Balyan RS, Banga RS et al (2002) Herbicide resistance management and evolution of zero-tillage—a success story. Research Bulletin, CCS Haryana Agricultural University, Hissar, 43 pp
23. Franke AC, Singh S, McRoberts N, Nehra AS, Godara S, Malik RK, Marshall G (2007) *Phalaris minor* seedbank studies: longevity, seedling emergence and seed production as affected by tillage regime. Weed Res 47:73–83
24. Norsworthy JK, Ward SM, Shaw DR, Llewellyn RS, Nichols RL, Webster TM et al (2012) Reducing the risks of herbicide resistance: best management practices and recommendations. Weed Sci 60(sp1):31–62
25. Green JM, Hazel CB, Forney D, Pugh LM (2008) New multiple herbicide crop resistance and formulation technology to augment the utility of glyphosate. Pest Manag Sci 64:332–339
26. Dale PJ (1994) The impact of hybrids between genetically modified crop plants and their related species: general considerations. Mol Ecol 3:31–36
27. Snow AA (2002) Transgenic crops—why gene flow matters. Nat Biotechnol 20:542
28. Nichols R, May L, Bourland F (2003) Special symposium—transgenic pest-resistant crops: status and testing issues. Crop Sci 43:1582–1583
29. Warwick SI, Legere A, Simard M-J, James T (2008) Do escaped transgenes persist in nature? The case of a herbicide resistance transgene in a weedy *Brassica rapa* population. Mol Ecol 17:1387–1395

30. Brookes G, Barfoot P (2011) GM crops: global socio-economic and environmental impacts, pp. 1996–2009. www.pgeconomics.co.uk/pdf/2011globalimpactstudy.pdf. Accessed 26 Nov 2013
31. Sudianto E, Beng-Kah S, Ting-Xiang N, Saldain NE, Scott RC, Burgos NR (2013) Clearfield rice: Its development, success, and key challenges on a global perspective. Crop Prot 49:40–51
32. Rao C, Dev M (2009) Biotechnology and pro-poor agricultural development. Econ Polit Wkly 44(52):56–64
33. Pannell DJ, Marshall GR, Barr N, Curtis A, Vanclay F, Wilkinson R (2006) Understanding and promoting adoption of conservation practices by rural landholders. Aust J Exp Agric 46:1407–1424
34. Raju KV, Wani SP, Anantha KH (2013) BHOOCHETANA: innovative institutional partnerships to boost productivity of rainfed agriculture in Karnataka, India. Resilient Dryland Systems Report No. 59. Patancheru 502 324, International Crops Research Institute for the Semi-Arid Tropics (ICRISAT), Andhra Pradesh, India, p 34
35. Wani SP, Sarvesh KV, Krishnappa K, Dharmarajan BK, Deepaja SM (eds) (2012) BHOOCHETANA: mission to boost productivity of rainfed agriculture through science-led interventions in Karnataka. Patancheru 502 324, International Crops Research Institute for the Semi-Arid Tropics, Andhra Pradesh, India, p 84
36. Somado EA, Guei RG, Keya SO (2008) NERICA: the new rice for Africa-a Compendium. Cotonou: WARDA. http://www.africarice.org/publications/nerica-comp/Nerica%20Compedium.pdf. Accessed 26 Nov 2013
37. Van Mele P (2006) Zooming-in, zooming-out: a novel method to scale up local innovations and sustainable technologies. Int J Agric Sustain *4*(2):131–142
38. Van Mele P, Wanvoeke J, Akakpo C, Dacko RM, Ceesay M, Beavogui L, Soumah M, Anyang R (2010) Videos bridging Asia and Africa: overcoming cultural and institutional barriers in technology-mediated rural learning. J Agric Educ Ext 16(1):75–87
39. Zossou E, Van Mele P, Vodouhe SD, Wanvoeke J (2009) The power of video to trigger innovation: rice processing. Int J Agric Sustain 7(2):119–129
40. Pontius J, Dilts R, Bartlett A (eds) (2002) From farmer field schools to community IPM. Ten years of IPM training in Asia. FAO Community IPM Program, Jakarta, Indonesia. FAO Regional Office for Asia and the Pacific, Bangkok 10200, Thailand, p 121
41. FAO (1996) The internet and rural development: recommendations for strategy and activity. Final Report and Executive Summary. http://www.fao.org/sd/CDdirect/CDDO/contents.htm. Accessed 26 Nov 2013

Index

B. S. Chauhan, G. Mahajan (eds.), *Recent Advances in Weed Management*,
DOI 10.1007/978-1-4939-1019-9, © Springer Science+Business Media New York 2014

S

T

W

Y